Molecular Biology and Biotechnology
6th Edition

Molecular Biology and Biotechnology

6th Edition

Edited by

Ralph Rapley
University of Hertfordshire, Hatfield, UK
Email: R.Rapley@herts.ac.uk

David Whitehouse
University of Hertfordshire, Hatfield, UK
Email: D.Whitehouse@herts.ac.uk

THE QUEEN'S AWARDS
FOR ENTERPRISE:
INTERNATIONAL TRADE
2013

Print ISBN: 978-1-84973-795-1

A catalogue record for this book is available from the British Library

Published by The Royal Society of Chemistry,
Thomas Graham House, Science Park, Milton Road,
Cambridge CB4 0WF, UK

Registered Charity Number 207890

Visit our website at www.rsc.org/books

Printed and bound by CPI Group (UK) Ltd, Croydon, CR0 4YY

Preface

There is no doubt that modern biological research and its subsequent applications rely heavily on the ability to analyse and manipulate a diversity of material at the molecular level. One of the exciting aspects of being involved in the field of molecular biology is the ever-accelerating rate of progress, both in the development and application of new methodologies. Since the 5th edition of this book there have been significant advances in biotechnology on many fronts. Notable advances in genome sequencing, proteomics and bioinformatics, for instance, have led to the characterization of molecular pathways, cellular interactions and even populations accelerating the understanding of disease processes, molecular diagnostics and targeted drug therapy. In an attempt to encompass the breadth of these advances, the new edition has been substantially revised and updated to provide a comprehensive overview. The book begins with several chapters on core technologies for example platform molecular biology and genomics, expression of recombinant proteins, proteomics and trangenesis. There follow chapters on antibody and protein engineering and biosensors. A series of chapters deal with more specialized topics reflecting current challenges in biotechnology including molecular analysis of yeast, molecular and microbial diagnostics, and the discovery and application of biomarkers. The concluding chapters address the important areas of tissue engineering, agricultural biotechnology and vaccine development. The editors believe the coverage of many of the key areas in molecular biotechnology will serve as a solid foundation for those embarking on or engaged in studies of this exciting field. Our intention is that this book shall continue to have a teaching function. As such, the 6th Edition of Molecular Biology and Biotechnology will be of particular interest to students of biology and chemistry, as well as to postgraduates and other scientific workers who require a sound introduction to this rapidly advancing and expanding area.

Ralph Rapley
David Whitehouse

Molecular Biology and Biotechnology, 6th Edition
Edited by Ralph Rapley and David Whitehouse
© The Royal Society of Chemistry 2015
Published by the Royal Society of Chemistry, www.rsc.org

Contents

Molecular Biology and Biotechnology, 6th Edition
Edited by Ralph Rapley and David Whitehouse
© The Royal Society of Chemistry 2015
Published by the Royal Society of Chemistry, www.rsc.org

CHAPTER 1

Molecular Biology Techniques and Bioinformatics

RALPH RAPLEY* AND NATALIA HUPERT

^aSchool of Life and Medical Sciences University of Hertfordshire, College Lane, Hatfield, Hertfordshire, AL10 9AB, United Kingdom
*Email: R.Rapley@herts.ac.uk

1.1 INTRODUCTION

The development of methods and techniques for studying processes at the molecular level has led to new and powerful ways of isolating, analysing, manipulating and exploiting nucleic acids. The completion of numerous genome projects has allowed the continued development of new exciting areas of biological sciences such as biotechnology, genome mapping, molecular medicine and gene therapy. In considering the potential utility of molecular biology techniques it is important to understand the basic structure of nucleic acids and gain an appreciation of how this dictates the function *in vivo* and *in vitro*. Indeed many techniques used in molecular biology mimic in some way the natural functions of nucleic acids such as replication and transcription. This chapter is intended to provide an overview of the general features of nucleic acid structure and function and describe some of the basic methods used in their isolation and analysis.

1.1.1 Primary Structure of Nucleic Acids

DNA and RNA are macromolecular structures composed of regular repeating polymers formed from nucleotides.[1] These are the basic building blocks of

Molecular Biology and Biotechnology, 6th Edition
Edited by Ralph Rapley and David Whitehouse
© The Royal Society of Chemistry 2015
Published by the Royal Society of Chemistry, www.rsc.org

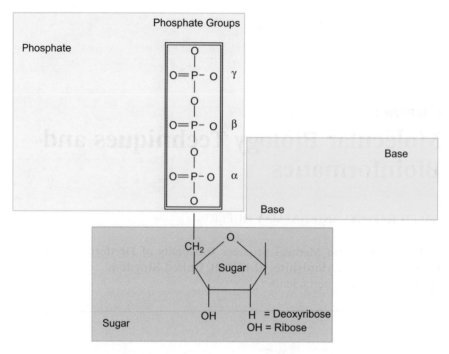

Figure 1.1   Representation of a deoxynucleoside triphosphate indicating the three components of a sugar, triphosphate and a base. The base can be either A, C, G or T. In RNA the 2′ carbon has an OH whereas it is deoxy in DNA.

nucleic acids and are derived from nucleosides which are composed of two elements: a five-membered pentose carbon sugar (2-deoxyribose in DNA and ribose in RNA), and a nitrogenous base (Figure 1.1). The carbon atoms of the sugar are designated 'prime' (1′, 2′, 3′, *etc.*) to distinguish them from the carbons of nitrogenous bases of which there are two types, either a purine or a pyrimidine. A nucleotide, or nucleoside phosphate, is formed by the attachment of a phosphate to the 5′ position of a nucleoside by an ester linkage. Such nucleotides can be joined together by the formation of a second ester bond by reaction between the phosphate of one nucleotide and the 3′ hydroxyl of another, thus generating a 5′ to 3′ phosphodiester bond between adjacent sugars; this process can be repeated indefinitely to give long polynucleotide molecules. DNA has two such polynucleotide strands, however since each strand has both a free 5′ hydroxyl group at one end, and a free 3′ hydroxyl at the other end, each strand has a polarity or directionality. The polarities of the two strands of the molecule are in opposite directions, and thus DNA is described as an 'anti-parallel' structure.

The purine bases (composed of fused five and six membered rings), adenine (A) and guanine (G) are found in both RNA and DNA, as is the pyrimidine (a single six-membered ring) cytosine (C). The other pyrimidines are each restricted to one type of nucleic acid: uracil (U) occurs exclusively in RNA, whilst thymine (T) is limited to DNA. Thus, it is possible to distinguish between RNA and DNA on the basis of the presence of ribose and uracil in RNA, and deoxyribose and thymine in DNA. However, it is the sequence of bases along a molecule that distinguishes one DNA (or RNA) from another.

1.1.2 Secondary Structure of Nucleic Acids

The two polynucleotide chains in DNA are usually found in the shape of a right handed double helix, in which the bases of the two strands lie in the centre of the molecule, with the sugar–phosphate backbones on the outside. A crucial feature of this double-stranded structure is that it depends on the sequence of bases in one strand being complementary to that in the other. A purine base attached to a sugar residue on one strand is always hydrogen bonded to a pyrimidine base attached to a sugar residue on the other strand. Moreover, adenine (A) always pairs with thymine (T) or uracil (U) in RNA, *via* two hydrogen bonds, and guanine (G) always pairs with cytosine (C) by three hydrogen bonds (Figure 1.2). When these conditions are met a stable double helical structure results in which the backbones of the two strands are, on average, a constant distance apart. Thus, if the sequence of one strand is known, that of the other strand can be deduced. The strands are designated as plus (+) and minus (−) and an RNA molecule complementary to the minus (−) strand is synthesised during transcription. The base sequence may cause significant local variations in the shape of the DNA molecule and these variations are vital for specific interactions between the DNA and various proteins to take place. Although the three dimensional structure of DNA may vary it generally adopts a double helical structure termed the B form or B-DNA *in vivo*.

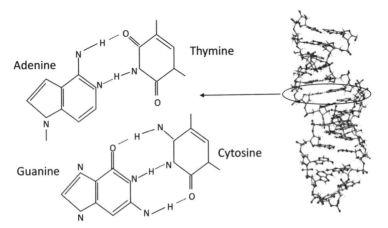

Figure 1.2   Representation of the four bases in DNA and their complementary base pairing, A–T and C–G through H bonds. The right hand panel indicates a DNA double helix.

1.1.3   Denaturation of Double Stranded DNA

The two anti-parallel strands of DNA are held together only by the weak forces of hydrogen bonding between complementary bases, and partly by hydrophobic interactions between adjacent, stacked base pairs, termed base-stacking. Little energy is needed to separate a few base pairs, and so, at any instant, a few short stretches of DNA will be opened up to the single-stranded conformation. However, such stretches immediately pair up again at room temperature, so the molecule as a whole remains predominantly double-stranded.

If, however, a DNA solution is heated to approximately 90 °C or above there will be enough kinetic energy to denature the DNA completely, causing it to separate into single strands. The temperature at which 50% of the DNA is melted is termed the melting temperature or T_m, and this depends on the nature of the DNA. If several different samples of DNA are melted, it is found that the T_m is highest for those DNAs which contain the highest proportion of cytosine and guanine, and T_m can actually be used to estimate the percentage $(C + G)$ in a DNA sample. This relationship between T_m and $(C + G)$ content arises because cytosine and guanine form three hydrogen bonds when base-paired, whereas thymine and adenine form only two. Because of the differential numbers of hydrogen bonds between A–T and C–G pairs, those sequences with a predominance of C–G pairs will require greater energy to separate or denature them. The conditions required to separate a particular nucleotide sequence is also dependent on environmental conditions such as salt concentration. If melted DNA is cooled, it is possible for the separated strands to reassociate, a process known as renaturation.

Strands of RNA and DNA will associate with each other, if their sequences are complementary, to give double-stranded, hybrid molecules. Similarly, strands of labelled RNA or DNA, when added to a denatured DNA preparation, will act as probes for DNA molecules to which they are complementary. This hybridisation of complementary strands of nucleic acids is a cornerstone for many molecular biology techniques and is very useful for isolating a specific fragment of DNA from a complex mixture. It is also possible for small single stranded fragments of DNA (up to 40 bases in length) termed oligonucleotides to hybridise to a denatured sample of DNA. This type of hybridisation is termed annealing and again is dependent on the base sequence of the oligonucleotide and the salt concentration of the sample.

1.2   ISOLATION AND SEPARATION OF NUCLEIC ACIDS

1.2.1   Isolation of DNA

The use of DNA for analysis or manipulation usually requires that it is isolated and purified to a certain extent.[2] DNA is recovered from cells by the gentlest possible method of cell rupture to prevent the DNA from fragmenting by mechanical shearing. This is usually in the presence of EDTA, which chelates the Mg^{2+} ions needed for enzymes that degrade DNA termed DNase. Ideally, cell walls, if present, should be digested enzymatically (*e.g.* lysozyme treatment of bacteria), and the cell membrane should be solubilised using detergent. If physical disruption is necessary, it should be kept to a minimum, and should involve cutting or squashing of cells, rather than the use of shear forces. Cell disruption (and most subsequent steps) should be performed at 4 °C, using glassware and solutions which have been autoclaved to destroy DNase activity (Figure 1.3).

After release of nucleic acids from the cells, RNA can be removed by treatment with ribonuclease (RNase) which has been heat treated to inactivate any DNase contaminants; RNase is relatively stable to heat as a result of its disulfide bonds, which ensure rapid renaturation of the molecule on cooling. The other major contaminant, protein, is removed by shaking the solution gently with water-saturated phenol, or with a phenol/chloroform mixture, either of which will denature proteins but not nucleic acids. Centrifugation of the emulsion formed by this mixing produces a lower organic phase, separated from the upper aqueous phase by an interface of denatured protein. The aqueous solution is recovered and

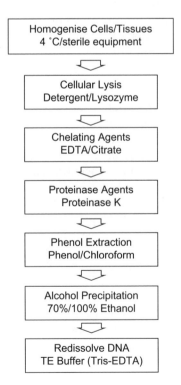

Figure 1.3 General steps involved in extracting DNA from cells or tissues.

deproteinised repeatedly, until no more material is seen at the interface. Finally, the deproteinised DNA preparation is mixed with two volumes of absolute ethanol, and the DNA allowed to precipitate out of solution in a freezer. After centrifugation, the DNA pellet is redissolved in a buffer containing EDTA to inactivate any DNases present. This solution can be stored at 4 °C for at least a month. DNA solutions can be stored frozen although repeated freezing and thawing tends to damage long DNA molecules by shearing.

The procedure described is suitable for total cellular DNA. If the DNA from a specific organelle or viral particle is needed, it is best to isolate the organelle or virus before extracting its DNA, since the recovery of a particular type of DNA from a mixture is usually rather difficult. Where a high degree of purity is required DNA may be subjected to density gradient ultra-centrifugation through caesium chloride, which is particularly useful for the preparation of plasmid DNA. It is possible to check the integrity of the DNA by agarose gel electrophoresis and determine the concentration of the DNA by using the fact that 1 absorbance unit equates to 50 µg/mL of DNA and thus:

$$50 \times A260 = \text{concentration of DNA sample } (\mu g/mL) \qquad (1.1)$$

Contaminants may also be identified in the sample by employing scanning UV-spectrophotometry from 200 nm to 300 nm. A ratio of 260 nm : 280 nm of approximately 1.8 indicates that the sample is free of protein contamination, which absorbs strongly at 280 nm.

1.2.2   Isolation of RNA

The methods used for RNA isolation are very similar to those described above for DNA; however, RNA molecules are relatively short, and therefore less easily damaged by shearing, so cell disruption can be rather more vigorous.[3] RNA is, however, very vulnerable to digestion by RNases which are present endogenously in various concentrations in certain cell types and exogenously on fingers. Gloves should therefore be worn, and a strong detergent should be included in the isolation medium to immediately denature any RNases. Subsequent deproteinisation should be particularly rigorous, since RNA is often tightly associated with proteins. DNase treatment can be used to remove DNA, and RNA can be precipitated by ethanol. One reagent that is commonly used in RNA extraction is guanidinium thiocyanate (GTC) which is both a strong inhibitor of RNase and a protein denaturant. It is possible to check the integrity of an RNA extract by analysing it by agarose gel electrophoresis. The most abundant RNA species are the rRNA molecules. For prokaryotes these are 16S and 23S and for eukaryotes the molecules are slightly heavier at 18S and 28S. These appear as discrete bands following agarose gel electrophoresis and importantly indicate that the other RNA components, such as mRNA, are likely to be intact. This is usually carried out under denaturing conditions to prevent secondary structure formation in the RNA. The concentration of the RNA may be estimated by using UV-spectrophotometry. At 260 nm, 1 absorbance unit equates to 40 μg/mL of RNA and therefore:

$$40 \times A260 = \text{concentration of RNA sample (μg/mL)} \qquad (1.2)$$

Contaminants may also be identified in the same way as for DNA by scanning UV-spectrophotometry, however in the case of RNA a 260 nm:280 nm ratio of approximately 2 would be expected for a sample containing no protein.

In many cases, it is desirable to isolate eukaryotic mRNA, which constitutes only 2–5% of cellular RNA, from a mixture of total RNA molecules. This may be carried out by affinity chromatography on oligo(dT)-cellulose columns. At high salt concentrations, the mRNA containing poly(A) tails binds to the complementary oligo(dT) molecules of the affinity column, and so mRNA will be retained; all other RNA molecules can be washed through the column by further high salt solution. Finally, the bound mRNA can be eluted using a low concentration of salt. Nucleic acid species may also be subfractionated by more physical means such as electrophoretic or chromatographic separations based on differences in nucleic acid fragment sizes or physico-chemical characteristics.

1.2.3 Enzymes Used In Molecular Biology

The discovery and characterisation of a number of key enzymes has enabled the development of various techniques for the analysis and manipulation of DNA. In particular, the enzymes termed type II restriction endonucleases have come to play a key role in all aspects of molecular biology.[4] These enzymes recognise certain DNA sequences, usually 4–6 bp (base-pairs) in length, and cleave them in a defined manner. The sequences recognised are palindromic or of an inverted repeat nature (Figure 1.4). That is, they read

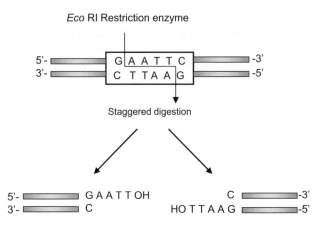

Figure 1.4 The cleavage of a DNA strand with a target site for the restriction enzyme *Eco*R1 indicating the ends of the DNA formed following digestion.

Table 1.1 Examples of restrictions enzymes with 4, 6, or 8 base recognition sequence.

Name	Recognition Sequence		Digestion Products	
Four Nucleotide Recognition Sequence				
*Hae*III	5′-GG↓CC-3′ 3′-CCGG-5′	5′-GG 3′-CC	CC-3′ GG-5′	**Blunt End Digestion**
*Hpa*II	5′-CCGG-3′ 3′-GGCC-5′	5′-C 3′-GGC	CGG-3′ C-5′	**Cohesive End Digestion**
Six Nucleotide Recognition Sequence				
*Bam*HI	5′-G↓GATTC-3′ 3′-GGCC-5′	5′-G 3′-CCTAG	GATCC-3′ G-5′	
*Eco*RI	5′-G↓AATTC-3′ 3′-CTTAAG-5′	5′-G 3′-CTTAA	AATCC-3′ G-5′	
*Hind*III	5′-A↓AGCTT-3′ 3′-TTCGAA-5′	5′-A 3′-TTCGA	AGCTT-3′ A-5′	
Eight Nucleotide Recognition Sequence				
*Not*I	5′-GC↓GGCCGC-3′ 3′-CGCCGGCG-5′	5′-GC 3′-CGCCGG	GGCCGC-3′ CG-5′	

the same in both directions on each strand. When cleaved they leave a flush-ended or staggered (also termed a cohesive-ended) fragment depending on the particular enzyme used. An important property of staggered ends is that those produced from different molecules by the same enzyme are complementary (or 'sticky') and so will anneal to each other. The annealed strands are held together only by hydrogen bonding between complementary bases on opposite strands. Covalent joining of the ends of each of the two strands may be carried out using the enzyme DNA ligase. This is widely exploited in molecular biology to enable the construction of recombinant DNA, *i.e.* the joining of DNA fragments from different sources. Approximately 500 restriction enzymes have been characterised that recognise over 100 different target sequences. A number of these, termed isoschizomers, recognise different target sequences but produce the same staggered ends or overhangs. A number of other enzymes have proved to be of value in the manipulation of DNA, as summarised in Table 1.1, and are indicated at appropriate points within the text (Figure 1.5).

1.2.4 Electrophoresis of Nucleic Acids

Electrophoresis in agarose or polyacrylamide gels is the most usual way to separate DNA molecules according to size. The technique can be used

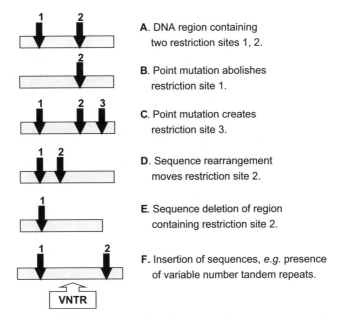

A. DNA region containing two restriction sites 1, 2.

B. Point mutation abolishes restriction site 1.

C. Point mutation creates restriction site 3.

D. Sequence rearrangement moves restriction site 2.

E. Sequence deletion of region containing restriction site 2.

F. Insertion of sequences, *e.g.* presence of variable number tandem repeats.

Figure 1.5 Indication of the application of restriction enzymes and the restriction fragment length generated with various situations.

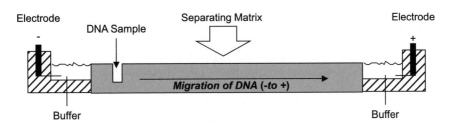

Figure 1.6 A schematic illustration of a typical horizontal gel electrophoresis set up for the separation of nucleic acids.

analytically or preparatively, and can be qualitative or quantitative. Large fragments of DNA such as chromosomes may also be separated by a modification of electrophoresis termed pulsed field gel electrophoresis (PFGE). The easiest and most widely applicable method is electrophoresis in horizontal agarose gels, followed by staining with ethidium bromide. This dye binds to DNA by insertion between stacked base pairs (intercalation), and it exhibits a strong orange/red fluorescence when illuminated with ultraviolet light. Very often electrophoresis is used to check the purity and intactness of a DNA preparation or to assess the extent of an enzymatic reaction during, for example, the steps involved in the cloning of DNA (Figure 1.6). For such checks 'mini-gels' are particularly convenient, since they need little preparation use small samples and provide results quickly.

Agarose gels can be used to separate molecules larger than about 100 bp. For higher resolution or for the effective separation of shorter DNA molecules polyacrylamide gels are the preferred method.

When electrophoresis is used preparatively, the piece of gel containing the desired DNA fragment is physically removed with a scalpel. The DNA may be recovered from the gel fragment in various ways. This may include crushing with a glass rod in a small volume of buffer, using agarase to digest the agarose leaving the DNA, or by the process of electroelution. In this method, the piece of gel is sealed in a length of dialysis tubing containing buffer, and is then placed between two electrodes in a tank containing more buffer. Passage of an electrical current between the electrodes causes DNA to migrate out of the gel piece, but it remains trapped within the dialysis tubing, and can therefore be recovered easily. An alternative to conventional analysis of nucleic acids by electrophoresis is through the use of microfluidic systems such as an Agilent Bioanalyser. These are automated machine based systems utilising carefully manufactured chip based units, frequently termed 'lab on a chip', where microlitre samples volumes may be used. Using complex miniaturised channels coupled with flow pumps and detection systems, nucleic acids can be separated and assayed rapidly. With the aid of advanced software, the analysis may provide much of the data required for nucleic acid and protein analysis such as sizing and quantitation.

1.3   MAPPING AND BLOTTING OF NUCLEIC ACIDS

1.3.1   DNA Blotting

Electrophoresis of DNA restriction fragments allows separation based on size to be carried out; however, it provides no indication as to the presence of a specific, desired fragment among the complex sample. This can be achieved by transferring the DNA from the intact gel onto a piece of nitrocellulose or nylon membrane placed in contact with it. This provides a more permanent record of the sample since DNA begins to diffuse out of a gel that is left for a few hours. First, the gel is soaked in alkali to render the DNA single stranded. It is then transferred to the membrane so that the DNA becomes bound to it in exactly the same pattern as that originally on the gel. This transfer, named a Southern blot after its inventor Ed Southern, can be performed electrophoretically or by drawing large volumes of buffer through both the gel and the membrane, thus transferring DNA from one to the other by capillary action.[5] The point of this operation is that the membrane can now be treated with a labelled DNA molecule, for example a gene probe. This single stranded DNA probe will hybridise under the right conditions to complementary fragments immobilised onto the membrane (Figure 1.7). The conditions of hybridisation, including the temperature and salt concentration are critical for this process to take place effectively. This is usually referred to as the stringency of the hybridisation and it is particular for each individual gene probe and for each sample of DNA. A series of washing steps with buffer is then carried out to remove any unbound probe and the

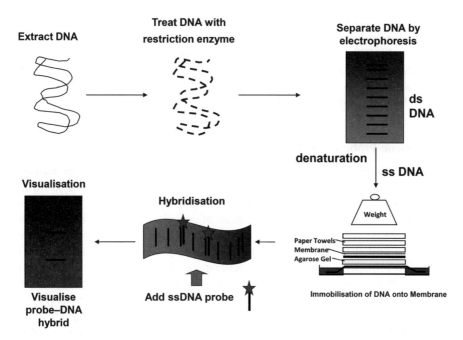

Extract DNA

Treat DNA with restriction enzyme

Separate DNA by electrophoresis

ds DNA

denaturation

ss DNA

Weight

Paper Towels
Membrane
Agarose Gel

Visualisation

Hybridisation

Immobilisation of DNA onto Membrane

Visualise probe–DNA hybrid

Add ssDNA probe

Figure 1.7 Steps involved in a Southern blot. Note that the denaturation step can be achieved using sodium hydroxide before the blotting stage with a nylon membrane.

membrane is developed after which the precise location of the probe and its target may be visualised. It is also possible to analyse DNA from different species or organisms by blotting the DNA and then using a gene probe representing a protein or enzyme from one of the organisms. In this way it is possible to search for related genes in different species. This technique is generally termed Zoo blotting.

1.3.2 RNA Blotting

The same basic process of nucleic acid blotting can be used to transfer RNA from gels onto similar membranes.[6] This allows the identification of specific mRNA sequences of a defined length by hybridisation to a labelled gene probe and is known as Northern blotting. With this technique it is possible to not only detect specific mRNA molecules but also to quantify the relative amounts of the specific mRNA. It is usual to separate the mRNA transcripts by gel electrophoresis under denaturing conditions since this improves resolution and allows a more accurate estimation of the sizes of the transcripts. The format of the blotting may be altered from transfer from a gel to direct application to slots on a specific blotting apparatus containing the nylon membrane. This is termed slot or dot blotting and provides a convenient means of measuring the abundance of specific mRNA transcripts without the need for gel electrophoresis; it does not however provide information regarding the size of the fragments.

A further method of RNA analysis that overcomes the problems of RNA blotting is termed the ribonuclease protection assay (RPA). This is a solution-based method where a probe that is complementary to the mRNA of interest is bound to form a hybrid that is resistant to digestion with RNAse. Thus while other single stranded RNA molecules are digested the intact hybrids can be further analysed by gel electrophoresis.

1.3.3   Gene Probe Derivation

The availability of a gene probe is essential in many molecular biology techniques yet in many cases is one of the most difficult steps.[7] The information needed to produce a gene probe may come from many sources but with the development and sophistication of genetic databases this is usually one of the first stages. There are a number of genetic databases throughout the world and it is possible to search these over the internet and identify particular sequences relating to a specific gene or protein. In some cases it is possible to use related proteins from the same gene family to gain information on the most useful DNA sequence. Similar proteins or DNA sequences but from different species may also provide a starting point with which to produce a so-called heterologous gene probe. Although in some cases probes are already produced and cloned it is possible, armed with a DNA sequence from a DNA database, to chemically synthesise a single stranded oligonucleotide probe. This is usually undertaken by computer controlled gene synthesisers, which link dNTPs together based on a desired sequence. It is essential to carry out certain checks before probe production to determine that the probe is unique, is not able to self-anneal or that it is self-complementary, all of which may compromise its use (Figure 1.8).

Where little DNA information is available to prepare a gene probe it is possible in some cases to use the knowledge gained from analysis of the corresponding protein. Thus it is possible to isolate and purify proteins and

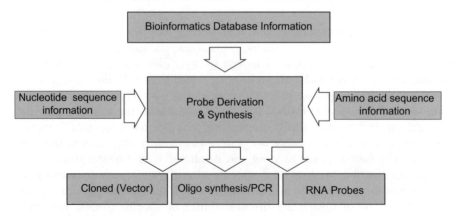

Figure 1.8   The various methods of producing a gene probe.

sequence part of the N-terminal end of the protein. From our knowledge of the genetic code, it is possible to predict the various DNA sequences that could code for the protein, and then synthesise appropriate oligonucleotide sequences chemically. Due to the degeneracy of the genetic code most amino acids are coded for by more than one codon, therefore there will be more than one possible nucleotide sequence that could code for a given poly-peptide. The longer the polypeptide, the greater the number of possible oligonucleotides which must be synthesised. Fortunately, there is no need to synthesise a sequence longer than about 20 bases, since this should hybridise efficiently with any complementary sequences, and should be specific for one gene. Ideally, a section of the protein should be chosen which contains as many tryptophan and methionine residues as possible, since these have unique codons, and there will therefore be fewer possible base sequences that could code for that part of the protein. The synthetic oligonucleotides can then be used as probes in a number of molecular biology methods.

1.4 LABELLING DNA GENE PROBE MOLECULES

An essential feature of a gene probe is that it can be visualised by some means. In this way, a gene probe that hybridises to a complementary se-quence may be detected and thus identify that desired sequence from a complex mixture.[7] There are two main ways of labelling gene probes. Fluorescent labelling is now a popular method for tagging nucleic acids and includes dyes such as Fluorescein amidite (FAM), digoxigenin-labelled nu-cleotides. Additionally fluorescent dyes such as 4′, 6-diamidino-2-phenyl-indole (DAPI) PicoGreen and RiboGreen, are commonly used. Radioactive labelling with 32 phosphorous (32P), or for certain techniques 35 sulfur (35S) and tritium (3H), can also be used. These may be detected by the process of autoradiography where the labelled probe molecule, bound to sample DNA, located for example on a nylon membrane, is placed in contact with an X-ray sensitive film. Following exposure the film is developed and fixed just as a black and white negative and reveals the precise location of the labelled probe and therefore the DNA to which it has hybridised.

Non-radioactive fluorescent labels are increasingly being used to label DNA gene probes and now many have similar sensitivities which, when combined with their improved safety, have led to their greater acceptance.

The labelling systems are termed either direct or indirect. Direct labelling allows an enzyme reporter such as alkaline phosphatase to be coupled dir-ectly to the DNA. Although this may alter the characteristics of the DNA gene probe they offer the advantage of rapid analysis since no intermediate steps are needed. However, indirect labelling is more popular at present. This relies on the incorporation of a nucleotide that has a label attached. At present three of the main labels in use are biotin, fluorescein and digoxy-genin. These molecules are covalently linked to nucleotides using a carbon spacer arm of 7, 14 or 21 atoms. Specific binding proteins may then be used as a bridge between the nucleotide and a reporter protein such as an enzyme. For example, biotin incorporated into a DNA fragment is recognised with a very high affinity by the protein streptavidin. This may either be coupled or conjugated to a reporter enzyme molecule such as alkaline phosphatase. This is able to convert a colourless substrate *para*-nitrophenol phosphate (PNPP) into a yellow coloured compound *para*-nitrophenol (PNP) and also offers a means of signal amplification. Alternatively, labels such as digoxygenin incorporated into DNA sequences may be detected by monoclonal antibodies, again conjugated to reporter molecules including alkaline phosphatase. Thus rather than the detection system relying on autoradiography, which is necessary for radiolabels, a series of reactions resulting in either a colour, light or chemiluminescence reaction takes place. This has important practical implications since autoradiography may take one to three days whereas colour and chemiluminescent reactions take minutes.

1.4.1 End Labelling of DNA Molecules

The simplest form of labelling DNA is by 5′ or 3′ end labelling. 5′ end labelling involves a phosphate transfer or exchange reaction where the 5′ phosphate of the DNA to be used as the probe is removed and in its place a labelled phosphate, usually 32P, is added. This is usually carried out by using two enzymes, the first, alkaline phosphatase, is used to remove the existing phosphate group from the DNA. Following removal of the released phosphate from the DNA a second enzyme polynucleotide kinase is added which catalyses the transfer of a phosphate group (32P labelled) to the 5′ end of the DNA. The newly labelled probe is then purified, usually by chromatography through a Sephadex column, and may be used directly (Figure 1.9).

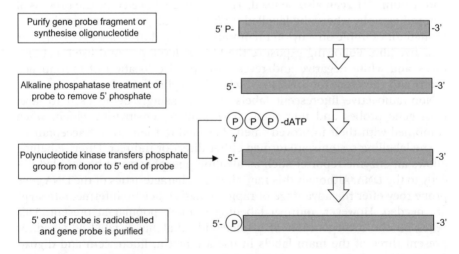

Figure 1.9 End labelling of a gene probe at the 5′ end with alkaline phosphatase and polynucleotide kinase.

Using the other end of the DNA molecule, the 3′ end, is slightly less complex. Here a new labelled dNTP (*e.g.* 32PadATP or biotin labelled dNTP) is added to the 3′ end of the DNA by the enzyme terminal transferase. Although this is a simpler reaction, a potential problem exists because a new nucleotide is added to the existing sequence and so the complete sequence of the DNA is altered which may affect its hybridisation to its target sequence. End labelling methods also suffer from the fact that only add one label is added to the DNA so they are of a lower activity in comparison to methods which incorporate label along the length of the DNA. Alternatively, fluorescent labels such as FAM may be used as an alternative to a radiolabel.

1.4.2 Random Primer Labelling

The DNA to be labelled is first denatured and then placed under renaturing conditions in the presence of a mixture of many different random sequences of hexamers or hexanucleotides. These hexamers will, by chance, bind to the DNA sample wherever they encounter a complementary sequence and so the DNA will rapidly acquire an approximately random sprinkling of hexanucleotides annealed to it. Each of the hexamers can act as a primer for the synthesis of a fresh strand of DNA catalysed by DNA polymerase since it has an exposed 3′ hydroxyl group. The Klenow fragment of DNA polymerase is used for random primer labelling because it lacks a 5′–3′ exonuclease activity. This is prepared by cleavage of DNA polymerase with subtilisin, giving a large enzyme fragment which has no 5′ to 3′ exonuclease activity, but which still acts as a 5′ to 3′ polymerase. Thus when the Klenow enzyme is mixed with the annealed DNA sample in the presence of dNTPs, including at least one which is labelled, many short stretches of labelled DNA will be generated. In a similar way to random primer labelling the polymerase chain reaction may also be used to incorporate radioactive or non-radioactive labels (Figure 1.10).

1.4.3 Nick Translation

A traditional method of labelling DNA is by the process of nick translation. Low concentrations of DNase I are used to make occasional single strand nicks in the double stranded DNA that is to be used as the gene probe. DNA polymerase then fills in the nicks, using an appropriate deoxyribonucleoside triphosphate (dNTP), at the same time making a new nick to the 3′ side of the previous one. In this way, the nick is translated along the DNA. If labelled dNTPs are added to the reaction mixture, they will be used to fill in the nicks, and so the DNA can be labelled to a very high specific activity.

1.5 THE POLYMERASE CHAIN REACTION (PCR)

There have been a number of key developments in molecular biology techniques; however, the one that has had the most impact in recent years has

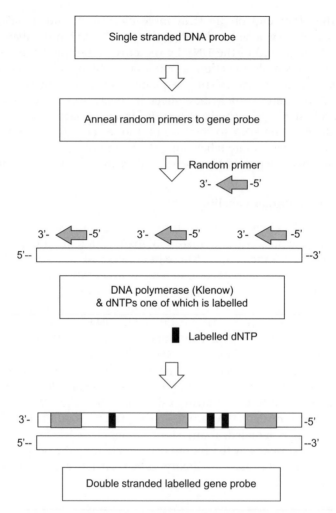

Figure 1.10 Random primer gene probe labelling. Random primers are incorporated and used as a start point for Klenow DNA polymerase to synthesise a complementary strand of DNA whilst incorporating a labelled dNTP at complementary sites.

been the polymerase chain reaction or PCR. One of the reasons for the adoption of the PCR is the elegant simplicity of the reaction and relative ease of the practical manipulation steps. Frequently this is one of the first techniques used when analysing DNA; it has opened up the analysis of cellular and molecular processes to those outside the field of molecular biology.[8,9]

The PCR is used to amplify a precise fragment of DNA from a complex mixture of starting material, usually termed the template DNA, and in many cases requires little DNA purification. In some respects, the polymerase chain reaction can be regarded as analogous to molecular cloning since it results in the generation of new DNA molecules based exactly upon the sequence of existing ones. PCR is a technique that is currently a mainstay of molecular biology. One of the reasons for the global adoption of the PCR is the elegant simplicity of the reaction and relative ease of the practical manipulation steps. Indeed, combined with the relevant bioinformatics resources for the design of oligonucleotide primers and for determination of the required experimental conditions, it provides a rapid means for DNA amplification, analysis and identification.

One problem with early PCR reactions was that the temperature needed to denature the DNA also denatured the DNA polymerase. However, the availability of a thermostable DNA polymerase enzyme isolated from the thermophilic bacterium *Thermus aquaticus* found in hot springs provided the means to automate the reaction. *Taq* DNA polymerase has an optimum temperature of 72 °C and survives prolonged exposure to temperatures as high as 96 °C and so is still active after each of the denaturation steps.

In contrast to conventional cell based cloning, PCR does require knowledge of the DNA sequences which flank the fragment of DNA to be amplified (target DNA). From this sequence information two oligonucleotide primers are chemically synthesised, each complementary to a stretch of DNA to the 3' side of the target DNA, one oligonucleotide for each of the two DNA strands (Figure 1.11). For many applications PCR has replaced the traditional DNA cloning methods as it fulfils the same function, the production of large amounts of DNA from limited starting material; however, this is achieved in a fraction of the time needed to clone a DNA fragment. Although not without its drawbacks the PCR is a remarkable development which has changed the approach of many scientists to the analysis of nucleic acids and continues to have a profound impact on core genomic and genetic analysis.

1.5.1 Steps in the PCR

The PCR consists of three well defined times and temperatures termed steps: (i) Denaturation at high temperature; (ii) annealing of primer and target DNA; and (iii) extension in the presence of a thermostable DNA polymerase. A single round of denaturation, annealing and extension is termed a 'cycle'. A typical PCR experiment consists of 30–40 cycles. In the first cycle the double stranded 'high molecular weight' template DNA is (i) denatured by heating the reaction mix to above 90 °C. Within the complex mass of DNA strands, the region to be specifically amplified (target) is thus made accessible to the primers. The temperature is then cooled to between 40–60 °C to allow the hybridisation of the two oligonucleotide primers, step (ii) which are present in excess, to bind to their complementary sites that flank the target DNA. The annealed oligonucleotides act as primers for DNA synthesis, since they provide a free 3' hydroxyl group for DNA polymerase. The DNA synthesis step (iii) is termed extension and carried out at 72 °C by a thermostable DNA polymerase, most commonly *Taq* DNA polymerase (Figure 1.12).

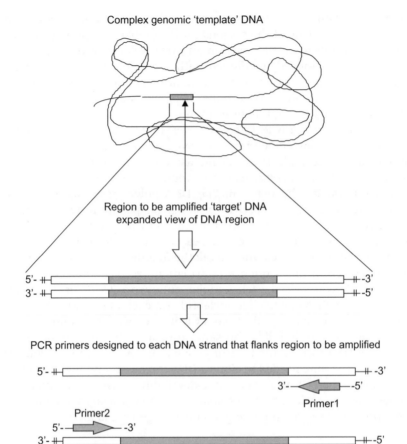

Complex genomic 'template' DNA

Region to be amplified 'target' DNA
expanded view of DNA region

PCR primers designed to each DNA strand that flanks region to be amplified

Primer2

Primer1

Primers are complementary to existing sequences necessitating
that some flanking sequence information is known

Figure 1.11   The location of PCR primers. PCR primers designed to sequences
adjacent to the region to be amplified allowing a region of DNA (*e.g.* a
gene) to be amplified from a complex starting material of genomic
template DNA.

DNA synthesis proceeds from both of the primers until the new strands
have been extended along and beyond the target DNA to be amplified. It is
important to note that since the new strands extend beyond the target DNA
they will contain a region near their 3′ ends, which is complementary to
the other primer (Figure 1.13). Thus, if another round of DNA synthesis is
allowed to take place not only will the original strands be used as templates
but also the new strands. The products obtained from the new strands will
have a precise length, delimited exactly by the two regions complementary to
the primers. As the system is taken through successive cycles of denaturation, annealing and extension all the new strands will act as templates and
so there will be an exponential increase in the amount of DNA produced.

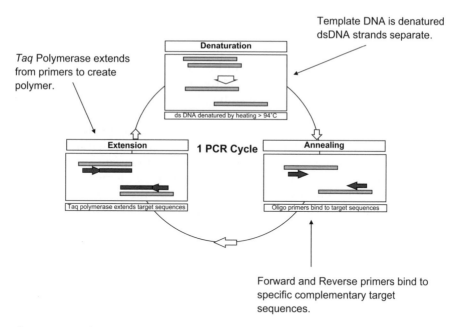

Figure 1.12 The three steps involved in one cycle of the PCR.

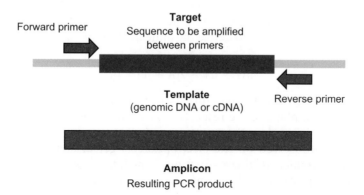

Figure 1.13 Terms associated with amplification components and resulting PCR products.

The net effect is to selectively amplify the target DNA and the primer regions flanking it leading to the production of millions of effectively identical copies.

For efficient annealing of the primers, the precise temperature at which the annealing occurs is critical and each PCR system has to be defined and optimised by the user. One useful technique for optimisation of annealing temperature is called Touchdown PCR where a programmable thermocycler machine is used to incrementally decrease the temperature until the optimum annealing is reached.[10] Reactions that are not optimised may give rise to other DNA products in addition to the specific target or may not produce any amplified products at all.[11] Another approach to reducing spurious non-specific amplification during the early stages of the reaction is termed 'Hot Start' PCR. In this method the reaction components are heated to the melting temperature before adding the polymerase. At one time Hot Start was achieved by introducing a physical wax barrier between the *Taq* polymerase and the remainder of the reaction components, which melted at the denaturation temperature allowing the *Taq* polymerase access to the reaction mix. More recently, modified polymerase enzyme systems have been developed that inhibit polymerisation at ambient temperature, either by the binding ligands such as antibodies or by the presence of bound inhibitors that dissociate only after a high-temperature activation step is performed.

1.5.2 PCR Primer Design

The specificity of the PCR lies in the design of the two oligonucleotide primers. These have to not only be complementary to sequences flanking the target DNA but must not be self-complementary or bind each other to form dimers since both prevent authentic DNA amplification. They also have to be matched in their GC content and have similar annealing temperatures and be incapable of amplifying unwanted genomic sequences. Manual design of primers is time consuming and often hit or miss, although formulae such as the following are still used to derive the annealing temperature or T_a for each primer:

$$4(G + C) + 2(A + T) = T_m \qquad (1.3)$$

Where T_m is the melting temperature of the primer/target duplex and G, C, A and T are the numbers of the respective bases in the primer. In general the T_a is set 3–5 °C lower than the T_m. On occasions, secondary or primer dimer bands may be observed on the electrophoresis gel in addition to the authentic PCR product. In these situations Touchdown or Hot start regimes may help. Alternatively, raising the T_a closer to the T_m can enhance the specificity of the reaction. The increasing use of bioinformatics resources such as PRIMER3, Generunner and Primer Design Assistant in the design of primers makes the design and selection of reaction conditions much more straightforward. These computer based resources allow the sequences to be amplified, primer length, product size, GC content *etc.* to be input and following analysis provides a choice of matched primer sequences. Indeed the initial selection and design of primers without the aid of bioinformatics would now be unnecessarily time-consuming. Some of these methods are useful when designing a multiplex PCR where more than one set of primers is included in a PCR. Here multiple amplicons are produced but the cycling conditions and design of the primers require careful optimisation. bioinformatic resources such as PrimerPlex are able to assist with the process of multiplex PCR design. Finally, before ordering or synthesising the primers it is wise to submit proposed sequences to a nucleotide sequence search program such as BLAST, which can be used to interrogate GenBank or other comprehensive public DNA sequence databases, to increase confidence that the reaction will be specific for the intended target sequence.[12]

1.5.3 Reverse Transcriptase PCR (RT-PCR)

RT-PCR is an extremely useful variation of the standard PCR which enables the amplification of specific mRNA transcripts from very small biological samples without the need for the rigorous extraction procedures associated with mRNA purification for conventional cloning purposes.[13] Conveniently, the dNTPs, buffer, *Taq* polymerase, oligonucleotide primers, reverse transcriptase (RT) and the RNA template are added together to the reaction tube. The reaction is heated to 37 °C thus allowing the RT to work and enables the production of a cDNA copy of the RNA strands that anneal to one of the primers in the mix. Following 'first strand synthesis' a normal PCR is carried out to amplify the cDNA product resulting in 'second strand synthesis' and subsequently a dsDNA product is amplified as usual. The choice of primer for the first strand synthesis depends on the experiment. If amplification of all mRNAs in the cell extract is required then an oligo dT primer that would anneal to all the poly A tails can be used. If a specific cDNA is sought, then a coding region specific primer can be used with success. The method is fast, accurate and simple to perform. It has many applications such as the assessment of transcript levels in different cells and tissues when combined with qPCR. When combined with allele specific primers, it also enables the amplification of cDNA from single chromosomes (Figure 1.14). RT-PCR is widely used as a diagnostic tool in microbiology and virology. A recent development in this area to try to quantitate rare alleles or viral nucleic acid copies is termed digital PCR (dPCR). Here the template is diluted or partitioned out to millions of single strands distributed in miniature chambers, emulsions and other surfaces. A PCR reaction is then performed such that a positive (1) or negative (0) will result. A determination of how many copies were in the original sample can then be assessed.[14] Here technological improvements have allowed the partition and amplification in titrated emulsions of oil which lowers the cost in comparison to other quantitative methods such as qPCR. This approach of solid phase PCR is also used in the template preparation for some of the next generation sequencing methods.[15]

1.5.4 Quantitative or Real Time PCR

One important evolution of the PCR method has been the development of Quantitative PCR (qPCR). This method has been gaining popularity for many applications because of the rapidity of the method compared to conventional PCR amplification whilst simultaneously providing a lower limit of detection and greater dynamic range.[16] As it is possible to track the reaction

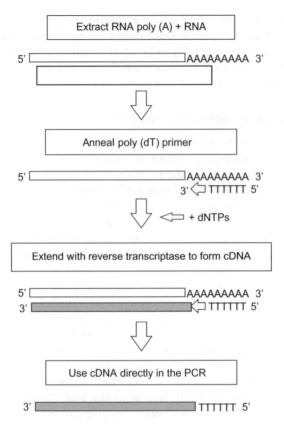

Figure 1.14   Reverse transcriptase PCR (RT-PCR). In RT-PCR mRNA is converted to complementary DNA (cDNA) using the enzyme reverse transcriptase. The cDNA is then used directly in the PCR.

in real time this allows assessments of any problems with the amplification process. Early quantitative PCR methods involved the comparison of a standard or control DNA template amplified with separate primers at the same time as the specific target DNA. These types of quantification rely on the reaction being exponential and so any factors affecting this may also affect the result. Other methods have involved the incorporation of a radiolabel through the primers or nucleotides and their subsequent detection following purification of the amplicon. In its simplest form a DNA binding dye such as SYBR Green is included in the reaction. As amplicons accumulate SYBR green binds the dsDNA proportionally. Fluorescence emission of the dye is detected following excitation. The binding of SYBR Green is non-specific but most qPCR machines will produce a melt curve that allows a degree of identity of the amplicon through its melting temperature profile. Indeed, refinement of this method, termed precision melt analysis, is able to differentiate between amplicons with one base difference and has value in

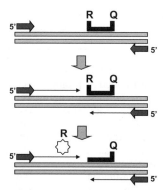

Figure 1.15 Representation of the 5′ nuclease assay or TaqMan approach of specific quantitative PCR where R is the reporter and Q is the quencher. Note *Taq* polymerase extends from the 3′ end of the primer in the second part and cleaves the R from Q in the lower panel producing a fluorescent signal.

mutation analysis. Subsequent DNA sequencing can provide a definitive confirmation. In order to detect specific amplicons during the PCR an oligonucleotide probe labelled with a fluorescent reporter and quencher molecule at either end can included in the reaction in the place of SYBR Green. This method is termed the 5′ fluorogenic exonuclease detection system or more commonly the TaqMan approach. When the oligonucleotide probe (TaqMan probe) binds to the target sequence the 5′ exonuclease activity of *Taq* polymerase degrades and releases the reporter from the quencher during extension. A signal is thus generated which increases in direct proportion to the number of starting molecules.[17] Thus the detection system is able to induce and detect fluorescence in real time as the PCR proceeds. Importantly it provides confirmation that the correct DNA sequence has been amplified (Figure 1.15). Refinements of the PCR process are also in development and promise rapid amplification and reporting. This is mainly due to advances in miniaturising the amplification system and using microfluidics on a small chip. Efficiency, multiplex throughput and low reagent costs are advantages of this system and similar methods are regularly employed in analysing nucleic acid and protein samples using Agilent Bioanalysers. At present, there is a move to provide a set of guidelines or information in the development and reporting of qPCR termed MIQE Guidelines or Minimum Information for Publication of Quantitative Real-Time PCR Experiments 17. There are numerous applications for the PCR as indicated in Table 1.2.

1.6 DNA SEQUENCING

The determination of the linear order or sequence of nucleotide bases along a length of DNA is one of the central techniques in molecular biology and

Table 1.2 Examples of applications of the PCR in various fields of biosciences. Certain techniques listed in the table are described in the text of chapters 2 and 3.

Field of Study	Applications	Specific Uses
DNA amplification	General molecular biology	Screening gene libraries
Bridge PCR	Next generation sequencing	Template cluster preparation
Production/labelling	Gene probe production	Use with blots/hybridisations
RT-PCR	RNA analysis	Active latent viral infections
Scenes of crime	Forensic Science	Analysis of DNA from blood
Microbial detection	Infection/disease monitoring	Strain typing/analysis RAPDs
Cycle sequencing	Sequence analysis	Rapid DNA sequencing possible
Referencing points in genome	Genome mapping studies	Sequence tagged sites STS
mRNA analysis	Gene discovery	Expressed sequence tags EST
Detection of known mutations	Genetic mutation analysis	Screening for Cystic fibrosis
Digital PCR (dPCR)	Quantification	Viral copy number analysis
Quantitative PCR (qPCR)	Quantification analysis	5′ nuclease (TaqMan assay)
Detection of unknown mutations	Genetic mutation analysis	Gel based PCR methods DGGE
Production of novel proteins	Protein engineering	PCR mutagenesis
Retrospective studies	Molecular archaeology	Dinosaur DNA analysis
Sexing or cell mutation sites	Single cell analysis	Sex determination of unborn
Studies on frozen sections	*In situ* analysis	Localisation of DNA/RNA

Abbreviations RT: reverse transcriptase, RAPDs: Rapid amplification polymorphic DNA, STS: Sequence tagged site, EST: Expressed sequence tags, DGGE: Denaturing gradient gel electrophoresis.

has played the key role in numerous genome mapping and sequencing projects. Two basic techniques were developed in the 1970s for efficient DNA sequencing, one based on an enzymatic method frequently termed Sanger sequencing, after its developer, and a chemical method termed Maxam and Gilbert sequencing, named for the same reason.[18–20] For large-scale DNA analysis Sanger sequencing and its variants are by far the most effective methods and many commercial kits are available for its use. However, there are certain occasions such as the sequencing of short oligonucleotides where the Maxam and Gilbert method is still more appropriate.

One absolute requirement for Sanger sequencing is that the DNA to be sequenced is in a single stranded form. Traditionally this demanded that the DNA fragment of interest be inserted and cloned into the specialised bacteriophage vector M13, which is naturally single stranded. Although M13 is still used, the advent of the PCR has provided the means not only to amplify a region of any genome or cDNA for which primer sequences are available, but also very quickly to generate the corresponding nucleotide sequence. This has led to an explosion in the accumulation of DNA sequence information and has provided much impetus for polymorphism discovery by resequencing regions of the genome from multiple individuals.

1.6.1 Sanger Chain Termination Sequencing

The Sanger method is simple and elegant and in many ways mimics the natural ability of DNA polymerase to extend a growing nucleotide chain based on an existing template.[18] Initially the DNA to be sequenced is allowed to hybridise with an oligonucleotide primer, which is complementary to a sequence adjacent to the 3′ side of DNA within a vector such as M13 (or within an amplicon in the case of PCR). The oligonucleotide will then act as a primer for synthesis of a second strand of DNA, catalysed by DNA polymerase. Since the new strand is synthesised from its 5′ end, virtually the first DNA to be made will be complementary to the DNA to be sequenced. One of the dNTPs required for DNA synthesis is labelled (or all four in some methods) with a fluorescent molecule or with 35S, and so the newly synthesised strand will be labelled. Alternatively, the primer may be labelled to provide the means of detection following the completion of the method.

1.6.2 Dideoxynucleotide Chain Terminators

The reaction mixture is then divided into four aliquots, representing the four dNTPs A, C, G and T. Using the adenine (A) tube as an example, in addition to all of the dNTPs being present in the mix, an analogue of dATP is added (2′3′-dideoxyadenosine triphosphate (ddATP)) which is similar to A except that it has no 3′ hydroxyl group. Since a 5′ to 3′ phosphodiester linkage cannot be formed without a 3′-hydroxyl group, the presence of the ddATP will terminate the growing chain. The situation for tube C is identical except that ddCTP is added; similarly, the G and T tubes contain ddGTP and ddTTP, respectively (Figure 1.16).

Since the incorporation of a ddNTP rather than a dNTP is a random event, the reaction will produce new molecules varying widely in length, but all terminating at the same type of base. Thus four sets of DNA sequence are generated, each terminating at a different type of base, but all having a common 5′ end (the primer). The four labelled and chain terminated samples are then denatured by heating and loaded next to each other on a polyacrylamide gel for electrophoresis. Electrophoresis is performed at approximately 70 °C in the presence of urea, to prevent renaturation of the DNA, since even partial renaturation alters the rates of migration of DNA fragments. Very thin, long electrophoresis gels are used for maximum resolution over a wide range of fragment lengths. After electrophoresis, the positions of labelled DNA bands on the gel are determined by

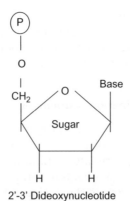

2'-3' Dideoxynucleotide

Figure 1.16 A 2'-3' Dideoxynucleotide. The base may be one of A, C, G or T (collectively known as ddNTPs).

autoradiography. Since every band in the lane from the ddATP sample must contain molecules which terminate at adenine, and that those in the ddCTP terminate at cytosine, *etc.*, it is possible to read the sequence of the newly synthesised strand from the autoradiogram, provided that the gel can resolve differences in length equal to a single nucleotide hence the ability to detect and characterise point mutations. Under ideal conditions, sequences up to about 300 bases in length can be read from one gel (Figure 1.17).

1.6.3 Sequencing Double Stranded DNA

It is also possible to undertake nucleotide sequencing from double stranded molecules such as plasmid cloning vectors and PCR amplicons directly. The double stranded DNA must be denatured prior to annealing with primer. In the case of plasmids, an alkaline denaturation step is sufficient. However, for PCR amplicons this is more problematic. Unlike plasmids, amplicons are short and reanneal rapidly. Denaturants such as formamide or dimethyl-sulfoxide (DMSO) have been used with to prevent the reannealing of PCR strands following their separation. Another strategy is to bias the amplification towards one strand by using a primer ratio of 100 : 1 which also overcomes this problem to a certain extent.

It is possible to physically separate and retain one PCR product strand by incorporating a molecule such as biotin into one of the primers. Following PCR the strand that contains the biotinylated primer may be removed by affinity chromatography with streptavidin coated magnetic beads, leaving the complementary PCR strand. This magnetic affinity purification provides single stranded DNA derived from the PCR amplicon and although somewhat time consuming, it does provide high quality single stranded DNA for sequencing.

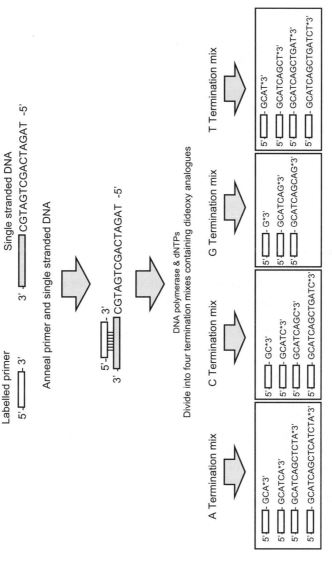

Figure 1.17 Overall scheme of the mechanism of chain termination sequencing.

1.6.4 Automated DNA Sequencing

Advances in fluorescent labelling chemistry have led to the development of high-throughput automated sequencing techniques.[21] Essentially most systems involve the use of dNTPs labelled with different fluorochromes (often referred to as dye terminators). The advantage of this modification is that since a different label is incorporated with each ddNTP it is unnecessary to perform four separate reactions. Therefore the four chain terminated products are run on the same track of a denaturing electrophoresis gel, each with their base specific dye such as Fluorscein or Tetramethyl-Rhodamine. These are excited by a laser and the dye then emits light at its characteristic wavelength. A diffraction grating separates the emissions which are detected by a charge coupled device (CCD) and the sequence interpreted by a computer. One advantage of this technique includes real time detection of the sequence. In addition the lengths of sequence that may be analysed are in excess of 500 bp. Capillary electrophoresis is increasingly being used for the detection of sequencing products. This is where liquid polymers in thin capillary tubes are used obviating the need to pour sequencing gels and requiring little manual operation. This substantially reduces the electrophoresis run times and allows high throughput to be achieved. A number of large-scale sequence facilities are now fully automated allowing the rapid acquisition of sequence data. Automated sequencing for genome projects is usually based on cycle sequencing using instruments such as the ABI PRISM 3700 DNA Analyzer.[22] This can be formatted to produce simultaneous reads in 384-well cycle sequencing reaction plates. The derived nucleotide sequences are downloaded automatically to databases and manipulated using a variety of bioinformatics resources (Figure 1.18).

1.6.5 Next Generation Sequencing

Advances in sequencing methods and developments in automation led to what is termed next generation sequencing (NGS).[23] This collection of methods, based on different platforms offers rapid, high-throughput analysis with the use of massively parallel systems that can determine the sequence of large numbers of different DNA strands at one time.[24] Currently the main NGS systems involve template preparation with some form of enrichment by PCR, usually followed by a sequencing by synthesis approach in a massively parallel fashion. Various platforms have been developed to undertake these processes. They include Illumina sequencing, Roche 454 sequencing, ABI's SOLiD sequencing, Ion torrent: Proton/PGM and Pacific Biosciences' real time sequencing system, among others. NGS has numerous applications in many fields including *de novo* sequencing of new genomes, whole genome sequencing (WGS), RNA-sequencing, targeted re-sequencing of specific genes, exome sequencing of protein coding regions, microbiome sequencing, ribonomics and epigenomics. One general feature of these systems or platforms is that they perform the methods in parallel, termed

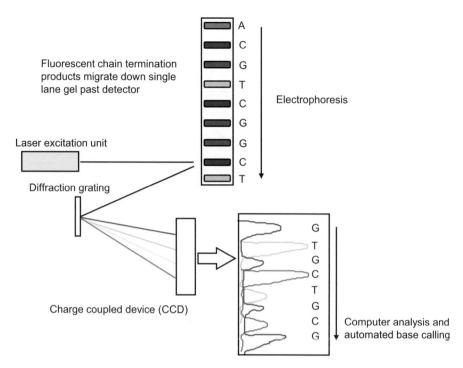

Figure 1.18 Representation of automated fluorescent dye terminator sequencing.

'massively parallel sequencing', so that large amounts of sequence data are produced simultaneously. Many of the methods also allow each fragment that is being sequenced to be uniquely tagged with a short identifying barcode in the initial sample preparation stage. The management, processing and analysis of enormous amounts of data in the reconstruction of large fragments of DNA derived from shorter reads also requires powerful computer hardware processing. At present the set up and capital costs for NGS systems are high in comparison to traditional Sanger sequencing although this will decrease as uptake of NGS rises. With the huge increase in data that NGS and massively parallel sequencing brings there is also a requirement for efficient bioinformatics resources. Table 1.3 indicates a list of a number of current and future NGS platforms.

1.6.6 Illumina Solex Sequencing Platform

The Illumina/Solexa method is similar to Sanger sequencing and is one example of a sequencing by synthesis method. Initially genomic DNA is fragmented into shorter fragments and adapters ligated to each end of the DNA. The DNA fragments are then immobilised onto a slide or flow cell containing a lawn of primers.[25] The DNA fragments hybridise at one end to one of the immobilised primers and then bend over and bind to a complementary primer also bound on the slide surface. The bridge formed is amplified by PCR (Bridge PCR) and produces DNA colonies or clusters in

Table 1.3 Comparison of a number of selected Next Generation Sequencing (NGS) methods.

Technique	Principal	Time (of run)	Length (of read)	Reads (per run)	Accuracy	Cost/10^6 bases	Advantage	Disadvantage
Sanger	Chain Termination	<3 Hours	400–900 bp	–	99.9%	$2400	Individual reads, highly accurate	Expensive, impractical for genomes
Illumina	Sequencing by Synthesis	1–10 Days	50–250 bp	≤3 million	98.0%	$0.15	High sequence yield	Expensive equipment
454	Pyrosequencing	24 Hours	700 bp	1 million	99.9%	$10	Read length, fast	Homopolymer errors
SOLiD	Sequencing by Ligation	7–14 Days	50 + 35/50 + 50 bp	1.2–1.4 billion	99.9%	$0.13	Accuracy	Short read assembly
Ion Torrent	Ion Semiconductor	2 Hours	≤400 bp	≤80 million	98.0%		Cheaper equipment	Homopolymer errors
Pacific Bio	Single molecule real time	0.5–2 hours	5000–8000 bp	400 Mb	99.9%	$0.33–$1	Long reads	Expensive equipment

each channel of the flow cell. The sequencing part of the technology involves the use of reversible dye terminators. Following denaturation one strand is removed and sequencing primers, polymerase and nucleotides are added. Each base is tagged with a different fluorophore and a $3'$ block and so as a complementary base is flowed across the cell it is incorporated and a laser is used to excite the fluorescence thus identifying the base. This is recorded in an image and happens in parallel at each cluster. Following this, the fluorophore is cleaved, the block removed, and the second cycle begins. All the information is processed to build up the sequence in each of the clusters. It took approximately 13 years and $3 billion to sequence the first human genome more than a decade ago. The current cost is estimated to be $10 000 although Illumina has announced the first $1000 genome sequence (Figure 1.19).

1.6.7 454 and Pyrosequencing

The 454 platform utilises the pyrosequencing format involves direct analysis of DNA fragments and this system allows the rapid sequencing of entire genomes by the 'shotgun' approach. First genomic DNA is randomly sheared and ligated to linker sequences that permit individual molecules captured on the surface of a bead to be amplified while isolated within an emulsion droplet by emulsion PCR (emPCR). Each bead is designed to contain one DNA fragment. A very large collection of such beads is arrayed in the 1.6 million wells of a fibre-optic slide.

Each PCR template is hybridised to an oligonucleotide and incubated with DNA polymerase, ATP sulfurylase, luciferase and apyrase. During the re-action the first of the four dNTPs are added and if incorporated release pyrophosphate (PPi), hence the name 'pyrosequencing'. The ATP sulfurylase converts the PPi to ATP, which drives the luciferase-mediated conversion of luciferin to oxyluciferin to generate light. Apyrase degrades the resulting component dNTPs and ATP. This is followed by another round of dNTP addition. A resulting pyrogram provides an output of the sequence.[26] The method provides short reads very quickly and is especially useful for the determination of mutations or SNPs. Current technology of 454 sequencing allows one gigabase or a third of the human genome to be derived in a single day in comparison to Sanger sequencing which would take over a year to complete.

1.6.8 ABI SOLiD Sequencing by Oligonucleotide Ligation and Detection

An alternative NGS platform developed by Applied Biosystems is the SOLID (Sequencing by Oligonucleotide Ligation and Detection) system. This pro-cess is similar to 454 technology in that it uses emulsion PCR (emPCR) to generate clonal amplification of single stranded DNA fragments (ligated to adapters) attached to beads. The beads are then attached to a glass at a high density.[27]

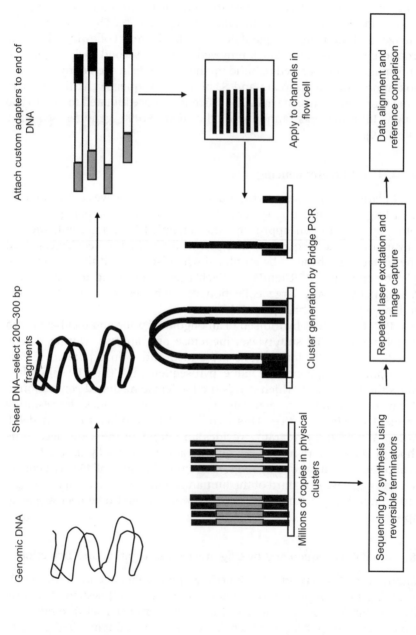

Figure 1.19 Schematic representation of the steps involved in the Illumina method of DNA sequencing.

A universal sequencing primer is hybridised to the adapter sequence on the beads containing the template. Ligase, together with a set of four fluorescently labelled di-base probes, is added to the reaction and compete for ligation to the sequencing primer. Specificity of the di-base probe is achieved by interrogating every 1st and 2nd base in each ligation reaction. Hybridisation, ligation, fluorescence detection and cleavage are performed for a predetermined number of cycles after which the extension product is removed and the template is reset with a primer offset by one base for a second round of ligation cycles.

1.6.9 Single Molecule Real Time Sequencing

A different approach that has recently been employed is the observation of DNA single molecule sequencing in real time.[28] This unique process is used by Pacific Biosciences with their SMRT platform. The method utilises phospholinked nucleotides where a fluorescent dye is attached through the phosphate group in the molecule, each of the four nucleotides having a different fluorescent dye. Thus during incorporation of the nucleotide into a growing chain by a DNA polymerase the phosphate is cleaved while a detector identifies which of the four nucleotides has been incorporated. The technology is based around a laser light illuminated nanophotonic containment cylinder termed a zero-mode waveguide within which the reaction takes place. Remarkably these cylinders are smaller than the wavelength of light and a sophisticated CCD detects flashes of light when a nucleotide is incorporated and records this as a movie.

There are many interesting future developments in DNA sequencing including Ion Torrent and nanoball systems and further sequencing by mass spectrometry, nanopore and tunnelling. There is no doubt that all the sequencing methods described and those in development will continue to have a profound effect on biosciences as did the original methods developed in the late 1970's. However, these methods generate vast amounts of data and it is now the developments in hardware and software that may provide the key to their success.[29]

1.7 BIOINFORMATICS

Modern molecular biology produces a tremendous amount of data from sequencing techniques. The development of computerised databases that store, organise and analyse biological information is now absolutely essential. In general, bioinformatics is a discipline of science that uses computational techniques to organise and analyse large amounts of biological data.[30]

1.7.1 Biological Databases

Databases were initially developed for different projects such as DNA sequencing, drug design, proteomic analysis and a range of biomedical issues.[31] Bioinformatic databases can be broadly classified into primary,

Table 1.4   A number of bioinformatic resources available over the world wide web.

---

1. Basic Local Alignment Search Tool. (2013). *Sequence comparisons database.* Available from http://blast.ncbi.nlm.nih.gov/.
2. ClustalW. (2013). *Multiple sequence alignment database.* Available from http:// www.ebi.ac.uk/Tools/msa/clustalw2/.
3. Expasy. (2010). *Protein analysis system.* Available from http://www.expasy.org/ tools.
4. Genbank. (2013). *Genetic sequence database.* Available from http://www.ncbi.nlm. nih.gov/genbank/.
5. GeneFisher. (2013). *PCR primer design.* Available from http://bibiserv.techfak. uni-bielefeld.de/genefisher2/.
6. Online Mendelian Inheritance in Man. (2013). *Compendium of human genetic disorders.* Available from http://www.ncbi.nlm.nih.gov/omim.
7. Pfam. (2013). *Collection of protein families.* Available from http://pfam.sanger. ac.uk/.
8. Protein Information Resource. (2009). *Protein sequences and functional information.* Available from http://pir.georgetown.edu/.
9. Protein Data Bank. (2013). *An information portal to biological macromolecular structures.* Available from http://www.rcsb.org/pdb/home/home.do.
10. PubMed. (2013). *Biomedical literature database.* Available from http://www.ncbi. nlm.nih.gov/pubmed.
11. Universal Protein Resource. (2013). *Catalogue of information on proteins.* Available from http://www.uniprot.org/.

---

secondary and specialised databases, depending on their contents. Primary databases contain raw sequence or structural data. Examples of these include GenBank and Protein Data Bank (PDB). Secondary databases contain information derived from the original primary databases. These are more sophisticated and require computational processing of raw data. Examples of secondary databases are SWISS-PROT and Protein Information Resource (PIR). Specialised databases have been developed that are concerned with a specific research area or a particular organism such as HIV database or FlyBase. An overview of some popular databases with their respective URLs is presented in Table 1.4.

1.7.2   Nucleotide Sequence Databases

There are three main sequence data resources that are available free on the world wide web: GenBank, European Molecular Biology Laboratory (EMBL) and DNA Data Bank of Japan (DDBJ). Data on nucleotide sequences are generated and submitted by scientists worldwide. These three databases closely collaborate and share information on a regular basis. GenBank, EMBL and DDBJ constitute International Nucleotide Sequence Database Collaboration (INSC). This means that the same nucleotide sequence is available to a user regardless of the database chosen. Despite the fact that these three databases contain identical raw sequence data, the output is presented in slightly different formats. The GenBank sequence format is described later in this chapter.

1.7.3 GenBank at National Centre for Biotechnology Information (NCBI)

GenBank is one of the fastest growing nucleic acid sequence repositories. This rapid growth is mainly because of the Expressed Sequence Tags (ESTs), short fragments of cDNA sequence, which are submitted to the genetic sequence databases. GenBank stores sequences that are derived from various organisms. These may be *Homo sapiens* (human), *Mus musculus* (house mouse), *Rattus norvegicus* (Norwegian rat) and *Zea Mays* (maize). Due to a high data volume GenBank files are seperated into divisions according to the sequence origin. Examples include viral (VRL), bacterial (BCT), high-throughput genomic (HTG) and high-throughput cDNA (HTC) divisions. In general, there are 17 such divisions in GenBank.[32]

Many biological databases contain hyperlinks to the relevant records in GenBank which has important information in the files. GenBank has a flat file structure and is divided into three distinct sections: header, features and sequence. The header section consists of a number of different subsections. The first line in the header section is the LOCUS number, which is a unique identifier of the sequence. There is also information on sequence length, molecule type, GenBank division and publication date. DEFINITION refers to a brief description of the sequence. The next line contains ACCESSION field that provides a code number for a sequence record. VERSION and GENE INDEX (GI) present the number of sequence revisions. If the sequence data is changed then the accession number remains the same but version and GI are changed. KEYWORDS field lists words/phrases associated with the sequence. SOURCE indicates the origin of an organism from which the data have been derived. There is information on publications related to a given sequence in the REFERENCE subsection. The features section contains SOURCE field that determines sequence length, scientific name of the source of the organism and the taxonomy identification number. GenBank records of the DNA sequence also contain CDS field, which refers to the boundaries of the gene's coding sequence. This subsection also contains information on exons locations and protein sequence after translational processes. The final section ORIGIN points to the start of the sequence.

1.7.4 Bioinformatics Data Formats

There are many other alternative sequence formats available and one important one is the FASTA format. FASTA is popular because of its simplicity as it only contains information about the sequence and has a format suitable for other databases. The first line is a sequence name that begins with (>) symbol. The actual sequence is written in one-letter code starting in the second line.

1.7.4.1 NCBI Databases Resources: BLAST, PubMed and OMIM

  a) Basic Local Alignment Search Tool (BLAST) is an important bioinformatic tool that allows searching sequence databases in order to identify matching sequences. The query sequence (sequence of interest) is compared with related sequences (subject sequences) that are found in the database and statistical significance of matches is calculated. BLAST is often used to aid in the identification of gene family members. There are several different BLAST programs and they are categorised according to the sequence to be analysed:

- BLASTn compares nucleotide sequence to nucleotide database
- BLASTp compares protein sequence to protein database
- BLASTx compares translated nucleotide sequence to protein database
- tBLASTn compares protein sequence to translated nucleotide database
- tBLASTx compares translated nucleotide sequence to translated nucleotide database

Once the appropriate BLAST is chosen a page is returned that allows pasting of the sequence in FASTA format to a query window and searching the database for similarity match.

BLAST produces graphical as well as textual representation of data. The graphical view includes coloured thin horizontal lines that correspond to the number of matches and their quality (red: most related; green and blue: moderately related, black: unrelated). The length of the line indicates the region of similarity between sequences. Each individual line is hyperlinked to its alignment sequence that can be viewed in the textual fragment of the BLAST results. Below the graphical view is the table with a list of entries matching the query sequence sorted according to Expect (E)-values in declining order. The E-value determines the statistical significance of the match. The lower the E-value the more significant the similarity match. The BLAST output also contains accession codes that are hyperlinked to the GenBank file and max scores that indicate how good the match was (the higher the score the better the sequence match). Query cover refers to the extent of sequence alignments and Identity represents the degree of sequences similarities. The alignment output is arranged into header, statistics and alignment subsections. The header subsection provides the name of the sequence and its identification number. The statistics subsection contains bit score (statistical measure of the sequence similarity), E-value, percentage of similarity, gap existence and strand (sense/antisense). The alignment presents the aligned sequences.

b) The NCBI also provides an access to PubMed. PubMed is one of the most popular textual databases with more than 23 million references from the peer-reviewed biomedical literature. PubMed takes advantage of the MeSH (Medical Subject Headings), which is a vocabulary thesaurus helping in the identification of relevant literature.

c) OMIM (Online Mendelian Inheritance in Man) is another comprehensive textual database available at NCBI that contains information about human genes that are responsible for various genetic diseases. The text contains a description of the disease, its clinical features, location of the gene on chromosome and reference details. OMIM is an ideal tool for the analysis of human genetic disorders.

1.7.5 Protein Databases: SWISS-PROT, TrEMBL, PIR, Pfam and PDB

One important protein sequence database is SWISS-PROT, which contains detailed descriptive information on proteins.[33] SWISS-PROT can be accessed *via* Expert Protein Analysis System (Expasy) server that offers various software tools for the analysis of proteins. Sequence data is obtained through TrEMBL (Translation from TREMBL), a database that contains translated nucleotide sequences derived from EMBL database. SWISS-PROT contains annotation data, which are checked and updated by external experts so the level of quality of this database is high. Essentially, annotations contain comments regarding function of the protein, posttranslational modifications, domain structure, binding sites, similarities to other protein sequences and protein-deficiency associated diseases. The majority of scientific data is collected from publications and review articles and is verified prior to submission. Each protein sequence has its corresponding annotation and this is very useful. In addition, sequence data contains links to other biological databases such as OMIM (catalogue of human genetic disorders) and PROSITE (dictionary of protein domains, families and functional sites) to facilitate the search process.

The PIR is another important bioinformatic tool that provides extensive protein sequence information and aids in the investigation and the analysis of proteins. PIR, SWISS-PROT and TrEMBL have been integrated and constitute the UniProt (Universal Protein Resource) database that covers a wider range of issues concerning proteins and simultaneously provides high quality of annotations, minimal redundancy and cross-references. UniProt is divided into UniProt Knowledgebase (UniProtKB), UniProt Reference Clusters (UniRef) and UniProt Archive (UniParc) databases. The UniProtKB comprises of detailed protein sequence information, including protein classifications and cross-references. The UniRef provides a cluster of closely related sequences that increase the speed of similarity searches and the UniParc contains information on history of protein sequences.[34]

There are also very useful databases that assist in the identification of protein domains such as Pfam database. At present, there are nearly 15 000 protein families stored in Pfam. The identification of domains within proteins is important as it may give rise to the discovery of functional properties of an unknown protein. PDB is a central repository of three-dimensional structures of large biological molecules. Structure of macromolecules (proteins and nucleic acids) is determined by nuclear magnetic resonance (NMR) or X-Ray crystallography. The PDB has a flat file structure containing the following information: protein name, authors, experimental design, secondary structure, compound details and cross-references. PDB also allows manipulating the image using software applications.

1.7.6 Multiple Sequence Alignments

An important resource that allows direct multiple sequence alignments of both nucleic acid and protein sequences is ClustalW. Using this allows the study of evolutionary processes and ancestral relationships.[35] ClustalW displays aligned sequences as either black and white or colour coded text. The colouring of residues depends on physicochemical characteristics of proteins. As a general rule, AVFPMILW residues are coloured red, DE are blue, RHK are magenta, STYHCNGQ are green and other remaining residues are grey. Below each alignment is a (*) symbol indicating identical match. Highly conserved sequences are displayed with a (:) symbol, whereas weakly conserved sequences are displayed with a (.) symbol. Gaps can also be present in the alignments and should be considered when designing probes or PCR primers for a particular gene. It is unthinkable nowadays to begin any biological research without the use of bioinformatics. Indeed, progress in sequencing, genome analysis, development of molecular biological and biotechnological methods all rely heavily on this area.

REFERENCES

1. J. D. Watson and F. H. Crick, *Nature*, 1953, **171**(4361), 964.
2. *The Nucleic Acids Protocols Handbook, 1st Ed*, ed. R. Rapley, Humana Press, Totowa, NY, 2000.
3. D. T. Gjerde and L. Hoang, *RNA Purification and Analysis: Sample Preparation, Extraction, Chromatography, 1st Ed*, John Wiley & Sons, Chichester, 2009.
4. H. O. Smith and K. W. Wilcox, *J. Mol. Bio.*, 1970, **51**, 379.
5. E. M. Southern, *J. Mol. Bio.*, 1975, **98**, 503.
6. J. C. Alwine, D. J. Kemp and G. R. Stark, *Proc. Natl. Acad. Sci. U. S. A.*, 1977, **74**, 5350.
7. A. de Muro and R. Rapley, *Gene Probes*, Humana Press, Totowa, NY, 2002, vol. 179.
8. R. K. Saiki, S. Scharf, F. Faloona, K. B. Mullis, G. T. Horn, H. A. Erlich and N. Arnheim, *Science*, 1985, **230**, 1350.
9. R. K. Saiki, D. H. Gelfand, S. Stoffel, S. J. Scharf, R. Higuchi, G. T. Horn, K. B. Mullis and H. A. Erlich, *Science*, 1988, **239**, 487.
10. R. H. Don, P. T. Cox, B. J. Wainwright, K. Baker and J. S. Mattick, *Nucleic Acids Res.*, 1991, **19**, 4008.
11. D. J. Sharkey, E. R. Scalice, K. G. Christy Jr., S. M. Atwood and J. L. Daiss, *Biotechnology*, 1994, **12**, 506.
12. *PCR Primer Design*, ed. A. Yuryev, Humana Press, Totowa, NY, 2007, vol. 402.
13. S. Corbet, J. Bukh, A. Heinsen and A. Fomsgaard, *J. Clin. Microbiol.*, 2003, **41**, 1091.
14. R. Sanders, J. Huggett, C. A. Bushell, S. Cowen, D. J. Scott and C. A. Foy, *Anal. Chem.*, 2011, **83**(17), 6474.

15. L. Biesecker, W. Burke, I. Kohane, S. E. Plon and R. Zimmern, *Nature Reviews*, 2012, **13**(1), 818.
16. R. Higuchi, C. Fockler, G. Dollinger and R. Watson, *Biotechnology*, 1993, **11**, 1026.
17. *PCR Technologies*, ed. T. Nolan and S. Bustin, CRC Press, Florida, USA, 2013.
18. F. Sanger, S. Nicklen and A. R. Coulson, *Proc. Natl. Acad. Sci. U. S. A.*, 1977, **74**(12), 5463.
19. A. M. Maxam and W. Gilbert, *Proc. Natl. Acad. Sci. U. S. A.*, 1977, **74**(2), 560.
20. L. T. C. Franca, E. Carrilho and T. B. L. Kist, *Q. Rev. Biophys.*, 2002, **35**(2), 169.
21. L. M. Smith, J. Z. Sanders, R. J. Kaiser, P. Hughes, C. Dodd, C. R. Connell, C. Heiner, S. B. Kent and L. E. Hood, *Nature*, 1986, **321**(6071), 674.
22. K. Murphy, K. Berg and J. Eshleman, *Clin. Chem.*, 2005, **51**(1), 35.
23. M. L. Metzker, *Nat. Rev. Genet.*, 2010, **11**, 31.
24. D. C. Koboldt, K. M. Steinberg, D. E. Larson, R. K. Wilson and E. R. Mardis, *Cell*, 2013, **155**, 27.
25. E. R. Mardis, *Hum. Genet.*, 2008, **9**, 387.
26. M. Ronaghi, S. Shokralla and B. Gharizadeh, *Pharmacogenomics*, 2007, **8**, 1437.
27. S. C. Schuster, *Nat. Methods*, 2008, **5**(1), 16.
28. M. A. Quail, M. Smith, P. Coupland *et al.*, *BMC Genomics*, 2012, **13**, 341.
29. C. Del Fabbro, S. Scalabrin, M. Morgante and F. M. Giorgi, *PLoS ONE*, 2013, **8**(12), e85024.
30. P. Hogeweg, *PloS Comput. Biol.*, 2011, **7**(3), e1002021.
31. J. D. Wren and A. Bateman, *Bioinformatics*, 2008, **24**(19), 2127.
32. D. A. Benson, I. Karsch-Mizrachi, D. J. Lipman, J. Ostell and D. L. Wheeler, *Nucleic Acid Res.*, 2004, **32**, 23.
33. B. Boeckmann, A. Bairoch, R. Apweiler, M. C. Blatter, A. Estreicher, E. Gasteiger, H. Huang, R. Lopez, M. Magrane, M. J. Martin, D. A. Natale, C. O'Donovan, N. Redaschi and L. S. Yeh, *Nucleic Acid Res.*, 2003, **31**(1), 365.
34. R. Apweiler, A. Bairoch, C. H. Wu, W. C. Barker, B. Boeckmann, S. Ferro, E. Gasteiger, H. Huang, R. Lopez, M. Magrane, M. J. Martin, D. A. Natale, C. O. Donovan, N. Redaschi and L. S. Yeh, *Nucleic Acids Res.*, 2004, **32**(90001), 115D.
35. M. A. Larkin, G. Blackshields, N. P. Brown, R. Chenna, P. A. McGettigan, H. McWilliam, F. Valentin, I. M. Wallace, A. Wilm, R. Lopez, J. D. Thompson, T. J. Gibson and D. G. Higgins, *Bioinformatics*, 2007, **23**(21), 2947.

CHAPTER 2

Genes and Genomes

DAVID B. WHITEHOUSE

School of Life and Medical Sciences, University of Hertfordshire,
College Lane, Hatfield, Hertfordshire, AL10 9AB, United Kingdom
Email: d.whitehouse@herts.ac.uk

2.1 INTRODUCTION

Groundbreaking discoveries in the 20th century include the double helix structure of DNA and cracking the genetic code. This knowledge paved the way to the development of recombinant DNA technology and "molecular biology" that has since enabled the identification, mapping, isolation and sequencing of genes and genomes. In the 50 years between the reporting of the DNA double helix in 1953 and the completion of the human genome sequence, dramatic progress in biotechnology has led to advances on many fronts including molecular cloning, genome wide mapping and mutation and polymorphism detection. Some of the enabling molecular methodologies and their applications are addressed in this chapter.

2.1.1 Genomics Background

'Genome' refers to the entire collection of hereditary material (DNA or RNA) in an organism or cell. The animal genome is subdivided into the nuclear genome (henceforth referred to as the genome) and the much smaller mitochondrial genome. The human genome is some 3 billion base pairs and consists of 23 linear DNA molecules that are packaged into 23 chromosomes and is estimated to encode about 22 000 genes, somewhat less than the estimated 35 000 genes in the grape (*Vitis sp.*) genome.[1] In contrast, the

Molecular Biology and Biotechnology, 6th Edition
Edited by Ralph Rapley and David Whitehouse
© The Royal Society of Chemistry 2015
Published by the Royal Society of Chemistry, www.rsc.org

mitochondrial genome consists of a single circular chromosome comprising of about 16 600 base pairs that encodes just 37 genes, of which 13 are protein coding. Modern genomics is the analysis of genes and genomes on a massive scale through application of high-throughput DNA technology platforms and bioinformatics that offer an unparalleled perspective on the organisation of the genetic material and genetic variation. There are potentially clear-cut benefits of genomic analysis to many disciplines, notably medical science *e.g.* for new diagnostics and biomarker discovery and agricultural bio-technology *e.g.* the development of new strains of GM crops. In the commercial sector, the potential for individual genetic testing has driven debates concerning and availability of personal genetic information.

2.1.2 Genomes and Chromosomes

Cellular genomes (*i.e.* non-viral) vary tremendously in size and complexity. The smallest known, from a bacterium contains about 600 000 DNA base pairs; in contrast, the eukaryote nuclear genome, which is the main focus of this chapter, often consists of billions of base pairs (Table 2.1). In many eukaryotes, notably representatives of plants, fungi, invertebrates, fish, amphibians and reptiles, genomic analysis is complicated by the polyploid nature of the genomes where nuclear DNA is packaged into several sets of chromosomes. In humans and the majority of the animal kingdom, including 'model species' such as mice and fruit flies, DNA is packaged into two sets of chromosomes *i.e.* the cells are said to be diploid. The human diploid genome consists of 23 pairs of chromosomes (22 autosomes and two sex chromosomes) that differ in size and staining properties, one set being inherited from each parent. When normal male cells undergoing metaphase are arrested, stained and examined under a light microscope, each pair of autosomes appears identical, but the sex chromosomes X and Y are distinct. In normal female cells, which have two X chromosomes, all 23 pairs of metaphase chromosomes appear identical.

2.1.3 Genomic Analysis

In prokaryotes, genomic analysis is simplified because the genomes consist of single chromosomes and simple gene structures, *i.e.* the genes comprise

Table 2.1 Approximate haploid genome sizes in various organisms.

Organism		Genome size
Bacteria	*Mycoplasma genitalium*	0.6 Mb
	Escherichia coli	4.6 Mb
Yeast	*Saccharomyces cerevisiae*	14 Mb
Nematode worm	*Caenorhabditis elegans*	100 Mb
Flowering plant	*Arabidopsis thalania*	115 Mb
Fruit fly	*Drosophilia melanogaster*	165 Mb
Puffer fish	*Fugu rubripes rubripes*	400 Mb
Human	*Homo sapiens*	3300 Mb

uninterrupted arrays of amino acid coding DNA sequences that extend from the transcription start site to the end of the coding region and there is relatively little intergenic non-coding DNA. In contrast, the DNA of the majority of eukaryote genes is broken up into small segments of coding DNA called exons and non-coding intervening sequences called introns. Furthermore, eukaryotic genes are often separated by vast lengths of intergenic non-coding DNA.

The first genome to be sequenced was from the virus-like bacteriophage PhiX174 and consists of 5386 nucleotides that encode 11 genes with no intergenic sequences.[2,3] Although the reporting of the PhiX174 sequence in 1977 preceded the publication of the human genome sequence by some 25 years, Sanger's pivotal discovery of the 'chain termination' DNA sequencing method (see Chapter 1) was vital to the Human Genome Project (HGP). 'Sanger sequencing' eventually led to the development of semi-automated high throughput DNA sequencing strategies that, together with the PCR and sophisticated bioinformatics tools, enabled the completion of the draft human genome sequence in 2001.[4,5] In the 21st century, further advances in genomic technologies and bioinformatics such as next generation sequencing (NGS, see Chapter 1), DNA microarray platforms and bioinformatics have enabled nucleotide sequencing and gene mapping of entire genomes for numerous species to become relatively commonplace.

2.1.4   Gene Identification

Gene mapping and identification in eukaryotes is complicated because the majority of nuclear genomic DNA is devoid of genes. It has been estimated that less than 2% of human nuclear DNA comprises approximately 22 000 expressed genes.[1] This appears to be too few genes to constitute the genetic blueprint for complex organisms such as vertebrates; however, because of post transcriptional processing the number of messenger RNA (mRNA) transcripts vastly exceeds the actual number of genes. Thus a relatively small number of genes produce a vast array of different mRNA transcripts that arise through various mechanisms such as alternative promoter usage and alternative exon splicing.[6,7] Further heterogeneity is introduced by post-translational modification of proteins during their maturation. Thus some 22 000 genes could give rise to ~1 000 000 proteins.

The 98% of the genome that is non-genic was once referred to as "junk DNA" because it appeared to be non-functional. However, there is growing evidence that as much as 90% of the genomic DNA may be actively transcribed giving rise to several classes of non-coding RNAs (ncRNA). At one time it was assumed that this uncoordinated transcriptional activity was due to background noise. However, it is becoming clear that some classes of ncRNAs are functionally active and could play key roles in molecular physiology and disease. This is a burgeoning area of research, which is yielding insights into cancer biology.[8]

2.1.5 Genetic Polymorphism

'Genetic polymorphism' is a precise term that relates to the presence of DNA variants, or alleles, at a locus; in human genetics a locus is said to be polymorphic if at least 2% of the individuals in a population are heterozygous. 'Locus' means a specific location of a DNA sequence in the genome, but need not be within a functional gene. Genetic polymorphisms arise from mutations that have increased in frequency in a population due the action of evolutionary forces such as natural selection or genetic drift. The most common type of polymorphism arises from single nucleotide substitutions (referred to as single nucleotide polymorphisms or SNPs). Normally SNPs are completely harmless, but when non-synonymous SNPs occur within coding regions, or in the associated non-coding regulatory DNA, there can be a loss of, or change of function leading to deleterious effects. A notable example is the β-globin allele, which is associated with sickle cell anaemia. The allele reaches polymorphic frequencies in several sub-Saharan African populations. The causative SNP is the replacement of A by T in codon 6 of the β-globin gene leading to the replacement of Glu by Val in the protein. Children who are homozygous for the Val allele will suffer from sickle cell anaemia.[9] Other types of molecular variants that sometimes reach polymorphic frequencies include short insertions and deletions. Of particular note are polymorphic minisatellite and microsatellite arrays that have been important markers in many genome projects. Polymorphism detection, which is discussed below, has become indispensable to genome wide analysis. Polymorphisms can be utilised as markers that together with others such as sequence tagged sites (STSs) and expressed sequence tags (ESTs) help to assemble long segments of DNA and map genes. For genetic mapping, panels of SNPs analysed using DNA microarrays have become the preferred method for genome wide association studies (GWAS) that seek to identify the susceptibility genes associated with complex (*i.e.* non-Mendelian) diseases.

2.2 RECOMBINANT DNA TECHNOLOGIES

A fleet of technologies have been developed to accelerate the pace and improve the precision of genetic and genome analysis. Innovations that stand out as particularly enabling include the discovery of DNA enzymes such as restriction endonucleases and the development of cloning vectors. Together these opened the way to the development of methods such as DNA sequencing, the polymerase chain reaction (PCR), genetic mapping and fragment assembly strategies, *etc.* that culminated in the success of the HGP. See Chapter 1 for a discussion of DNA sequencing and PCR methodologies.

2.2.1 Molecular Cloning Outline

Molecular cloning involves the manipulation and amplification of small segments of DNA leading to the accumulation of large amounts of identical copies.

Traditional DNA cloning vectors such as pUC18 were derived from circular DNA molecules called bacterial plasmids and bacteriophages such as lambda phage and M13. Cloning vectors replicate efficiently in specialised strains of bacterial host cells resulting in the production of multiple copies of the recombinant DNA fragment that was inserted into the vector.[10] Before a DNA fragment can be inserted into a cloning vector, it must be prepared so that the DNA sequences of its ends are compatible with the ends of the 'cloning site' in vector molecule DNA sequence. This is achieved using commercially available restriction enzymes (Type II restriction endo-nucleases) that cleave double stranded DNA at specific nucleotide sequences. In nature, restriction enzymes constitute part of the bacterial defence system since they digest double stranded DNA from harmful bacteriophages and other foreign DNA as it enters the cell. More than 3500 restriction enzymes have been identified, some of which are commercially available and used for routine molecular biology applications.[11] The unique DNA sequences, *i.e.* restriction or recognition sites, that restriction enzymes bind to are usually palindromic; most restriction enzymes are homodimers.

If a plasmid vector and 'foreign' DNA fragment are cut with the same re-striction enzyme that leaves small overhangs of DNA bases, it is relatively straightforward to join, or ligate, the two molecules because their 'sticky ends' of complementary single stranded DNA will base pair (Figure 2.1). The resulting molecule is a recombinant vector that is capable of growing in the host cell.

2.2.2   Cloning Vectors

Many cloning vectors are engineered to have 'multiple cloning sites' or 'polylinkers' that consist of several different restriction sites for the most widely used restriction enzymes; for example, the pUC18 polylinker has 10 restriction sites and each site occurs only once in the plasmid (Figure 2.2). Thus pUC18 will be linearised as a single molecule if it is cut with any of the polylinker specific restriction enzymes. Under the correct conditions the linearised vector is able to ligate to a fragment of 'foreign' DNA, which has been prepared using the same restriction enzyme. Upon successful ligation, which is achieved using the enzyme DNA ligase, the vector molecule con-taining the recombinant DNA insert is re-circularised. Cloning vectors also have an origin of DNA replication and one or more selectable marker genes, typically these are antibiotic resistance genes.[10]

The choice of vector depends on the size of the insert to be cloned and the purpose of the experiment (Table 2.2). Plasmid and some lambda phage vectors are used for cloning small inserts of up to a few kilobases (kb) of DNA, cosmid vectors (a hybrid between a plasmid and specific lambda se-quences) can replicate efficiently with up to about 50 kb of inserted DNA; bacterial artificial chromosomes (BACs) and yeast artificial chromosomes (YACs) are capable of replicating significantly larger inserts of up to about 500 kb and more than 1000 kb of DNA, respectively. BACs and YACs are

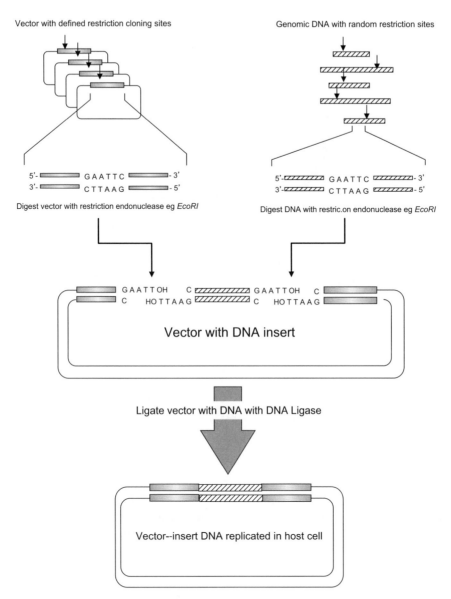

Figure 2.1   Overview of restriction endonuclease digestion and DNA ligation in DNA
          cloning.

essential tools in eukaryotic genome projects where vast lengths of DNA
need to be cloned, mapped and arranged into a series of overlapping DNA
segments, called 'contigs'. Whereas YACs are capable of holding more 1000 kb
of insert DNA they have a tendency to become unstable causing the insert
DNA to rearrange; BACs on the other hand are far more stable and have
become the mainstay cloning vectors for many genome projects.

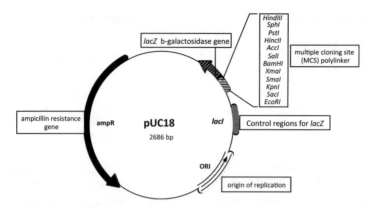

Figure 2.2 Map indicating the important features of the plasmid cloning vector pUC18.

Table 2.2 Examples of vectors generally available for cloning DNA fragments.

Vector	Host Cell	Vector Structure	Insert Range (kb)
M13	*E. coli*	Circular virus	1–4
Plasmid	*E. coli*	Circular plasmid	1–5
Phage λ	*E. coli*	Linear virus	2–25
Cosmids	*E. coli*	Circular plasmid	35–45
BACs	*E. coli*	Circular plasmid	50–500
YACs	*S. cerevisiae*	Linear chromosome	100–2000

BAC: bacterial artificial chromosome, YAC: yeast artificial chromosome

2.2.3   The Cloning Process

The process in which plasmid vector DNA is introduced into the host cell is known as bacterial transformation. Since 'naked' vector DNA is hydrophilic and the bacterial cell wall is normally impermeable to such molecules, the host cells must be made 'competent' by exposing them to conditions that make them permeable to DNA. Treatment with a solution of calcium chloride in early log phase of growth causes the cells to become permeable to chloride ions. When these competent cells are mixed with DNA and heat shocked at 42 °C, the swollen cells are able to take up naked DNA molecules; only a single DNA molecule is permitted to enter a cell. Thus individual colonies of transformed bacteria produce single recombinant vector molecules. An alternative to $CaCl_2$ transformation is electroporation during which the permeability of the bacterial cell wall is increased following the application of an electric field. Following transformation, the cells are grown on selective media so that only transformants will survive to form colonies. For example, if the vector has an ampicillin resistance gene, which is absent in the host strain, only cells containing the vector will be able to form colonies on media containing ampicillin.

2.2.3.1 Library Screening.   There are several methods of screening for colonies that contain recombinant vectors. For example, where the cloning site in the vector lies within an antibiotic resistance gene, successful integration of the insert will lead to inactivation of the resistance gene and recombinant colonies can be identified by the technique of replica plating (Figure 2.3). In this method, the pattern of colonies on the original Petri dish is printed onto a nutrient agar plate containing the selective antibiotic and the position of the recombinant colonies, *i.e.* those that fail to grow on the selective antibiotic, is noted. With some vectors, a method called blue/white selection can be performed. In blue/white selection, successful insertion of a foreign DNA molecule in the vector disrupts an enzyme gene (the *LacZ* gene of beta galactosidase) that otherwise forms a blue product when the transformed colonies are exposed to the substrate X-gal (5-bromo-4-chloro-3-indolyl β-D-galactopyranoside). Thus recombinant colonies are white and non-recombinants are blue (Figure 2.4).

2.2.3.2 Identifying Clones.   Bacterial clones containing the sought after recombinant vectors can be identified either by their capacity to hybridise to specific radioactive isotope labelled or biotin labelled cDNA or genomic DNA probes, or by immunodetection of expressed recombinant protein products (using specialised expression vectors, such as pET, that allow protein expression from the insert DNA). Both approaches are technically straightforward and involve transferring the colonies from a master agar

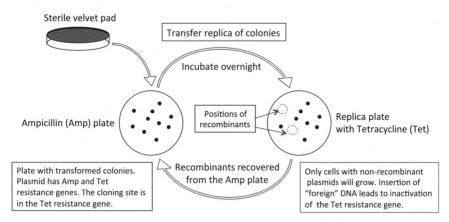

Figure 2.3   Replica plating to detect recombinant plasmids. A sterile velvet pad is pressed on to the surface of an agar plate, picking up some cells from each colony growing on that plate. The pad is then pressed on to a fresh agar plate containing the selective antibiotic, thus inoculating it with cells in a pattern identical with that of the original colonies. Clones of cells that fail to grow on the second plate (owing to the loss of antibiotic resistance) can be recovered from their corresponding colonies on the first plate.

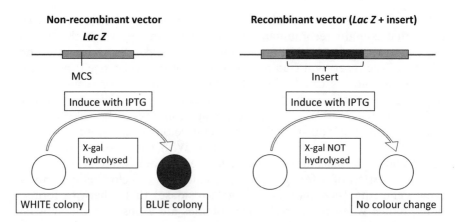

Figure 2.4 Principle of blue/white selection for the detection of recombinant vectors. In the presence of the inducer IPTG the β-galactosidase (*LacZ*) gene is transcribed. The recombinant DNA insert disrupts the expression of the *LacZ* gene which encompasses the multiple cloning site (MCS), hence the substrate X-gal is not hydrolysed and the recombinant colonies remain white.

plate onto carefully orientated nitrocellulose or nylon membranes and then treating the membrane bound bacterial colonies to induce cell lysis. The DNA (or protein if an expression vector is used) from the lysed colonies is then immobilised on the membrane, which is used for the hybridisation with a labelled probe. Recombinant colonies can be detected on X-ray film by radiolabelled or biotin labelled DNA probes (the latter are detected using streptavidin/enzyme conjugates and a chemiluminescent substrate), either by autoradiography or enzyme generated chemiluminescence. The X-ray film must then be aligned with the agar master plate to allow the positive colonies to be identified and picked (Figure 2.5). For detecting recombinant proteins in expression vectors, antibody probes are employed and in a manner analogous to a direct ELISA test. The antibody probe is conjugated to an enzyme such as horseradish peroxidase or alkaline phosphatase and chemiluminescence or an insoluble coloured reaction product reveals the position of the recombinant colonies.

2.2.4 DNA Libraries

A 'DNA library' is a collection of recombinant clones, or DNA molecules generated from a specific source. DNA extracted from nucleated cells is used to make a 'genomic' library and a cDNA library is constructed from cDNA that has been reverse transcribed from mRNA.

The purpose of a standard genomic library is to produce a set of clones from total genomic DNA from a single organism that collectively contain

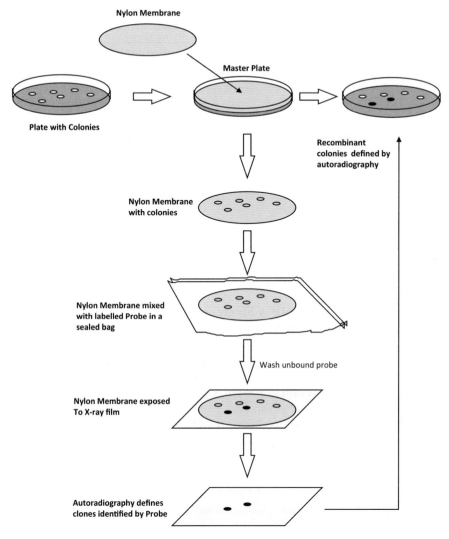

Figure 2.5 Method for detection of recombinant clones by colony hybridisation with labelled gene probes. The bacteria immobilised on the nylon (or nitrocellulose) membrane are lysed in alkali conditions to make the plasmid DNA accessible to the probe.

enough DNA fragments to give complete coverage of the genome under study; the tissue source of the genomic DNA is irrelevant since the genome is identical in all nucleated cells. Specialised genomic DNA libraries such as chromosome specific libraries, prepared from chromosomes sorted by flow cytometry,[12] can be used to shorten the path between the starting DNA and generation of a genome map and sequence. The first step in the preparation of a genomic DNA library is to break up large segments of chromosomal DNA into smaller pieces that can be inserted into a cloning vector; this can

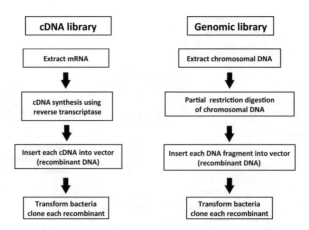

Figure 2.6   Comparison of the general steps involved in the construction of genomic and cDNA libraries.

be achieved either physically, *e.g.* by sonication, or enzymatically using restriction endonucleases.

The second type is the cDNA library that is made from mRNA that has been reverse transcribed by the enzyme reverse transcriptase. Reverse transcriptase produces complementary DNA (cDNA) copies of mRNA, which are then inserted into a vector. In contrast to a genomic library, a cDNA library is representative of the expressed genes in a particular cell or tissue type. Thus a skeletal muscle cDNA library contains sequences expressed in the muscle tissue at the time the mRNA was harvested. cDNA libraries are particularly useful for cloning genes from sources where there is biological information. For example, mammalian skeletal muscle produces high levels of phosphoglucomutase (PGM1) enzyme activity, thus *PGM1* cDNA clones are well represented in skeletal muscle cDNA libraries.[13] Essential features of both types of library are illustrated in Figure 2.6.

2.3   GENOME ANALYSIS

2.3.1   Mapping Genomes and Genes

Genome and gene maps provide information on the location of genes and other markers that enables the user to navigate to specific features and regions of interest. The two main approaches for locating genic and other DNA fragments on chromosomes are genetic and physical mapping. Genetic or linkage mapping aims to determine the genetic distance between a pair of polymorphic markers, or a marker and a disease gene locus by estimating the recombination frequency between them. Linkage is the propensity for genes at two nearby loci to remain on the same chromatid at meiosis, *i.e.* they tend not to be separated by recombination. A linkage map is constructed by analysing recombination frequencies between several pairs of polymorphic marker loci. The genetic distance between the markers, which

is determined by their recombination frequency, is measured in cen-
timorgans (cM) rather than base pairs. For any pair of marker loci, the lower
the recombination frequency, the shorter the genetic distance between the
markers. Linkage mapping in humans is statistically complex and relies on
the availability of DNA samples from several multi-generation pedigrees,
with affected individuals if a Mendelian disease is being mapped.[14] In the
past, linkage analysis in humans was a rather sluggish process because of
the dearth of suitable polymorphic markers. However, the introduction of
DNA technology and the discovery of tandem repeat polymorphisms such as
TA repeats augmented the use of linkage analysis, which became a major
mapping tool in the HGP. Since less than 300 well-chosen marker loci permit
full coverage of the human genome, linkage mapping is an effective approach to discovering the approximate location of an unmapped gene or
Mendelian disease. Where the lack of pedigree information prevents classical linkage mapping, other genetic mapping techniques can be employed
such as affected sib-pair analysis (ASP) and the transmission disequilibrium
test (TDT).[14] In contrast to linkage and association mapping, physical or
genome mapping provides the locations of sets of markers separated by
their physical distances, *i.e.* number of base pairs or cytogenetic locations
rather than recombination frequencies. Physical mapping is addressed in
Section 2.4.1.

2.3.2 Tools for Genetic Mapping

2.3.2.1 RFLPs and Minisatellites. The main tools for genetic mapping are
polymorphic genetic markers. The first DNA polymorphisms were detected
using the Southern blotting technique for analysing genomic DNA digested
with restriction enzymes.[15] Restriction fragment length polymorphisms
(RFLPs) were revealed using a specific genomic or cDNA probe that hybridised with digested DNA from the locus of interest. Southern Blotting involves
the transfer of DNA fragments, generated by restriction enzyme digestion,
from an electrophoresis slab gel made of agarose to an immobilising
membrane such as nitrocellulose or nylon. The immobilised DNA is then
hybridised with a locus specific radioactive or biotin labelled DNA probe that
could have originated from a cDNA or a genomic DNA clone or a PCR product
(see Chapter 1 for a discussion of PCR). When the autoradiograph or
fluorograph of the probed restriction fragments from a set of unrelated
individuals is examined, person to person variation in the resulting band
pattern reflects the presence of an RFLP.[16] The majority of RFLPs consist of
two alleles (*i.e.* cut and uncut) and result from single nucleotide substitutions
leading to alteration of a restriction site (Figure 2.7).

A second type of RFLP is caused by different lengths of the DNA fragments
(variable number of tandem repeats polymorphism, VNTR) rather than the
presence or absence of restriction sites. VNTRs are extremely polymorphic
blocks of repetitive DNA where each repeat is between 10 and 100 nucleotides (minisatellite DNA). Hypervariable VNTR bands were first detected on

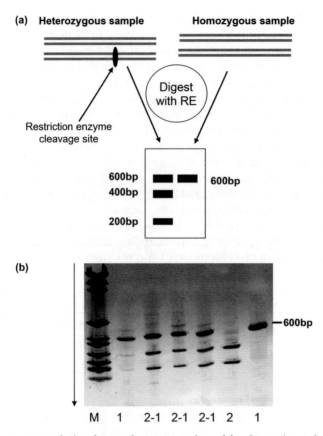

Figure 2.7 RFLP analysis of a 600 bp PCR product. (a) Schematic analysis of products from heterozygous and homozygous individuals digested with a
restriction endonuclease. Three bands arise from the heterozygous individual, one that represents the uncleaved fragments and the other, the
cleaved fragments. (b) An *Sfa*II RFLP revealed in six individuals following
digestion of a 600 bp PCR product.

Southern blots by hybridisation with a labelled 'core' probe after digestion of
human DNA digested with a restriction enzyme, often *Hin*F1 that cuts the
DNA either side of the minisatellite arrays.[17] DNA core probes such as 33.6
and 33.15 that are derived from the human myoglobin gene sequence (*MB*
on 22q13.1) recognise a common core sequence of about 10 to 15 base pairs
that is shared between many different minisatellite loci. Such 'multilocus'
probes revealed tremendous person-to-person variation in the complex
patterns of bands that became known as DNA fingerprints.[18] DNA fingerprints have had a role in forensic medicine where, apart from identical
twins, they can unambiguously identify individuals.

However, DNA fingerprints were not amenable to Mendelian segregation
analysis or gene mapping because it was not possible to assign alleles due to
the complexity of the electrophoretic band patterns. A considerable advance

was the development of single locus VNTR probes, which were isolated from specific cloned minisatellites and able to detect individual mini satellite loci. Single locus probes (SLPs) produce simple patterns in which heterozygous samples display two bands and homozygotes, a single band, thus they became widely used for gene mapping because the allelic bands could be readily traced through families (Figure 2.8).

The overall length of a single VNTR locus can exceed several kilobases and thus VNTRs were generally analysed using Southern blotting techniques rather than PCR. The average heterozygosity of polymorphic VNTR loci is

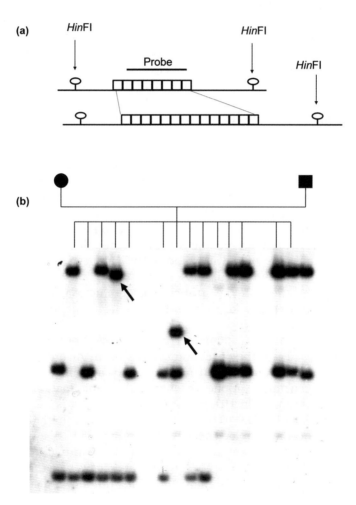

Figure 2.8   (a) Schematic illustration of a heterozygous VNTR locus. The invariant *Hin*F1 sites reveal a length polymorphism resulting from the repeat number variation in the two minisatellite alleles. (b) Southern blot analysis of hypervariable minisatellite B6.7 using a single locus probe on *Hin*F1 digests of genomic DNA from a large family. The arrows indicate new mutations not found in either parent.

about 70%; however, loci with the highest heterozygosities tend to be unstable.[19] It has been estimated that there are between 15 000 and 20 000 VNTR loci in the human genome, but since they tend to cluster near the ends of chromosomes they are of limited use for genome mapping. Interestingly, the repeat blocks in many VNTR loci vary slightly within the sequences of the repeats giving rise to repeat types along the length of the two alleles. These variants are detectable using PCR based methods and are the basis of the minisatellite variant repeats (MVR) typing system, an elegant means of producing highly informative individual genetic band patterns that somewhat resemble barcodes.[20]

2.3.2.2 Microsatellites. Microsatellites (also referred to as Short Tandem Repeats or STRs) have a much simpler structure than minisatellites, usually comprising between two to seven nucleotides per repeat; they are also relatively stable. Because the overall length of the tandem repeat is relatively small, microsatellites are amenable to PCR amplification. In general, STRs produce simple band patterns on electrophoresis gels; tri and tetra nucleotide microsatellites tend to be preferred over dinucleotide repeat polymorphisms that have been associated with a greater tendency to produce confusing secondary bands, sometimes called 'stutter' bands. At one time STRs were the most widely used source of marker polymorphisms in the HGP, and they still form the basis of modern forensic DNA profiling, for example the CODIS system. Although STRs tend to be less polymorphic than minisatellites, they are found scattered throughout the human genome and thus are better mapping tools. The human genome contains over a million STR loci;[4] by the mid 1990s thousands of polymorphic STRs had been used to generate linkage maps of the human chromosomes.

Interestingly some of the triplet repeat microsatellite sequences have been found to play a key role in the aetiology of neurological genetic disorders such as Huntington disease (HD, OMIM 143100), myotonic dystrophy (DM, OMIM 160900) and fragile X syndrome (FMR1, OMIM 309550). In these cases the onset and severity of the disorder is related to an increase in the number of triplet repeats that tend to expand from generation to generation, a phenomenon called 'anticipation' in medical genetics.[14] For example in HD, the onset of symptoms is associated with the increasing length of a CAG triplet repeat located within the Huntington gene (*HTT*). The CAG triplet encodes the amino acid Glutamine, which gives rise to a poly-glutamine tract within the huntingtin protein. In healthy individuals, the number of CAG repeats ranges between six and 26, whereas in patients with more than 40 repeats, full-blown disease is to be expected.[21]

Although the molecular mechanisms leading to microsatellite variation are not completely understood, it is thought that replication slippage is the most common mechanism leading to the gain or loss of one or more repeat units.[22] Other mutational mechanisms, such as those involving unequal crossing over or duplication events, have also been considered.[23]

The tendency to mutate can also depend on the chromosomal environment of a particular locus and whether or not it is transcribed.[24–26] Microsatellite loci tend to exhibit higher mutation rates than single nucleotide substitution loci.[26]

2.3.2.3   Single Nucleotide Polymorphisms (SNPs).   The most widespread type of genetic variation is the result of single nucleotide substitution. Commonly referred to as single nucleotide polymorphisms or SNPs, these variant sites are found scattered throughout the genomes of most species. A SNP is caused by a single nucleotide replacement such as GAATTC to GTATTC that occurs with a frequency of at least 1% in the population. SNP is a new name for an old phenomenon, namely the type of point mutation that can be detected by Southern blotting when performing conventional RFLP analysis and indeed the type of mutation that was indirectly detected in gene products (enzyme polymorphisms) using zone electrophoresis before the advent of DNA technology.[27] It is clear that SNPs are extremely abundant: there are an estimated 15 million SNPs in the human genome.[28] In a study to investigate genetic backgrounds in different populations, 1.6 million human SNPs were genotyped in 1184 individuals.[29] SNPs are not distributed evenly throughout the genome and their frequency can vary up to tenfold.[28,30] Unsurprisingly because of selection pressure, SNPs are less abundant in coding regions than non-coding regions; other factors that affect their distribution are local recombination and mutation rates.[31] The majority of SNPs are located in repeat regions which are difficult to analyse, but even so SNPs are the best choice of polymorphic markers that offer coverage of the whole genome. It has been estimated that there are 5.3 million common SNPs, each with a frequency of 10 to 15%, that account for the bulk of DNA differences. Such SNPs occur on average once every 600 base pairs within the human genome in non-coding DNA[32] thus the level of heterozygosity in the human genome resulting from SNPs is extraordinarily high. Although most single nucleotide polymorphisms result from single nucleotide base differences, many SNP detection methods can detect small insertions and deletions equally well. Although SNPs have a use in classical gene mapping by linkage analysis, they have attracted great interest as marker polymorphisms for large scale GWAS that attempt to find associations between 'susceptibility' loci and complex traits such as obesity.[33] Most SNPs are 'anonymous' genetic markers used to narrow down the regions and identify variants that have phenotypic effects. On rare occasions, the SNPs themselves may contribute to the phenotype. SNPs within genes, especially those in coding regions that lead to missense mutations, may well change the function, regulation or expression of a protein.[9,34] SNPs have also been discovered in the splice sites of genes and these can result in variant protein products with different exon arrays.[35] A number of SNPs have been discovered in the controlling region of genes and some of these are reported to affect expression and regulation of proteins.[36,37] It should be emphasised that the majority of SNPs occur in non-coding sequences that are separated from functional genes by vast genomic distances and are unlikely to have functional significance.

The US National Center for Biotechnology Information (NCBI) curates and maintains a SNP database (dbSNP) which contains sequence details of several million SNPs, together with short insertions and deletions, from a variety of species. The structural simplicity of SNPs compared with microsatellites renders them ideal for full-scale automation and unambiguous scoring of alleles. However, in contrast to microsatellite loci, because most SNPs represent single base changes or small insertions or deletions, they can have only two alleles and thus their heterozygosity cannot exceed 0.5; microsatellites may have multiple alleles at individual loci with heterozygosities sometimes approaching 1.0. The low polymorphism information content of individual SNP loci compared with microsatellites is offset by their relative abundance.

2.3.3   Mutation and Polymorphism Detection

For many years, electrophoretic methods have prevailed for polymorphism typing and mutation detection. With the completion of the HGP there has been an intensifying requirement for faster and improved methods to discover and characterise polymorphic markers in the effort to create linkage and genome maps to assist the identification of genes for complex (*i.e.* non-Mendelian) traits and diseases.

Whereas DNA sequencing platforms, particularly next generation sequencing, are now the prominent methods for mutation discovery,[38] a variety of alternative approaches were developed to discover and genotype SNPs and STR polymorphisms.[39] These methods can be categorised according to their platform technologies *e.g.* electrophoretic or non-electrophoretic as well as the molecular features they assess, such as the differential conformation of DNA strands or the melting properties of PCR products. In addition, many methods have been developed to genotype known SNPs; these often use allele specific primers and probes and real-time qPCR amplification (such as the 5′-nuclease assay, see Chapter 1) or microarray technology (Table 2.3). Following a brief introduction to high throughput electrophoresis based methods, a conformation based approach (single strand conformation polymorphism), a DNA melting based approach (denaturing high performance liquid chromatography) and DNA microarray technologies are discussed.

A range of gel based and capillary electrophoresis methods and instruments has been introduced that are suitable for large-scale high-throughput genotyping of STR and SNP loci. For instance PCR products labeled with infrared fluorescent dyes can be separated by electrophoresis on thin denaturing gels (*e.g.* 0.25 mm), which have a capacity of up to 64 samples per gel and run times between 1 and 3.5 hours depending on the resolution required, and results recorded and analysed on instruments such as the

Table 2.3 Methods for discovering (D) and genotyping (G) SNPs and STR polymorphisms. SNPs (single nucleotide polymorphisms) can be assessed by the melting and conformational properties of double stranded (ds) and single stranded (ss) DNA, respectively. STRs (short tandem repeats) are genotyped by analysis of fragment sizes. NGS: Next Generation Sequencing.

Type	Basis of Detection	Technology	Application	Throughput
SNP	Melting of dsDNA	dHPLC	D,G	Medium
		DGGE	D	Low
		qPCR Melt-curve analysis[a]	D	Low
	Conformation of ssDNA	SSCP	D	Low
		Capillary electrophoresis SSCP	D	Medium
	Probe hybridisation	DNA microarrays	G,D	Very high
		MALDI-TOF	G,D	High
		Allele specific PCR[a]	G	Low
		5'nuclease assay[a]	G	Low
	Primer extension	Dideoxy sequencing[a]	D	Medium
		NGS (some platforms)[a]	D	Very high
STR	Fragment size analysis	Gel electrophoresis	G	Medium
		Capillary electrophoresis	G	High

[a]See Chapter 1 for details of PCR and DNA sequencing technologies.

LI-COR 4300 (Lincoln, NE). Gel based methods have now largely been replaced by capillary electrophoresis (CE). CE was introduced as a workable technique in the 1990s and has since undergone many developments and sophistications. In contrast to gel-based methods, capillary electrophoresis can be fully automated as the products of a PCR reaction can be introduced robotically to an instrument with multiple capillaries allowing large numbers of genotypes to be characterised in parallel and in a short time; therefore relatively few highly skilled personnel are needed. Once the polymorphic STRs are amplified using fluorophore labelled PCR primers the PCR products are loaded into the capillaries and electrophoresed. They are identified by laser-stimulated fluorescence as they pass by a detector that records the electrophoretic mobilities. There are a number of CE instruments useful for genotyping (fragment) analysis with varying capacities. An example of a high-throughput system is the 48 capillary Genetic Analyzer 3730xL (Applied Biosystems), which has the capacity to electrophorese over 4000 samples a day (yielding more than 90 000 STR genotypes) and operate unattended for 24 h. Dedicated software is used to process and determine the allele sizes and record the genotypes of the samples. SNP genotyping by CE is more complex since the allelic PCR products are normally the same length. There are a number of proprietary and non-proprietary approaches to accomplish SNP genotyping. Some are based on mini-sequencing (see Chapter 1), others on ligation reactions and allele-specific reagents.

2.3.3.1 *Single Stranded Conformation Polymorphism (SSCP).* SSCP was first described in 1989,[40] and has been widely applied in genetic analysis both for the discovery and genotyping of SNPs.[41] The principle of SSCP analysis is to denature PCR products to make them single stranded and then to separate the single strands by gel electrophoresis under non-denaturing conditions during which single stranded fragments of DNA will adopt unique and specific conformations as they attempts to fold into the most stable structure. A mutated strand (allele) differing by just one nucleotide can adopt a different conformation to it is wild-type counterpart. The structural differences in the conformers can be identified by electrophoresis where a homozygous sample displays two bands and a heterozygote, four bands, one for each denatured strand (Figure 2.9). For mutation detection, the resolving power of the technique is improved by screening PCR fragments that are less than 300 base pairs in length.[42] Provided that PCR-SSCP analysis is conducted under appropriate conditions, for example by repeating each electrophoresis experiment at two running temperatures or by varying the concentration of denaturant (such as formamide) in the gel, it is an efficient method for discovering new alleles.[43–45] SSCP has been adapted for capillary electrophoresis; in this semi-automated format,

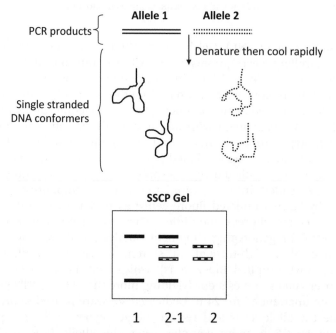

Figure 2.9   Principle of SSCP analysis. In the presence of heat and formamide the dsDNA PCR products from a heterozygote are denatured for several minutes to form single stranded DNA (ssDNA). Immediately before loading onto a non-denaturing electrophoresis gel, the samples are cooled on ice to encourage the formation of ssDNA conformers. Each conformer has a unique electrophoretic mobility as shown on the gel diagram.

PCR fluorophore labelled primers generate products that are recorded and analysed by capillary electrophoresis and laser induced fluorescence under non-denaturing conditions.[46] It is also amenable to be multiplexed where several PCR products labelled with different fluorophores are analysed simultaneously in the same electrophoretic gel lane or capillary.[47]

2.3.3.2 Denaturing High-Performance Liquid Chromatography (DHPLC).

Denaturing high-performance liquid chromatography is a fast and reliable method for SNP detection and discovery and is an alternative to melt-curve analysis that can be performed following real time qPCR (see Chapter 1) or denaturing gradient gel electrophoresis (DGGE).[48–50] The DHPLC technique is based on the separation of homoduplexes and hetero-duplexes formed between reference and mutant DNA molecules by iron-pair reversed-phase high-performance liquid chromatography under partially denaturing conditions. For DNA fragments between about 100 and 1500 base pairs, dHPLC is capable of detecting all single base substitutions as well as small insertions and deletions.

To perform the analysis, DNA fragments are amplified by the PCR from both reference and test samples. The two PCR products are then mixed together and fully denatured at 95 °C before being slowly reannealed by reducing the temperature to about 25 °C. Three classes of duplexes are formed: homoduplexes of both the reference and test DNAs and hetero-duplexes that form from the annealing of reference and mutant DNA strands. Any nucleotide sequence differences between the between the reference and test samples will produce mismatches in the heteroduplexes. Mismatched heteroduplexes and homoduplexes are then introduced to an iron-pair of the reversed-phase high-performance liquid chromatography flow path where they bind to the HPLC capillary. The temperature is then increased to a threshold level, usually between 50–60 °C, whereupon the DNA duplexes become partially denatured. Initially the duplexes are retained in the HPLC capillary through ionic binding interactions between the negatively charged phosphate backbone of the partially denatured DNA fragments and the beads in the cartridge that are coated with the positively charged ion pair reagent TEAA. The column is then eluted with a TAEE/acetonitrile gradient under conditions such that the heteroduplexes with a mismatched base pair(s) elute before the more stable homoduplexes. Note that for SNP detection purposes, the two homoduplexes are indistinguishable. The eluted fragments then pass through the UV detector and the absorbance is measured and the data is analysed by computer. Two sets of peaks are observed for each duplex (Figure 2.10). The method is rapid, taking on average 7 minutes to analyse a genotype and amenable to multiplexing through the use of laser induced fluorescence detection that is analogous to high throughput DNA sequencing strategies.[51] From a practical perspective the method requires the use of *Pfu* polymerase instead of *Taq* polymerase for generating PCR products. *Pfu* has proof reading properties, which minimise the introduction of confusing PCR induced mutations.

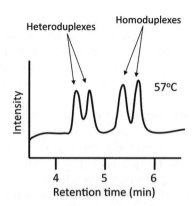

Figure 2.10   Schematic representation of dHPLC analysis of PCR products showing that the homoduplex and heteroduplex DNA species are differentially eluted when the column is held at the tm of the PCR fragment, in this case 57 °C. N.B. to detect homozygous point mutations, the mutant PCR product is mixed with reference DNA.

2.3.3.3 DNA Microarrays.   The desire for rapid access to comprehensive genomic and gene expression data has stimulated the development of DNA microarrays which together with NGS and CE platforms are key technologies for high-throughput genomic analysis. DNA microarrays provide a radically different approach to large-scale characterisation of polymorphic sites and other genomic features[52] and currently represent the extreme end of miniaturisation. A microarray consists of many thousands of DNA oligonucleotide probes, designed to encompass the features of interest, attached to a glass or silicon 'chip' that is about the size of a microscope coverslip, or to arrays of microspheres (beads). The probes can be applied to the solid supports either by printing or by *in situ* synthesis. They can then be hybridised with test amplicons from genomic DNA or cDNA, which have been labelled with one or two fluorescent dyes depending on the experimental protocol. Unhybridised material is washed away and the result is recorded using a confocal laser scanner. The data are collected, analysed and displayed using dedicated computer programs. The investment in equipment and skills needed to design and prepare microarrays is a disincentive for many research laboratories; this has encouraged some biotech companies to commercialise pre-prepared microarrays for a variety of applications (see www.labome.com) as well as the required instrumentation and software.

The first DNA microarrays were designed to measure mRNA transcript levels from thousands of genes in a single experiment.[53] Such functional genomics studies enable the physiological state of cells and overall gene expression pattern to be correlated. Gene expression arrays have many applications including transcriptional profiling of human diseases, such as cancers, which promise to improve the understanding of neoplasia, improved diagnostics, response to drugs and identification of novel therapeutic targets.[54] Microarray data from functional genomics experiments can be accessed from the Genome Expression Omnibus (GEO) public database, which curates gene expression data from high-throughput platforms.

DNA microarrays are not limited to gene expression studies (Figure 2.11). The first genotyping arrays were devised to identify key mutations in highly variable medically important genes and genomes such as the tumour suppressor gene *TP53* (OMIM 191170)[55] and the human immunodeficiency virus (HIV).[56] More recently genotyping chips have been designed to record SNP profiles on a 'whole genome' basis. Information from post-genomic projects such as the HapMap and 1000 Genome initiatives to map haplotypes and characterise human variation in different population groups have been used to identify SNPs and provide coverage of expressed genes.[57] Commercial

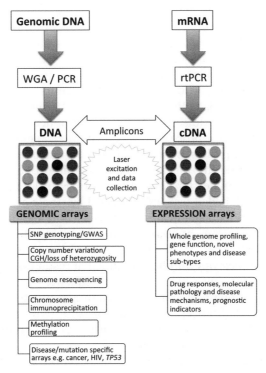

Figure 2.11 A summary of the major steps when using DNA microarrays for genomic and expression studies and some applications. For genomic microarrays, WGA (whole genome amplification) can be achieved by PCR or non-PCR based routes. In expression studies, mRNA is converted to complementary DNA (cDNA) using reverse transcriptase followed by PCR (rtPCR). Fluorophore labels, such as cyanine dyes, are used to indicate the levels of hybridisation between the test (or target) DNA and the probes can be incorporated into the target DNA prior to hybridisation, or once hybridisation has occurred.

microarrays such as the high density Affymetrix SNP 6.0, or Illumina's 1M bead array simultaneously analyse many hundreds of thousands of human SNPs and other features such as copy number variants in a genomic DNA sample. The technology has proved instrumental for GWAS to investigate the genetics of complex diseases and conditions such as cardiovascular disease, cancers (dedicated cancer genome arrays are available), obesity, diabetes *etc.* and other applications such as predicting responses to drugs and investigating the genetic structure of populations. Microarrays have also been, or are being developed to monitor SNP profiles and copy number variation in economically and scientifically important species such as pigs,[58] cattle,[59] maize[60] and mice.[61]

The current estimate of around 10–15 million SNPs in the human genome suggests that around 10% have been covered by the currently available high-density arrays, which have been designed to include SNPs with minor-alleles with frequencies not less than 0.025–0.05. Genomic mapping microarrays preferably include probes for tag-SNPs, which are markers for regions of high linkage disequilibrium, SNPs for autosomes, X and Y chromosomes, recombination hotspots *etc.* SNP arrays are designed for biallelic polymorphisms, thus three classes of genotype *e.g. AA*, *AB* and *BB* must be detected for each locus. Different DNA microarray platforms utilise different chemistries.[52] Following whole genome 'one primer' PCR based amplification and fluorescence labelling of the test DNA, Affymetrix SNP arrays work through differential hybridisation of labelled amplicons to 25 bp arrayed SNP specific probes in a process that is somewhat analogous to Southern blot probing (longer probes are also included for detection of copy number variants). In contrast, the Illumina Infinium microsphere system recommends non-PCR amplification and utilises arrays of longer 50 bp probes: hybridisation of unlabelled test DNA followed by single base extension with two colour immunohistochemical staining is used to identify the genotypes. Although DNA microarray analysis provides a highly efficient GWAS platform for SNP genotyping,[52,62] it can also be used to detect other genomic features. For example: copy number variants that result from genomic events such as duplication, insertion or loss of sequences and loss of heterozygosity;[63] methylation profiles of CpGs including promoter regions, to study their epigenetic significance;[64] outcomes of ChiP (chromosome immunoprecipitation) experiments to identify features such as transcription factor binding sites[65] and histone modifications.[66]

An alternative technology platform for SNP detection is matrix assisted laser desorption/ionisation (MALDI-TOF) mass spectrometry (MS).[67] Whilst MS is limited by the number of SNPs that can be genotyped and is not capable of genome wide analysis, it offers the advantage exquisite precision and therefore can be used to confirm the results of SNP profiling by GWAS as well as for fine mapping. In the Sequenom MassARRAY iPLEX system up to 36 SNPs can be multiplexed by PCR and genotyped simultaneously, amplicons undergo single base extensions to introduce large mass differences between alleles that enables unambiguous genotyping by medium-throughput (*i.e.* not whole genome) MALDI-TOF MS. This high resolution MS platform is finding useful applications in several areas including agricultural biotechnology.[68]

2.4 GENOME PROJECTS BACKGROUND

The HGP was conceived in the mid 1980s, but launched in 1990. This led to an upsurge in DNA sequencing activity and an avalanche of sequence data. By 1995 the first cellular (as opposed to viral) genome of a bacterium had been completed and this was followed by a stream of sequenced genomes, from prokaryotes initially, and then eukaryotes. Two independent drafts of the human genome euchromatic DNA sequence were published simultaneously in 2001 as a result of the publicly funded International Human Genome Project and the privately funded initiative led by Celera Genomics.[4,5] This led to estimates that the human genome consists of between 30 500 and 35 500[4] or 26 000 and 38 000[5] expressed genes. Today these are considered overestimates, the actual number being closer to 22 000 genes.[1] Rapid and powerful advances in cloning and sequencing technologies and computational biology have transformed genome sequencing initiatives from massive long term endeavours to relatively low cost, quick factory-style undertakings. By early 2014, complete genome sequences had been recorded for 18 850 species including 906 eukaryotes.[69] It is estimated that the cost of the HGP, which involved a massive international effort over more than 10 years, was in the order of $500 million. In 2001 it was predicted that the cost of sequencing a human genome would, in the foreseeable future, be just $1000. In January 2014 this prediction appeared to have been fulfilled with the introduction of Illumina's next generation sequencing platform, the HiSeqX 10 desktop sequencing machine, which utilises its proprietary 'sequencing by synthesis' (SBS) technology and has the claimed ability to sequence more than 18 000 human genomes a year at a cost of around $1000 dollars a genome (www.illumina.com). Potentially this could bring human genomic analysis into the realms of personal medicine.

2.4.1 Mapping and Sequencing Strategies

The HGP aimed to produce four types of map: physical, linkage, DNA sequence and expressed gene. Physical and linkage maps provided essential anchor points and frameworks to align DNA sequences and assign genes. A high-resolution physical map based on the analysis of overlapping DNA clones represents the actual distance in DNA base pairs between genetic markers and other landmarks. But the ultimate physical map is the DNA sequence itself. Low-resolution physical maps were generated from techniques such as somatic cell hybridisation and fluorescence *in situ* hybridisation; these methods were also applicable for assigning genes to chromosomes and will be addressed in Section 2.5.

There are two approaches to genome sequencing: whole genome shotgun sequencing (WGS) and the more labour intensive hierarchical shotgun sequencing (HS). In simple organisms such as bacteria and viruses where the chromosomes are haploid and very little repeat sequence occurs, or for sequencing individual human genes, WGS works well.[70,71] In contrast, for eukaryotic genomes where in many cases repeat sequences abound, including the human genome (>50% repeats), and there is considerable genetic variation within the diploid genome, it has been argued that HS offers considerable advantages over WGS sequencing and this was the approach adopted by the publicly funded International Human Genome Sequencing Consortium.[4]

2.4.1.1 Hierarchical Sequencing. This approach relies on the production of a set of large-insert clones (typically 100–200 kb each) that cover the entire genome.[4] Since the clones are all independent and ultimately positioned on a physical map in an order that represents each chromosome, repeated sequences are far less troublesome leading to fewer gaps than encountered by the WGS approach. For many genome projects, high capacity vectors such as BACs are advantageously used for generating large-insert clones since they are less likely to rearrange than alternatives such as YACs. Long-range physical maps are generated from the production of 'contigs'. Contigs are a set of overlapping DNA fragments that have been obtained from independent clones and positioned relative to one another so that they form a contiguous array. To obtain contigs, genomic libraries must be prepared from high molecular weight DNA that has either been partially digested with restriction enzymes or randomly sheared. In contrast to complete digestion, partial digestion of DNA (or random shearing) will result in a set of overlapping fragments (Figure 2.12). Thus partial digestion ensures that when each DNA fragment is cloned into a vector it has ends that will overlap with other clones. Thus when the overlaps are identified and aligned the clones can be positioned, or ordered, so that a physical map is produced.

Large-insert clones are broken down further into sets of smaller overlapping subclones and sequenced using the shotgun sequencing approach. In order to position the overlapping ends into a contig representing the large insert clone, it is preferable to undertake DNA sequencing of both ends of the individual subclones (double barrelled shotgun sequencing). Eventually the entire DNA sequence of the large-insert clone is obtained by computer-based alignment of individual sub-clone DNA end sequences. In order to minimise the overlaps and identify the large-insert clones for further sequencing, *in silico* restriction enzyme mapping can be undertaken to produce a 'fingerprint clone contig'. In the HGP, fingerprint clone contigs were mapped to human chromosomal locations by each chromosome workgroup using resources such as panels of human radiation hybrids (RH), fluorescence *in situ* hybridisation (FISH) with human chromosomes and existing

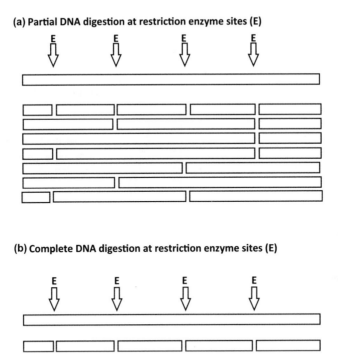

(a) Partial DNA digestion at restriction enzyme sites (E)

(b) Complete DNA digestion at restriction enzyme sites (E)

Figure 2.12 Comparison of partial and complete digestion of DNA molecules at restriction enzymes sites (E).

genetic maps. Radiation hybrids are panels of human–hamster cell hybrids formed by fusing human cells containing radiation-generated fragments of human chromosomes with hamster cells that are used for constructing genetic maps that are complementary to both recombination maps and physical maps based on contigs.[72]

In order to define a common way for all research laboratories to order clones and connect physical maps together, an arbitrary molecular technique based on the polymerase chain reaction has been developed to generate sequence tagged sites (STS). These are small unique sequences between 200–300 base pairs that are amplified by PCR.[73] The uniqueness of the STS is defined by the PCR primers that flank the STS. If the PCR results in amplification then the STS is present in the clone being tested. In this way, defining STS markers that lie approximately 100 kb apart along a contig map allows the ordering of those contigs. Thus, all groups working with clones have publicly available defined landmarks with which to order clones produced in their DNA libraries (Figure 2.13). STSs may also be generated from polymorphic markers that may be traced through families along with other DNA markers and located on a genetic linkage map. Polymorphic STSs may thus serve as markers on both physical and genetic linkage maps for each chromosome and therefore provide a means for aligning the two types of map.

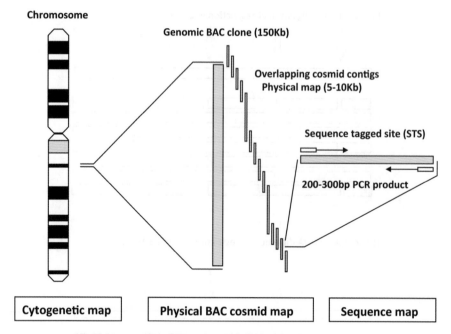

Chromosome

Genomic BAC clone (150Kb)

Overlapping cosmid contigs
Physical map (5-10Kb)

Sequence tagged site (STS)

200-300bp PCR product

| Cytogenetic map | Physical BAC cosmid map | Sequence map |

Figure 2.13 Scheme for the use of STS markers in the hierarchical physical mapping a human chromosome using BAC clones.

In addition to the human genome, the hierarchical sequencing approach has been used to sequence several genomes including those of the yeast *Saccharomyces cerevisiae* and the nematode worm *Caenorhabditis elegans*.[74,75] High quality BAC clone based physical genome maps in one species can be of great value to genome projects in other species where some conservation of genomic sequence and gene order might be expected. For example, the outputs of the HGP have provided anchors for ordering BAC clones generated in several other species such as mouse, rat and cattle, allowing simplification of clone alignments and physical map building as well as the generation of comparative maps in the respective species.[76-78] The International Human Genome Sequencing Consortium sequence was reported as finished in 2004 (Build 35) and contains 2.85 billion nucleotides interrupted by 341 gaps. It covers approximately 99% of the euchromatic genome.[79]

2.4.1.2 Whole Genome Shotgun Sequencing (WGS). In contrast to the hierarchical BAC by BAC approach which relies on the availability of genetic and physical maps for success, WGS is based on the strategy of sequencing a vast number of random genomic clones followed by a computer based analysis of the DNA sequences which identifies matching sequence in different clones. This enables the assembly of a chromosomal DNA sequence, in principle without other map resources. As with the HS approach, overlapping clones are required, but since the clones are destined for direct sequence analysis, only vectors that contain small to medium inserts are used. Thus once the overlaps have been identified the entire sequence is assembled. WGS was the approach adopted by the privately funded human genome initiative.[5]

Although WGS remains somewhat controversial for sequencing complex genomes of 'higher' organisms because of the problems associated with repeat sequences and heterozygosity, it is a widely used approach. The number of complex genomes sequenced by this WGS is increasing and includes the fruit fly *Drosophila*, mosquito (anopheles), mouse, puffer fish, dog and grapevine.[80–85] However, in some cases, such as the silk worm genome project, the WGS method has resulted in many seemingly irresolvable gaps in the genome and so the BAC-based hierarchical ordering of clones was used to close the gaps.[86] Advances in computational analysis of WGS sequences suggests that the problems caused by repeat sequences could be overcome, thus the approach can be expected to gain more ground in future genome projects.[87]

2.5   GENE DISCOVERY AND LOCALISATION

A major goal of all genome projects is to identify, map and characterise the genes. These objectives are prerequisites for advances in biotechnology, medicine and genetics.

2.5.1   Laboratory Approaches

Some well-established methods for mapping genes predate the DNA era including somatic cell hybridization and family based linkage analysis. Somatic cell hybridization was a forerunner of the radiation hybrid mapping panels that played a significant role in the HGP for high-resolution gene mapping. Somatic cell mapping panels are derived from human–rodent cell hybrids that are formed by fusing human cells with rodent cells. The hybrids tend to lose human chromosomes at random, which allows the establishment of clones that retain different human chromosomes. This enables the assignment of a DNA probe, or an isozyme detected by its activity, to a particular human chromosome.[88] Both somatic cell hybridization and linkage analysis require knowledge of the gene to be mapped and thus cannot be used alone for gene discovery. However, both approaches can be used to provide low resolution localisation for a gene: in the case of the somatic cell hybrids, to a particular chromosome and for linkage analysis, to a linkage group, which may be associated with a chromosome in some instances.[14] A limitation of linkage analysis is that the gene being mapped and the marker genes must be heterozygous in key parent individuals; this was frequently a stumbling block before the availability of polymorphic DNA markers such as microsatellites and SNPs.

Several gene mapping and identification techniques have been developed for use with cloned DNA. A gene can be localised to a chromosome by fluorescence *in situ* hybridisation (FISH). FISH is carried out on R-banded metaphase chromosomes using a labelled genomic clone. The clone hybridised to its complementary sequence on a chromosome and its position revealed by fluorescence confocal microscopy. By alternating the microscope between the hybridised clone and R-band display, the gene can be assigned to a particular chromosome band to a resolution of a few megabases.[13]

A high-resolution development of FISH sometimes called Fibre FISH can be used to order cloned genes on DNA fibres prepared from chromatin from interphase nuclei. The fibres are made to extend and using this approach it is possible to hybridise three genomic clones simultaneously enabling their relative positions and order to be determined.[89] This approach can resolve the order of clones that are only about 2–7 kb apart.

Identifying genes *de novo* in cloned DNA presents a different type of challenge. Several strategies have been developed with varying degrees of success. In some species, expressed genes are associated with upstream unmethylated GpC dinucleotide rich sequences. This has enabled the use of the restriction endonuclease *Hpa*II to scan cloned DNA for CpG dinucleotide rich regions. Because this enzyme only cleaves the unmethylated cytosine of the CpG dinucleotide, cleavage at such sites may indicate the presence of an expressed gene. Upon digestion with *Hpa*II, these regions form tiny fragments and are known as *Hpa*II tiny fragments (HTF) or CpG islands.[90]

Considerable progress in gene localisation and discovery has been made with a PCR based method known as EST mapping. An EST or Expressed Sequence Tag is a small PCR product that has been generated from a cDNA sequence, thus reflecting an expressed gene in the cell or tissue from which the mRNA was prepared. Public domain ESTs are available from the dbEST database.[91] Release 130101 of dbEST contains more than 74 million ESTs from a large number of species. EST mapping used with large insert genomic clones can provide information on genome organisation, including gene density and localisation.[92] ESTs in many species reflect genes of unknown function. For example, a recent study of the brown planthopper, a serious pest of rice plants, generated a library of more than 37 000 nuclear genome ESTs some of which were unrelated to any gene sequences in the databases. These and others could be used to search for genes of relevance to understanding the biology of the pest species and identifying potential target genes for developing novel insecticides against planthoppers.[93,94]

A further gene isolation system, which uses adapted vectors, termed exon trapping or exon amplification may be used to identify exon sequences. Exon trapping requires the use of a specialised expression vector that will accept fragments of genomic DNA containing sequences for splicing reactions to take place. Following transfection of a eukaryotic cell line a transcript is produced that may be detected by using specific primers in a RT-PCR. This indicates the nature of the foreign DNA by virtue of the splicing sequences present.[95]

2.5.2 Bioinformatics Approaches

Parallel developments in bioinformatics have been essential for the success of the genome projects. In the closing stages of the HGP, some institutions were generating up to 175 000 individual DNA sequence reads per day, information that needed to analysed, submitted to databases, annotated and checked for quality. A multiplicity of DNA and protein sequences from different species is submitted daily by laboratories world wide to databases such as GenBank (USA, DNA sequences), EMBL-BANK (Europe, DNA sequences), DDBJ (Japan, DNA sequences) and Swissprot (Europe, protein sequences). All of these ever expanding and annotated databases are available to the public and the three DNA sequence databases update and share information on a daily basis. These massive primary sequence databases have stimulated the emergence of a large number of specialised genomic and protein databases dedicated to various subjects such as a particular species group, biological features such as disease markers, protein families, protein domains, *etc.* Nucleic Acids Research online Molecular Biology Database Collection is a public repository that lists more than 1000 databases.[96]

Many algorithms such as BLAST[97] have been developed to search databases for matching sequences of nucleotides or amino acids, and there are now unparalleled opportunities for characterising genes and for studying groups of related genes *in silico*. It has been estimated that the majority of new cDNA and EST sequences will show similarity to proteins of known function and that many of the sequences will show similarity to each other. Further precision for identifying gene families is afforded by automated approaches to access databases such as Pfam to search for protein domains.[98]

Other computational approaches have been devised to predict features of genes in otherwise anonymous genomic sequences. Programs such as Genescan that predict exons, genes, promoter regions, polyA tails and other features associated with expressed DNA have been used to indicate possible genic regions. 'Cloning' genes using database resources and dedicated algorithms is now a reality and biocomputing has become an applied science which enables investigators to identify genes *in silico*. Integrating and annotating the expanding genomic data delivered by next generation sequencing and DNA microarrays is a continuing challenge for bioinformaticians and database curators.

In addition to the DNA, expression and protein sequence databases, there has been a parallel development of textbases. Two examples are OMIM (Online Mendelian Inheritance in Man; www.omim.org), which catalogues salient clinical synopses, genomic features and comprehensive references for more than 18 000 human genetic traits (the majority of which are known at the molecular level) and the Human Genome Nomenclature database (HGNC; www.genenames.org), which assigns names to all known human genes and curates a searchable database. There are also many dedicated databases, such as the cancer knowledgebase canSAR (www.cansar.icr.ac.uk), that serve various research communities.

2.6 CONCLUSION

In medicine a major challenge is unravelling the genetic basis of common diseases that will affect the majority individuals in all populations. The HGP has produced the genomic resources to tackle these multifactorial disorders including cardiovascular disease, cancers, obesity, diabetes *etc.* Complex traits do not exhibit Mendelian inheritance patterns observed in single gene disorders such as cystic fibrosis. Instead the action of several genes, each with a small effect, and environmental influences, modify the risk of disease. Various types of association studies are used in an attempt to define the genes that underlie the propensity for some individuals to develop such diseases. Population and family based association studies such as GWAS and affected sib pair analysis make use of the gene maps, polymorphic marker resources and genome technologies in an attempt to pinpoint small regions of chromosomes associated with the disease phenotype. Chromosomal regions showing positive association with disease can then be resequenced[6] and combined with functional studies in an attempt to identify the causative mutations.[99] It is anticipated that advances on several fronts will lead to improved understanding of molecular pathology, discovery of new drug targets and better diagnostics. The translation of genomic scale laboratory based approaches to end users such as hospitals and doctors offices will require the development of fast, high precision and low-cost technology platforms. Innovative instruments such as Oxford Nanopore Technologies nanopore real time 'hand held' DNA sequencer[100] and point-of-care diagnostic platforms such as QuantuMDX's continuous flow nanowire biosensors[101] should become available in the near future. The convenience and levels of multiplexing of such miniaturised technologies could greatly exceed those of current diagnostic systems such as the Multiplex Ligation-dependent Probe Amplification (MLPA) assays.[102]

The outcomes of animal and plant genome projects will undoubtedly yield social, humanitarian and economic benefits. Genetic modification (GM) of food species is highly likely to grow in importance as world food supplies become limited. Genetically modified crops, principally corn, soybean, rice, rapeseed and cotton have already been produced with a range of GM characters such as resistance to herbicides, insects, viruses and drought, delayed fruit ripening (tomato), altered oil and vitamin content *etc.* In the next few decades, transgenic animals that produce increased yields of meat, disease resistance, optimised fat content and ability to thrive in extreme environments could be produced. In the pharmaceutical industry, transgenic technologies hold great promise for the lowering the cost production of therapeutic proteins that could be routinely harvested from cows milk and plants. The growing effort of genome sequencing, gene discovery and manipulation promises to overcome the previously encumbering difficulties associated with locating genes with the potential for species improvement, new antibiotic targets *etc.*[103] However, before implementation of these GM developments can be implemented, full risk assessments will be necessary to understand the possible impacts on human health and the environment.

REFERENCES

1. M. Pertea and S. L. Salzberg, *Genome Biol.*, 2010, **11**, 206.
2. F. Sanger, S. Nicklen and A. R. Coulson, *Proc. Natl. Acad. Sci. U. S. A.*, 1977, **74**, 5463.
3. F. Sanger, G. M. Air, B. G. Barrell, N. L. Brown, A. R. Coulson, J. C. Fiddes, C. A. Hutchison III, P. M. Slocombe and M. Smith, *Nature*, 1977, **265**, 687.
4. International Human Genome Sequencing Consortium, *Nature*, 2001, **409**, 860.
5. J. C. Venter, M. D. Adams, E. W. Myers, P. W. Li, R. J. Mural, G. G. Sutton, H. O. Smith, M. Yandell, C. A. Evans, R. A. Holt, *et al.*, *Science*, 2001, **291**, 1304.
6. P. Carninci, *Trends Genet.*, 2006, **22**, 501.
7. P. Carninci, T. Kasukawa, S. Katayama, J. Gough, M. C. Frith, N. Maeda, R. Oyama, T. Ravasi, B. Lenhard, C. Wells, *et al.*, *Science*, 2005, **309**, 1559.
8. J. Sana, P. Faltejskova, M. Svoboda and O. Slaby, *J. Transl. Med.*, 2012, **10**, 103.
9. F. B. Piel, A. P. Patil, R. E. Howes, O. A. Nyangiri, P. W. Gething, T. N. Williams, D. J. Weatherall and S. I. Hay, *Nat. Commun.*, 2010, **1**, 104.
10. J. Sambrook and P. MacCallum, *Molecular Cloning: A Laboratory Manual*, Cold Spring Harbor Laboratory Press, NY, Third Edition, 2001.
11. R. J. Roberts, T. Vincze, J. Posfai and D. Macelis, *Nucleic Acids Res.*, 2010, **38**(Database issue), D234.
12. K. E. Davies, B. D. Young, R. G. Elles, M. E. Hill and R. Williamson, *Nature*, 1981, **293**, 374.
13. D. B. Whitehouse, W. Putt, J. U. Lovegrove, K. Morrison, M. Hollyoake, M. F. Fox, D. A. Hopkinson and Y. H. Edwards, *Proc. Natl. Acad. Sci. U. S. A.*, 1992, **89**, 411.
14. *An Introduction to Genetic Analysis*, ed. A. J. F. Griffiths, S. R. Wessler, S. B. Carroll and J. Doebley Gelbart, W. H. Freeman and Company, New York, 10th edn, 2012.
15. E. M. Southern, *Trends Biochem. Sci.*, 2000, **25**, 585.
16. R. Leach, R. DeMars, S. Hasstedt and R. White, *Proc. Natl. Acad. Sci. U. S. A.*, 1986, **83**, 3909.
17. A. J. Jeffreys, V. Wilson and S. L. Thein, *Nature*, 1985, **314**, 67.
18. A. J. Jeffreys, V. Wilson and S. L. Thein, *Nature*, 1985, **316**, 76.
19. K. Tamaki, C. A. May, Y. E. Dubrova and A. J. Jeffreys, *Hum. Mol. Genet.*, 1999, **8**, 879.
20. K. Tamaki, C. H. Brenner and A. J. Jeffreys, *Forensic Sci. Int.*, 2000, **113**, 55.
21. R. A. C. Roos, *Orphanet J Rare Dis*, 2010, **5**, 40.
22. C. Schlötterer and D. Tautz, *Nucleic Acids Res.*, 1992, **20**, 211.

23. J. M. Hancock, in *Microsatellites: Evolution and Applications*, ed. D. Goldstein and C. Schlötterer, Oxford University Press, New York, 1999, pp. 1–9.
24. H. Ellegren, *Nat. Genet.*, 2000, **24**, 400.
25. J. D. Hawk, L. Stefanovic, J. C. Boyer, T. D. Petes and R. A. Farber, *Proc. Natl. Acad. Sci. U. S. A.*, 2005, **102**, 8639.
26. T.-M. Kim, P. W. Laird and P. J. Park, *Cell*, 2013, **155**, 858.
27. H. Harris and D. A. Hopkinson, *Handbook of Enzyme Electrophoresis in Human Genetics (with Supplements)*, Oxford American Publishing Co., New York, l976.
28. The 1000 Genomes Project Consortium, *Nature*, 2010, **467**, 1061.
29. International HapMap 3 Consortium, *Nature*, 2010, **467**, 52.
30. International SNP Map Working Group, *Nature*, 2001, **409**, 928.
31. D. E. Reich, S. F. Schaffner, M. J. Daly, G. McVean, J. C. Mullikin, J. M. Higgins, D. J. Richter, E. S. Lander and D. Altshuler, *Nat. Genet.*, 2002, **32**, 135.
32. L. Kruglyak and D. A. Nickerson, *Nat. Genet.*, 2001, **27**, 234.
33. Y. F. Pei, L. Zhang, Y. Liu, J. Li, H. Shen, Y. Z. Liu, Q. Tian, H. He, S. Wu, S. Ran, Y. Han, R. Hai, Y. Lin, J. Zhu, X. Z. Zhu, C. J. Papasian and H. W. Deng, *Hum. Mol. Genet.*, 2014, **23**, 820.
34. F. J. de Serres, *Environ. Health Perspect.*, 2003, **111**, 1851.
35. A. Woolfe, J. C. Mullikin and L. Elnitski, *Genome Biol.*, 2010, **11**, R20.
36. E. M. El-Omar, M. Carrington, W. H. Chow, K. E. McColl, J. H. Bream, H. A. Young, J. Herrera, J. Lissowska, C. C. Yuan, N. Rothman, *et al.*, *Nature*, 2000, **404**, 39.
37. A. Ligers, N. Teleshova, T. Masterman, W. X. Huang and J. Hillert, *Genes Immun.*, 2001, **2**, 45.
38. X. Wu, C. Ren, T. Joshi, T. Vuong, D. Xu and H. T. Nguyen, *BMC Genomics*, 2010, **11**, 469.
39. *Single Nucleotide Polymorphisms, Methods and Protocols*, ed. A. A. Komar, Humana Press, Totowa, New Jersey, 2009.
40. M. Orita, H. Iwahana, H. Kanazawa, K. Hayashi and T. Sekiya, *Proc. Natl. Acad. Sci. U. S. A.*, 1989, **86**, 2766.
41. T. Tozaki, N.-H. Choi-Miura, M. Taniyama, M. Kurosawa and M. Tomita, *BMC Med. Genet.*, 2002, **3**, 6.
42. K. Hayashi and D. W. Yandell, *Hum. Mutat.*, 1993, **2**, 338.
43. S. P. Yip, W. Putt, D. A. Hopkinson and D. B. Whitehouse, *Ann. Hum. Genet.*, 1999, **63**, 129.
44. W. Yan, H. Zhou, Y. Luo, J. Hu and J. G. H. Hickford, *PLoS One*, 2014, **9**, e88691.
45. S. P. Yip, D. A. Hopkinson and D. B. Whitehouse, *Biotechniques*, 1999, **27**, 20.
46. T. Tahira, A. Suzuki, Y. Kukita and K. Hayashi, *Methods Mol. Biol.*, 2003, **212**, 37.
47. W. Choi, G. W. Shin, H. S. Hwang, S. P. Pack, G. Y. Jung and G. Y. Jung, *Electrophoresis*, 2014, **35**, 1196.

48. D. Guzowski, A. Chandrasekaran, C. Gawel, J. Palma, J. Koenig, X. P. Wang, M. Dosik, M. Kaplan, C. C. Chu, S. Chavan *et al.*, *J. Biomol. Tech.*, 2005, **16**, 154.

49. D. Liu, Y. Zhang, Y. Du, G. Yang and X. Zhang, *DNA Seq.*, 2007, **18**, 220.

50. M. Shi, Y. Hou, J. Yan, R. Bai and X. Yu, *J. Forensic Sci.*, 2007, **52**, 235.

51. A. Premstaller, W. Xiao, H. Oberacher, M. O'Keefe, D. Stern, T. Willis, C. G. Huber and P. J. Oefner, *Genome Res.*, 2001, **11**, 1944.

52. T. LaFramboise, *Nucleic Acids Res.*, 2009, **37**, 4181.

53. V. Trevino, F. Falciani and H. A. Barrera-Saldaña, *Mol. Med.*, 2007, **13**, 527.

54. D. R. Rhodes, S. Kalyana-Sundaram, V. Mahavisno, R. Varambally, J. Yu, B. B. Briggs, T. R. Barrette, M. J. Anstet, C. Kincead-Beal, P. Kulkarni, *et al.*, *Neoplasia*, 2007, **9**, 166.

55. S. A. Ahrendt, S. Halachmi, J. T. Chow, L. Wu, N. Halachmi, S. C. Yang, S. Wehage, J. Jen and D. Sidransky, *Proc. Natl. Acad. Sci. U. S. A.*, 1999, **96**, 7382.

56. R. Gonzalez, B. Masquelier, H. Fleury, B. Lacroix, A. Troesch, G. Vernet and J. N. Telles, *J. Clin. Microbiol.*, 2004, **42**, 2907.

57. T. J. Hoffmann, M. N. Kvale, S. E. Hesselson, Y. Zhan, C. Aquino, Y. Cao, S. Cawley, E. Chung, S. Connell, J. Eshragh, M. Ewing, J. Gollub, M. Henderson, E. Hubbell, C. Iribarren, J. Kaufman, R. Z. Lao, Y. Lu, D. Ludwig, G. K. Mathauda, W. McGuire, G. Mei, S. Miles, M. M. Purdy, C. Quesenberry, D. Ranatunga, S. Rowell, M. Sadler, M. H. Shapero, L. Shen, T. R. Shenoy, D. Smethurst, S. K. Van den Eeden, L. Walter, E. Wan, R. Wearley, T. Webster, C. C. Wen, R. A. Li Weng, A. Whitmer, S. C. Williams, C. Wong, A. Zau, C. Finn, P.-Y. Schaefer, Kwok and N. Risch, *Genomics*, 2013, **98**, 79.

58. W. Luo, S. Chen, D. Cheng, L. Wang, Y. Li, X. Ma, X. Song, X. Liu, W. Li, J. Liang, H. Yan, K. Zhao, C. Wang, L. Wang and L. Zhang, *Int. J. Biol. Sci.*, 2012, **8**, 870.

59. M. Gautier, D. Laloë and K. Moazami-Goudarzi, *PLoS One*, 2010, **5**, e13038.

60. R. A. Swanson-Wagner, S. R. Eichten, S. Kumari, P. Tiffin, J. C. Stein, D. Ware and N. M. Springer, *Genome Res.*, 2010, **20**, 1689.

61. Y. Kayashima, H. Tomita, S. Zhilicheva, S. Kim, H.-S. Kim, B. J. Bennett and N. Maeda, *PLoS One*, 2014, **9**, e88274.

62. K. J. Lindquist, E. Jorgenson, T. J. Hoffmann and J. S. Witte, *Genet. Epidemiol.*, 2013, **37**, 383.

63. C. R. Coughlin II, G. H. Scharer and T. H. Shaikh, *Genome Med.*, 2012, **4**, 80.

64. J. Liu, Z. Zhang, M. Bando, T. Itoh, M. A. Deardorff, J. R. Li, D. Clark, M. Kaur, K. Tatsuro, A. D. Kline, C. Chang, H. Vega, L. G. Jackson, N. B. Spinner, K. Shirahige and I. D. Krantz, *Nucleic Acids Res*, 2010, **38**, 5657.

65. G. M. Euskirchen, J. S. Rozowsky, C.-L. Wei, W. H. Lee, Z. D. Zhang, S. Hartman, O. Emanuelsson, V. Stolc, S. Weissman, M. B. Gerstein, Y. Ruan and M. Snyder, *Genome Res.*, 2007, **17**, 898.

66. J. A. McCann, E. M. Muro, C. Palmer, G. Palidwor, C. J. Porter, M. A. Andrade-Navarro and M. A. Rudnicki, *BMC Genomics*, 2007, **8**, 322.

67. I. G. Gut, *Hum. Mutat.*, 2004, **23**, 437.

68. S. D. Pant, C. P. Verschoor, F. S. Schenkel, Q. You, D. F. Kelton and N. A. Karrow, *Gene*, 2014, **537**, 302.

69. Genomes OnLine Database, Release v.5, 2014, http://www.genomesonline.org.

70. R. C. Gardner, A. J. Howarth, P. Hahn, M. Brown-Luedi, R. J. Shepherd and J. Messing, *Nucleic Acids Res.*, 1981, **9**, 2871.

71. S. L. Chissoe, A. Bodenteich, Y. F. Wang, Y. P. Wang, D. Burian, S. W. Clifton, J. Crabtree, A. Freeman, K. Iyer, L. Jian, *et al.*, *Genomics*, 1995, **27**, 67.

72. G. Gyapay, K. Schmitt, C. Fizames, H. Jones, N. Vega-Czarny, D. Spillett, D. Muselet, J. F. Prud'homme, C. Dib, C. Auffray, *et al.*, *Hum. Mol. Genet.*, 1996, **5**, 339.

73. E. A. Stewart, K. B. McKusick, A. Aggarwal, E. Bajorek, S. Brady, A. Chu, N. Fang, D. Hadley, M. Harris, S. Hussain, *et al.*, *Genome Res.*, 1997, **7**, 422.

74. H. W. Mewes, K. Albermann, M. Bahr, D. Frishman, A. Gleissner, J. Hani, K. Heumann, K. Kleine, A. Maierl, S. G. Oliver, *et al.*, *Nature*, 1997, **387**, 7.

75. C. *elegans* Sequencing Consortium, *Science*, 1998, **282**, 2012.

76. S. G. Gregory, M. Sekhon, J. Schein, S. Zhao, K. Osoegawa, C. E. Scott, R. S. Evans, P. W. Burridge, T. V. Cox, C. A. Fox, *et al.*, *Nature*, 2002, **418**, 743.

77. M. Krzywinski, J. Wallis, C. Gosele, I. Bosdet, R. Chiu, T. Graves, O. Hummel, D. Layman, C. Mathewson, N. Wye, *et al.*, *Genome Res.*, 2004, **14**, 766.

78. International Bovine BAC Mapping Consortium, *Genome Biol.*, 2007 **8**, R165.

79. International Human Genome Sequencing Consortium, *Nature*, 2004, **431**, 931.

80. M. D. Adams, S. E. Celniker, R. A. Holt, C. A. Evans, J. D. Gocayne, P. G. Amanatides, S. E. Scherer, P. W. Li, R. A. Hoskins, R. F. Galle, *et al.*, *Science*, 2000, **287**, 2185.

81. R. A. Holt, G. M. Subramanian, A. Halpern, G. G. Sutton, R. Charlab, D. R. Nusskern, P. Wincker, A. G. Clark, J. M. C. Ribeiro, R. Wides, *et al.*, *Science*, 2002, **298**, 129.

82. Mouse Genome Sequencing Consortium, *Nature*, 2002, **420**, 520.

83. S. Aparicio, J. Chapman, E. Stupka, N. Putnam, J. M. Chia, P. Dehal, A. Christoffels, S. Rash, S. Hoon, A. Smit, *et al.*, *Science*, 2002 **297**, 1301.

84. E. F. Kirkness, V. Bafna, A. L. Halpern, S. Levy, K. Remington, D. B. Rusch, A. L. Delcher, M. Pop, W. Wang, C. M. Fraser, *et al.*, *Science*, 2003, **301**, 1898.

85. R. Velasco, A. Zharkikh, M. Troggio, D. A. Cartwright, A. Cestaro, D. Pruss, M. Pindo, L. M. Fitzgerald, S. Vezzulli, J. Reid, *et al.*, *PLoS ONE*, 2007, **2**, e1326.
86. S. Zhan, J. Huang, Q. Guo, Y. Zhao, W. Li, X. Miao, M. R. Goldsmith, M. Li and Y. Huang, *BMC Genomics*, 2009, **10**, 389.
87. S. Istrail, G. G. Sutton, L. Florea, A. L. Halpern, C. M. Mobarry, R. Lippert, B. Walenz, H. Shatkay, I. Dew, J. R. Miller, *et al.*, *Proc. Natl. Acad. Sci. U. S. A.*, 2004, **101**, 1916.
88. H. N. Bu-Ghanim, C. M. Casimir, S. Povey and A. W. Segal, *Genomics*, 1990, **8**, 56.
89. N. Hornigold, M. van Slegtenhorst, J. Nahmias, R. Ekong, S. Rousseaux, C. Hermans, D. Halley, S. Povey and J Wolfe, *Genomics*, 1997, **41**, 385.
90. C. A. Sargent, I. Dunham and R. D. Campbell, *EMBO J.*, 1989, **8**, 2305.
91. M. S. Boguski, T. M. Lowe and C. M. Tolstoshev, *Nat. Genet.*, 1993 **4**, 332.
92. L. L. Qi, B. Echalier, S. Chao, G. R. Lazo, G. E. Butler, O. D. Anderson, E. D. Akhunov, J. Dvořák, A. M. Linkiewicz, A. Ratnasiri, *et al.*, *Genetics*, 2004, **168**, 701.
93. H. Noda, S. Kawai, Y. Koizumi, K. Matsui, Q. Zhang, S. Furukawa, M. Shimomura and K. Mita, *BMC Genomics*, 2008, **9**, 117.
94. D. R. Price, H. S. Wilkinson and J. A. Gatehouse, *Insect Biochem. Mol. Biol.*, 2007, **3**, 1138.
95. G. M. Duyk, S. W. Kim, R. M. Myers and D. R. Cox, *Proc. Natl. Acad. Sci. U. S. A.*, 1990, **87**, 8995.
96. M. Y. Galperin, *Nucleic Acids Res.*, 2008, **36**(Database issue), D2.
97. S. F. Altschul, T. L. Madden, Al. A. Schaffer, J. Zhang, Z. Zhang, W. Miller and D. J. Lipman, *Nucleic Acids Res.*, 1997, **25**, 3389.
98. R. D. Finn, J. Mistry, B. Schuster-Böckler, S. Griffiths-Jones, V. Hollich, T. Lassmann, S. Moxon, M. Marshall, A. Khanna, R. Durbin, *et al.*, *Nucleic Acids Res.*, 2006, **34**(Database Issue), D247.
99. W. Bodmer and C. Bonilla., *Nat. Genet.*, 2008, **40**, 695.
100. E. Pennisi, *Science*, 2014, **343**, 829.
101. J. O'Halloran, *International Patent Application*, PCT/GB2009/005008, 2009.
102. I. Bergval, S. Sengstake, N. Brankova, V. Levterova, E. Abadía, N. Tadumaze, N. Bablishvili, M. Akhalaia, K. Tuin and A. Schuitema, *et al.*, *PLoS One*, 2012, **7**, e43240.
103. W. Liu, J. S. Yuan and C. N. Stewart Jr., *Nat. Rev. Genet.*, 2013; **14**, 781.

CHAPTER 3

Protein Expression

STUART HARBRON

Principal Consultant, The Enzyme Technology Consultancy, Patent
Specialist, Romax Technology Limited, Visiting Lecturer, University of
Hertfordshire, Berkhamsted, Hertfordshire, UK
Email: stuart.harbron@experts.co.uk

3.1 INTRODUCTION

Knowledge of the full sequence of many genomes has led to the identification of thousands of genes encoding proteins with unknown or poorly
known activity, which can only be elucidated by expression of the genes
and analysis of the expressed protein by functional assay.[1] Producing recombinant proteins in forms that are either suitable for elucidating function
for investigative purposes or in amounts useful for therapeutic applications
is a key challenge. To produce the protein in good yield and in the right form
"requires the success of three individual factors: expression, solubility and
purification."[2] Important aspects of processes and techniques leading to
expression of proteins of interest are the main focus of this chapter.

3.2 HOST-RELATED ISSUES

Commercially available systems permit molecular cloning and protein expression in commonly used host organisms. Bacterial expression systems
are highly attractive in this respect for a number of reasons, including:

- Their rapid growth rates;
- Their ability to use relatively inexpensive substrates;

Molecular Biology and Biotechnology, 6th Edition
Edited by Ralph Rapley and David Whitehouse
© The Royal Society of Chemistry 2015
Published by the Royal Society of Chemistry, www.rsc.org

- Their well-characterised genetics;
- The availability of a large number of cloning vectors; and
- A variety of mutant host strains.

The most popular hosts are *Escherichia coli, Bacillus subtilis*, yeast, and cultured cells of higher eukaryotes such as insect or mammalian cells. *E. coli* is frequently used because the very large body of information available makes it relatively well understood, and there are well-characterised protocols for manipulating this microbe. Even so "the production of soluble proteins in *E. coli* remains a hit-or-miss affair."[2] For example, there are many proteins for which *E. coli* is not the ideal host for expression, including proteins having more than 500 amino acids, those which are highly hydrophobic, proteins having many cysteines (because the reducing environment in *E. coli* prevents the formation of disulfide bonds), and those requiring post-translational modification or other treatments. Furthermore, if the protein of interest is from a eukaryotic organism, as it often is for a protein having a commercial interest, then there are immediately three problems associated with expressing it in a prokaryotic system such as *E. coli*, and these problems relate to the difference in the mechanism of gene expression between the two systems.

First, bacteria are not capable of processing RNA to remove introns. Fortunately, this can be more-or-less easily overcome by generating double-stranded DNA copies of mRNA molecules isolated from the eukaryotic organism by using the mRNA as a template with a reverse transcriptase. This double-stranded copy, or cDNA, will not contain introns and can act as the coding sequence in expression vectors. The problem of introns has also been addressed by synthesising fragments of the gene chemically and subsequent ligation, but this pre-supposes that the amino acid sequence of the protein of interest is known.[3–6]

Secondly, the RNA polymerase of a prokaryotic host will not bind to and transcribe the gene encoding the protein of interest unless it has an appropriate promoter sequences upstream of the coding region. To be useful as tools for protein expression the promoter must be strong, have a low basal expression, be easily transferred, be easily and economically induced, and be unaffected by commonly used ingredients in culture media. Basal transcription, which is transcription in the absence of the inducer, can be dealt with through the use of a suitable repressor: this is especially important if the expression target introduces cellular stress, which would select for plasmid loss. Either thermal or chemical triggers can be used to initiate promoter induction, and some commonly used systems are listed in Table 3.1.[7]

Finally, prokaryotic ribosomes will not bind to the mRNA produced by transcription unless there is a ribosome-binding site (RBS) on the mRNA, just before the coding region. Initiation of translation can be a significant limiting factor in expression of cloned genes.[15] Translation initiation from the translation initiation region of the transcribed messenger RNA requires

Table 3.1 Some promoter systems for *E. coli.*[7]

Expression level	System	Induction	Cost	Commercial construct
++	λ P$_L$ promoter[8]	Δt	0	Invitrogen pLEX
+ $\rightarrow$ ++	*lac* promoter[9]	IPTG	+++	
++	*trc, tac* promoter[10]	IPTG	+++	GE Lifesciences pTrc, pGEX
+ $\rightarrow$ +++	*araBAD* promoter (P$_{BAD}$)[11]	L-arabinose	+	Invitrogen pBAD
+ $\rightarrow$ +++	*rhaP$_{BAD}$*[12]	L-rhamnose	+++	Invitrogen rhaPbad
++ $\rightarrow$ +++	*tetA* promoter/ operator[13]	Anhydroteracycline	+	
++++	T7 RNA polymerase[14]	IPTG	+++	Agilent pET

an RBS and a translation initiation codon.[16] Efficiency of translation initiation is influenced by the codon following the initiation codon, and abundant adenine seems to lead to highly expressed genes.[17] In addition to an initiation codon, AUG, other nucleotides, particularly in the 5′ untranslated leader of the mRNA, are needed to create suitable secondary and tertiary structures in mRNA and facilitate interaction between the mRNA and the ribosome, such as the Shine–Dalgarno (S–D) sequence.[18] Other sequences that increase the level of translation include translation enhancers such as the Epsilon sequence in the g10L ribosome-binding site of phage T7.[19,20] Translation termination is preferably mediated by the stop codon UAA in *E. coli.* Increased efficiency of translation termination is achieved by insertion of consecutive stop codons or the UAAU stop codon.[21] To obtain expression of foreign genes in *E. coli* it is necessary to incorporate ribosome-binding motifs into the recombinant DNA molecule. Furthermore, some sequences (such as the S–D sequence) must be located at an optimal distance from the translation start codon. This is most readily achieved by construction of fusion genes where an entire untranslated leader and 5′ coding sequence from a naturally occurring gene is present.

3.3 VECTORS

To counter some of the issues related to the capabilities of the host, expression vectors have been developed. These are DNA constructs that are stably maintained and propagated in a host and which contain promoter and ribosome-binding sites positioned just before one or more sites for restriction endonucleases to allow the insertion of foreign DNA. These regulatory sequences, such as that from the *lac* operon of *E. coli*, are usually derived from genes which, when induced, are strongly expressed in bacteria.

Expression vectors vary in their complexity, ease of manipulation, and the length of DNA sequence they can accommodate (the insert capacity). Vectors are derived from naturally occurring entities such as bacterial plasmids, bacteriophages or combinations of their constituent elements, such as cosmids. For applications to do with the expression of proteins, plasmids are the most important of these.

A plasmid is an autonomously replicating, extrachromosomal circular DNA molecule, distinct from the normal bacterial genome and non-essential for cell survival under non-selective conditions. Genes carried by plasmids often include those for conferring antibiotic resistance, to enable conjugation, or for the metabolism of 'unusual' substrates. These are attractive candidates for modification for use as vectors, particularly if they are replicated at a high rate and are not easily 'lost' from the host in non-selective conditions. It is clear from the previous section that a number of key elements are more-or-less essential to the design of these vectors.[22,23]

One of the more successful plasmids is pBR322 (Figure 3.1), which has been widely used and has a number of desirable key features, which are further discussed below:

- It is small (much smaller than a natural plasmid);
- It has a relaxed origin of replication;
- Two genes coding for resistance to antibiotics; and
- Single recognition sites for a number of restriction enzymes at various points around the plasmid.

The small size means that it is resistant to damage by shearing, and is efficiently taken up by bacteria, a process termed transformation.

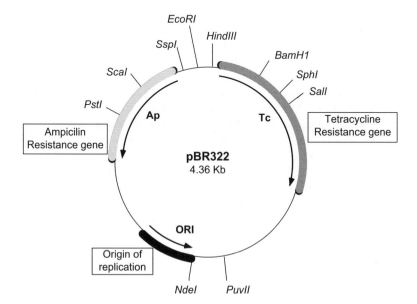

Figure 3.1 Map and important features of pBR322, including restriction sites.

A relaxed, as opposed to stringent, origin of replication means that it is not tightly linked to cell division, and plasmid replication will happen far more frequently than chromosomal replication, leading to a large number of plasmid molecules per cell,[22] and any vector with a replication origin in *E. coli* will replicate (together with any incorporated DNA) more-or-less efficiently. In stringent regulation, replication is in synchrony with cell division. The replicon may also have associated *cis* acting elements.[24] The origin of replication is most commonly ColE1, as in pBR322 (copy number 15–20) or pUC (copy number 500–700), or p15A, as in pACYC184 (copy number 10–12). These multi-copy plasmids are stably replicated and maintained under selective conditions and plasmid-free daughter cells are rare.[25] Different replicon incompatibility groups and drug resistance markers are required when multiple plasmids are employed for the co-expression of gene products. Derivatives containing ColE1 and p15A replicons are often combined in this context since they are compatible plasmids,[26] meaning that they may be stably maintained in the same cell.[27]

One of the antibiotic resistance genes allows cells that contain the plasmid to be selected: if cells are plated on medium containing an appropriate antibiotic, only those that contain plasmid will grow to form colonies. The other resistance gene can be used, as described below, for detection of those plasmids that contain inserted DNA. The most common drug resistance markers in recombinant expression plasmids confer resistance to ampicillin, kanamycin, chloramphenicol or tetracycline.

Recognition sites for restriction enzymes are used to open or linearise the circular plasmid. Linearising a plasmid allows a fragment of DNA to be inserted and the circle closed. The variety of sites not only makes it easier to find a restriction enzyme which is suitable for both the vector and the foreign DNA to be inserted, but, since some of the sites are placed within an antibiotic resistance gene, the presence of an insert can be detected by loss of resistance to that antibiotic. This is termed insertional inactivation.

The protocol utilised for using a plasmid such as pBR322 to introduce DNA encoding the protein of interest into the host cell is summarised below.

First a fragment of DNA encoding the protein of interest and digested with *Bam*H1 is isolated and purified, or produced *via* PCR. Plasmid pBR322 is also treated with *Bam*H1, and both are deproteinised to inactivate the restriction enzyme. Since *Bam*H1 cleaves to give sticky ends, the plasmid and digested DNA fragments can be ligated using T4 DNA ligase. This yields a plasmid containing a single fragment of the DNA as an insert, but the mixture will also contain products, such as plasmid which has recircularised without an insert, dimers of plasmid, fragments joined to each other, and plasmid with an insert composed of more than one fragment. Most of these unwanted molecules are be eliminated during subsequent steps. The products of such reactions are usually identified by agarose gel electrophoresis.

Secondly, host *E. coli* is transformed using the ligated DNA plasmid. Bacteria termed competent can be induced to take up DNA from their surroundings by prior treatment with Ca^{2+} at 4 °C followed by a brief increase in temperature, termed heat shock. Plasmid DNA added to the suspension of competent host cells will thus be imported during this process. Small, circular molecules are taken up most efficiently, whereas long, linear molecules will not enter the bacteria.

Thirdly, after a brief incubation to allow expression of the antibiotic resistance genes the cells are plated onto medium containing the antibiotic (*e.g.* ampicillin). Any colonies that grow are obviously derived from cells that contain plasmid, since this carries the gene for resistance (to ampicillin).

Fourthly (Figure 3.2), to distinguish between those colonies containing plasmids with inserts of the DNA encoding the protein of interest and those that simply contain recircularised plasmids, the colonies are replica plated, using a sterile velvet pad, onto plates containing tetracycline in their medium. The plasmid carries the tetracycline resistance gene, but the *Bam*HI site lies within this gene, which means that the plasmid will show insertional inactivation in the presence of insert, but will be intact in those plasmids that have merely recircularised. Thus colonies that grow on ampicillin but not on tetracycline must contain plasmids with inserts. Since replica plating gives an identical pattern of colonies on both sets of plates, it is straightforward to recognise the colonies with inserts, and to recover them from the ampicillin plate for further growth. This illustrates the importance of a second gene for antibiotic resistance in a vector.

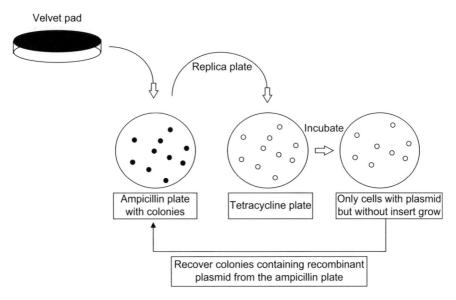

Figure 3.2   Replica plating to detect recombinant plasmids. A sterile velvet pad is pressed on to the surface of an agar plate, picking up some cells from each colony growing on that plate. The pad is then pressed on to a fresh agar plate, thus inoculating it with cells in a pattern identical with that of the original colonies. Clones of cells that fail to grow on the second plate (*e.g.* owing to the loss of antibiotic resistance) can be recovered from their corresponding colonies on the first plate.

This fourth step can be omitted if, prior to ligation in the first step, the mixture is treated with the enzyme alkaline phosphatase, which removes 5'-phosphate groups essential for ligation. Following ligation between the 5'-phosphate of insert and the 3'-hydroxyl of plasmid, only recombinant plasmids and chains of linked DNA fragments will be formed. It does not matter that only one strand of the recombinant DNA is ligated, since the nick will be repaired by bacteria transformed by the modified plasmid. Including this step increases the yield of recombinant plasmid containing inserts.

A variety of plasmids based on pBR322 have been developed, including a series of plasmids termed pUC (Figure 3.3), which include the gene encoding β-galactosidase and some of the more commonly-used restriction sites concentrated into a region termed the multiple cloning site or MCS. When the pUC plasmid has been used to transform the host cell, *E. coli*, the gene is 'switched on' by adding the inducer IPTG (isopropyl-β-D-thiogalactopyrano-side) and the enzyme β-galactosidase is produced. This enzyme hydrolyses a colourless substance called X-gal (5-bromo-4-chloro-3-indolyl-β-galactopyr-anoside) leading to the precipitation of a blue insoluble material. However, disruption of the gene by the insertion of DNA encoding the protein of interest means that X-gal is not hydrolysed. This means that a host cell having a pUC plasmid carrying DNA encoding the protein of interest will be white or colourless in the presence of X-gal, whereas a host cell having an intact non-recombinant pUC plasmid will be blue since its gene is fully functional and not disrupted. This approach, termed blue/white selection, allows rapid initial identification of recombinant host cells and has been included in a number of later vector systems (Figure 3.4).

These approaches detect not only host cells containing a plasmid carrying the DNA encoding the protein of interest, but also host cells in which

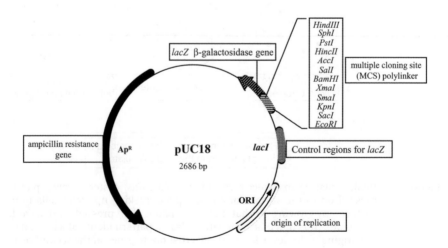

Figure 3.3   Map and important features of pUC18, including restriction sites.

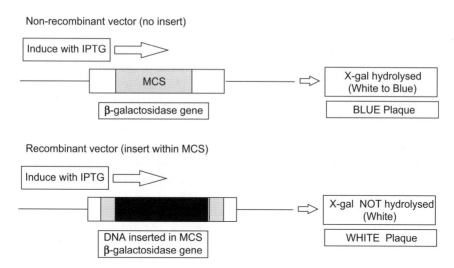

Non-recombinant vector (no insert)

| Induce with IPTG |

MCS

β-galactosidase gene

X-gal hydrolysed
(White to Blue)

BLUE Plaque

Recombinant vector (insert within MCS)

| Induce with IPTG |

DNA inserted in MCS
β-galactosidase gene

X-gal NOT hydrolysed
(White)

WHITE Plaque

Figure 3.4 Principle of blue/white selection for the detection of recombinant vectors.

insertional inactivation of antibiotic resistance genes has happened as a result of the misincorporation of the DNA insert.

3.3.1 Expression Vector Constructs

Fortunately, not only have plasmid vectors been developed which contain promoter and ribosome-binding sites positioned just before one or more restriction sites that allow the facile insertion of the DNA encoding the protein of interest, they have also been made available commercially. Indeed, a wide range of expression constructs based on different promoters is available.[28,29] Table 3.1 shows some of the commonly used systems including the T7 based pET expression system (commercialised by Agilent and Invitrogen, amongst others) and one based on the araBAD promoter (*e.g.*, Invitrogen pBAD), which are discussed below.

3.3.1.1 pBAD. Regulation of the arabinose operon in *E. coli* is directed by the product of the *araC* gene,[30] which controls the synthesis rate of the AraE, AraF and AraG proteins, required for arabinose uptake, as well as the AraB, AraA and AraD enzymes, required for its catabolism. That means that the intrinsic state of the *ara*-specific promoters is off and AraC turns them on, whereas the set state of the *lac* operon promoter is on and the *lac* repressor turns it off. In addition, pBAD is catabolite repressed, which means that growing the culture in the presence of glucose will further repress expression.

Whilst expression of a cloned gene from plasmids containing the araBAD promoter can be modulated over several orders of magnitude in cultures grown in the presence of sub-saturating concentrations of arabinose,[11] individual cells are either fully induced or uninduced.[31] Cells having the natively controlled arabinose transport gene (araE) are either induced or uninduced, the relative fraction of which is controlled by the concentration of arabinose. The population-averaged variation in expression from pBAD as a function of inducer concentration is proportional to the percentage of cells that are fully induced (*vs.* uninduced) rather than the level of expression in individual cells. This all-or-none phenomenon, which can have undesirable effects on the expression of heterologous genes, can be eliminated in *E. coli* by expression of araE from arabinose-independent promoters. In these arabinose-transport engineered cells, all cells in the population have approximately the same induction level.[32] Strains capable of transporting L-arabinose, but not metabolising it, such as a recA, endA strain are therefore most suitable.

Expression plasmids based on the araBAD promoter are designed for tight control of background expression and precise control of the expression levels of the target protein.[11] This contrasts to the all-or-nothing induction achieved by most other bacterial expression systems.[33] A pBAD plasmid, derived from pBR322 and shown in Figure 3.5, is a 4.1 kb construct having the following elements:

- *araC*: ORF encoding *araC* protein.
- *amp^R*: codes for ampicillin resistance.
- *ori*: *col E1* origin of replication.
- *pBAD*: *araBAD* promoter.
- *rrnB*: transcription termination region.
- *MCS*: multiple cloning site, has the sequence encoding the protein of interest.

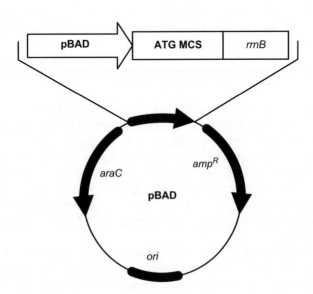

Figure 3.5 pBAD plasmid.

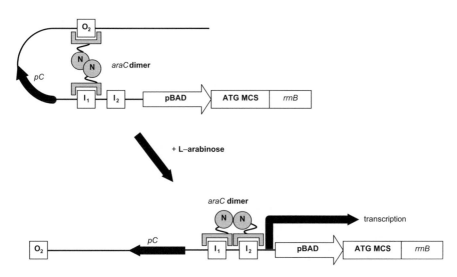

Figure 3.6 Control of pBAD system.

The AraC dimer binds three sites in the arabinose operon, I_1, I_2 and O_2 (Figure 3.6). In the absence of arabinose, the AraC dimer contacts the O2 site located within the *araC* gene, 210 base pairs upstream from pBAD. The other half of the *araC* dimer contacts the I_1 site in the promoter region forming the DNA loop as shown. Transcription from pBAD and the *araC* promoter (pC) is thus inhibited by the loop. Upon binding of arabinose, the *araC* dimer changes its conformation so that it binds to the I_2 site of pBAD instead of the O_2 site. This removes the loop structure, and transcription by RNA polymerase initiates. Binding of the *araC* dimer to the I_1 and I_2 sites is stimulated by cAMP receptor protein (CRP) which means that background expression from ara-BAD can be reduced by glucose mediated catabolite repression.[11]

The pBAD Expression System is the basis for the TOPO® one-step cloning and expression system from *Invitrogen*. Here, the expression vector is lin-earised, so that it has single 3′-thymidine (T) overhangs and bound topo-isomerase I. This means that PCR product having single 3′-deoxyadenosine (A) as a result of Taq polymerase activity inserts to ligate efficiently with the vector. Topoisomerase I from the Vaccinia virus binds to duplex DNA at specific sites and cleaves the phosphodiester backbone after 5′-CCCTT in one strand and forms a covalent bond between the 3′-phosphate of the cleaved strand and a tyrosyl residue of topoisomerase I. This bond between the DNA and enzyme is subsequently attacked by the 5′-hydroxyl of the original cleaved strand, reversing the reaction and releasing topoisomerase. This process results in the PCR product becoming incorporated into the expression vector. The vector also codes for His-patch thioredoxin which serves to both increase translation efficiency, and in some cases, solubility (see Section 3.4.1), as well as providing a purification-facilitating tag (see Section 3.4.2).

3.3.1.2 pET. This construct, derived from pBR322, includes hybrid promoters, multiple cloning sites for the incorporation of different fusion partners and protease cleavage sites, and has been developed for a variety of expression applications.[34,35] It is said to be the most powerful system developed for the production of recombinant proteins when *E. coli* is the host.[36] The pET plasmid, shown in Figure 3.7, is a 5.4 Kb construct having the following elements:

- *lacI*: codes for the *lac* repressor protein.
- *amp^R*: codes for ampicillin resistance.
- *ori*: *col E1* origin of replication.
- *lacO*: codes for the *lac* operon.
- P_{T7}: codes for the T7 promoter, which is specific only for T7 RNA polymerase.
- T_{T7}: codes for the T7 terminator.
- *MCS*: multiple cloning site, has the sequence encoding the protein of interest.

A key feature is the use of P_{T7}, a 20-nucleotide sequence that is not recognised by the *E. coli* RNA polymerase, nor does T7 RNA polymerase occur in the prokaryotic genome sequence; thus in the absence of T7 RNA polymerase, P_{T7} is not activated and the protein of interest is not produced. When P_{T7} is activated, as described below, being a viral promoter it transcribes rapidly, with a maximum speed of around 230 nucleotides per second, some five times faster than *E. coli* RNA polymerase.

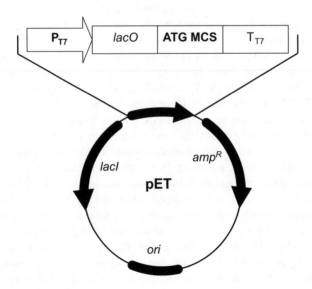

Figure 3.7 pET plasmid.

The pET expression system from Agilent is further engineered to take advantage of the features of the T7 bacteriophage gene 10 that promotes high-level transcription and translation.[37] This RBS dramatically enhanced translation efficiency, often more than 40-fold greater than that obtained using a 'consensus' RBS.

Expression requires a host strain lysogenised by a DE3 phage fragment, encoding the T7 RNA polymerase, under the control of the IPTG-inducible *lacUV5* promoter, and Figure 3.8 shows a schematic of the process. A copy of the *lacI* gene is present on the *E. coli* genome and on pET. *LacI* is a weakly expressed gene and a 10-fold enhancement of the repression is achieved when the overexpressing promoter mutant *LacIq* is employed.[38] The *lac* repressor protein, *lacI*, represses the *lacUV5* promoter of the host cell and the *T7/lac* hybrid promoter encoded by the expression plasmid. In the absence of an inducer then, the *lacI* tetramer binds to the *lac* operator on both the host cell genome and the plasmid. This prevents the host cell from producing T7 RNA polymerase, and prevents the plasmid from producing the protein of interest.

When the inducer, typically IPTG, is introduced, it binds and triggers the release of tetrameric *lacI* from the *lac* operator on both the genome and the plasmid, which triggers the expression of T7 RNA polymerase in the host cell. Transcription of the target gene from the T7/lac hybrid promoter is thus initiated.

There can be low background expression from pET expression plasmids; this may be reduced by co-expression of T7 lysozyme, a natural inhibitor of T7 RNA polymerase, using either plasmid pLysS or pLysE. These plasmids

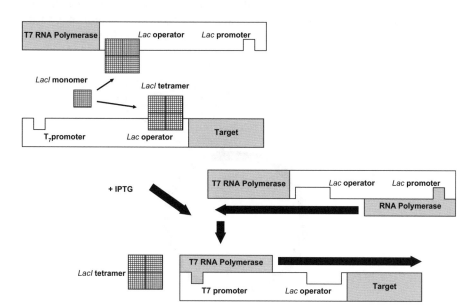

Figure 3.8 Control of pET system.

harbour the T7 lysozyme gene in silent (pLysS) and expressed (pLysE) orientations, with respect to the cognate tetracycline responsive (Tc) promoter.[39]

Although the *lacUV5* promoter is less sensitive to regulation by the cAMP-CRP (cAMP receptor protein) complex than the *lac* promoter, incorporation of 1% glucose in the cultivation medium reduces cAMP levels and enhances repression of the promoter significantly. Host strains deficient in the *lacY* gene, which encodes lactose permease, improves control of target protein expression.[40]

Lucigen have developed two variants of pET shown in Figure 3.9, having the following elements:

- *ROP*: repressor of priming.
- *kan^R*: codes for ampicillin resistance.
- *ori*: *col E1* origin of replication.
- RBS: the ribosome binding site: (1) *lac* operon or (2) T7 gene 10 leader.
- P: the promoter: (1) T7 promoter or (2) rha pBAD.
- T: the terminator.
- *Tr*: transcription terminators prevent transcription into or out of the insert.
- MCS: multiple cloning site, has the sequence encoding the protein of interest.

Thus the kanamycin resistance gene replaces the ampicillin resistance gene, and the vector includes a repressor of priming (for low copy number); the transcription terminators prevent transcription into or out of the insert. Neither vector includes the *lacI* repressor protein, which means that these are about half the size of a typical pET vector, improving cloning efficiency

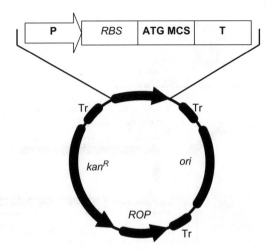

Figure 3.9 Improved pET plasmid.

and simplifies downstream manipulations such as site-directed mutagenesis. One vector is under the control of the T7-lac promoter, and the
absence of the *lacI* gene is instead provided by the BL21(DE3) cells. The other
vector utilises the rhaPBAD promoter and does contain the lacZ alpha gene
fragment, which means that blue/white colony screening is not possible.
However, only minimal colony screening is necessary because the background from empty vector is typically <10%.[41] In the absence of rhamnose
the transcriptional activity of rhaPBAD is very low, which allows stable clone
construction for potentially toxic gene products.[42] Two activators, RhaR and
RhaS, which bind rhamnose positively control transcription.[43] RhaR activates its own transcription as well as that of RhaS, which in turn activates
transcription from rhaPBAD. This cascade means adjustment of expression
levels can be controlled by controlling the concentration of rhamnose.[44] This
feature is useful for proteins that are potentially toxic to the host cells, or
that are difficult to express in soluble form. Transcription is also controlled
by the cAMP-dependent transcriptional activator protein and is subject to
catabolite repression. In the presence of glucose, cAMP levels remain low
and rhaPBAD remains inactive, even when rhamnose is available.

These vectors form the basis of Lucigen's Expressioneering Technology,
which is an *in vivo* recombinational approach which allows cloning of PCR
product into the expression vector with no further sample clean-up.[41] The
PCR product is mixed with a linearised pRham vector and transformed into a
modified *E. coli*, which leads to the formation of the clone in the pRham
vectors and subsequent expression of the protein of interest. The high
transformation efficiency of thee cells makes them ideal for cloning, and
their recA-, endA- genotype allows recovery of high quality plasmid DNA if
required for cloning into other hosts.

3.4 EXPRESSION PROBLEMS

Most problems relating to expression of heterologous proteins are probably
to do with the differences between codon usage in eukaryotes and *E. coli.*[7]
Codon usage in *E. coli* is reflected by the level of cognate amino-acylated
tRNAs available, and minor or rare codons tend to be genes expressed at low
levels. Codons rare in *E. coli* are often abundant in heterologous genes from
eukaryotes,[45] which tend to be the proteins of commercial interest. What
this means is that attempts to express genes containing rare codons can
result in translational errors due to ribosomal stalling at positions on the
messenger RNA that require incorporation of amino acids coupled to minor
codon tRNAs. Insufficient tRNA pools can lead to translational stalling,
premature translation termination, translation frameshift and aminoacid
misincorporation.[46]

The most problematic codons[29] are shown in Table 3.2. A number of approaches have been used in an attempt to reduce problems associated with
codon bias, such as co-transforming the host with a plasmid harbouring a
gene cognate with the problematic codons.[47] Several plasmids are available

Table 3.2   Some rare codons in *E. coli*.

Codon	Amino acid	Frequency (per 1000) (*E. coli*)	tRNA gene	Frequency (per 1000) (*H. sapiens*)
AGG	Arg	2.1	*arg*U	12.0
CGA	Arg	2.4	*arg*W	6.2
AGA	Arg	2.4	*arg*U	12.2
CCC	Pro	2.4	*pro*L	19.8
CUA	Leu	3.4	*leu*W	7.2
AUA	Ile	5.0	*ile*X	7.5
CGG	Arg	5.0	*arg*W, *arg*X	11.4
GGA	Gly	8.2	*gly*T	16.5
AAG	Lys	8.8	*lys*	31.9

for rare tRNA co-expression, most of which are based on the p15A replication origin, which enable maintenance in the presence of the ColE1 origin. Commercially available *E. coli* strains for protein expression include Rosetta™ 2 host strains from Merck Millipore, which are BL21 derivatives designed to enhance the expression of eukaryotic proteins that contain codons rarely used in *E. coli*. These strains supply tRNAs for seven rare codons (AGA, AGG, AUA, CUA, GGA, CCC, and CGG) on a compatible chloramphenicol-resistant plasmid. The tRNA genes are driven by their native promoters. Agilent offers BL21-CodonPlus-RIL chemically competent cells that carry extra copies of the *argU*, *ileY*, and *leuW* tRNA genes. The tRNAs encoded by these genes recognise the AGA/AGG, AUA, and CUA codons, respectively. The presence of these additional tRNA genes resolves the issue of codon bias for organisms whose genome is AT-rich.

Another issue affecting efficient protein expression[29] is the stability of mRNA in *E. coli*. Exononucleases RnaseII and PNPase, and the endonuclease RnaseE act to degrade mRNA, which can have a half-life of between seconds and 20 minutes. Strains are available, such as the BL21 Star strain from Invitrogen, which have a mutation in *rne131*, the gene encoding RnaseE; stability of mRNA transcripts in this strain are thus significantly improved. Another approach is to introduce 5′ and 3′ stabilising structures to reduce damage from exonuclease attack, such as by fusion to a sequence encoding green fluorescent protein.[48]

A further hazard is faced when the expressed protein is to be secreted,[49] and issues include the premature aggregation in the cytoplasm and proteolysis in the cell envelope. The former can be prevented through the exploitation of the SRP or Tat pathways. Protease digestion of the protein can be avoided using *E. coli* BL derivates that are *lon*[50] and *ompT* protease deficient.

A further problem reducing efficiency of protein expression is inclusion body formation. Whether these form through a passive event occurring through hydrophobic interactions between unfolded polypeptide chains or by specific clustering mechanisms is unknown.[51] However, inclusion bodies do not appear to be inert aggregates,[52] but may be an unbalanced equilibrium between *in vivo* protein aggregation and solubilisation,[49] hence formation may be minimised by reducing culture temperature to slow host metabolism and expression rate, or addition of sorbitol, betaine, sucrose or raffinose to the culture medium. Alternatively, the DNA encoding the protein of interest may be modified to include solubility-enhancing tags (see Section 3.4.1) or by coexpression of plasmid-encoded chaperones;[23] molecular chaperones may also be used as part of a refolding strategy.[53] The idea here is that as a newly synthesised polypeptide is produced on the ribosome, it associates with a trigger factor chaperone, which prevents hydrophobic patches on the polypeptide from interacting.[54] Once the trigger factor is removed, the folding proceeds to yield the protein in a natural state. Other chaperones enhance protein degradation (DnaK and GroEL).

Over-expression of toxic proteins that are difficult or impossible to produce in bacteria can be achieved using modified strains, such as OverExpress C41 and C43 from Avidis, which are based on *E. coli* DE3.[55]

Finally, *E. coli* accumulates lipopolysaccharide (LPS), or endotoxin, which is pyrogenic in humans and other mammals; this is problematic if the protein of interest is intended for therapeutic use, and it must be purified in a second step to become endotoxin-free.[56]

3.5 FUSION PROTEINS

Recalling the comment that production of proteins requires the success of three individual factors: expression, solubility and purification,[1] it is clear from what has been covered so far that protein expression, whilst technically reasonably straightforward, does not easily yield useful protein products—the protein may be inactive, precipitated, incomplete, *etc.*

Recent advances in genomics, proteomics, and bioinformatics have facilitated the use of recombinant DNA technology in order to evaluate any protein of interest, without prior knowledge of the protein's cellular location or function. Tagging the protein of interest using recombinant DNA techniques allows the facile modification of proteins of interest potentially offering improvements in solubility and purification.[57] The approach has made it clear that some affinity tags also increase solubility of the protein.[58–62]

Fusion tags developed for recombinant protein production include affinity tags for protein detection or purification,[7,63–65] solubility enhancement,[66] tag removal[7] or technical approaches.[67–69]

3.5.1 Solubility-Enhancing Tags

Fluorescence of *E. coli* cells expressing GFP fused to a gene corresponding to a protein of interest has been found to correlate well with the solubility of the protein when expressed alone.[70] The approach has also been used to identify solubilising interaction partners for 'insoluble' targets such as integration host factor β.[71] Solubility-enhancing and purification-assisting fusion partners[1,60] that have been described or are in use are shown in Table 3.3, and some work better than others, or work best on certain types of protein.

Numerous solubility tags are listed in Table 3.3, but most work has utilised just a few of these, notably glutathione-S-transferase (GST), maltose-binding protein (MBP), N-utilisation substance A (NusA), and thioredoxin (Trx). GST and MBP also function as purification-facilitating (affinity) tags; GST binds to glutathione resin,[77] and MBP binds strongly to amylose resin (see Section 3.5.2).[79]

GST, from *Schistosoma japonicum*, is a 26 kDa protein. It is considered, at best, to be a poor solubility enhancer[87,88] in *E. coli*. Hydrophobic regions or charged residues on proteins fused to GST may contribute to insoluble expression. GST fusion proteins that are larger than 100 kDa may also be insoluble. In both cases, they can be purified after solubilisation with mild detergents.[90] The GST-tag often protects against intracellular proteolysis and

Table 3.3 Solubility-enhancing and purification-facilitating tags.

Tag	Protein	Source organism	Affinity matrix	Reference
BAP	Biotin acceptor peptide		Avidin	72
CBP	Calmodulin-binding peptide		Calmodulin	73
DsbC	Disulfide bond C	*Escherichia coli*		74
FLAG	FLAG tag peptide		Anti-FLAG antibody	75
GB1	Protein G B1 domain	*Streptococcus* sp.		76
GST	Glutathione S-transferase		Glutathione	77
		Schistosoma japonicum		59
His$_6$	Hexahistidine tag		Metal chelates	78
MBP	Maltose-binding protein		Amylose	79
		Escherichia coli		80, 81
NusA	N-Utilisation substance	*Escherichia coli*		61
SET	Solubility-enhancing tag	Synthetic		82
Skp	Seventeen kilodalton protein	*Escherichia coli*		83
Strep-II	Streptavidin-binding peptide		Streptavidin	84
SUMO	Small ubiquitin-modifier	*Homo sapiens*		85
T7PK	Phage T7 protein kinase	Bacteriophage T7		74
Trx	Thioredoxin	*Escherichia coli*		30
ZZ	Protein A IgG ZZ repeat domain	*Staphylococcus aureus*	IgG	86

stabilises the recombinant protein in the soluble fraction as monomers or homodimers.[91–94]

MBP is a 42 kDa protein encoded by the *mal*E gene of *E. coli* K12.[25] MBP fused to either the N- and C-terminus has been shown to increase the expression and folding of eukaryotic fusion proteins expressed in bacteria[58,88] and a significant body of evidence exists to show that N-terminal MBP fusions can frequently produce soluble proteins when the unfused partners are insoluble.[88,89,95,96] Some maltodextrin-binding proteins from other bacteria provide an even greater enhancement of solubilisation than MBP[58]. In addition, there are also commercial vectors for the expression of MBP fusion proteins in the non-reducing environment of the *E. coli* periplasm (*e.g.*, New England Biolabs pMAL), addition of which may improve the folding of disulfide bond-containing proteins.

NusA is a transcription termination anti-termination factor and is a 55 kDa hydrophilic protein that promotes the soluble expression even of hydrophobic fusion proteins in *E. coli*.[68] In *E. coli*, wild-type NusA pauses in DNA transcription by RNA polymerase,[97] which slows translation and permits more time for critical folding events to occur.[61] NusA provides solubility enhancement comparable to MBP,[98–100] and a comparison of normally aggregation-prone proteins tagged with either MBP or NusA showed that both MBP and NusA enhanced solubility overall while not affecting the structure of the aggregation-prone proteins.[80] MBP and NusA have been used to enhance the solubilisation of ScFv antibodies in *E. coli*.[101,102] NusA has no independent purification-facilitating functionality.

Thioredoxin A (TrxA), an 11.6 kDa protein from *E. coli*, demonstrates high cytoplasmic solubility and thermal stability, which may be conferred to TrxA fusion proteins. It has been used to increase soluble protein expression of recombinant proteins,[109] and has been reported in several studies to be nearly as efficient as MBP in promoting solubility,[87,98] although other studies have shown it to be less effective.[85]

Other solubility tags include the Small Ubiquitin-related Modifier (SUMO) modifier protein, which is involved in post-translational modification. When used as an N-terminal carrier protein, SUMO promotes folding folding and structural stability, which leads to enhance functional production.[85,103–108] In some cases it appears to be as effective as MBP.[85] An additional benefit of using the SUMO tag is the availability of a specific SUMO protease (*S. cerevisiae* UlpI), which recognises and removes the SUMO tag at a Gly–Gly motif.

A fragment of the bacteriophage T7 protein kinase gene (T7PK) functions not only as a solubility enhancer, but also appears to enhance overall levels of expression.[85]

Membrane proteins account for approximately one-third of the human proteome and they form a majority of current drug targets.[110–113] However, standard over-expression and purification techniques are often unsuitable for this class of proteins, which presents a major obstacle to research progress. Mistic (membrane-integrating sequence for translation of integral membrane protein constructs)[114] has been applied to enhancing expression and membrane integration of eukaryotic integral membrane proteins in *E. coli* when linked to the N-terminus.[115,116] Mistic's mechanism of action remains unclear; however, in some cases the use of Mistic and its orthologs have significantly increased levels of eukaryotic integral membrane protein expression in *E. coli.*[117]

An improved approach for expression and purification of PglB, a bacterial bacterial oligosaccharyl transferase, has recently been reported.[118] The gene was incorporated into several vector-fusion protein constructs, and by screening at each step, expression levels were increased from tens of micrograms to several milligrams of active protein per litre of *E. coli* culture. The stability of the expressed protein was also increased from several hours to greater than six months post-purification. The best constructs used the MBP tag, the T7-tag and the GB1 tag.

Even with the expression systems noted above, problems still remain. Low expression yield and poor refolding efficiency of small recombinant proteins expressed in *E. coli* hinders the large-scale purification of such proteins for structural and biological investigations.[76] MBP and NusA are rather large partners and attention has focussed on smaller partners. The advantages of using a smaller partner include:

- Reduced energy demands on the host cell.
- Diminished steric hindrance.

One such is the N-terminal fragment of translation initiation factor IF2, which has a molecular weight of 17.4 kDa and has been used for production of recombinant streptavidin.[62] Other smaller tags include a His_6-tagged N-terminal fragment (52 residues) of *Staphylococcal nuclease* R (HR52), which has been used for the expression of small peptides.[119] This system also simplifies the purification protocol due to a one-step affinity purification procedure, and dramatically increases the final yield because of the smaller size of the fusion partner. However, its high hydrophobicity can interfere with the purification and refolding of hydrophobic peptides. The small size (<30 amino acids) of SET (Solubility Enhancement Tags; Stratagene) also means that in some cases removal of the tag for structural studies may be unnecessary. These comprise highly acidic amino acid sequences which are believed to reduce folding interference with the protein of interest.[82] Similarly, two other small protein tags, GB1 and ZZ, have been used with some success to enhance the expression and solubility of peptides and small proteins.[76,86,120] Using GB1[76] with a small cysteine-rich toxin, mutant myotoxin a (MyoP20G), the highly expressed fusion protein was refolded using an unfolding/refolding protocol, which could be monitored using heteronuclear single quantum coherence (HSQC) NMR spectroscopy. The final product yielded well-resolved NMR spectra, with a topology corresponding to the natural product. This system seems suitable for highly hydrophobic and cysteine-rich small proteins. A further fusion partner,[121] the Fh8 tag (8 kDa), has recently been evaluated as a solubility enhancing tag in *E. coli* and compared to other commonly used fusion partners. Six difficult-to-express target proteins (RVS167, SPO14, YPK1, YPK2, Frutalin and CP12) were fused to eight fusion tags (His, Trx, GST, MBP, NusA, SUMO, H and Fh8). The Fh8 partner improved protein expression and solubility as much as Trx, NusA or MBP fusion partners. Proteins cleaved from the Fh8 fusions were soluble and obtained in similar or higher amounts than proteins from the cleavage of other partners as Trx, NusA or MBP. The Fh8 fusion tag therefore acts as an effective solubility enhancer, and its low molecular weight potentially gives it an advantage over larger solubility tags by offering a more reliable assessment of the target protein solubility when expressed as a fusion protein.

The advent of recombinatorial cloning and high-throughput expression techniques furnishes a toolbox of solubility tags for the protein expressionist, any one of which might prove the best tool for a given task. For example, reduction of inclusion body formation to enhance overall yield of an expressed protein has been studied using RpoA (DNA-directed RNA polymerase α-subunit)[118] and Tsf (an elongation factor)[119] as fusion proteins. The proteins were fused to nine aggregation-prone human proteins, leading to enhanced yields. These are believed to shield active surfaces of heterologous proteins associated with non-specific protein–protein interactions that lead to the formation of inclusion bodies. In further study,[122] the industrial biocatalysts uridine phosphorylase from *Aeropyrum pernix* K1 and (+)-γ-lactamase and (−)-γ-lactamase from *Bradyrhizobium japonicum* USDA 6 were expressed in *E. coli* by using the pET System and then examined. Using a histidine tag as a fusion partner for protein expression did reduce the formation of inclusion bodies in these examples, suggesting that removing the fusion tag can promote the solubility of heterologous proteins.

3.5.2 Purification-Facilitating Tags

Purification-facilitating tags, or affinity tags, enable different proteins to be purified using a common method, and obviate the need for an understanding of the dark arts associated with conventional chromatographic purification.[63] Additionally, affinity purification reduces the number of unit operations and produces high yields, imbuing the approach with a high degree of economic favourability. Some commonly use purification-enhancing tags are listed in Table 3.3.

As mentioned above, GST and MBP also function as purification-facilitating (affinity) tags; GST binds to glutathione resin,[77] and MBP binds strongly to amylose resin[79] (see 3.5.1). GST has been utilised for single-step purification of fusion proteins from crude lysate by affinity for immobilised glutathione and eluted under non-denaturing conditions with 10 mM reduced glutathione.[77] MBP fusion proteins have been utilised for single-step purification by affinity to cross-linked amylose.[79] MBP fusion proteins bound to immobilised amylose are eluted under non-denaturing conditions with 10 mM maltose. In either case, the fusion tag can be removed during or after purification (see below).

His-tags are widely used, and pET and pBAD vectors are available with a His_6-encoding sequence. A schematic for the expression and purification of a His-tagged protein of interest is shown in Figure 3.10. Purification is based on the use of metal ions complexed with a resin-immobilised chelating agent; cellular extract containing the expressed protein is applied to the column, the column is washed to remove unbound moieties, and the His-tagged protein of interest is eluted by applying a solution containing the metal ion to the column. A similar approach can be adopted for a fusion between the protein of interest and FLAG, an octapeptide recognised by the M1 monoclonal antibody. The FLAG-tagged protein may be purified on resin-immobilised M1 mAb in the presence of calcium, which is needed for binding. After subsequent washing steps, the FLAG-tagged protein of interest is eluted by applying to the column a solution containing a chelating agent.

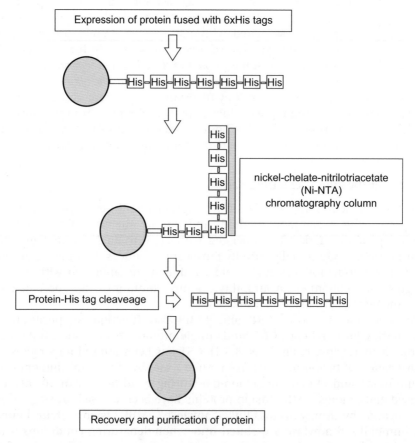

Figure 3.10 Recovery of proteins using (His_6-tag) and (Ni-NTA) chromatography columns.

Another approach[123] has been to use tandem affinity tags to allow the expressed protein to be obtained in greater purity, even though this does increase the size of the expressed product. In a recent study, an 16.5 kDa tag tag, CHiC (choline-binding His-tag), consisting of an N-terminal 6xHis-tag, a choline binding domain followed by a proteolytic site specific for the tobacco etch virus endopeptidase (TEV) was evaluated. The polypeptides used in the study from *Streptococcus pneumoniae* were PcsB, a peptidoglycan hydrolase, and ECL1, involved in splitting of the septum during cell division. CHiC-tagged PcsB and ECL1 were expressed in *E. coli* and sequentially purified as described above (Figure 3.9). After TEV digestion, the CHiC-tag, TEV-protease and undigested fusion protein were easily separated from the target protein in a single purification step. The approach yielded 4–7 mg of recombinant PcsB and ECL1 per litre of cell culture with a purity estimated to be at least 95%. In addition, it was found that the tag increased the solubility of PcsB.

Disadvantages of processes that use affinity tagging to purify the expressed protein include the cost of the affinity matrix that binds the tag and the consequential need for a chromatographic step. Obviously, an approach that dispensed with the chromatographic step and utilised a different method for separating the tagged protein from other proteins would offer advantages. Annexin B1, a member of the annexin superfamily from *Cysticercus cellulosae*, can be expressed in a soluble form in *E. coli*, and purified by a Ca2+-triggered precipitation step, which is fast and easy to perform, followed by resolubilisation of purified protein. Intein is a protein-intervening sequence that catalyses its efficient and precise excision from a host protein in a process mediated by pH changes or the addition of thiols, and which has been engineered for protein purification. Using an annexin B1-intein hybrid as a fusion tag,[124] the expressed B1-Int-protein can be precipitated using calcium and recovered by centrifugation, and the fusion partner subsequently removed by thiol-induced cleavage of the intein. The efficacy of the approach was demonstrated by expression and purification of human IL2 and single-chain plasminogen activator, urokinase-type (scuPA).

3.5.3 Tag Removal

Deleterious effects of an affinity tag on the properties of the target protein include:

- A change in conformation;[125]
- Lower yields;[126]
- Inhibition of enzyme activity;[127,128]
- Alteration in biological activity;[129]
- Undesired flexibility in structural studies[130] and toxicity.[131]

It is therefore desirable to remove the tag. Clearly, a strategy for the removal of the tag needs to be developed at the outset if the aim is to produce a 'native' (*i.e.*, tagless) protein for human use, when it is necessary to remove not only the cleaved fusion partner but also both enzyme(s) used to cleave it. One such approach is the TAGZyme system (Qiagen), a system based on recombinant exoproteases such as DAPase, which allow the efficient and precise removal of N-terminal His tags from proteins. One or more amino acid residue that the exoprotease is not able to remove can be included between the His-tag and the protein of interest to ensure that only the His-tag is removed. The DAPase has a C-terminal His$_6$ tag, which means that after treatment of the fusion protein with the enzyme, the mixture can be re-applied to the column containing the chelating agent, and the cleaved protein of interest will not now bind.

Alternatively, it can be arranged that the fusion product contain a linker sequence that can be cleaved, typically by an endoprotease, such as entero-kinase. Again, enterokinase-encoding regions can be included in pBAD and other expression systems.

GST expression vectors, such as the pGEX series from GE Healthcare, commonly include specific protease cleavage sites between the tag and partner proteins. Thus, the GST-tag can be readily removed from GST fusion proteins after or during purification, where affinity for immobilised glu-tathione greatly simplifies purification of the protein of interest from cleavage products.

A number of vectors for generating MBP fusion proteins are commercially available, including the pMAL series from New England Biolabs and pIVEX series from Roche, which contain a specific protease cleavage site in the region between MBP and the multiple cloning site.

In principle, self-cleaving elastin-like protein (ELP or inteins) tags, allow target protein to be purified by simple ELP-mediated precipitation steps, followed by self-cleavage and removal of the ELP tag. However, inteins usually cause some pre-cleavage during protein expression, significantly decreasing final yields. By splitting the intein into two ELP-tagged segments, each incapable of pre-cleavage alone, tight control of the tag cleaving re-action is obtained. Four different sizes of target proteins were purified using this approach with final yields comparable to or higher than a contiguous intein–ELP approach.[132]

3.6 CELL-FREE SYSTEMS

This chapter has focused on approaches for expressing protein in a suitable host cell, and the problems attendant to this process have been rehearsed in some detail; an alternative approach, is the use of cell-free systems.[133] These contain of all necessary components for translation, requiring RNA as a template. Cell-free protein expression can be performed in two steps (RNA transcription followed by translation) or combined into single step; the former may have higher yields but is less convenient, as the latter DNA (PCR product or expression clones) can be used as a template,[1] simplifying the overall process. Systems have been described from *Escherichia coli*, wheat germ extract, and mammalian cell extracts, including human and rabbit.[134–138]

A system based on the wheat embryo system has been developed by the Centre for Eukaryotic Structural Genomics, in co-operation with Ehime University and CellFree Sciences.[139] This automated platform for producing proteins for NMR-based structural proteomics is able to carry out as many as 384 small-scale screening reactions per week. A desktop robot is also available which, according to the manufacturers, performs transcription, translation and batch affinity purification in around 35 hours. It can run on either a 6-well format or a 24-well format to express up to 6 or 24 genes of interest. The robot produces 2.5 to 3.0 mg of crude proteins per run.

Other systems are available, including Expressway Cell-Free Expression Systems from Invitrogen, an *E. coli*-based in-vitro system that is able to produce up to milligram quantities of active recombinant protein in a tube reaction format.

3.7 HIGH-THROUGHPUT APPROACHES

As noted above, whilst one tag may be the ideal choice for a given protein, or even a given group of proteins, it does not follow that it will function equally well with every protein: thus it remains unknown which to use for the protein of interest, there are no rules of thumb.

Technologies have been developed that have made the generation of new expression vectors relative easy,[88,91] allowing parallel cloning into multiple vectors having different tags to be an almost routine matter. This means that the suitability of a range of solubility-enhancing tags can be assessed and compared in a single experiment.[68,88,89] Furthermore, the approach can be extended to look at larger numbers, and more diverse types, of target proteins.[88,98]

Fortunately, a number of high-throughput technologies have been developed fulfilling the need for systematic and high-throughput cloning methodologies for the study of protein function. ORFs are captured in a common configuration so that the same basic reagents and steps can be used on all of them. Cloning steps are chosen which assure molecular conservation, avoiding amplification, so that once a sequence-verified ORF is introduced into the system, it will not require re-sequencing after any transfer steps. Differences in approaches relate to the mechanism for transferring ORFs from one plasmid to another, and all the methods aim to provide simple, rapid, reliable, and highly efficient approaches; the best methods being automated.[1]

Commercial systems include Promega's *Flexi*®, Invitrogen's *Gateway* system, Merck Millipore's *Radiance* and *In-Fusion* from Clontech. The *Gateway* system not only allows exchange of ORFs between libraries, synthesised genes, PCR products, or indeed many other sources, *via* an 'entry' clone, the clone itself can be used to develop expression clones using a standardised procedure,[140] which is based around bacteriophage λ site-specific

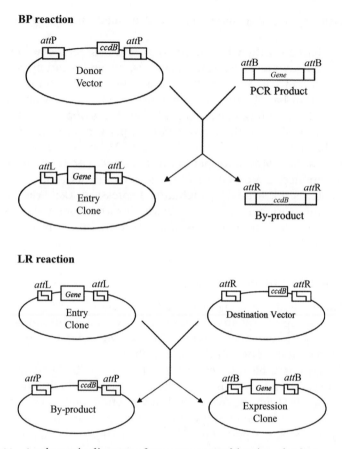

Figure 3.11 A schematic diagram of one-step recombination cloning.

recombinatorial system (see Figure 3.11) and does not use restriction enzymes. It is based on the BP reaction (recombination between *att*B and *att*P sites) and the LR reaction (recombination between *att*L and *att*R sites), in which specific recombination sites are used for gene cloning and transfer from one vector to another. In the first reaction, the BP reaction, mediated by an enzyme mixture, the PCR product (or cloned gene) is flanked by the site-specific *att*B sites, which recombine with the *att*P sites in the donor vector. The resultant product is an entry clone containing the cloned sequence flanked by new *att*L recombination sites. In the second reaction, the LR Reaction, the entry clone having the *att*L sites recombines with a destination vector having *att*R sites to create an *att*B-containing expression clone (see Figure 3.11).

Furthermore, the expression clone may contain multiple ORFs in a fixed and known orthogonality. This means it can be used to produce, in at least a semi-automated manner, permutations and combinations of ORFs, and such approaches have been used to explore metabolomic features of proteins from a particular pathway, for example. But it may also be used to generate combinations of ORFs corresponding to proteins of interest and fusion proteins, enabling production of a range of expression clones that can be used to produce fusion protein for simplified purification and subsequent analysis. In one study, 75 different ORFs were transferred into expression vectors in combination with four different fusion tags. The efficiency and usefulness of the exercise was evaluated in respect of yield and solubility.[141]

A similar approach has been used for high-throughput expression and screening of membrane proteins, a particular challenge hindering routine structure determination. The approach was used for 49 *E. coli* integral membrane proteins, and 71% of these could be produced at sufficient levels to allow milligram amounts of protein to be relatively easily purified.[142]

In another study of 36 prokaryotic P-type transporters, a wide ensemble of modified constructs was generated and tested for expression in *E. coli*, membrane localisation, detergent extraction, and homogeneity. The choice of promoter, the choice of source organism providing the cloned gene, and, most importantly, the position of the affinity tag had a big effect on successful production. Following the initial screening, material from nine of the 36 targets was suitable for crystallisation or other structural studies.[143] The expression of 20 membrane proteins, both peripheral and integral, in three prokaryotic (*E. coli*, *Lactobacillus lactis*, *Rhodobacter sphaeroides*) and three eukaryotic (*Arabidopsis thaliana*, *Nicotiana benthamiana*, Sf9 insect cells) hosts were investigated in another study.[110] The proteins tested were of various origins (bacteria, plants and mammals), functions (transporters, receptors, enzymes) and topologies (between zero and 13 transmembrane segments). The Gateway system was used to clone all 20 genes into appropriate vectors for the hosts to be tested. Culture conditions were optimised for each host, and specific strategies were tested, such as the use of Mistic fusions in *E. coli*. All but three of the proteins were produced at adequate yields for functional and, in some cases, structural studies.

3.8   OTHER HOSTS

The focus in this chapter has been on the use of *E. coli* as the host system, but the use of bacterial expression systems is somewhat limited for eukaryotic proteins on account of the need for post-translational modifications to produce a correct glycosylation pattern. For such proteins, eukaryotic system may be a better choice. One such system is based on a monkey COS cell line having a defective region of the SV40 genome stably integrated into the COS cell genome. Inserting an expression vector having the SV40 origin of replication and the protein of interest into the COS cell initiates viral replication leading to a high level of expression of the protein. A disadvantage of this system is the lysis of the COS cells; another is the limited insert capacity of the vector. Other eukaryotic cloning vectors[144] for expression of the protein of interest, including *Saccharomyces cerevisiae*,[145–147] *Pichia* pastoris,[147,148] insect[149] and mammalian[145] cell lines are available.

A recent approach has developed a stable mammalian cell expression system based on the piggyBac transposon.[150] This doxycycline-inducible system is cell line-independent and requires only a single transfection/selection step, and yields stably transfected mammalian cell cultures for large-scale protein production. The approach was used for large-scale production (140–750 mg scale) of an endoplasmic reticulum-resident fucosyltransferase and two potential anticancer protein therapeutic agents.

3.9 CONCLUSION

Molecular cloning techniques have evolved a long way from simple usage to express proteins from heterologous DNA incorporated into pBR322; now many expression systems and many hosts are available to ensure that the "expression" part of the "expression, solubility and purification"[2] target is met. But there is more to protein expression than merely producing small amounts of it—the more recent challenge has been to provide a reasonable quantity of it in a state that at least approximates to the 'natural' state, for proteomic applications, for pharmaceutical research and for commercial use. This has been met by the development of solubility-enhancing and purification-facilitating tags on the one hand, and on the development of high-throughput approaches on the other. This means that is now possible quickly to screen which of these tags might be most suitable for expressing the protein of interest, potentially rendering the purification of a protein of interest to an almost routine procedure. Molecular cloning for protein expression is thus a key technique for the future understanding and exploitation of proteomics.

REFERENCES

1. F. Festa, J. Steel, X. Bian and J. Labaer, *Proteomics*, 2013, **13**, 1381.
2. D. Esposito and D. K. Chatterjee, *Curr. Opin. Biotechnol.*, 2006, **17**, 353.
3. M. H. Caruthers, S. L. Beaucage, C. Becker, W. Efcavitch, F. F. Fisher, G. Galluppi, R. Coldman, P. Dettaseth, F. Martin, M. Matteucci and Y. Stabinsley, in *Genetic Engineering*, ed. J. K. Setlow and A. Hollaender, Plenum Press, New York and London, 1982, p. 119.
4. S. Narang, *J. Biosci.*, 1984, **6**, 739.
5. M. D. Edge, A. R. Green, G. R. Heathcliffe, P. A. Meacock, W. Shuch, D. B. Scanlon, T. C. Atkinson, C. R. Newton and A. F. Markham, *Nature*, 1981, **191**, 756.
6. D. G. Yanasura and D. J. Henner, in *Gene Expression Technology, Methods in Enzymology*, ed. D. V. Goeddel, Academic Press, San Diego, 1990, 185, p. 54.
7. K. Terpe, *Appl. Microbiol. Biotechnol.*, 2006, **72**, 211.
8. C. M. Elvin, P. R. Thompson, M. E. Argall, P. Hendry, N. P. Stamford, P. E. Lilley and N. E. Dixon, *Gene*, 1990, **87**, 123.
9. B. Gronenborn, *Mol. Gen. Genet.*, 1976, **148**, 243.

10. J. Brosius, M. Erfle and J. Storella, *J. Biol. Chem.*, 1985, **260**, 3539.
11. L. M. Guzman, D. Belin, M. J. Carson and J. Beckwith, *J. Bacteriol.*, 1995, **177**, 4121.
12. A. Haldimann, L. Daniels and B. Wanner, *J. Bacteriol.*, 1998, **180**, 1277.
13. A. Skerra, *Gene*, 1994, **151**, 131.
14. F. W. Studier and B. A. Moffatt, *J. Mol. Biol.*, 1986, **189**, 113.
15. H. A. de Boer and A. S. Hui, in *Gene Expression Technology, Methods in Enzymology*, ed. D. V. Goeddel, Academic Press, San Diego, 1990, 185, p. 103.
16. H. P. Sørensen, B. S. Laursen, K. K. Mortensen and H. U. Sperling-Petersen, *Recent Res. Dev. Biophys. Biochem.*, 2002, **2**, 243.
17. C. M. Stenstrom, H. Jin, L. L. Major, W. P. Tate and L. A. Isaksson, *Gene*, 2001, **263**, 273.
18. S. Ringquist, S. Shinedling, D. Barrick, L. Green, J. Binkley, G. D. Stormo and L. Gold, *Mol. Microbiol.*, 1992, **6**, 1219.
19. P. O. Olins and S. H. Rangwala, in *Gene Expression Technology, Methods in Enzymology*, ed. D. V. Goeddel, Academic Press, San Diego, 1990, 185, p. 1–15.
20. P. O. Olins, C. S. Devine and S. H. Rangwala, in *Expression Systems and Processes for rDNA Products*, ed. R. T. Hatch, C. Goochee, A. Moreira and Y. Alroy, ACS, Washington, DC, 1991, p. 17.
21. E. S. Poole, C. M. Brown and W. P. Tate, *EMBO J.*, 1995, **14**, 151.
22. F. Baneyx, *Curr. Opin. Biotechnol.*, 1999, **10**, 411.
23. P. Jonasson, S. Liljeqvist, P. A. Nygren and S. Stahl, *Biotechnol. Appl. Biochem.*, 2002, **35**, 91.
24. G. del Solar, R. Giraldo, M. J. Ruiz-Echevarria, M. Espinosa and R. Diaz-Orejas, *Microbiol. Mol. Biol. Rev.*, 1998, **62**, 434.
25. D. Summers, *Mol. Microbiol.*, 1998, **29**, 1137.
26. M. P. Mayer, *Gene*, 1995, **163**, 41.
27. K. G. Hardy, *Plasmids*, IRL Press, Oxford, 1987.
28. G. Hannig and S. C. Makrides, *Trends Biotechnol.*, 1998, **16**, 54.
29. H. P. Sørensen and K. K. Mortensen, *J. Biotechnol.*, 2005, **115**, 113.
30. E. Englesberg, C. Squires and F. Meronk Jr., *Proc. Natl. Acad. Sci. U. S. A.*, 1969, **62**, 1100.
31. D. A. Siegele and J. C. Hu, *Proc. Natl. Acad. Sci. U. S. A.*, 1997, **94**, 8168.
32. A. Khlebnikov, T. Skaug and J. D. Keasling, *J. Ind. Microbiol. Biotechnol.*, 2002, **29**, 34.
33. R. M. Morgan-Kiss, C. Wadler and J. E. Cronan Jr., *Proc. Natl. Acad. Sci. U. S. A.*, 2002, **99**, 7373.
34. J. W. Dubendorff and F. W. Studier, *J. Mol. Biol.*, 1991, **219**, 45.
35. F. W. Studier, A. H. Rosenberg, J. J. Dunn and J. W. Dubendorff, *Methods Enzymol.*, 1990, **185**, 60.
36. S. Zhu, C. Gong, L. Ren, X. Li, D. Song and G. Zheng, *Appl. Microbiol. Biotechnol.*, 2013, **7**, 837.
37. P. O. Olins, C. S. Devine, S. H. Rangwala and K. S. Kavka, *Gene*, 1988, **73**, 227.

38. M. P. Calos, *Nature*, 1978, **274**, 762.
39. F. W. Studier, *J. Mol. Biol.*, 1991, **219**, 37.
40. A. Khlebnikov and J. D. Keasling, *Biotechnol. Prog.*, 2002, **18**, 672.
41. E. Steinmetz, *BioTechniques*, 2011, **50**, 263.
42. A. Haldimann, L. Daniels and B. Wanner, *J. Bacteriol.*, 1998, **180**, 1277.
43. S. M. Egan and R. F. Schleif, *J. Mol. Biol.*, 1993, **234**, 87.
44. M. J. Giacalone, A. M. Gentile, B. T. Lovitt, N. L. Berkley, C. W. Gunderson and M. W. Surber, *BioTechniques*, 2006, **40**, 355.
45. J. F. Kane, *Curr. Opin. Biotechnol.*, 1995, **6**, 494.
46. C. Kurland and J. Gallant, *Curr. Opin. Biotechnol.*, 1996, **7**, 489.
47. G. Dieci, L. Bottarelli, A. Ballabeni and S. Ottonello, *Protein Expr. Purif.*, 2000, **18**, 346.
48. I. Arechaga, B. Miroux, M. J. Runswick and J. E. Walker, *FEBS Lett.*, 2003, **547**, 97.
49. G. Georgiou and L. Segatori, *Curr. Opin. Biotechnol.*, 2005, **16**, 538.
50. T. A. Phillips, R. A. van Bogelen and F. C. Neidhardt, *J. Bacteriol.*, 1984, **159**, 283.
51. A. Villaverde and M. M. Carrio, *Biotechnol. Lett.*, 2003, **25**, 1385.
52. M. M. Carrio, R. Cubarsi and A. Villaverde, *FEBS Lett.*, 2000, **471**, 7.
53. A. Mogk, M. P. Mayer and E. Deuerling, *ChemBioChem*, 2002, **3**, 807.
54. E. Deuerling, H. Patzelt, S. Vorderwulbecke, T. Rauch, G. Kramer, E. Schaffitzel, A. Mogk, A. Schulze-Specking, H. Langen and B. Bukau, *Mol. Microbiol.*, 2003, **47**, 1317.
55. B. Miroux and J. E. Walker, *J. Mol. Biol.*, 1996, **260**, 289.
56. D. Petsch and F. B. Anspach, *J. Biotechnol.*, 2000, **76**, 97.
57. C. L. Young, Z. T. Britton and A. S. Robinson, *Biotechnol. J.*, 2012, **7**, 620.
58. R. B. Kapust and D. S. Waugh, *Protein Sci.*, 1999, **8**, 1668.
59. P. A. Nygren, S. Stahl and M. Uhlen, *Trends Biotechnol.*, 1994, **12**, 184.
60. K. Terpe, *Appl. Micobiol. Biotechnol.*, 2003, **60**, 523.
61. G. D. Davis, C. Elisee, D. M. Newham and R. G. Harrison, *Biotechnol. Bioeng.*, 1999, **65**, 382.
62. H. P. Sørensen, H. U. Sperling-Petersen and K. K. Mortensen, *Protein Expr. Purif.*, 2003, **32**, 252.
63. J. Arnau, C. Lauritzen, G. E. Petersen and J. Pedersen, *Prot. Expr. Purif.*, 2006, **48**, 1.
64. D. S. Waugh, *Trends Biotechnol.*, 2005, **23**, 316.
65. B. Brizzard, *BioTechniques*, 2008, **44**, 693.
66. D. Walls and S. T. Loughran, *Methods Mol. Biol.*, 2011, **681**, 151.
67. J. Koehn and I. Hunt, *Methods Mol. Biol.*, 2009, **498**, 1.
68. R. C. Stevens, *Struct. Fold Des.*, 2000, **8**, R177.
69. A. Jacquet, V. Daminet, M. Haumont, L. Garcia, S. Chaudoir, A. Bollen and R. Biemans, *Protein Expr. Purif.*, 1999, **17**, 392.
70. G. S. Waldo, B. M. Standish, J. Berendzen and T. C. Terwilliger, *Nat. Biotechnol.*, 1999, **17**, 691.

71. H. Wang and S. Chong, *Proc. Natl. Acad. Sci. U. S. A.*, 2003, **100**, 478.
72. P. J. Schatz, *Biotechnology*, 1993, **11**, 1138.
73. P. Vaillancourt, C. F. Zheng, D. Q. Hoang and L. Breister, *Methods Enzymol.*, 2000, **326**, 340.
74. Z. Zhang, Z. H. Li, F. Wang, M. Fang, C. C. Yin, Z. Y. Zhou, Q. Lin and H. L. Huang, *Protein Expr. Purif.*, 2002, **26**, 218.
75. A. Einhauer and A. Jungbauer, *J. Biochem. Biophys. Methods*, 2001, **49**, 455.
76. Y. Cheng and D. J. Patel, *Biochem. Biophys. Res. Commun.*, 2004, **317**, 401.
77. D. B. Smith and K. S. Johnson, *Gene*, 1988, **67**, 31.
78. V. Gaberc-Porekar and V. Menart, *J. Biochem. Biophys. Methods*, 2001, **49**, 335.
79. C. di Guan, P. Li, P. D. Riggs and H. Inouye, *Gene*, 1988, **67**, 21.
80. S. Nallamsetty and D. S. Waugh, *Protein Expr. Purif.*, 2006, **45**, 175.
81. J. D. Fox and D. S. Waugh, *Methods Mol. Biol.*, 2003, **205**, 99.
82. Y. B. Zhang, J. Howitt, S. McCorkle, P. Lawrence, K. Springer and P. Freimuth, *Protein Expr. Purif.*, 2004, **36**, 207.
83. D. K. Chatterjee and D. Esposito, *Protein Expr. Purif.*, 2006, **46**, 122.
84. S. Vossand and A. Skerra, *Protein Eng.*, 1997, **10**, 975.
85. J. G. Marblestone, S. C. Edavettal, Y. Lim, P. Lim, X. Zuo and T. R. Butt, *Protein Sci.*, 2006, **15**, 182.
86. Y. Zhao, Y. Benita, M. Lok, B. Kuipers, P. van der Ley, W. Jiskoot, W. E. Hennink, D. J. Crommelin and R. S. Oosting, *Vaccine*, 2005, **23**, 5082.
87. M. Hammarström, N. Hellgren, S. van den Berg, H. Berglund and T. Hard, *Protein Sci.*, 2002, **11**, 313.
88. M. R. Dyson, S. P. Shadbolt, K. J. Vincent, R. L. Perera and J. McCafferty, *BMC Biotechnol.*, 2004, **4**, 32.
89. I. Kataeva, J. Chang, H. Xu, C. H. Luan, J. Zhou, V. N. Uversky, D. Lin, P. Horanyi, Z. J. Liu, L. G. Ljungdahl, J. Rose, M. Luo and B.-C. Wang, *J. Proteome Res.*, 2005, **4**, 1942.
90. W. Kaplan, P. Husler, H. Klump and J. Erhardt, *Protein Sci.*, 1997, **6**, 399.
91. M. W. Parker, M. Lo Bello and G. Federici, *J. Mol. Biol.*, 1990, **213**, 221.
92. K. Lim, J. X. Ho, K. Keeling and G. L. Gilliland, *Protein Sci.*, 1994, **3**, 2233.
93. X. Ji, P. Zhang, R. N. Armstrong and G. L. Gilliland, *Biochemistry*, 1992, **31**, 10169.
94. Y. Maru, D. E. Afar, O. N. Witte and M. Shibuya, *J Biol. Chem.*, 1996, **271**, 15353.
95. D. Busso, B. Delagoutte-Busso and D. Moras, *Anal. Biochem.*, 2005, **343**, 313.
96. S. Braud, M. Moutiez, P. Belin, N. Abello, P. Drevet, S. Zinn-Justin, M. Courcon, C. Masson, J. Dassa, J. B. Charbonnier, J.-C. Boulain, A. Menez and M. Gondry, *J Proteome Res.*, 2005, **4**, 2137.

97. I. Gusarov and E. Nudler, *Cell*, 2001, **107**, 437.
98. A. Dummler, A. M. Lawrence and A. de Marco, *Microb. Cell. Fact.*, 2005, **4**, 34.
99. A. Schrodel, J. Volz and A. de Marco, *J. Biotechnol.*, 2005, **120**, 2.
100. P. Turner, O. Holst and E. N. Karlsson, *Protein Expr. Purif.*, 2005, **39**, 54.
101. H. Bach, Y. Mazor, S. Shaky, A. Shoham-Lev, Y. Berdichevsky, D. L. Gutnick and I. Benhar, *J. Mol. Biol.*, 2001, **312**, 79.
102. L. Zheng, U. Baumann and J. L. Reymond, *J. Biochem. (Tokyo)*, 2003, **133**, 577.
103. M. P. Malakhov, M. R. Mattern, O. A. Malakhova, M. Drinker, S. D. Weeks and T. R. Butt, *J. Struct. Funct. Genomics*, 2004, **5**, 75.
104. T. R. Butt, S. C. Edavettal, J. P. Hall and M. R. Mattern, *Protein Expr. Purif.*, 2005, **43**, 1.
105. X. Zuo, S. Li, J. Hall, M. R. Mattern, H. Tran, J. Shoo, R. Tan, S. R. Weiss and T. R. Butt, *J. Struct. Funct. Genomics*, 2005, **6**, 103.
106. X. Zuo, M. R. Mattern, R. Tan, S. Li, J. Hall, D. E. Sterner, J. Shoo, H. Tran, P. Lim, S. G. Sarafianos, L. Kazi, S. Navas-Martin, S. R. Weiss and T. R. Butt, *Protein Expr. Purif.*, 2005, **42**, 100.
107. C. M. Guzzo and D. C. Yang, *Protein Expr. Purif.*, 2007, **54**, 166.
108. J. E. Dominy, C. R. Simmons, L. L. Hirschberger, J. Hwang, R. M. Coloso and M. H. Stipanuk, *J. Biol. Chem.*, 2007, **282**, 25189.
109. E. R. LaVallie, Z. Lu, E. A. Diblasio-Smith, A. Collins-Racie and J. M. McCoy, *Methods Enzymol.*, 2000, **326**, 322.
110. F. Bernaudat, A. Frelet-Barrand, N. Pochon, S. Dementin and P. Hivin, *et al.*, *PLoS ONE*, 2011, **6**, e29191.
111. J.-J. Lacapere, E. Pebay-Peyroula, J.-M. Neumann and C. Etchebest, *Trends Biochem. Sci.*, 2007, **32**, 259.
112. G. von Heijne, *J. Intern. Med.*, 2007, **261**, 543.
113. K. McLuskey, A. W. Roszak, Y. Zhu and N. W. Isaacs, *Eur. Biophys. J.*, 2010, **39**, 723.
114. T. P. Roosild, J. Greenwald, M. Vega, S. Castronovo, R. Riek and S. Choe, *Science*, 2005, **307**, 1317.
115. T. P. Roosild, M. Vega, S. Castronovo and S. Choe, *BMC Struct. Biol.*, 2006, **6**, 10.
116. G. Kefala, W. Kwiatkowski, L. Esquivies, I. Maslennikov and S. Choe, *J. Struct. Funct. Genomics*, 2007, **8**, 167.
117. H. Dvir and S. Choe, *Protein Expr. Purif.*, 2009, **68**, 28.
118. M. B. Jaffee and B. Imperiali, *Protein Expr. Purif.*, 2013, **89**, 241.
119. Y. Cheng, D. Liu, Y. Feng and G. Jing, *Protein Pept. Lett.*, 2003, **10**, 175.
120. W. J. Bao, Y. G. Gao, Y. G. Chang, T. Y. Zhang, X. J. Lin, X. Z. Yan and H. Y. Hu, *Protein Expr. Purif.*, 2006, **47**, 599.
121. S. J. Costa, A. Almeida, A. Castro, L. Domingues and H. Besir, *Appl. Microbiol. Biotechnol.*, 2013, **97**, 6779.

122. S. Zhu, C. Gong, L. Ren, X. Li, D. Song and G. Zheng, *Appl. Microbiol. Biotechnol.*, 2013, **97**, 837.
123. G. A. Stamsås, L. S. Håvarstein and D. Straume, *J. Microbiol. Methods*, 2013, **92**, 59.
124. F.-X. Ding, H.-L. Yan, Q. Mei, G. Xue, Y.-Z. Wang, Y.-J. Gao and S.-H. Sun, *Appl. Microbiol. Biotechnol.*, 2007, **77**, 483.
125. A. Chant, C. M. Kraemer-Pecore, R. Watkin and G. G. Kneale, *Protein Expr. Purif.*, 2005, **39**, 152.
126. A. Goel, D. Colcher, J. S. Koo, B. J. Booth, G. Pavlinkova and S. K. Batra, *Biochim. Biophys. Acta*, 2000, **1523**, 13.
127. K. M Kim, E. C. Yi, D. Baker and K. Y. Zhang, *Acta Crystallogr. Sect. D Biol. Crystallogr.*, 2001, **57**, 759.
128. S. Cadel, C. Gouzy-Darmon, S. Petres, C. Piesse, V. L. Pham, M. C. Beinfeld, P. Cohen and T. Foulon, *Protein Expr. Purif.*, 2004, **36**, 19.
129. I. Fonda, M. Kenig, V. Gaberc-Porekar, P. Pristovaek and V. Menart, *Sci. World J.*, 2002, **15**, 1312.
130. D. R. Smyth, M. K. Mrozkiewicz, W. J. McGrath, P. Listwan and B. Kobe, *Protein Sci.*, 2003, **12**, 1313.
131. E. G. E. de Vries, M. N. de Hooge, J. A. Gietema and S. de Jong, *Clin. Cancer Res.*, 2003, **9**, 912.
132. C. Shi, Q. Meng and D. W. Wood, *Appl. Microbiol. Biotechnol.*, 2013, **97**, 829.
133. Y. Endo and T. Sawasaki, *Curr. Opin. Biotechnol.*, 2006, **17**, 373.
134. F. Festa, S. M. Rollins, K. Vattem, M. Hathaway, *et al.*, *Proteomics Clin. Appl.*, 2013, **5–6**, 372.
135. E. D. Carlson, R. Gan, C. E. Hodgman and M. C. Jewett, *Biotechnol. Adv.*, 2012, **30**, 1185.
136. N. Ramachandran, J. V. Raphael, E. Hainsworth and G. Demirkan, *et al.*, *Nat. Methods*, 2008, **5**, 535.
137. N. Goshima, Y. Kawamura, A. Fukumoto and A. Miura, *et al.*, *Nat. Methods*, 2008, **5**, 1011.
138. D. Busso, R. Kim and S. H. Kim, *J. Biochem. Biophys. Methods*, 2003, **55**, 233.
139. D. A Vinarov and J. L. Markley, *Exp. Rev. Proteomics*, 2005, **2**, 49.
140. F. Katzen, *Expert Opin. Drug Discov.*, 2007, **2**, 571.
141. U. Korf, T. Kohl, H. van der Zandt, R. Zahn, S. Schleeger, B. Ueberle, S. Wandschneider, S. Bechtel, M. Schnölzer, H. Ottleben, S. Wiemann and A. Poustka, *Proteomics*, 2005, **5**, 3571.
142. S. Eshaghi, M. Hedrén, M. Ignatushchenko, A. Nasser, T. Hammarberg, A. Thornell and P. Nordlund, *Prto. Sci.*, 2005, **14**, 676.
143. O. Lewinson, A. T. Lee and D. C. Rees, *J. Mol. Biol.*, 2008, **377**, 62.
144. T. Brown, *Gene Cloning and DNA*, Blackwell, Oxford, 5th edition, 2006.
145. L. Borsig, E. G. Berger and M. Malissard, *Biochem. Biophys. Res. Commun.*, 1997, **240**, 586.

146. L. J. Byrne, K. J. O'Callaghan and M. F. Tuite, *Methods Mol. Biol.*, 2005, **308**, 51.
147. B. Prinz, J. Schultchen, R. Rydzewski, C. Holz, M. Boettner, U. Stahl and C. Lang, *J. Struct. Funct. Genomics*, 2004, **5**, 29.
148. D. Daly and M. T. Hearn, *J. Mol. Recognit.*, 2005, **18**, 119.
149. T. A. Kost, J. P. Condreay and D. L. Jarvis, *Nat. Biotechnol.*, 2005, **23**, 567.
150. Z. Li, I. P. Michael, D. Zhou, A. Nagy and J. M. Rini, *PNAS*, 2013, **110**, 5004.

CHAPTER 4

Proteins and Proteomics

NATALIA HUPERT AND DAVID B. WHITEHOUSE*

School of Life and Medical Sciences, University of Hertfordshire, College Lane, Hatfield, Hertfordshire, AL10 9AB, United Kingdom
*Email: d.whitehouse@herts.ac.uk

4.1 INTRODUCTION

Proteomics is the large-scale and high-throughput study of proteins.[1] The term 'proteome' was first used in 1995 and derives from PROTEins expressed in a genOME.[2] A proteome is the population of all proteins expressed by the genome in a specific cell type, tissue or biological fluid at a particular time, and under specific conditions. 'Proteomics' was coined in 1997 to introduce uniformity of terminology with 'genomics', the study of the genome,[3] and has the potential to address aspects of biology that are beyond the reach of genomics and transcriptomics. For instance, it can be used to identify precisely the expression and cellular locations of specific proteins that could not be predicted from mRNA levels, identify posttranslational modifications that could not be ascertained from gene sequences and characterise many aspects of protein function and protein interactions. Any tissue, cell type or extracellular fluid is amenable to proteomic analysis.[4] In contrast to the genome, which comprises the set of all genes in a species, the proteome in different cells, tissues and fluids is dynamic and constantly changing in response to metabolic requirements, developmental stage and environmental stresses.[5] It should also be noted that the total number of proteins far exceeds the total number of genes. It is believed that the Human Genome comprises around 20 000 protein coding genes with some 180 000 exons that occupy about 1% of the genome.[6] Through the action of alternative splicing

Molecular Biology and Biotechnology, 6th Edition
Edited by Ralph Rapley and David Whitehouse
© The Royal Society of Chemistry 2015
Published by the Royal Society of Chemistry, www.rsc.org

(it has been estimated that between 74% and 99% of human multi-exon genes are alternatively spliced), mutation and other molecular phenomena, ~150 000 mRNA transcripts could be produced from 20 000 genes.[7] This gives rise to an equivalent number of primary protein products. However, the number of protein species is greatly increased through the action of post-translational events such as heteromer formation and posttranslational modifications such as glycosylation and phosphorylation. Therefore, it is reasonable to assume that 20 000 genes could give rise to perhaps a million or more different proteins.[8] Individual cell types and tissues will contain a restricted repertoire of the total proteome, but taken together the numerical challenge for proteomics is enormous.

4.1.1 Outline of the Proteomics Workflow

The aim of a basic proteomics experiment is to isolate and characterise a single protein from a complex biological sample. The first step in the pro-teomics workflow is the extraction of proteins from the biological material of interest. This is followed by their separation, normally by gel electrophoresis or liquid chromatography. The preferred separation method in many pro-teomics laboratories is two-dimensional electrophoresis (2DGE), where proteins are first separated by their isoelectric points and then by their molecular weights.[9] Proteins of interest are cut from the gel, digested with a proteinase and the resulting peptides are analysed by mass spectrometry. Protein identification can be achieved by matching the masses of the ex-perimental peptide fragments with those from known proteins recorded in sequence databases. When peptide matching is unsuccessful, a peptide can be further fragmented and analysed to allow the *de novo* identification of amino acid sequences that can be used to search DNA sequence databases. A generalised proteomics workflow is shown in Figure 4.1. Proteomics

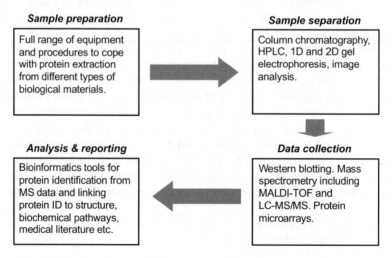

Figure 4.1 A typical proteomics laboratory workflow.

depends on sophisticated instrumentation capable of processing multiple samples with a very high degree of precision and powerful bioinformatics tools. However, its roots can be traced to traditional protein biochemistry and the separation methods that were employed to identify specific proteins in complex biological mixtures.[10] There are many applications for proteomics ranging from the identification and characterisation of proteins and their interactions to probing metabolic pathways and the assessment of protein expression profiles in health and disease. An important practical outcome is the identification of new biomarkers of complex diseases, such as cancer, that could lead to improvements in *in vitro* diagnosis, prognosis and identification of novel drug targets. Following a brief introduction to basic protein chemistry, this chapter describes each stage of the proteomics workflow and some of its applications.

4.2 PROTEIN CHEMISTRY

Proteins are large complex molecules that are composed of linear arrays of amino acids. Each amino acid contains a central (alpha) carbon atom to which an amino group (-NH$_2$) and a carboxyl group (-COOH) are attached (Figure 4.2). Also, attached to the alpha carbon atom are a hydrogen atom and a side group (R-group). The R-group varies between different amino acids giving each its own unique chemical properties. For example, some amino acids are positively or negatively charged while others are hydrophilic or hydrophobic. The R-groups are important in determining the three-dimensional structure and function of proteins. 20 amino acids are encoded in DNA and these make up the vast majority of proteins in living things; the number of protein structures and hence functions is enormous.[11]

4.2.1 Protein Structure

Amino acids are linked together by peptide bonds. The peptide bond is formed by a condensation reaction when a carboxyl group of one amino acid reacts with amino group of another and a water molecule is released (Figure 4.3). A peptide bond formed between two amino acids gives rise to a

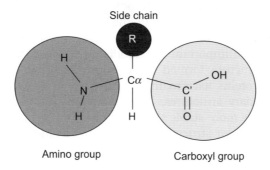

Figure 4.2 The general structure of an amino acid.

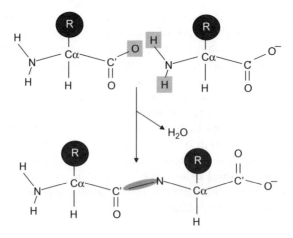

Figure 4.3 Formation of a peptide bond, also called an amide linkage, occurs when the carboxyl group of one amino acid reacts with the amino group of another. This condensation reaction is accompanied by the elimination of water.

dipeptide and the joining together of three amino acids creates a tripeptide *etc*. Historically, peptide chains with less than 20 amino acids were referred to as 'oligopeptides', 'polypeptides' were associated with 20–100 amino acids and proteins with more than 100 amino acids.[12] Protein structure is organised into four categories: primary, secondary, tertiary and quaternary.[68]

4.2.1.1 Primary Structure. The primary structure is the linear sequence of amino acids in the protein. One of the first proteins for which the primary structure was determined was the hormone insulin (51 amino acids). Proteins made up of approximately the same ratios of amino acids, but in different sequences, have different secondary and tertiary structures and perform different functions.

4.2.1.2 Secondary Structure. Primary structures can fold or twist into more complex secondary structures (Figure 4.4). There are two main types of regular secondary structures: α-helix and β-sheet. α-helix is formed as a result of a right-handed twist of the polypeptide chain along its axis. The C=O group of one amino acid and the N-H group of another are held together by hydrogen bonds, which are made every fourth amino acid. Hydrogen bonds are important for stabilising α-helixes. The less common β-sheet is flat structure formed by hydrogen bonding between adjacent strands. Some proteins have mainly α-helixes, or β-sheets, whereas others have both secondary structures; in general proteins with high levels of β-sheets tend to be more stable than those without. Primary sequence that does not form a regular structure is referred to as random coil.

4.2.1.3 Tertiary Structure. The tertiary structure is the three-dimensional shape of the folded polypeptide chain and is stabilised by different types

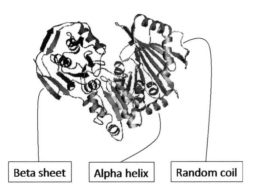

| Beta sheet | Alpha helix | Random coil |

Figure 4.4 Secondary structures of proteins: examples of α-helix, β-sheet and random coil in a 3D ribbon model of the Phosphoglucomutase (PGM1) enzyme, EC 5.4.2.2.

of chemical bonds including covalent such as disulfide bridges that can form between adjacent cysteine residues, ionic and hydrogen bonds, as well as hydrophobic interactions. Because many of the bonds stabilising the tertiary structure are relatively weak, the majority of proteins are liable to denature at high temperatures or when exposed to denaturants such as urea and guanidinium thiocyanate, or extremes of pH.

Proteins are divided into two main classes based on their 3D structure: globular and fibrous. The majority of proteins are globular; these have irregular shapes and tend to be soluble in aqueous solvents and are most amenable to proteomic analysis. Fibrous proteins are tough molecules found only in animals and are mainly insoluble in water. Examples of fibrous proteins include collagen, elastin, keratin and actin.

4.2.1.4 Quaternary Structure. Proteins comprising more than one polypeptide chain are said to have a quaternary structure; an example is the haemoglobin heterotetramer, which consists of four globin polypeptide chains (in adults, two α-chains and two β-chains). Many proteins form complexes with other types of molecules. For example, they may combine with carbohydrates to form glycoproteins or lipids to form lipoproteins[13] or nucleic acids to form ribonucleoproteins and ribosomes.

4.2.2 Protein Function

There is a close relationship between structure and function. Globular proteins include enzymes; peptide hormones; and defence proteins such as antibodies, complement, cytokines, proteinases and their regulators; transporter proteins that can be cytoplasmic, membrane bound or circulate in the plasma; membrane bound receptor proteins that enable cells to respond to external stimuli. Fibrous proteins such as elastin and the collagens are structural proteins associated with connective tissue and

Table 4.1 The primary functions of proteins.

Function	Examples
Catalysis	Enzymes
Communication	Peptide hormones, receptors
Cellular signalling	Cytokines *e.g.* interferons, interleukins
Defence	Antibodies, complement proteins
Transport	Haemoglobin, transferrin
Structural framework	Collagens, elastin
Osmotic pressure	Albumin
Antioxidants	Albumin, superoxide dismutases

the maintenance of the shape and function of different organs.[14] An overview of some of the most important functions of proteins is shown in Table 4.1.

4.3 PROTEOMICS WORKFLOW: SAMPLE PREPARATION

Success in proteomics experiments depends on appropriate sample preparation.[15] The challenges are different to those encountered when working with nucleic acids since proteins represent a vast diversity of structures including fibrous proteins and globular proteins, protein complexes and those with posttranslational modifications. Therefore, a proteomics laboratory requires a full range of procedures to cope with a huge variety of biological materials and protein structures.

4.3.1 Cell Lysis and Protein Extraction

Biochemical analysis often requires a protein in its pure form. Regardless of whether the starting material is brain, skeletal muscle or plant cells, it is often important to recover the proteins in their native state. There are several approaches to break up tissues and lyse cells including sonication (using an ultrasonic probe), osmotic shock, mechanical disruption (homogenisation) and exposure to chaotropes and detergents capable of breaking plasma membranes. The resulting homogenate or cell/tissue extract should contain all the macromolecules including proteins and nucleic acids as well as subcellular particles such as ribosomes and organelles.[16]

4.3.1.1 Centrifugation. A crude homogenate may be centrifuged at low speed to produce a clear extract containing soluble proteins that is suitable for electrophoresis or liquid chromatography without further treatment. Centrifugation may also be used to fractionate the extract: at a low speed, large and dense components such as nuclei, fragments of membrane, *etc.* are pelleted at the bottom of the tube, as the speed is increased, smaller sub-cellular components are pelleted (see Figure 4.5).

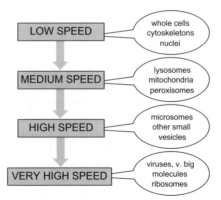

Figure 4.5 Differential centrifugation: repeated centrifugation with increasing relative centrifugal forces enables fractionation of sub-cellular components according to size and density.

Differential ultracentrifugation is a widely used procedure for separating organelles, membrane fragments, ribosomes and large molecules. Centrifugation in a sucrose gradient is useful for sub cellular fractionation of ribosomes, mitochondria and other organelles: a shallow density gradient, which increases in concentration toward the bottom of the tube serves to stabilise the cell components and prevent convective mixing. The sub-cellular particles form bands as they migrate through the gradient at different rates.[17] An alternative method is buoyant density centrifugation, usually done in a caesium chloride solution. During ultracentrifugation the caesium chloride solution forms a density gradient along which the components move until they reach the region equal to their own density. Components with the highest buoyant density accumulate towards the bottom of the tube, and those with the lowest buoyant density towards the top. Buoyant density gradient centrifugation, also known as equilibrium centrifugation, is a highly efficient fractionation method.[18] On completion of centrifugation; the fractions can be collected through a hole pierced in the base of the tube.

4.3.1.2 Removing Contaminants. Protein extracts often contain substances such as salts and detergents that are capable of interfering with downstream operations[19] such as the resolution of bands or spots on electrophoresis gels. There are several techniques to remove contaminants and 'clean up' samples such as gel filtration (desalting), dialysis and precipitation. Gel filtration and dialysis remove contaminants by size exclusion and dilution, whereas precipitation, which involves lowering the solubility of proteins by the addition of salts such as ammonium sulfate or organic compounds such as TCA, enables the precipitated proteins to be isolated from soluble contaminants.[20] In contrast to gel filtration and dialysis, precipitation allows the concentration of protein samples and is thus especially useful when a relatively dilute source of proteins such as body fluid is used; it also limits the number of proteins lost in a sample.[21]

4.3.1.3 Protein Assays. Following the 'clean up' stage it is essential to determine the total protein concentration in the sample; it is important to load the appropriate concentration of protein to ensure good 2DGE images *etc.*[22] The most extensively used are reagent based colourimetric assays, such as the Bradford and Lowry methods, in which the sample is mixed with an aqueous reagent that results in a colour change. The intensity of the colour is measured by spectrophotometry and a standard curve of known protein concentrations against their absorbance values makes it possible to determine the concentration of proteins in an unknown sample.

The Bradford assay involves a chemical reaction between Coomassie Brilliant Blue G-250 and several amino acid residues, primarily arginine and to a lesser extent lysine, tyrosine, histidine, tryptophan and phenylalanine. The reaction leads to a shift in absorbance of the dye from 465 nm to 595 nm. The Bradford assay is a straightforward method and can be used with reducing agents to stabilise proteins. However, some detergents such as sodium dodecyl sulfate (SDS) may compromise the results.[23] The Lowry assay involves two chemical reactions with the Folin phenol reagent (phosphormolybdic-phosphotungstic reagent): first copper reacts with proteins in an alkaline solution followed by reduction of the Folin reagent. The reactions produce a blue colour and the absorbance is read at 750 nm.[24] A drawback of this method is the possibility of interference by some reagents such as barbital, chloride, citrate, TRIS (tris(hydroxymethyl)aminomethane) and EDTA (ethylenediaminetetraacetic acid). The Lowry assay has been modified for use with samples containing reducing agents and detergents; this is known as RC DC (reducing agent compatible [RC], detergent compatible [DC]) protein assay. Both the Bradford and Lowry assays have advantages and disadvantages and that need to be considered.

4.3.2 Protein Separation

The protein extract is next subjected to a separation step to produce single protein suitable for mass spectrometry. Techniques for separation of single proteins from complex mixtures include column chromatography, which is often used to part-purify a sample, and gel electrophoresis.

4.3.2.1 Column Chromatography. Complex mixtures of proteins can be fractionated according to their ionic properties (ion exchange), relative sizes (size exclusion or gel filtration) or their affinity with a ligand (affinity chromatography). The proteins are separated as they pass through a column packed with the appropriate porous stationary phase. Small volumes of the column eluate are then collected using a fraction collector and tested for the presence of protein. Figure 4.6 illustrates size exclusion (gel

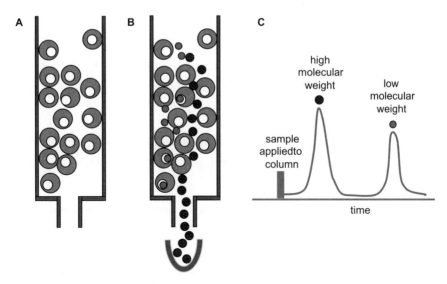

Figure 4.6 Gel filtration, or size exclusion column chromatography: (A) The column is packed with beads of cross-linked dextran (*e.g.* Sephadex™) that consists of a network of pores large enough to trap small proteins but too small for large proteins that are therefore excluded. (B) When the extract is applied, the large proteins are eluted first and smaller proteins are retarded. (C) The eluted fractions can be monitored by UV spectrophotometry.

filtration) chromatography. The matrix for ion-exchange chromatography consists of either negatively (*e.g.* carboxymethyl groups) or positively (*e.g.* diethylaminoethyl groups) charged beads. Proteins are thus separated according to their charge. Positively charged proteins bind to negatively charged beads and negatively charged proteins bind to positively charged beads. Proteins are eluted from the column by gradually changing the buffer pH or ionic strength. In gel filtration chromatography (Figure 4.6), the matrix consists of beads that contain pores and proteins are separated according to their size. Small proteins enter the pores and therefore migrate through a column at a slower rate than large proteins than cannot enter the pores. Large proteins are thus eluted first. In affinity chromatography, the matrix is covalently linked to a ligand that is specific for the protein of interest (*e.g.* an antibody). Proteins are separated according to their specific binding properties. Unbound proteins are washed out, whereas proteins that bind to the ligand are retained on the column and eluted by changing ionic strength or pH of the buffer.[25]

From the perspective of isolating single protein species or peptides, the most useful format is high-performance liquid chromatography (HPLC). HPLC offers more separation strategies than traditional column chromatography and has several additional advantages including speed (the samples are forced through the columns under great pressure), the requirement for much smaller starting volumes, higher sensitivity of detection and greater resolving power. In addition once the detector records a protein peak, it can be directly introduced into an ioniser and then a mass spectrometer (MS).[26] In contrast to gel-based methods, HPLC is suitable for automating sample separation and data collection.

4.4 ELECTROPHORETIC METHODS

Gel electrophoresis is the most widely used separation technique in proteomics laboratories. Gel based techniques can be grouped into three categories:

(1) Zone electrophoresis where separation of proteins is based on their net charge at certain buffer pH and the sieving effect of the gel.
(2) Gel isoelectric focusing (IEF) in which native, or denatured, proteins migrate through a pH gradient generated by polyampholytes until they reach their isoelectric point (where their net electrostatic charge is zero).
(3) SDS gel electrophoresis where proteins are denatured, coated with sodium dodecyl sulfate and separated according to their molecular masses.[27] In contrast to zone and SDS gel electrophoresis, during IEF the proteins concentrate into very narrow bands. Thus, IEF is suited to resolving protein isoforms that would otherwise be refractory to analysis by zone electrophoresis. The most widely used method in proteomics is two-dimensional gel electrophoresis (2DGE), which combines IEF and SDS. In the first dimension, the proteins are separated according to their isoelectric points on an IEF gel strip, which is then applied to an SDS slab gel in which the proteins are separated according to their molecular masses in the second dimension.[28]

4.4.1 Zone Electrophoresis

Zone (or 'native') electrophoresis using slabs of starch gel as the support medium[29] became a mainstay technique for biochemical genetics. Starch gel electrophoresis allowed the accurate comparison of electrophoretic mobilities of specific proteins in different samples and was invaluable for the assessment of person-to-person variation in enzymes leading to many seminal discoveries.[30] Zone electrophoresis has undergone refinements and developments including the introduction of polyacrylamide gels, which advantageously permits an increased resolution of the separated protein bands. The principle of zone electrophoresis is the separation of proteins according to their electrostatic charges at certain pH values and the sieving effects of the support medium. When placed in an electric field with a buffer pH usually between 7.5 and 8.6, most soluble proteins carry a net negative charge and migrate towards the anode. On completion of electrophoresis the gel is stained to reveal the position of the separated proteins. Non-specific protein stains such as Amido Black and Coomassie Brilliant Blue R250 enable abundant proteins to be visualised. Functional detection methods utilising enzyme activity and chromogenic or fluorogenic substrates can be used to visualise specific enzymes, whereas Western blotting (described in Section 4.5.6) can be used for immunodetection.

4.4.2 Polyacrylamide Gel Isoelectric Focusing (IEF)

IEF, which is another method for separating proteins in their native state, is conducted in a gel that contains a stable pH gradient,[31] which allows proteins to be separated according to their isoelectric points. In contrast to zone electrophoresis where the proteins tend to diffuse as soon as they are applied to the gel, IEF is a 'concentration process', which means the operator is not constrained to apply the protein extracts in narrow zones. As IEF proceeds all proteins carrying a net charge will migrate along the pH gradient, acidic proteins will tend to migrate towards the anode and basic proteins towards the cathode (Figure 4.7). As individual proteins approach the pH value of their isoelectric points, they begin to lose their net charges. On reaching their isoelectric points their net charges will be zero. Once focused into narrow bands, any diffusion away from the pI (isoelectric point) will restore a charge on the proteins, which will immediately 'refocus' to their pI thus maintaining the tight resolution of the bands for considerable periods. In the first version gel IEF the pH gradient was established using mixtures of soluble polyampholytes such as Ampholines® and Pharmalytes®. Polyampholytes are mixtures of small amphoteric molecules that are weak acids or weak bases that differ in their acid dissociation constants (pKa); they resemble oligomers of amino acids.[32] In an IEF gel the polyampholytes migrate towards the electrodes according to their charges. Commercial polyampholytes were available in a wide range of pH values ranging from pH 3.5–10 to sub pH unit ranges such as pH 4.2–4.9. IEF is typically conducted in thin polyacrylamide sheets using horizontal flatbed equipment with efficient cooling. A disadvantage of soluble polyampholytes is that the pH gradient will only retain its linear shape for a limited period of time after which cathodal drift will cause the gradient to stretch in the acidic region of the gel.[33] For this reason the use immobilised pH gradients (IPG) has gained a foothold and is preferred for IEF procedures in proteomics; IPGs eliminate gradient instability whilst providing a high buffering capacity. IPGs are formed by making a gel with acrylamide monomers that are derivatised with buffering molecules that have carboxylic acid groups or tertiary amines that differ considerably in acid dissociation constants. Thus the pH gradient, which is covalently attached to the gel matrix, is prepared by mixing two Immobilines® such as pH 4.0 and pH 6.0 in a gradient maker so that their ratios in the gel will produce of stable linear pH gradient between pH 4.0 and pH 6.0.

4.4.3 SDS-PAGE

Sodium dodecyl sulfate polyacrylamide gel electrophoresis (SDS-PAGE), which was introduced in the late 1960s, separates proteins according to their

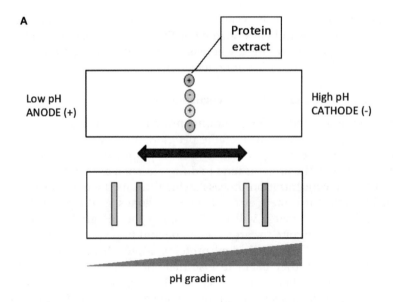

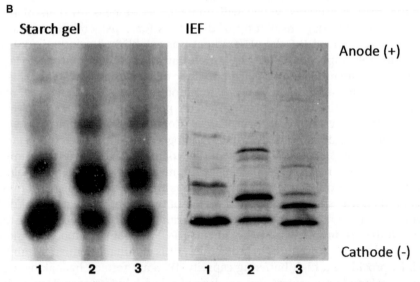

Figure 4.7   Polyacrylamide gel isoelectric focusing (IEF). (A) The proteins in a
complex mixture separate along the pH gradient according to their isoelectric points (pI). (B) PGM1 isozymes from three human red blood cell
samples separated by both starch gel electrophoresis (left panel) and IEF
(right panel). The improved resolution of IEF reveals additional genetic
variation in the isozyme patterns.

molecular mass rather than their intrinsic charge characteristics, hence it is
a useful method for estimating the molecular masses of unknown proteins.[34] SDS is an anionic detergent with chaotropic properties that binds to
proteins causing them to lose their native conformations. In the absence of

SDS, proteins migrate at a rate determined by their intrinsic electrostatic charge and shape. Prior to electrophoresis, the proteins should be heat denatured and the charge on their R-groups masked through the binding of SDS. Studies have indicated that the number of SDS molecules that bind a protein is about half of the number of amino acids and that SDS binds to all proteins in a uniform manner.[35] The high levels of bound SDS completely overwhelm the intrinsic electrostatic charges causing all of the proteins to become negatively charged; thus they migrate towards the anode. The mobility differences observed are due principally to the sieving effect of the gel: smaller proteins will migrate fastest. Because of their constant charge/mass ratio, the electrophoretic mobility of SDS treated proteins is a function of their masses thereby enabling the molecular mass of unknown proteins to be estimated. In most experiments, sets of marker proteins of known mass are electrophoresed and used produce a calibration graph of distance against log MW on which the mobility of the unknown protein can be plotted.

Successful separation of proteins depends on the acrylamide/bisacrylamide concentration in the gel; this is dictated by the molecular masses of proteins to be resolved. For instance proteins with high molecular mass should be separated in gels with lower acrylamide concentrations and *vice versa*. Proteins with a molecular mass of 60–200 kDa require 5% gels; proteins with a molecular mass of 16–70 kDa require 10% gels; proteins with a molecular mass of 12–45 kDa require 15% gels. Provided a constant ratio of acrylamide to bisacrylamide is used, as the total acrylamide concentration increases the size of the pores decreases and smaller molecules are separated more readily.[36]

In a typical SDS-PAGE experiment the gel consists of two phases: a stacking gel and a resolving gel. The stacking gel is a narrow strip of low concentration of acrylamide (often 3%), which allows the SDS treated proteins to concentrate in a tight band by the process of isotachophoresis before they enter the resolving gel.[37] The stacking gel usually has a pH of 6.8 (Tris-HCl), whereas the resolving gel has a pH of 8.8 (Tris-HCl). The tank, or running buffer, which connects the gels to the electrodes has a pH of 8.3 (Tris-Glycine). Isotachophoresis is well understood; under the influence of an electric current, glycine ions in the running buffer migrate into the stacking gel where the pH is reduced to 6.8, during which the negatively charged glycine looses its net charge. This decreases the migration rate of glycine in the electric field. In contrast, the chloride ions (from Tris-HCl) travel at a higher speed in the electric field. The difference in the mobility of ions results in the formation of two distinct fronts: the chloride front and the glycine front. The samples become trapped between the chloride and glycine fronts and they begin to concentrate into tight bands. When the protein samples enter the resolving gel (pH 8.8), the glycine becomes negatively charged again and starts migrating through the gel at a greater speed; leaving the SDS coated proteins to electrophorese at their own rates.[38]

SDS-PAGE has many applications in protein chemistry, proteomics and molecular biology. For example, the steps in a protein purification effort could be monitored, protein structure can be interrogated and even allelic variation of polymorphic proteins could be assessed as in the case of many mucin genes.[39]

4.5   2D GEL ELECTROPHORESIS

Since the early work on proteomics, electrophoresis has been the mainstay for isolating and preparing proteins for mass spectrometry. Proteins can be separated in one dimension (SDS-PAGE) according to their molecular weights; however some proteins are too alike in size and mobility and may not be separated on a gel. Thus the resolving power of SDS-PAGE is limited. A much more powerful and high-resolution technique is two-dimensional poly-acrylamide gel electrophoresis (2DGE), where proteins are first separated by their isoelectric points and then by molecular weights.[9] Therefore the technique combines high-resolution charge separation using IPGs in the first dimension and molecular mass separation using SDS-PAGE in the second dimension (Figure 4.8). The 2DGE procedure commences with the IEF step, which takes place in gel strips containing immobilised pH gradients (IPG).

4.5.1   The First Dimension: IPG Strip Selection

Dehydrated IPG strips are commercially available (*e.g.* ReadyStrip, BioRad) and have pH gradients that range from a broad, *e.g.* pH 3–10, to narrow, such as pH 3.5–4.5. Typically, a broad pH range should be chosen in order to obtain an overview of proteins in the sample. It is then possible to focus in on a region of interest by narrowing the pH range. Narrow pH range strips provide better separation of proteins and may reveal greater number of spots

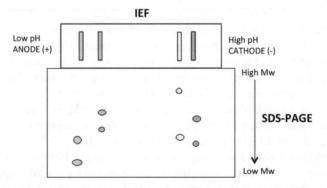

Figure 4.8   Schematic representation of two-dimensional electrophoresis (2DGE). The proteins are separated in the first dimension according to their isolectric points and in the second dimension according to their molecular weights.

in comparison to broad pH range strips. IPG strips are available in several sizes *e.g.* 24 cm, 18 cm, 13 cm, 11 cm and 7 cm. The choice depends on the size of the SDS gel to be used in the 2D separation. Generally, longer IPG strips provide greater separation and loading capacity for larger samples.

4.5.2 Sample Loading Methods

There are two ways of sample application: the protein sample can either be mixed with a rehydration buffer or it can be applied to the IPG strip (already rehydrated) through sample wells. The rehydration loading method is easier and it allows loading and separating greater amounts of proteins.

4.5.3 IPG Strip Rehydration

The selection of the most suitable rehydration buffer is generally dictated by the solubility of the proteins in the sample. Typically, a rehydration solution includes urea, detergent such as CHAPS, a reducing agent such as β-mercaptoethanol or dithiothreitol (DTT), carrier ampholyte mixtures (Pharmalytes®) and an optional dye. Urea is used to solubilise and denature proteins, whereas CHAPS prevents protein aggregation. The reducing agent is included to cleave disulfide bonds and the carrier ampholyte mixtures enhance protein separation.[40] The amount of the sample applied to a rehydration buffer depends on the method chosen to visualize the results. For instance, the recommended amount of proteins for silver staining is 5–20 μg, whereas for Coomassie Blue staining, 50–100 μg of the sample is used. The proteins will not be visible if too little protein is loaded or be over-stained if too much is loaded.

4.5.4 IPG Strip Equilibration

Following IEF, the IPG strip has to be equilibrated. This involves incubation in the SDS buffer followed by reduction and oxidation steps. A reducing agent such as DTT breaks disulfide bonds of proteins. This is followed by incubation with an oxidising agent such as iodoacetamide, which alkylates reactive sulfhydryl groups and thus prevents the generation of oxidation artefacts during the SDS-PAGE step. The equilibration solution also contains urea and glycerol, which reduce electroendosmosis and improve transfer of proteins from the first to the second dimension.[41] Pre-prepared IPG strips of the desired pH range are typically used for the first dimension of 2DGE. Several strips can be focused side by side using a flatbed IEF apparatus allowing for replicates or different samples to be focused simultaneously.

4.5.5 The Second Dimension: SDS-PAGE

Following IEF and the second equilibration step, the IPG strips are carefully removed and washed in SDS gel tank buffer (*e.g.* Tris HCl) to remove any equilibration solution. The top of an SDS-PAGE gel is then rinsed with water to remove any acrylamide and the IPG strip carefully placed along the gel so that there is uniform contact and no air bubbles. At this stage, MW markers are placed in a well next to the IPG strip. Finally, molten agarose is used to seal the top of the gel and strip. Once solidified, the gel cassette is placed in a gel tank, buffer added and the proteins are electrophoresed at a constant voltage. After the extract has been separated by 2DGE the gel has to be stained to reveal the positions of the protein spots. Several different detection methods are available. The method chosen should be sensitive enough to detect the proteins of interest in the sample being analysed. It should also be compatible with mass spectrometry (MS), and wherever possible be non-toxic and environmentally friendly.[42] The most commonly used stains are Coomassie Blue, fluorescent and silver stains. All have advantages and disadvantages. Coomassie Blue staining is easy to perform, cheap and is compatible with MS, but it has a low sensitivity of detection; fluorescence based stains have the highest sensitivity of detection and are most are suitable for use with MS, but they are costly and need to be visualised using a fluorescence imager; silver staining is more difficult to perform and shows higher sensitivity of detection than Coomassie Blue. However, MS analysis of silver stained proteins is more problematic than the other methods, although several strategies are available to prepare silver stained proteins for MALDI-TOF MS.

4.5.6   Specific Protein Detection: Western Blotting

Where specific antibodies are available, protein detection can be facilitated by western blotting without proceeding to mass spectrometry. Proteins are transferred electrophoretically from the gel to a nitrocellulose or polyvinylidene fluoride (PVDF) membrane. Before immunodetection using a specific antibody, the membrane is incubated with a 'blocking' solution (containing agents such as BSA or Tween 20) to prevent non-specific binding of antibody. The membrane is then incubated with a primary antibody that is specific for the protein under investigation. The primary antibody may be labelled with an enzyme such as horseradish peroxidase (HRP) or a fluorescent dye such as rhodamine or Cy3 and used for 'direct detection'.[43] However, indirect detection is generally more successful and widely used. In this case, the primary antibody is not labelled. Once the antibody, for example a mouse-anti-human protein IgG, has reacted with the antigen, unbound antibody is washed away and the membrane incubated with a second antibody, for example a rabbit-anti-mouse IgG, conjugated to an enzyme or fluorophore, which binds to the primary antibody. The protein spots are visualised following incubation with the enzyme substrate leading to the production of an insoluble coloured product, or a chemiluminescence signal that can be recorded with a CCD camera. Horseradish peroxidase (HRP) and alkaline phosphatase (ALP) are both used as enzyme conjugate labels for western blot detection. Alternatively, a fluorescence imager can be used to record fluorescent labels. In each case, clearly visible bands (if a 1D gel is used) or spots are revealed that corresponds to the position of the specific protein.[44]

4.6 PROTEIN IDENTIFICATION: MASS SPECTROMETRY

Mass spectrometry is the best way to identify proteins. Following trypsin digestion of a protein recovered from a 2D gel, the masses of the tryptic peptides are determined and compared with those generated *in silico* from protein sequence databases. An accurate identification of a protein is possible even if the masses of only a few tryptic peptides are known.

4.6.1 Matrix Assisted Laser Desorption/Ionisation-Time of Flight (MALDI-TOF)

A variety of established mass spectrometry methods are used to identify proteins, but perhaps the most widely used is matrix-assisted laser desorption ionisation-time-of-flight (MALDI-TOF) MS. A plug containing the protein spot to be analysed is cut out of the 2D gel and digested with trypsin to yield peptides for mass spectrometric analysis. Trypsin cuts the protein sequence after lysine (K) and arginine (R), except when the residue is followed by a proline (P). Peptide fragments comprising of 6–20 residues are optimal for MALDI-TOF. The peptides are then mixed with an organic compound such as alpha-cyano-4-hydroxcinnamic acid, *i.e.* the matrix substance, and spots of matrix plus peptides are then dried on the surface the metal MALDI plate.[45] The plate is inserted into the mass spectrometer where a pulsed laser beam excites the crystallised matrix molecules that absorb most of the laser light energy, thereby protecting peptide mixtures from fragmentation. This leads to matrix desorption and ionisation of the peptides by proton transfer from the matrix.[46] Hence sample ions that are produced in the ion source have a single positive charge (although ions carrying charges $+2$ and $+3$ can also be generated) and travel into the gas phase (Figure 4.9).

The ionised peptides are accelerated in an electric field and travel through a flight tube to the detector (Figure 4.10). The time taken to reach the detector depends on the mass and charge (m/z) of the peptide ion, since z is almost always 1, the m/z ratio can be taken as the mass of the peptide. Ions with larger m/z ratios fly more slowly than smaller ions and thus arrive at the detector after smaller ions. The MS gathers the mass data and generates a spectrum of ion intensity plotted against m/z ratio. The identification of the proteins through the determination of peptide masses and database searching is referred to as peptide-mass fingerprinting. Advantageously, in MALDI-TOF MS the matrix absorbs most of the laser energy thus ionisation conditions are relatively gentle, which means that peptides are not fragmented and high molecular weight species, even whole proteins, can be analysed. However, because of the requirement for mixing the analyte with

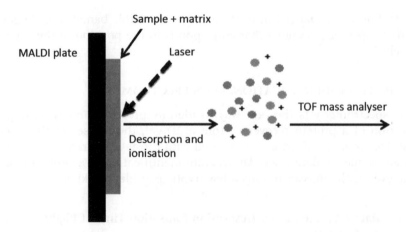

Figure 4.9 Matrix-assisted laser desorption and ionisation (MALDI).

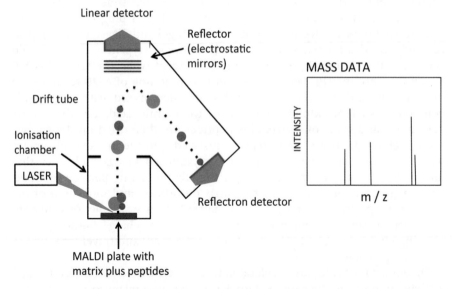

Figure 4.10 Schematic representation of MALDI time of flight (TOF) mass spectrometry. Adapted from www.proteomicsnijmegen.nl.

the matrix, MALDI-TOF is not readily suited to automation and the throughput is largely determined by the size of the MALDI plate.

4.6.2 Liquid Chromatography Coupled Tandem Mass Spectrometry (LC-MS/MS)

In contrast to MALDI-TOF MS, liquid chromatography tandem mass spectrometry (LC-MS/MS) allows for 'in-line' processing of samples and is therefore amenable to automation and high-throughput operations. LC-MS/MS can be used to identify various molecules from complex biological samples as well as to determine the amino acid sequences of peptides that are refractory to peptide mass fingerprinting.[47] *De novo* sequencing is indispensible if the genome of the organism being studied has not yet been sequenced. In addition, LC-MS/MS is used to identify and characterise posttranslational modifications by analysing the sequence of the peptide and looking for additional chemical groups such as carbohydrates (glycosylation) or phosphates (phosphorylation).

The instrumentation for LC-MS/MS consists of three elements: an HPLC, an electrospray ionisation source and a mass analyser. In a typical experiment, the protein is subjected to enzymatic cleavage to generate a mixture of peptides. The peptide fragments are then separated using HPLC and transferred to a fine capillary tube that has a high voltage, *i.e.* the electrospray ionisation source. This results in the formation of highly charged droplets consisting of both sample and a solvent. As the droplets exit the capillary, the solvent starts to evaporate so the droplets gradually decrease in size. Finally, highly charged analyte molecules are ejected from the droplets and travel into the mass analyser (Figure 4.11). The LC-MS/MS mass analyser consists of four parallel metal rods (a quadrupole) to which both constant and oscillating electric fields are applied. Peptide ions, which were produced in the ion source, migrate along the middle of the quadrupole and are separated according to their *m/z* ratio. Only ions of particular *m/z* values reach the detector. The separation of peptides with various *m/z* ratios is possible by altering the electric field in the quadrupoles[48] The system typically comprises of three quadrupoles (Figure 4.12). In the first (Q1), a single specific ion (the precursor ion) is selected according to its *m/z* ratio and travels to the second quadrupole (Q2). In Q2 the precursor ion collides with a collision gas (*e.g.* argon) in the process called collision-induced dissociation (CID) and is further fragmented. This fragmentation procedure breaks the

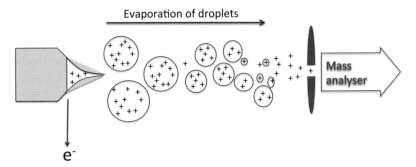

Figure 4.11   Electrospray ionisation. The large initial droplets in the aerosol evaporate and reduce in size until all the solvent has evaporated leaving the charged analytes to form a charged gas phase before they enter the mass analyser.

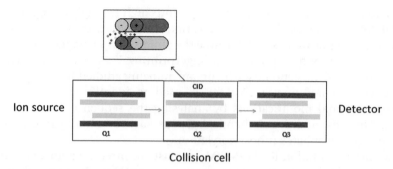

Figure 4.12 Tandem quadrupole system. The Q1 and Q2 are mass spectrometers and Q2 is a collision cell.
Adapted from Ho *et al.*[69]

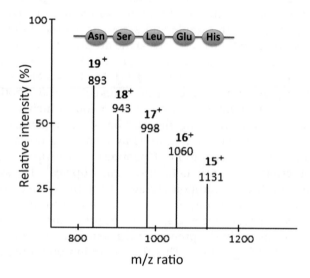

Figure 4.13 Example of an ESI mass spectrum.
Adapted from Ho *et al.*[69]

peptide bonds, producing various fragments that differ by a single amino acid, low energy conditions during CID, means that the peptide ions fragment in predictable patterns.[47] The mass analysis of fragmented peptides takes place in the third quadrupole (Q3) to produce the mass spectrum, which is use to deduce the the *de novo* amino acid sequence of the peptide or characterise posttranslational modifications. The mass spectrum is a graphical representation of the relationship between relative abundance of ions and their *m/z* ratios. All the peaks that are presented in the mass spectrum shown in Figure 4.13 refer to an individual peptide carrying multiple charges. The number of charges on the protein depends on the peptide's molecular weight. By knowing the charge of each peak and its *m/z* ratio it is possible to determine the mass of the molecule. For example, a peptide with *m/z* ratio of 1131 carries 15 charges. Therefore, $m/15 = 1131$ so the mass equals 16965 Daltons (Da). 15 must be subtracted to obtain the peptide's mass, which is 16950 Da in this example. The mass of the remaining peaks can be determined in the same way. An average value for all peaks is calculated to measure the masses of peptides accurately, the more accurate the mass of the peptides the less number of hits in the database. The *de novo* amino acid sequence is determined by analysing the differences in mass of several small peptides. Posttranslational modifications can also be detected by identifying an increase in the mass of a particular amino acid.

4.7 PROTEIN IDENTIFICATION: BIOINFORMATICS

Mass spectrometry experiments generate mass spectra of tryptic peptides (*e.g.* from MALDI TOF MS), mass data that can be used for *de novo* peptide sequencing or that represents posttranslational modifications (MSMS). Following the determination of the peptide mass fingerprint or *de novo* peptide sequence, bioinformatics is used to identify the protein of interest. Proteins can be identified by matching a list of experimental tryptic peptide masses (the mass fingerprint) to the list of all peptide masses generated by simulated tryptic cleavage in the database.[49] The aim of database search is to find a profile match. In the real world, tryptic digestion is not always perfect and may result in partial digestion of proteins due to the failure of trypsin to cut after Lys and Arg residues, especially where there is an adjacent Pro. In addition, peptides produced by MALDI-MS are highly charged, increasing their relative masses. For these reasons it is important that the database allows for some degree of freedom in matching the fragmentation patterns. The probability of successful identification can be improved by inputting as much information as possible to the search engine. For instance, in addition to the mass of the peptide and its sequence, the molecular weight, isoelectric point and the species from which the protein was obtained can aid identification. This approach is the best suited to species whose genomes have been sequenced and mapped.[50]

Mascot (2013, identification and characterisation with MS/MS data, available from: www.matrixscience.com) is a widely used suite of programs dedicated to the analysis of MS data and can be used for identification, quantification and further characterisation of proteins using mass data generated by different MS platforms.

Expasy (2010, Expert Protein Analysis System, available from www.expasy.org) is a comprehensive suite of tools for the analysis of protein sequences and structures as well as MS and 2-DGE data analysis. Expasy offers a range of nine programs for peptide analysis. For example, AACompIdent program is capable of identifying the protein of interest based on its amino acid composition by comparing the experimental data with known amino acid compositions held on SWISS-PROR/TrEMBL databases. Information such as the name of the species or keywords can narrow the database search results. TagIdent identifies the protein by comparing an unknown sequence tag (up to six residues) with proteins held on UniProt Knowledgebase database (SWISS-PROT and TrEMBL). It can generate a list of candidate proteins that match the isoelectric point and molecular weight of an unknown protein. MultiIdent uses peptide mass fingerprinting data together with isoelectric point, molecular weight, amino acid composition, sequence tag and species name to identify the protein.

ProFound (2012, protein identification and characterisation database, available from: prowl.rockefeller.edu) is also a valuable tool, which searches protein sequence databases utilising peptide mass fingerprinting data. It takes advantage of a Bayesian algorithm that ranks the protein sequences found in the database according to their probability of generating the peptide mass fingerprints.

Where it is not possible to match peptides in the protein sequence databases, the *de novo* amino acid sequences obtained by MS/MS can be analysed using tblastn to search for homologues in DNA sequence databases such as GenBank. In order to further the analysis of structure and function of a given protein, laboratory studies are required. Recombinant DNA technology is often used to sufficient quantities of proteins for such studies. Basic principles of DNA recombinant technology are discussed in Chapters 1 and 2.

4.8   EXPRESSION PROTEOMICS

2DGE has become a widely used technique for the analysis of protein expression patterns, primarily due to its relative simplicity. There are, however, drawbacks to this method such as poor reproducibility between replicates and low sensitivity of detection using Coomassie Blue. Furthermore, 2DGE is capable of visualising only the proteins with molecular weights between 10 and 100 kDa and it generally fails to separate membrane proteins because of the formation of protein aggregates during IEF.[51] Gel based methods circumventing some of these problems have been developed, such as Difference Gel Electrophoresis (DIGE).

4.8.1   Difference Gel Electrophoresis (DIGE)

The DIGE method involves labelling of protein samples with fluorescent dyes before 2DGE. Frequently used fluorophores include Cy2, Cy3 or Cy5; these cyanine dyes carry an N-hydroxysuccinimidyl ester reactive group that forms covalent bonds with the ε-amino group of lysine residues.[1] A maximum of three protein samples can be analysed in a DIGE experiment; treated and control samples are labelled with fluorescent dyes that have different excitation wavelengths. They are then adjusted to equalise total protein concentration, mixed together and separated by 2DGE and the individual protein spots are detected using a fluorescence scanner. If there are no differences in individual protein concentrations between the samples, excitation of the 'multiplexed' spots will yield images of intermediate colour. Any difference in expression between the samples will result in the appearance or disappearance, or gradation, of one of the fluorescent dyes in the merged image. Thus DIGE is a straightforward method to compare specific protein levels from two or three samples in a single gel[52] that ensures improved sensitivity of detection and reproducibility in comparison with the standard 2DGE approaches.[53] The analysis of protein expression patterns is carried out using computer programs dedicated to 2D-DIGE analysis such as De-Cyder.[54] There are some limitations, however: for example 2D-DIGE is not suitable for extremely acidic or basic proteins, or hydrophobic proteins or those smaller than about 15 kDa. In addition, the fluorescent dyes are expensive compared with silver or Coomassie blue staining and there is a requirement for lysine residues.[52]

4.8.2 Isotope Coded Affinity Tags (ICAT)

A useful non-gel based method for protein expression studies is ICAT (isotope coded affinity tags). The ICAT method is capable of a greater degree of sensitivity and reproducibility than DIGE.[1] It quantifies differences in protein levels in two samples by MS.[55] ICAT is a chemical isotope labelling method that uses the ICAT reagent, which consists of three components: a reactive group specific for thiol groups, *e.g.* on cysteine residues; an isotope-coded light or heavy linker; and a biotin moiety for affinity chromatography.[56] As with DIGE, two samples are compared simultaneously. For ICAT, the samples are labelled with either light or heavy ICAT reagents. The two samples are then mixed together and digested with trypsin; next, a streptavidin affinity column is used to isolate tryptic peptides with the coded tags from unlabelled peptides. ICAT labelled peptides are analysed using LC-ESI-tandem mass spectrometry[57] and the resulting mass data are analysed by searching the database such as proICAT.[58] ICAT is a powerful method because the peptides with low expression levels can still be detected on 2D gels.[59] The limitations of this approach include the inability to detect acidic proteins, proteins that lack cysteine residues and posttranslational modifications that are not on cysteine containing tryptic peptides.[60]

4.8.3 Protein Microarrays

An important development of protein science in the era of functional genomics has been the introduction of protein microarrays that have the potential for high-throughput studies of protein interactions. Protein microarrays allow screening of thousands of proteins in a single experiment and are analogous to DNA microarrays (see Chapter 2). The types of analysis amenable to microarrays include protein–protein, protein–DNA/RNA and protein–small molecule interactions as well as enzymic reactions such as phosphorylation.[61] A solid support medium with arrays of specific proteins or ligands capable of binding to their complementary molecules is incubated with an extract containing the molecules of interest. A characteristic signal pattern is produced when the query molecules that are fluorescently labelled react with proteins on the array or an enzyme substrate produce a fluorescent or coloured product. The intensity of the signal is directly proportional to the amount or activity of the molecule captured on the microarray.[62] Fluorescent spots are recorded and the resulting patterns interpreted.

There are two main types of protein microarrays: the forward phase arrays (FFA) and reverse phase arrays (RPA). In FFA, capture molecules, usually antibodies, are fixed on a solid surface. Each microscopic spot comprises one type of antibody. After the appropriate incubation time only the specific analyte from the mixture of other molecules binds the antibody. The protein interaction is detected following the addition of a second labelled antibody that binds to the specific analyte. In RPA, unknown protein mixtures are immobilised on the solid phase and a fluorescently labelled specific antibody is applied. Interactions are detected only for complementary molecules.[62] The FFA is conceptually more similar to DNA microarrays, whereas the RPA is more similar to western blots.

In contrast to DNA microarrays that have reached a high level of sophistication, protein microarray technology continues to pose many challenges. Unlike oligonucleotide probes, the behaviour of proteins arrayed on a solid phase is not nearly so predictable. For instance, proteins tend to cross-react with non-specific molecules, some classes of proteins may become unstable and change their physicochemical properties and the experimental conditions may affect protein conformation and hence activity or binding affinity. There are also some biological challenges. For instance many genes produce several protein variants from differentially spliced mRNAs, such variants may differ in their antigenicity and biological activity; posttranslational modifications can add an additional layer of complexity. Discriminating between various isoforms of a single protein is a major challenge for protein microarray technology. In theory, it is possible to produce specific protein binders for each query molecule but this would be very expensive. Furthermore, in contrast to DNA microarray technology, it is not possible to amplify low abundance proteins.[63]

Despite obvious problems associated with the analysis of proteins, microarrays have several advantages. They offer the potential advantage of being scalable, flexible, automatable and easy to perform.[63] Furthermore, microarray technology allows the operator to control environmental parameters such as pH, temperature and concentration of the labelled probe.[64]

4.9 PRACTICAL APPLICATIONS

Proteomic studies can provide fundamental information on cellular and molecular physiology in health and disease by assessing changes in protein expression in normal and abnormal samples. Such studies could pave the way to the identification of novel drug targets which in turn could accelerate new drug development.[65] The discovery of drug targets is dependent on the analysis of the structure and function of a given protein. By knowing the 3D structure of the protein, small molecules can be designed, for example, to inhibit the active site of an enzyme. Signal transduction pathways and cellular processes facilitated by enzymes could thus be blocked. This is an important approach in the drug development process and it is expected that power of proteomics could contribute to the development of both conventional and personalised drugs. For example, proteomic studies have enabled the identification of HIV-1 protease, which is responsible for cleavage of the HIV protein. The survival of the virus depends on this enzyme and thus an inactivation of the HIV-1 protease would to destroy HIV. Typically, an *in silico* approach is used to identify active sites from structural data. Virtual ligand screening is capable of determining the quality of the fit of the drug molecule to various sites in the protein in order to enhance or disable its functionality.[66]

The identification of alterations in the structure and function of the protein as well as changes in protein expression patterns and protein–protein interactions are important for diagnosis and prognosis. The high-throughput identification and and quantification of proteins in healthy and diseased tissues and fluids promises to improve *in vitro* diagnosis on a broad front. Proteomics holds great promise for the rapid discovery of protein biomarkers and drug targets. For example, a proteomics based approach was used to identify biomarkers for Alzheimer's disease.[67] Alzheimer's disease is associated with increased levels of β-secretase, which stimulates the formation of β-amyloid protein. This protein accumulates to form plaques in the brain, which results in dementia. Therefore, effective targeting of β-secretase could usefully inhibit the build up of β-amyloid protein between nerve cells.

4.10 CONCLUSIONS

Proteomics is the high throughput and automated study of protein expression, structure and function; it offers unparalleled opportunities to delve into protein transactions and the cellular mechanisms they underpin. Compared to genomics and transcriptomics, proteomics addresses biology directly and can provide insights to dynamic aspects of pathology and cell physiology at the molecular level. A proteomics experiment frequently sets out to identify (ID) all the proteins expressed in cell type, tissue or fluid under defined conditions. The first step in the workflow, sample preparation, is crucial to the success of the experiment. Thereafter separation of the sample by electrophoresis or liquid chromatography enables the isolation of specific proteins that can be identified by immunodetection or more frequently, mass spectrometry. Thus a typical proteomic workflow might include separation by 2DGE, gel staining and image analysis followed by spot excision and tryptic digestion. The masses of the tryptic peptides are measured by MALDI-TOF mass spectrometry and the resulting peptide mass spectra searched in databases to find sufficient matches to ID the protein. If protein identification is unsuccessful using MALDI, peptide ions can be selected and fragmented using MS/MS to produce *de novo* amino acid sequences. The aim of proteome profiling and ID (which is equivalent to mapping the genome) is to catalogue all the proteins in a particular sample and optionally analyse posttranslational modifications. In contrast, expression proteomics sets out to identify proteins that are expressed at different levels in test samples and controls. The test samples may be from individuals affected by a specific pathology or from experiments to monitor the effect of pharmaceuticals, *etc*. Whilst standard 2DGE can be used for expression proteomics, there are drawbacks such as poor reproducibility and the need for multiple replicates of the gels. Two very effective methods have been devised to overcome these drawbacks: gel based DIGE and liquid chromatography based ICAT;[1,54,57] both are widely used for expression studies.[70] A third category of proteomics, sometimes called functional proteomics, is being developed to assess protein–protein and protein–ligand interactions. Interaction proteomics is currently based on the relatively new protein microarray technology. It is anticipated that protein microarrays, particularly in the reverse-phase format, will provide valuable insights into many aspects of pathology and cell physiology through the understanding of the networks of protein interactions that underline cellular responses.

The application of proteomics technologies is yet to be fully realised though there are indications that there will be significant benefits. Through expression proteomic studies it is hoped that new biomarkers, which are defined as 'measurable characteristics that reflect physiological, pharmacological, or disease processes' will be discovered and employed to improve clinical outcomes for many types of pathology.[71] These could include diagnostic and prognostic markers as well as those that could predict responses to drug therapy, *etc*. In the field of drug discovery, since most drug targets are proteins it is anticipated that new targets will be identified through proteomics studies. Although at a relatively early stage in the cycle of value creation, it is anticipated that proteomics technologies have the potential for industrial scale uses in medical, pharmaceutical and agricultural biotechnology.

BIOINFORMATICS SOURCES CITED IN THE TEXT

Nijmegen Proteomics Facility. (2013). *Maldi-TOF MS*. Available from: www.proteomicsnijmegen.nl

REFERENCES

1. K. Chandramouli and P. Y. Qian, *Hum. Genomics Proteomics*, 2009, **1**, 1.
2. V. C. Wasinger, S. J. Cordwell, A. Cerpa-Polijak, J. X. Yan, A. A. Gooley, M. R. Wilkins, M. W. Duncan, R. Harris, K. L. Williams and I. Humphery-Smith, *Electrophoresis*, 1995, **16**, 1090.
3. *Proteome Research: New Frontiers In Functional Genomics*, ed. M. R. Wilkins, K. L. Williams, R. D. Appel and D. F. Hochstrasser, Springer-Verlag, Berlin, 1997.

4. M. A. Alaoui-Jamali and Y. J. Xu, *J. Zhejiang University SCIENCE B*, 2006, 7, 411.
5. S. J. Fey and P. M. Larsen, *Curr. Opin. Chem. Biol.*, 2001, 5, 26.
6. D. W. Nebert, G. Zhang and E. S. Vesell, *Annu. Rev. Pharmacol. Toxicol.*, 2013, 53, 355.
7. J. M. Johnson, J. Castle, P. Garrett-Engele, Z. Kan, P. M. Loerch, C. D. Armour, R. Santos, E. E. Schadt, R. Stoughton and D. D. Shoemaker, *Science*, 2003, 302, 2141.
8. J. Godovac-Zimmermann and L. R. Brown, *Mass Spectrom. Rev.*, 2001, 20, 1.
9. R. L. Gundry, M. Y. White, C. I. Murray, L. A. Kane, Q. Fu, B. A. Stanley and J. E. Van Eyk, *Curr. Protocol. Mol. Biol.*, 2009, Unit 10.25, 1.
10. S. D. Patterson and R. H. Aebersold, *Nat. Genet.*, 2003, 33, 311.
11. M. Boyle and K. Senior, *Human Biology*, Collins, London, 2008.
12. F. H. Martini and J. L. Nath, *Fundamentals of Anatomy and Physiology*, Pearson Education, Inc., San Francisco, 2009.
13. W. K. Purves, D. Sadava, G. H. Orians and H. C. Heller, *Life: The Science Of Biology*, W.H. Freeman and Company, Gordonsville, 2004.
14. E. N. Marieb, *Human Anatomy and Physiology*, Benjamin Cummings, New York, 2001.
15. *Proteomics of Human Body Fluids: Principles, Methods, and Applications*, ed. V. Thongboonkerd, Humana Press, New York, 2007.
16. B. Alberts, A. Johnson, J. Lewis, M. Raff, K. Roberts and P. Walter, *Molecular Biology of the Cell*, Garland Science, New York, 2002.
17. J. L. Cole, J. W. Lary, T. P. Moody and T. M. Laue, *Methods Cell Biol.*, 2008, 84, 143.
18. B. Alberts, D. Bray, K. Hopkin, A. Johnson, J. Lewis, M. Raff, K. Roberts and P. Walter, *Essential Cell Biology*, Garland Science, New York, 2010.
19. L. Jiang, L. He and M. Fountoulakis, *J. Chromatogr. A*, 2004, 1023, 317.
20. A. Bodzon-Kulakowska, A. Bierczynska-Krzysik, T. Dylag, A. Drabik, P. Suder, M. Noga, J. Jarzebinska and J. Silberring, *J. Chromatogr. B*, 2006, 849, 1.
21. D. Martins de Souza, B. M. Oliveira, E. Castro-Dias, F. V. Winck, R. S. Horiuchi, P. A. Baldasso, H. T. Caetano, N. K. D. Pires, S. Marangoni and J. C. Novello, *Briefings Funct. Genomics Proteomics*, 2008, 7, 312.
22. T. Berkelman, *Mol. Biol.*, 2008, 424, 43.
23. M. M. Bradford, *Anal. Biochem.*, 1976, 72, 248.
24. O. H. Lowry, N. J. Rosebrough, A. L. Farr and R. J. Randell, *J. Biol. Chem.*, 1951, 193, 265.
25. W. W. Ward and G. Swiatek, *Curr. Anal. Chem.*, 2009, 5, 1.
26. J. J. Pitt, *Clin. Biochem. Rev.*, 2009, 30, 19.
27. F. Chevalier, *Proteome Sci.*, 2010, 8, 1.
28. A. Gorg, W. Weiss and M. J. Dunn, *Proteomics*, 2004, 4, 3665.
29. O. Smithies, *Biochem. J.*, 1955, 61, 629.
30. H. Harris and D. A. Hopkinson, *Handbook of Enzyme Electrophoresis in Human Genetics (with Supplements)*, Oxford American Publishing Co., New York, 1976.

31. H. J. Isaaq and T. D. Veenstra, *BioTechniques*, 2008, **44**, 697.
32. R. J. Braun, N. Kinkl, M. Beer and M. Ueffing, *Anal. Bioanal. Chem.*, 2007, **389**, 1033.
33. T. Rabilloud and C. Lelong, *J. Proteomics*, 2011, **74**, 1829.
34. A. L. Shapiro, E. Viñuela and J. V. Maizel, *Biochem. Biophys. Res. Commun.*, 1967, **28**, 815.
35. P. G. Righetti, A. V. Stoyanov and M. Y. Zhukov, *The Proteome Revisited: Theory and Practice of All Relevant Electrophoretic Steps*, Elsevier Science, Amsterdam, 2001.
36. S. R. Gallagher, *Curr. Protocol. Mol. Biol.*, 2006, 10.2.1, 1.
37. D. M. Hawcroft, *Electrophoresis*, Oxford University Press, New York, 1997.
38. Science Squared, How SDS-PAGE Works, available from: http://bitesize-bio.com/580/how-sds-page-works, 2008.
39. D. M. Swallow, S. Gendler, B. Griffiths, G. Corney, J. Taylor-Papadimitriou and M. E. Bramwell, *Nature*, 1987, **328**, 82.
40. T. Berkelman and T. Stenstedt, *2-D Electrophoresis: Principles and Methods, Amersham Biosciences AB*, Uppsala, Sweden, 2001.
41. A. Gorg, O. Drews, C. Luck, F. Weiland and W. Weiss, *Electrophoresis*, 2009, **30**, 1.
42. R. Westermeier and R. Marouga, *Biosci. Rep.*, 2005, **25**, 19.
43. T. Mahmood and P. C. Yang, *North Am. J. Med. Sci.*, 2012, **4**, 429.
44. C. Moore, *Introduction to Western Blotting*, MorphoSys, Oxford, 2009.
45. J. Gobom, M. Schuerenberg, M. Mueller, D. Theiss, H. Lehrach and E. Nordhoff, *Anal. Chem.*, 2001, **73**, 434.
46. J. K. Lewis, J. Wei and G. Siuzdak, in *Encyclopedia Of Analytical Chemistry*, ed. R. A. Meyers, John Wiley & Sons Ltd., Chichester, 2001, pp. 5880–5894.
47. C. Delahunty and J. R. Yates, *Methods*, 2005, **35**, 248.
48. W. Li, J. Zhang and F. L. S. Tse, *Handbook of LC-MS bioanalysis: Best Practices, Experimental Protocols and Regulations*, John Wiley & Sons, Inc., New Jersey, 2013.
49. R. Aebersold and M. Mann, *Nature*, 2003, **422**, 198.
50. J. Xiong, *Essential Bioinformatics*, Cambridge University Press, Cambridge, 2006.
51. F. Delom and E. Chevet, *Proteome Sci.*, 2006, **4**, 1.
52. G. Van den Bergh and L. Arckens, *Expert Rev. Proteomics*, 2005, **2**, 243.
53. G. Zhou, H. Li, D. DeCamp, S. Chen, H. Shu, Y. Gong, M. Flaig, J. W. Gillespie, N. Hu, P. R. Taylor, M. R. Emmert-Buck, L. A. Liotta, E. F. Petricoin III and Y. Zhao, *Mol. Cell. Proteomics*, 2002, **1**, 117.
54. R. Marouga, S. David and E. Hawkins, *Anal. Bioanal. Chem.*, 2005, **382**, 669.
55. A. S. Haqqani, J. F. Kelly and D. B. Stanimirovic, *Methods Mol. Biol.*, 2008, **439**, 225.
56. L. Monteoliva and J. P. Albar, *Briefings Funct. Genomics Proteomics*, 2004, **3**, 220.

57. M. Sethuraman, M. E. McComb, H. Huang, S. Huang, T. Heibeck, C. E. Costello and R. A. Cohen, *J. Proteom. Res.*, 2004, **3**, 1228.
58. X. J. Li, H. Zhang, J. A. Ranish and R. Aebersold, *Anal. Chem.*, 2003, **75**, 6648.
59. F. Schmidt, S. Donahoe, K. Hagens, J. Mattow, U. E. Schaible, S. H. Kaufmann, R. Aebersold and P. R. Jungblut, *Mol. Cell. Proteomics*, 2004, **3**, 24.
60. H. Zhou, J. A. Ranish, J. D. Watts and R. Aebersold, *Nat. Biotechnol.*, 2002, **20**, 512.
61. A. J. Nijdam, M. R. Zianni, E. E. Herderick, M. M.-C. Cheng, J. R. Prosperi, F. A. Robertson, E. F. Petricoin III, L. A. Liotta and M. Ferrari, *J. Proteome. Res.*, 2009, **8**, 1247.
62. L. A. Liotta, V. Espina, A. I. Mehta, V. Calvert, K. Rosenblatt, D. Geho, P. J. Munson, L. Young, J. Wulfkuhle and E. F. Petricoin III, *Cancer Cell*, 2003, **3**, 317.
63. P. Cutler, *Proteomics*, 2003, **3**, 3.
64. G. MacBeath, *Nat. Biotechnol.*, 2001, **19**, 828.
65. M. Abhilash, *Internet J. Genomics Proteomics*, 2008, **4**, 1.
66. S. Tyagi, Raghvendra, U. Singh, T. Kalra, K. Munjal and Vikas, *Int. J. Pharm. Sci. Rev. Res.*, 2010, **3**, 87.
67. A. Hye, S. Lynham, M. Thambisetty, M. Causevic, J. Campbell, H. L. Byers, C. Hooper, F. Rijsdijk, S. J. Tabrizi, S. Banner, C. E. Shaw, C. Foy, M. Poppe, N. Archer, G. Hamilton, J. Powell, R. G. Brown, P. Sham, M. Ward and S. Lovestone, *Brain*, 2006, **129**, 3042.
68. P. Bradley and J. Calvert, *Catch Up Biology For The Medical Sciences*, Scion Publishing Limited, Bloxham, 2006.
69. C. S. Ho, C. W. K. Lam, M. H. M. Chan, R. C. K. Cheung, L. K. Law and L. C. W. Lit, *Clin. Biochem. Rev.*, 2003, **24**, 3.
70. C. Fu, J. Hu, T. Liu, T. Ago, J. Sadoshima and H. Li, *J. Proteome Res.*, 2008, **7**, 3789.
71. P. C. Guest, M. G. Gottschalk and S. Bahn, *Genome Med.*, 2013, **5**, 17.

CHAPTER 5
Transgenesis

ELIZABETH J. CARTWRIGHT,* WEI LIU AND XIN WANG

University of Manchester, Room 5.001 AV Hill Building, Oxford Road, Manchester, M13 9PT, UK
*Email: elizabeth.j.cartwright@manchester.ac.uk

5.1 INTRODUCTION

5.1.1 From Gene to Function

The goal of the Human Genome Project was met in 2004 when the sequence of every gene in the human genome was released and each was mapped to its precise chromosomal location.[1,2] In the following decade we have built on this information and have made major steps in understanding our genetic make-up. Through the '1000 Genomes Project' we are increasing our understanding of human genetic variation;[3] the ENCODE (Encyclopaedia of DNA elements) project is set to reveal the functional and regulatory elements of the human genome, that act at the protein and RNA levels, to control cells and gene activity (http://genome.ucsc.edu/ENCODE/); and numerous genome wide association studies (GWAS) are identifying potential genetic risk factors for human disease.

It is now vital that we fully address the next big challenge—to identify the function of the 21 000 protein-encoding genes in human health and disease. A number of technologies and model organisms are currently used to analyse and understand mammalian gene function including the techniques that will be the focus of this chapter—transgenesis technologies.

This chapter will outline the transgenic techniques that are currently used to modify the genome in order to extend our understanding of the *in vivo*

Molecular Biology and Biotechnology, 6th Edition
Edited by Ralph Rapley and David Whitehouse
© The Royal Society of Chemistry 2015
Published by the Royal Society of Chemistry, www.rsc.org

function of these genes in normal mammalian development and physiology, as well as in pathogenesis.

Transgenesis is a general term which covers numerous ways of modifying the genome of intact organisms, for example, it is possible to eliminate a gene of interest or introduce extra pieces of DNA (overexpress) in a whole organism or a specific tissue. Transgenesis can be carried out in numerous organisms including plants, the fruit fly (*Drosophila*), worm (*Caenorabditis elegans*), frog (*Xenopus*), zebrafish, large mammals including sheep and small mammals such as the rat and mouse. Each of these model organisms is valuable in its own right; the fruit fly and the worm have been extensively used to provide information regarding the basic functional processes of organisms such as cell proliferation, metabolic pathways and developmental processes. The zebrafish is particularly suited to use in studies to elucidate vertebrate gene function, and large mammals have even been used as 'bio incubators' to produce recombinant proteins. Although historically the rat has been used in many physiological experiments the ability to modify its genome is much more limited than in the mouse. Subsequently, it is the mouse that has become by far the most popular organism in which to study mammalian gene function and in particular human disease.

This chapter will outline the most commonly used transgenic technologies applied to the mouse genome including overexpression transgenesis, gene targeting and conditional gene modifications, and it will introduce technologies that have the potential to significantly speed up the process of *in vivo* gene modification. In addition, it will outline how we determine *in vivo* function in the mice that are generated by these transgenic techniques, and finally, we address some of the ethical issues which must be considered when using animals as a model system for research.

5.2 TRANSGENESIS BY DNA PRONUCLEAR INJECTION

Transgenesis by DNA pronuclear injection is the oldest and still the most widely used method of introducing foreign genes into the mammalian germ line to subsequently generate mammalian transgenic animals (including mice, rats and livestock). The basis of this technique, which was developed in the early 1980s by, amongst others, Gordon, Ruddle, Brinster, Palmiter and colleagues,[4-6] is the microinjection of DNA of interest (the transgene) into the pronucleus of fertilised oocytes (single-cell embryos) leading to the ectopic expression (overexpression) of the transgene in the genome.

5.2.1 Generation of a Transgenic Mouse

This is a multi-step process (see Figure 5.1) which begins with the construction of the transgene, involves collection of fertilised oocytes, microinjection of the transgene DNA into the pronucleus of the oocytes, reimplantation of those eggs into a foster mother, identification of pups carrying the transgene and establishment of the transgenic line.

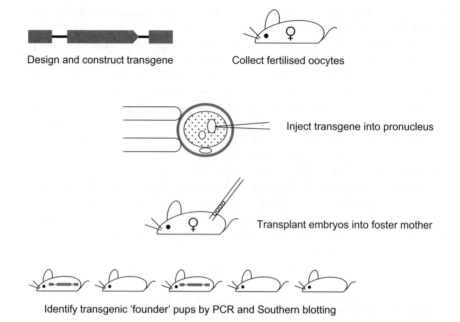

Design and construct transgene   Collect fertilised oocytes

Inject transgene into pronucleus

Transplant embryos into foster mother

Identify transgenic 'founder' pups by PCR and Southern blotting

Figure 5.1 Generation of transgenic mice by DNA pronuclear injection. Fertilised oocytes are collected from superovulated females. Transgene DNA is injected into one of the pronuclei. Embryos are re-implanted into pseudopregnant foster females and the resulting pups are screened for integration of the transgene. Founder (transgene positive) mice are then mated with wild-type mice to test for germ line transmission and transgene activity.

5.2.1.1 Step 1: Construction of the Transgene. The transgene can be any DNA sequence/gene of interest to be added to the mouse genome, *i.e.* it will be present in addition to the normal complement of mouse genes. It is possible to use overexpression of a transgene in several ways: (i) to determine the function of a gene or (ii) to utilise its properties. For example, transgenes are often generated by linking a gene of interest with the regulatory regions of other known genes in order to express the gene of interest and its product at a higher level of expression, or in a specific tissue or stage of development. Such experiments can be used to elucidate the normal function of the gene. For example, Figure 5.2 shows the transgene constructs used to overexpress the plasma membrane calcium pump isoform 4 (PMCA4) in the cardiomyocytes (under the regulation of the MLC2v promoter) and in the vascular smooth muscle (driven by the SM22α promoter)—analysis of these transgenic lines has revealed that PMCA4 is involved in contractility in the heart and in the maintenance of peripheral blood pressure and vascular tone.[7,8] It is possible to use DNA pronuclear injection experiments to determine the function of regulatory elements of genes by fusing them to a reporter gene such as LacZ or GFP, whose expression can be readily detected.[9,10] It is also possible to use a

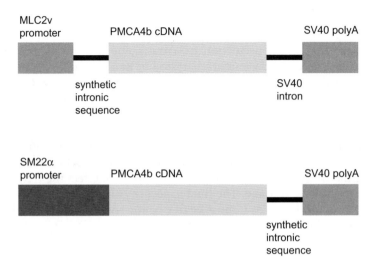

Figure 5.2 Construction of transgene. Examples of two transgenes designed to overexpress isoform 4 of the plasma membrane calcium pump (PMCA4b) in different tissues. Both constructs contain the same essential elements, *i.e.* a promoter sequence, cDNA of the gene of interest, a region of intronic sequence and a polyA tail. The top panel will result in the overexpression of PMCA4b in the heart, whereas the transgene in the lower panel was designed to overexpress PMCA4b in vascular smooth muscle cells.

transgene to express the wild-type form of a gene in a mouse that carries an endogenous mutant form *i.e.* it can be used to rescue a genetic defect.[11]

Transgene design has a major influence on the successful expression of the microinjected DNA; there are a number of elements that must be incorporated into the transgene construct, in addition to the protein coding sequence of the gene of interest, that are critical for gene expression. Essentially a transgene should also contain a promoter, an intron and a transcriptional stop sequence. The promoter is the regulatory sequence which determines in which cells and at what time the transgene is active and is usually derived from sequences upstream of, and including, the transcriptional start site which contain the necessary regulatory elements for transcription. The protein coding sequence is usually derived from the cDNA of the gene of interest and will contain both the ATG start codon and a translational stop codon. It is important to include an intron in the transgene construct as this increases transgene activity;[12] however, this does not need to be an intron from the gene of interest. It is not clear why the inclusion of an intron has a positive effect on transgene expression but it is hypothesised to be due to a functional link between transcription and splicing. The transgene must also contain a transcriptional stop signal including a polyA addition sequence (AAAUAA). A number of transgenes have included both the intron and transcriptional stop signal at the end of the coding sequence (as in Figure 5.2—SM22α-PMCA4b transgene), although they can be included separately (see Figure 5.2—MLC2v-PMCA4b transgene).

It is clear that the construction of a transgene is a multi-step cloning process which is most conveniently carried out in a bacterial plasmid. It has been shown that it is important to remove all of the prokaryotic plasmid sequences from the transgene prior to microinjection as its inclusion may inhibit its expression.[13] It is also clear that DNA that has been linearised prior to microinjection, rather than being left in its circular or supercoiled state, will integrate with greater efficiency.[14] Interestingly, although a large transgene (>30 kb) may present technical difficulties when cloning and isolating the DNA, their large size does not appear to effect the frequency of integration into the genome. The final aspect to consider is the purity of the DNA to be microinjected as it must be free of any contaminants which might be harmful to the oocytes to be injected.

5.2.1.2 Step 2: Collection of Fertilised Oocytes. To improve the efficiency of the generation of the transgenic mice it is essential to have a large number of viable fertilised eggs (oocytes) available for microinjection. Normally female mice will generate six–ten eggs when naturally ovulating; however, it is possible to recover 20–30 eggs from a single female by inducing superovulation. The consequence is that less female mice are required for the production of a suitable number of quality eggs. Young female mice are injected with a combination of hormones which mimic natural mouse hormones; these ensure that large numbers of mature eggs are released from the ovaries simultaneously. Treatment with follicle stimulating hormone is followed 44–48 hours later by treatment with leutinising hormone. Fertilised oocytes are collected from the oviducts of the females following overnight mating with stud males. The oocytes will be collected at 0.5 days *post coitum* (dpc) as this allows time for the sperm to complete fertilisation following mating. In addition, at this stage the pronuclei from both gametes will be visible for several hours after embryo collection, which is necessary to enable successful injection of the transgene DNA.

5.2.1.3 Step 3: Pronuclear Injection of DNA Transgene. Microinjection of the transgene requires specialist, expensive equipment, and highly trained personnel. The whole microinjection process is visualised under an inverted microscope which is attached to two micromanipulators; one to manipulate the glass pipette used to hold and secure the oocyte during the injection process (known as the holding pipette), the other micromanipulator controls the fine glass pipette used to inject the DNA transgene into the pronucleus (see Figure 5.3). For an excellent and detailed technical description of the microinjection process see Hogan *et al.*[15]

It is usual that between 60–80% of the microinjected embryos will survive the process (viability can be tested by culturing the embryos to the two-cell stage) and be suitable for re-implantation into a surrogate female.

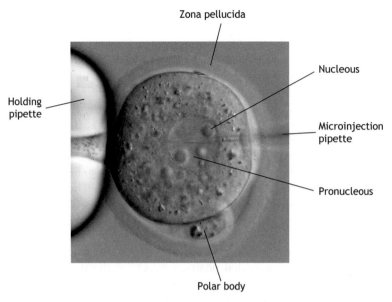

Zona pellucida

Nucleous

Holding
pipette

Microinjection
pipette

Pronucleous

Polar body

Figure 5.3 Microinjection of DNA into the pronucleus. The oocyte is secured by gentle suction on to the holding pipette. The microinjection pipette is advanced through the zona pellucida, it pierces the plasma membrane and is then pushed into the pronucleus. The oocyte has two pronuclei, either of which can be injected; the male pronucleus tends to be the largest and therefore easiest to visualise and inject. The DNA is then expelled from the injection pipette causing the pronucleus to swell.

Surrogate/recipient females are rendered pseudopregnant by mating with either a vasectomised or genetically sterile male. Of these re-implanted embryos between 10–25% will survive to term. It is this high attrition rate of embryos at each stage which necessitates the use of such large numbers of fertilised eggs at the beginning of the whole procedure.

The transgene will integrate randomly into the genome, into any chromosome, including the sex chromosomes, and often in multiple copies in a head-to-tail array.[16] Interestingly, it appears that integration most often occurs at just one site in the genome, even if multiple copies integrate.

Experience has shown that when microinjecting linearised DNA approximately 25% of the resulting pups carry the transgene, these are known as founder transgenics. Due to the precise timing of injection and therefore integration of the transgene into the pronucleus, prior to DNA replication, a high proportion (approximately 70%) of the transgenic mice carry the transgene in every cell of the body. It is essential that the transgene integrates into the germ cells to be able to transmit the gene modification to the next generation. In those mice in which integration of the transgene occurred after DNA replication the transgene will integrate into a proportion of the somatic and germ cells only; these mice are known as mosaics.

5.2.1.4   Step 4: Generation of a Transgenic Line.   Founder transgenic mice are identified by DNA analysis including PCR and Southern blotting. It is common for DNA to be isolated from small pieces of tissue taken from the ears of the mice or from a small section of the end of the tail. The presence or absence of the transgene can be identified by PCR analysis using transgene-specific primers which will not amplify the endogenous gene locus. Southern blot analysis is then used to verify the integration of the transgene and to identify the number of copies that have inserted into the genome.

Once founders have been identified they must be bred to wild-type mice to determine if the transgene can be passed to the next generation, in other words, they will be tested for germ-line transmission. The resultant pups must then be screened by PCR to identify heterozygous (*Tg*/+) transgenic mice from their wild-type (+/+) littermates—these would be expected in a 50:50 ratio.

One of the main drawbacks of generating transgenic mice by pronuclear injection of DNA is the so-called position effect *i.e.* the transgene can integrate at any chromosomal location and the site of integration itself could affect its expression. For example, a transgene could insert into a region of the chromosome that suppresses gene activity or could even insert into a functioning endogenous gene. Both of these scenarios would lead to a phenotype that is not solely due to the effect of the transgene. It is therefore essential that the phenotype of at least three independent founder lines carrying the same transgene are analysed to determine the effect of over-expression of the gene of interest.

5.2.2   Summary of Advantages and Disadvantages of Generating Transgenic Mice by Pronuclear Injection of DNA

Although the generation of transgenic mice by DNA pronuclear injection has several disadvantages, and the list of potential technical problems is lengthy, it remains a very rewarding technique which provides essential information regarding the biological function of a gene. Table 5.1 summarises the major advantages and disadvantages of this process.

5.3   GENE TARGETING BY HOMOLOGOUS RECOMBINATION IN EMBRYONIC STEM CELLS

Gene targeting was developed to ablate (knockout) the function of a specific gene. Although gene targeting is still most commonly used to generate this type of mutation the method is now the basis of many techniques which can produce a whole variety of DNA modifications to the mouse genome including the deletion/re-arrangement of large regions of chromosomes containing multiple genes,[17,18] the introduction of subtle mutations into the genome,[19,20] the knock-in of a gene of interest,[21-23] and conditional gene targeting (see Section 5.4).

Table 5.1 Advantages and disadvantages of generating transgenic mice by pro-nuclear injection of DNA.

Advantages
Conceptually straightforward.
Cloned DNA from any species can be microinjected.
Integration of transgene is relatively efficient.
Process is relatively quick—taking as little as 3 months to generate a construct, microinject DNA, analyse genotype of founder pups and test for germ-line transmission of the transgene.

Disadvantages
Microinjection requires specialist, expensive equipment and highly trained personnel.
Integration site is random which may affect the level and pattern of expression. Therefore need to analyse the phenotype (effect of the gene deletion) of mice from several founder lines.
The endogenous gene remains intact in the genome which may complicate molecular and phenotypic analysis.

The technique of gene targeting in the mouse has enabled major advances in our understanding of mammalian gene function; in normal developmental, biochemical and physiological processes, as well as in disease processes. The technique was established by bringing together two important findings: (i) that homologous recombination can be used to specifically modify a target mammalian gene and (ii) that embryonic stem cells isolated from pre-implantation stage mouse embryos can be maintained in culture and then used to contribute to the germ line of another mouse. The importance of this technique to biomedical research was acknowledged when Mario R. Capecchi, Martin J. Evans and Oliver Smithies jointly received The Nobel Prize in Physiology or Medicine 2007 for their discoveries of the *'principles for introducing specific gene modifications in mice by the use of embryonic stem cells.'*

5.3.1 Basic Principles

5.3.1.1 Homologous Recombination. Homologous recombination is used to introduce modifications into a gene by exchanging endogenous DNA sequences in the chromosome with cloned DNA sequences (targeting vector). Gene targeting by homologous recombination in mammalian cells was first reported in 1985,[24,25] allowing for the stable insertion of DNA into the genome at a predictable site.

5.3.1.2 Embryonic Stem Cells. Embryonic stem (ES) cells are isolated from the inner cell mass of pre-implantation stage embryos at 3.5 days *post coitum*; at this stage embryos are known as blastocysts. The crucially important factor about the progenitor cells of these early embryos is that they are *pluripotent*—they have the potential to differentiate into any cell type, including the germ cells, of the subsequent embryo. It was in 1986 that Martin Evans and colleagues demonstrated that these ES cells could be maintained in culture, where they could be genetically modified, and then used to re-populate the germ line of another mouse, thereby generating chimaeric offspring containing genetic material from both the recipient mouse and the ES cells.[26] The chimaeric mice, when mated, could pass on the genetic mutation to the next generation with the mutation being inherited according to Mendel's laws of genetics.

Oliver Smithies, Mario Capecchi and their colleagues brought together the homologous recombination and ES cell technologies to develop the process known as *Gene Targeting*.[27,28]

It is essential that ES cells to be used in gene targeting experiments maintain their pluripotency in order to contribute to the germ line of chimaeric mice. It has been found that the presence of the cytokine leukaemia inhibitory factor (LIF) is essential to ensure that ES cells do not differentiate *in vitro*. For this reason ES cells are generally grown on a feeder layer of fibroblasts which secrete LIF into the culture medium. When ES cells are grown in optimal culture conditions they grow as discrete colonies of densely packed cells, with no obvious sign of differentiation (see Figure 5.4). Other culture conditions have been shown to be essential in maintaining the ES cells in their undifferentiated state, including growing at a high cell density and ensuring that the colonies are thoroughly disaggregated when passaging (splitting) the cells, since clumps of cells are likely to differentiate.

Historically the majority of ES cells lines in use have been derived from the 129 strain of mouse which has an agouti coat colour genotype (for a summary of mouse coat colour genetics refer to [15]); this is useful when identifying chimaeric mice (see Step 5 in Section 5.3.2.5 below). Another interesting feature of many of the ES cell lines currently in use is that they have an XY (male) genotype which leads to the production of mainly male chimaeras (even when XY ES cells are injected into a female blastocyst, the resulting chimaeric pup will tend to be male).

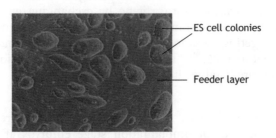

Figure 5.4   ES cell colonies growing on a layer of fibroblast feeder cells. Healthy, undifferentiated ES cells form densely packed colonies.

5.3.2 Generation of a Knockout Mouse

As with transgenesis by pronuclear injection, gene targeting by homologous recombination in embryonic stem cells is a multi-step process. It begins with the generation of the targeting vector which is transferred by electroporation into the ES cells. The ES cells are cultured and analysed for presence of the homologously recombined DNA sequence; the targeted ES cells are then injected into blastocyst stage embryos. The resultant pups are screened for the level of chimaerism (percentage contribution from the targeted ES cells), and the chimaeric mice are then tested for their ability to pass the targeted mutation through the germ line and to generate mice heterozygous for the mutation. Heterozygous mice can then be mated to generate the homozygous knockout mouse (see Figure 5.5).

5.3.2.1 Step 1: Generation of a Targeting Vector. When designing and constructing a targeting vector a number of factors must be considered

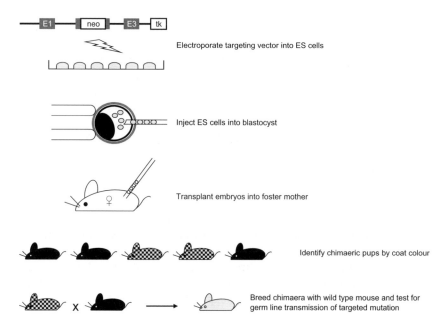

Electroporate targeting vector into ES cells

Inject ES cells into blastocyst

Transplant embryos into foster mother

Identify chimaeric pups by coat colour

Breed chimaera with wild type mouse and test for germ line transmission of targeted mutation

Figure 5.5 Generation of gene knockout mice by gene targeting in ES cells. The targeting vector is electroporated into the ES cells. ES cells that have undergone homologous recombination are injected into blastocyst stage embryos and these embryos are then transplanted to pseudopregant foster mothers. Chimaeric offspring can be identified by their coat colour; these pups will carry the targeted mutation carried by the injected ES cells. Chimaeric offspring can then be mated to wild type mice to determine whether they transmit the targeted mutation through the germ line to give pups heterozygous for the mutation. The heterozygous offspring can then be intercrossed to mice homozygous for the mutation.

which will influence the type of mutation to be introduced, the efficiency of targeting, and the ease with which successful targeting can be detected.

5.3.2.1.1 DNA Homologous With the Chromosomal/Gene Site of Interest.
For successful and efficient targeting the vector must contain at least 5–10 kb of isogenic DNA homologous with the sequence to be targeted. This homologous sequence is divided between the short-arm of homology (1–1.5 kb) and a long-arm of homology (4–8 kb); this enables easy screening of the ES clones. It is ideal to identify gene targeted colonies by PCR designed to span the short-arm of homology. It is known that the efficiency of homologous recombination is decreased when there are base pair differences between the donor and recipient DNA.[29] For this reason it is now common practice for the DNA used to construct the targeting vector to originate from the same mouse strain as the ES cells (*i.e.* isogenic DNA).

5.3.2.1.2 Positive and Negative Selection Cassettes.
Since gene targeting by homologous recombination occurs at low frequencies (typically 10^{-5}–10^{-6} of ES cells treated with construct DNA) and the targeting construct is much more likely to insert randomly into the genome it is essential to be able to screen ES cell colonies quickly and efficiently for successful targeting. For this reason, the vast majority of targeting vectors will be designed to insert a positive selection cassette into the gene of interest. For example, the neomycin phosphotransferase gene (neo) is often used as a positive marker, which when expressed in the ES cell genome will render the cells resistant to treatment with the antibiotic neomycin sulphate (G418). A negative selection marker can also be used to enrich for gene targeted colonies. The selection marker is cloned outside of the homologous sequence in the targeting vector and will therefore not insert into the genome when homologous recombination occurs, but will insert into the genome if random integration of the targeting vector occurs. For example, the herpes simplex virus thymidine kinase gene (HSVtk) when expressed in ES cells will produce a toxic product in the presence of gancyclovir (a thymidine analog), killing ES cells expressing this gene (see Figure 5.6).

5.3.2.2 *Step 2: ES Cell Transfection.*
The most efficient method for introducing the targeting vector into the ES cells is by electroporation. The linearised vector DNA is electroporated into a large number of ES cells in a single cell suspension; the cells are then plated onto fresh feeder cells. Twenty-four hours after electroporation the selection process can begin which will kill cells which have not incorporated the targeting vector by homologous recombination. The ES cells are cultured in media containing the drugs used for selection for 7–10 days; this will enrich the population

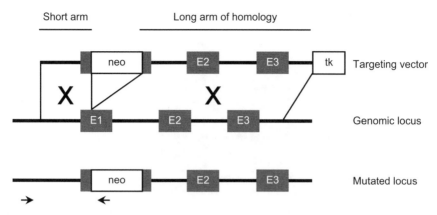

Figure 5.6 Gene targeting strategy. The upper panel depicts a conventional targeting vector in which the positive selection cassette (neo) has been cloned
into the first exon (E1) of the target gene and the negative selection
cassette has been cloned at the end of the targeting construct after exons
2 and 3 (E2 & E3). The middle panel represents the genomic locus and
homologous recombination will take place between the genomic locus
and the two arms of homologous sequence in the targeting vector. The
crossover points are depicted by the X. The lower panel depicts the
mutated locus when homologous recombination has occurred. The two
small arrows represent the PCR primers which will be used to amplify
the sequence across the short arm of homology to identify ES cells which
have undergone homologous recombination.

with cells that have undergone homologous recombination, however, it
must be noted that this process is not 100% efficient.

*5.3.2.3 Step 3: Identification of ES Cells Targeted by Homologous
Recombination.* To identify the ES cells that have undergone gene targeting
by homologous recombination discrete colonies are identified and picked.
The colonies are dissociated into single cells by treatment with trypsin, divided between two wells on duplicate microtitre plates and cultured. The
purpose of dividing the cells between duplicate plates is to allow one plate
of cells to be used to prepare DNA to identify targeted ES cells and the cells
from the second plate can be used to inject into blastocysts. Genomic DNA
is prepared from each ES cell clone which is then screened by PCR to
identify clones in which homologous recombination has occurred. It is
likely that several hundred ES cell clones will require screening so it is
important that a robust and easy PCR strategy is used—it is for this reason
that it is usual to amplify the region across the short arm of homology (see
Figure 5.6). Positive clones must then be further analysed, usually by Southern blot and DNA sequencing, to verify that all regions of the targeting
vector have undergone the desired recombination event.

5.3.2.4 Step 4: Injection of ES Cells into Blastocysts. Blastocysts, which
are 3.5-day-old embryos, are collected from the uterus of the donor female.

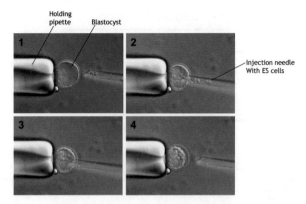

Figure 5.7 Injection of targeted ES cells into blastocysts. (1) The blastocyst is held on the holding pipette by gentle suction. (2) The injection needle containing ES cells is advanced into the blastocyst cavity (blastocoel), (3) where the ES cells are released and (4) the injection needle is removed.

It is usual when using ES cells from the 129 strain of mouse to collect blastocysts from a C57Bl/6 mother; this mouse line has a black coat colour (which is helpful when identifying chimaeric mice—see step 5 below). ES cells carrying the desired mutation are treated to give a single cell suspension. The ES cells are drawn up into the injection pipette by gentle suction and the blastocyst to be injected is held by suction onto the holding pipette. The injection pipette is advanced into the cavity of the blastocyst, which is known as the blastocoel, and 10–15 ES cells are released (see Figure 5.7). After injection the embryos are cultured for a few hours to allow them to slowly re-expand before being transferred to the uterus of a pseudopregnant foster mother. Pups should be born 17 days later.

5.3.2.5 Step 5: Identification of Chimaeric Mice and Breeding to Generate Homozygous Mutant (Knockout) Mice. Approximately one week after mouse pups are born their coat colour becomes apparent. At this stage it is possible to identify agouti from non-agouti coat colour. It is therefore possible to identify chimaeric mice by their coat colour if ES cells from the 129 mouse strain (agouti) have contributed to the development of a C57Bl/6 embryo (non-agouti). Embryos in which the ES cells had made no contribution would appear as wild-type C57Bl/6 (black), whereas those pups in which the 129 ES cells had made a contribution would contain a certain level of agouti coat colouring. Chimaeric mice therefore contain some cells carrying the targeted mutation on one allele and other cells which are wild type.

To generate a gene knockout mouse it is essential that some of the germ cells carry the targeted mutation. To test for germ line transmission of the mutation chimaeric mice are bred to wild-type mice; should germ line transmission occur a proportion of the pups will be heterozygous for the

Table 5.2 Advantages and disadvantages of generating a knockout mouse by gene targeting in embryonic stem cells.

Advantages

The integration site and therefore the gene modification is highly specific.

A variety of mutations can be achieved including null mutations (gene knockout), deletion/rearrangement of large regions of chromosomes, site-specific mutations, gene knock-in.

Recessive alleles can be studied.

Disadvantages

Microinjection requires specialist, expensive equipment and highly trained personnel.

Process is very time consuming, taking 1.5–2 years to generate a targeting vector, target ES cells, identify homologous recombination events, microinject ES cells, test chimaeric pups for germ line transmission of mutation.

Process is expensive as it is labour intensive, requires expensive equipment and the mouse husbandry costs will be high.

Embryonic lethality—if the target gene is essential for development of the embryo then it will not be possible to study the role of the gene in the adult mouse.

Sometimes difficult to determine if the phenotype observed is primarily due to the deletion of the target gene or is a secondary consequence of the deletion of the target gene on a downstream pathway.

targeted mutation. Heterozygous mice can then be bred to produce mice homozygous for the targeted mutation—gene knockout mice.

5.3.3 Summary of Advantages and Disadvantages of Generating Gene Knockout Mice

As will be evident from the section above, the generation of a knockout mouse by gene targeting in ES cells is not a process to be undertaken lightly. It is labour intensive, technically demanding and fraught with problems; however, with good experimental design and know-how this technique will provide enormous amounts of invaluable information about the function of the target gene. Table 5.2 summarises the major advantages and disadvantages of this process.

5.4 CONDITIONAL GENE TARGETING

As has been described above, gene targeting has rapidly become a highly developed technology that facilitates the study of gene function. However, 30% of germline mutations result in embryonic lethality[30] thereby precluding analysis of gene function in the adult tissues. Moreover, gene targeting disrupts the function of the gene within all tissues, creating a complex phenotype in which it may be difficult to distinguish direct function of the gene of interest in a particular tissue from any secondary effects. To resolve these critical problems, Gu and colleagues developed the conditional inactivation of gene expression using the Cre/loxP system.[31]

Figure 5.8   Scheme of Cre binding to the loxP site. Cre specifically recognises and binds to inverted repeats surrounding an asymmetric 8 bp core region.

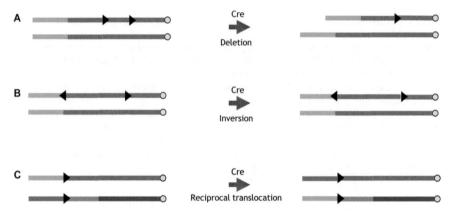

Figure 5.9   Cre-mediated recombination. The orientation of the loxP sites influences the type of recombination that occurs. (A) Deletion of DNA occurs between two loxP sites in the same orientation; following recombination one loxP site remains in the genome. (B) Inversion of DNA occurs between two loxP sites in the opposite direction. (C) Reciprocal translocation occurs between loxP sites placed in non-homologous chromosomes. The arrowheads indicate the orientation of the loxP sites.

Cre (Cause REcombination) recombinase is a 38 kDa enzyme isolated from the P1 bacteriophage. The loxP site is a 34 bp sequence comprising two 13 bp inverted repeats separated by an asymmetric 8 bp core which indicates the orientation of the loxP site (see Figure 5.8). The Cre recombinase efficiently binds the two loxP sites and catalyses the recombination, resulting in the deletion, inversion or translocation of the sequences between the two recombination sites, according to the orientation of two loxP sites (see Figure 5.9).

Cre is not the only recombinase that has been described for use in conditional gene targeting; Flp recombinase from *Saccharomyces cerevisiae* is also used.[32] Similar to the Cre/loxP system, Flp recognises two FRT sites (34 bp DNA sequence) and catalyses the DNA recombination between the two recombination sites. The Flp recombinase is not as effective as Cre at 37 °C; its optimum function is at 25–30 °C.[33] The use of the Flp/FRT system is not as widespread as that of the Cre/loxP system, due to the limited availability of mouse lines expressing the Flp recombinase. This section will therefore focus on the description of the Cre/loxP recombination system.

5.4.1 Generation of a Conditional Knockout Mouse using the Cre/loxP System

The generation of conditional knockout mice is a multi-step process, which involves mating two separate mouse lines generated by DNA pronuclear injection and/or gene targeting by homologous recombination in ES cells.

(i) Homologous recombination in embryonic stem (ES) cells is used to generate mutant mice (Flox mice) in which the essential exon/s of the gene is flanked by two loxP sites. These sites do not interfere with the normal expression of the gene but constitute a binding domain for the Cre recombinase which will excise the DNA fragment between the loxP sites.

(ii) Generating inducible and/or tissue-specific Cre expressing mouse lines by pronuclear injection or knock-in method using embryonic stem cells.

(iii) Crossing the Flox mouse with the specific Cre-expressing mouse to achieve the deletion of the gene in a particular cell lineage or tissue type at a certain time.

5.4.1.1 Step 1: Generation of Flox Mouse Models. The *tri-loxP* targeting vector has been successfully used to generate numerous Flox mouse lines. For example, to produce *mkk4* Flox mice ($mkk4^{flox/flox}$), a thymidine kinase neomycin resistance (Neo-TK) cassette containing two loxP sites was inserted behind exon 4, while a third loxP site was placed in front of exon 4 (see Figure 5.10). Exon 4 of this gene contains the important kinase function domain; the removal of this exon causes a frame shift in the open reading frame, resulting in the loss of expression of MKK4. The targeting construct was electroporated into ES cells and then subjected to selection using the antibiotic G418. The resulting positive clones were transfected with a Cre-expression plasmid to remove the Neo-TK selection cassette; it is important to remove this cassette because it may interfere with the normal expression of the target gene. The ES cells were counter-selected with 1-(2-deoxy-2-fluoro-β-D-arabinofuranosyl)-5-iodouracil (FIAU) which results in the death of cells which retained the positive/negative selection cassette. The ES clones that retained two loxP sites flanking exon 4 were identified by Southern blotting and were subsequently used to generate the *mkk4* Flox mice.[34]

5.4.1.2 Step 2: Generation of the Cre-Expressing Mouse Lines. Over the past few years many tissue-specific Cre mouse lines have been developed which have greatly facilitated the study of gene function. Established Cre lines can be identified using the Jackson Laboratory Cre Repository database (http://cre.jax.org). The core part of this approach is to choose a tissue-specific promoter which drives Cre expression specifically and efficiently. Three methods are widely used for generating Cre mouse models.

5.4.1.2.1 Standard Transgenesis By Pronuclear Injection. The generation of a Cre-expressing mouse in which a transgene vector consisting of

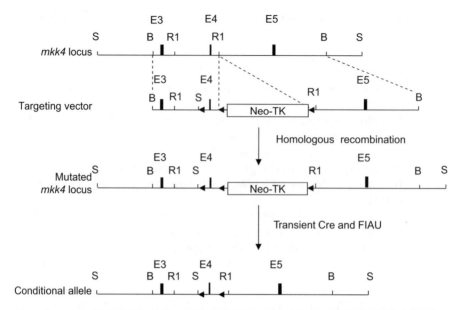

Figure 5.10 Conditional disruption of the *mkk4* gene. The figure shows the *mkk4*
wild-type locus, the targeting vector containing the three *loxP* sites,
represented by black triangles, the predicted structure of the mutated
mkk4 locus, the selected conditional allele obtained by Cre-mediated
recombination. In the conditional *mkk4* allele, *loxP* sites are located
either side of exon 4. Restriction enzyme sites are indicated (B, BamHI;
R, EcoRI; S, SpeI).

Cre driven by a tissue-specific promoter is inserted into the genome by
standard transgenesis is the most popular method. It is a relatively rapid
procedure with a straightforward concept, and in many cases has generated very efficient Cre lines. However, this method involves random integration of the Cre-transgene into the genome which sometimes results in
unwanted Cre expression or low Cre expression levels which prohibits any
further analysis of the function of the gene of interest.

5.4.1.2.2 Bacterial Artificial Chromosome (BAC) Transgenesis. BAC
clones contain large fragments of mouse genomic DNA including most
regulatory elements of the chosen promoter; thus introducing Cre into the
BAC presents an advantage that Cre expression pattern reflects truthful
endogenous gene expression. The main drawbacks of this technique are
that an additional loxP site exists in the BAC vector which needs to be removed prior to pronuclear injection; and that other genes and their promoter regions may exist within the BAC vector and may therefore interfere
with the desired Cre expression pattern.[35]

**5.4.1.2.3 Knock-In Approach Using Homologous Recombination in ES
Cells.** The knock-in strategy requires homologous recombination in ES

cells to insert the Cre recombinase construct downstream of the endogenous targeted promoter. The advantage of this approach is that Cre expression is closely correlated to the expression profile of the chosen promoter, however, as has already been highlighted in this chapter this is a long and expensive procedure.

Many genes have different functions, from developmental stages to adulthood. Using an early promoter (*i.e.* the promoter from a gene that is expressed very early in development) to drive Cre activity in a tissue-specific gene-knockout system sometimes results in embryonic lethality. To overcome this limitation, inducible gene targeting was first developed by Kühn and colleagues.[36] This method can induce gene inactivation in mice at a given time point using modification of the Cre recombinase system. For example, the Cre-ER line is widely used to achieve such temporal control of genetic recombination; in this mouse line Cre has been fused to a mutated form of the ligand binding domain of the estrogen receptor.[37] The mutant ER does not bind its natural ligand at physiological concentrations but binds to the synthetic ligand tamoxifen and several of its metabolites (*e.g.* 4-hydroxytamoxifen). Thus, Cre-ER activity can be induced following tamoxifen injection, thereby allowing the temporal (time-specific) deletion of the loxP flanked gene. At present, a number of tissue-specific inducible Cre lines have been reported, such as the αMHC-MerCreMer line for the study of heart function[38] and the Nestin-CreERT2 line used in brain function studies.[39] The availability of these elegant and novel Cre lines has provided us with a unique opportunity to precisely study gene function and regulation.

5.4.1.3   Step 3: Breeding Strategies to Generate Temporal- and/or Tissue-specific Gene Knockout Mice.   Two rounds of breeding are required to produce temporal- and/or tissue-specific gene-knockout mice (see Figure 5.11). The first breeding step between the floxed mouse and the cre-expressing mouse results in progeny which carry a single floxed allele and the Cre vector (Gene A$^{fl/+}$:Cre). The second breeding step crosses these mice with homozygous flox mice (Gene A$^{fl/fl}$); pups with a number of different genotypes will be born, including the desired conditional knockouts (Gene A$^{fl/fl}$:Cre).

5.4.2   Chromosomal Engineering using the Cre-loxP System

Chromosomal abnormalities cause many human genetic disorders and foetal loss, such as DiGeorge syndrome, which is due to the deletion of a large region within chromosomal 22 and is characterised by congenital heart defects, immunodeficiency and abnormal facial appearance. Due to the restriction on working with human material and the lack of experimental models, progress in dissecting the genetic basis underlying congenital diseases has been slow. In 1999, Lindsay and colleagues made a revolutionary step when they successfully created the DiGeorge mouse model bearing the 1.2 Mb chromosomal deletion using homologous recombination in ES cells and the Cre-loxP technique.[18] Due to the unique properties of the Cre-loxP

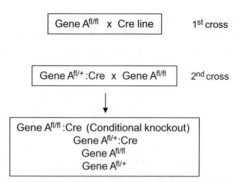

Figure 5.11   Breeding scheme to generate tissue/temporal-specific knockout mice. The floxed mouse is crossed with the Cre-expressing mouse, which after two breeding steps will result in a conditional gene knockout.

system, the manipulation of chromosomal rearrangements (deletion, inversion or translocation) in order to model human diseases has become feasible and this will greatly empower scientists in studying human inherited diseases.

5.4.3   Summary of Advantages and Disadvantages of Conditional Gene Targeting

Conditional gene targeting is an exceptionally elegant extension of the classical gene targeting approach which enables the expression of a gene modification at a given time and/or in a particular cell type or tissue. The system can be exploited to introduce subtle mutations into the genome and to make large chromosomal modifications. However, its disadvantages must be fully considered before embarking on the study of gene function using this system. Table 5.3 summarises the main advantages and disadvantages of generating a conditional knockout mouse.

5.5   TETRACYCLINE INDUCIBLE GENE SWITCH SYSTEM

Conditional gene expression technologies are powerful tools for gene function studies, however, the most successful systems (Cre/LoxP and Flp/FRT) introduce permanent genetic modifications. By contrast, the tetracycline inducible gene switch system allows multiple switches in gene expression and potentially reversible phenotypic changes.

5.5.1   Basic Principles

Tetracyclines are a family of broad-spectrum antibiotics which are well characterized and widely applied for the treatment of bacterial infections. They act by inhibiting gene expression; however, bacteria have evolved a self-protection system against tetracyclines to ensure survival and normal

Table 5.3 Advantages and disadvantages of generating a conditional knockout mouse.

Advantages

Avoids embryonic lethality and complicated secondary effects which can be induced by conventional gene targeting.

Gene function can be studied in a specific cell type or tissue at any given time point.

Chromosomal abnormalities causing human diseases can be modelled in mice.

Disadvantages

Two transgenic lines (Flox mice and Cre mice) are required to generate conditional knockout mouse models.

Efficiency of gene deletion varies depending on the gene locus position and the Cre activity.

Time-consuming and expensive.

---

function. The transcription of tetracycline resistance genes is regulated by TETracycline Repressor protein (TetR). In the anti-tetracycline system, in the absence of tetracycline, binding of TetR to the 19bp TETracycline Operator (TetO) sequence (TCCCTATCAGTGATAGAGA) inhibits transcription of the resistance genes. However, if tetracycline enters the bacterial cells, TetR binds to tetracycline thus allowing expression of the resistance proteins.[40]

5.5.2 Generation of Tetracycline Inducible Transgenic Mouse

The inducible system has been developed, based on the natural and effective binding characteristics of TetR to TetO, such that highly sensitive and fine regulation of the transgene of interest can be achieved. Mouse models generated by this approach have been utilised to efficiently study gene function.

Two separate transgenic lines must be generated and then cross bred to obtain a tetracycline inducible transgenic mouse. The mouse lines are the TetR driver transgenic mouse and the TetR responder transgenic mouse, which are both generated by the DNA pronuclear injection technique.[41–43]

5.5.2.1 Step 1: Generation of the Driver Transgenic Mouse. Two versions of the Tet system exist and are known as Tet-Off and Tet-On, based on their response to tetracyclines (see Figure 5.12):

1. _Tetracycline transactivator protein (tTA) for Tet-Off system._ tTA protein is comprised of the full length prokaryotic TetR protein and, most commonly, the activation domain of the herpes simplex virus (HSV) VP16 protein.[44] A cell or tissue specific promoter is then fused upstream to drive tTA protein expression (*e.g.* the transgenic tTA driver carrying the α-MHC promoter is used for cardiomyocyte specific tTA protein expression[45]).

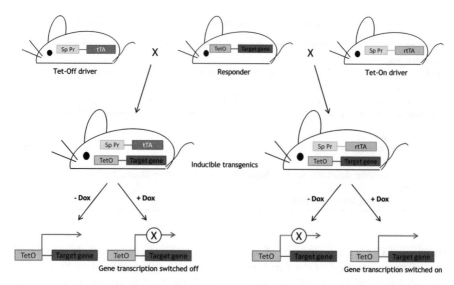

Figure 5.12 Generation of transgenic mice using the tetracycline inducible gene switch system. A tissue or cell-specific promoter (Sp Pr) is fused to tTA or rtTA to generate a Tet-Off driver or Tet-On driver transgenic mouse, respectively. In the responder transgenic line, TetO is fused to the target gene of interest. The breeding of the responder transgenic mouse with one of the driver transgenic mice creates a double transgenic mouse. In the Tet-Off system, transcription of the target gene is inhibited in the presence of Dox; whilst the Tet-On system requires Dox to be present for activation of the target gene.

2. *Reverse tetracycline transactivator (rtTA) for Tet-On system.* rtTA is also a fusion protein comprised of a cell or tissue specific promoter, the TetR motif, and the VP16 activation domain. However, in rtTA, a four amino acid substitution in the TetR moiety alters its protein structure and TetO binding character, leading to a completely reverse response to tetracycline.[46]

5.5.2.2 Step 2: Generation of the TetR Responder Transgenic Mouse. The second transgenic mouse carries another essential component, the TetR responsive element, in which TetO sequences are fused to a minimal promoter to control transgene expression. The specific interaction of TetO with the TetR region in tTA or rtTA results in transcriptional regulation of the target transgene[44] (see Figure 5.12).

5.5.2.3 Step 3: Generation of Tetracycline Inducible Transgenic Mouse. The next step in the generation of these binary transgenic mouse models is to crossbreed the driver transgenic mouse and the responder transgenic lines.[47] Finally, administration of the tetracycline (often the derivative doxycycline) permits transcription of the target gene to be switched on or off in specific tissues or cells.

Expression of the target gene is dependent on both the activity of tTA/rtTA controlled by specific promoter and whether or not tetracycline is present. In the tTA dependent/Tet-off system tTA binds to TetO to activate transcription of the target gene whilst the addition of tetracycline inactivates tTA and causes downregulation of transgene expression. Hence the term *Tet-Off*. The rtTA system acts in reverse; without tetracycline rtTA cannot recognize the TetO sequences, resulting in transcriptional inhibition of the target gene. Conversely, transcription is stimulated by rtTA in the presence of tetracycline, giving rise to the term *Tet-On*[46] (see Figure 5.12).

5.5.3 Summary of Advantages of Generating Transgenic Mice by Tetracycline Inducible Gene Switching

Tetracycline inducible gene switch technology controls gene expression at the transcriptional level. The tetracycline transgenic mouse models allow expression of the target transgene in a time specific and tissue or cell specific manner and importantly it enables reversibility and quantitative control of gene expression. These features make this type of transgenic model important for the study of disease progression and the potential for the disease state to be reversed.

5.6 GENOME EDITING BY ENGINEERED ENDONUCLEASES

The field of gene customisation continues to evolve rapidly and in recent years we have seen the development of a number of innovative approaches designed to speed up the process of generating genetically modified mice by using *engineered endonucleases*. The technology is considered to have such potential to increase our understanding of gene function that it was elected as 'Method of the Year 2011' by *Nature Methods*.[48]

Engineered endonucleases can be used to generate numerous types of genetic modification including gene knockout and knockin; reporter genes can be added to the genome and allelic mutations generated. One of these endonuclease systems even has the capability of generating mutations in multiple genes in one organism.[49]

Zinc finger nucleases (ZFNs) were the first to be utilised and are therefore the best characterised of the nucleases, but the development of transcription activator-like effector nucleases (TALENs) has the potential to make a bigger impact on this field of research. This is mainly due to their simpler DNA-binding code and the relative ease with which they can be designed. TALENs technology is also proving to be a powerful tool for gene editing not only in laboratory animals but also other large domestic species as well as human genomes, thus providing a new gene targeting platform for modified model organisms and gene therapeutics.

The nucleases comprise a sequence-specific DNA-binding domain linked to a non-specific DNA cleavage domain. Both methods work by directly microinjecting the site-specific nucleases either into the pronucleus (as a supercoiled DNA plasmid) or into the cytoplasm (as an *in vitro*-transcribed mRNA) of one-cell fertilized embryos. Double-strand breaks (DSB) are then induced at the target locus which stimulates either non-homologous end joining (NHEJ) or homology-directed repair (HDR). The injected embryos, thereafter, are implanted into the uterus of pseudopregnant females to obtain genetically modified pups.

The CRISPR/Cas system differs from ZFNs and TALENs in that it is based on the use of RNA-guided DNA endonucleases. The CRISPR/Cas system was identified in bacteria which had evolved an RNA-based adaptive immune system that uses CRISPR (clustered regularly interspaced short palindromic repeat) and Cas (CRISPR-associated) proteins to detect and destroy foreign DNA. This characteristic has been harnessed to produce an endonuclease system which can be used to generate multiple gene modifications within the genome.[50]

5.7 PHENOTYPIC ANALYSIS OF GENETICALLY MODIFIED MICE

Genetically modified mice, generated by any of the techniques described within this chapter, are merely a sophisticated *tool* with which to work when determining the function of a gene. The mutant mouse must be fully analysed at the molecular, biochemical, and *in vivo* levels to determine the function of that gene.

Initially all mice will be analysed to determine their *genotype*, for example, PCR or Southern blot analysis will be carried out to determine whether the mouse carries the desired gene mutation and whether it is heterozygous or homozygous for the mutation.

It is the effect that the *genotype* has on the *phenotype*, the characteristics and traits of the mouse, which is of key interest.

Having ascertained the genotype of the mouse it is essential to determine that the introduced genetic modification has resulted in the desired altered gene regulation; for example, that disruption of the target gene or the introduction of a DNA construct has resulted in the ablation or over-expression of the gene product, respectively. RNA analysis by reverse transcriptase-PCR (RT-PCR) or real-time PCR, Northern blotting, or *in situ* hybridisation can be used to establish the expression pattern of the transgene or confirm disruption/deletion of the gene of interest. Modified levels/localisation of the gene product (protein) can then be identified by Western blot analysis and immunohistochemistry.

Having identified the direct effect that the gene modification has had on the gene of interest it is important to investigate the downstream effects of the gene modification by determining expression levels of genes in the same pathways, or known interaction partners *etc.* DNA microarray technologies are extremely valuable in identifying novel genes whose expression may be influenced by modification of the gene of interest.[51]

The functional tests that must be carried to determine the effect of the gene modification will depend on each particular gene, for example, it would be important to measure intracellular Ca^{2+} levels in a mouse model in which a calcium handling protein is disrupted, or the activity of an enzyme should be tested where the gene encoding that enzyme has been modified.

To determine the role of the gene of interest in normal development, physiology and pathology there are numerous assessments that can be carried out. The first thing to assess is whether the gene of interest is vital to embryonic development and therefore whether its modification leads to embryonic lethality. The simplest assessment is to analyse the genotyping results; for example, are homozygous gene knockout offspring born when heterozygous mice are crossed? If the gene modification does lead to a lethal phenotype it will then be necessary to assess the point of lethality and ultimately how and why the modification leads to embryonic death.[52] It is interesting to note that analysis of large numbers of mouse models carrying a homozygous gene deletion has revealed that over 30% are embryonic lethal.[30]

There is a vast array of tests that can be carried out to determine the function of the gene in normal physiology ranging from very simple observations of the gross appearance of the mouse to extremely sophisticated screens.[53] For example, a simple first-line phenotyping screening protocol named SHIRPA is widely used to identify and characterise phenotypic changes associated with a gene mutation.[30,54] This series of simple tests and observations is designed to assess the general health of the mouse by measuring body weight and temperature; behavioural characteristics such body posture, gait, locomotion, presence of tremor, and coat colour; and the sensory functions of the mouse. The effect of the gene mutation on other organs systems including the heart can also be measured using specialised tests and equipment.[7,55–57] Simple breeding tests can be used to determine if the gene modification affects fertility; which can then be followed up, for example, with in depth analysis of sperm function.[58]

It is therefore the analysis of the functional, biochemical and molecular changes that occur due to a gene modification that will help us to determine the function of the genes in the genome.

5.8 ETHICAL AND ANIMAL WELFARE CONSIDERATIONS

The mouse has become an extremely popular model for the study of mammalian gene function for a number of reasons. Although it has long been known that the mouse has a similar physiological structure to humans we now know that approximately 99% of mouse genes have human homologs.[59] Mice are mammals and subsequently undergo similar developmental processes to humans; they also have many physiological and behavioural similarities. In addition, mice develop some of the same diseases as humans including cancer, cardiovascular disease, hypertension, obesity, diabetes, glaucoma, neurological defects, deafness, osteoporosis and asthma. For these reasons mice are considered by many as an essential biomedical tool in our armoury to combat human disease through increased understanding of disease mechanisms and development of therapeutic agents.

It is not only the physiological and genetic aspects of the mouse that make it a suitable model for studying gene function; on a practical level the mouse is very convenient to work with as it has a short gestation period (19–21 days), reaches breeding age within 6–8 weeks and colonies are relatively inexpensive to maintain.

It is clear however that the mouse is not without its disadvantages when using it as a model in which to study human gene function and the development of human disease. It is obvious that even though the two species share 99% gene homology that millions of years of evolution have resulted in major differences between the two species.

The mouse does not naturally develop all diseases exhibited by humans, for example, one of the most important diseases of the elderly population is Alzheimer's disease; a disease which mice do not normally develop. However, genetic manipulation of the mouse genome through gene targeting and transgenesis has enabled many aspects of this disease to be modelled.[60] In other instances reproducing mutations in the mouse genome which are known to cause disease in humans does not lead to mimicking of the human phenotype.[61,62] A recent development in extending the use of the mouse as a model for human disease is by 'humanising' the mouse, this is a process which includes replacing large segments of the mouse genome with the syntenic human sequence (syntenic regions are areas of chromosomes in which genes occur in the same order in different species), allowing both the gene itself and surrounding regulatory regions to be replaced.[63] This new adaptation of transgenic technology promises to enable more accurate models of human disease to be developed in the mouse. It is clear therefore that the mouse is not a perfect model in which to study mammalian gene function and for use as a model for human disease; however, the application of transgenic techniques and genetic mutation in the mouse genome has greatly informed our knowledge of gene function.

As research scientists using an animal system as a model it is essential that we consider the welfare of our animal models and must closely weigh the benefit we gain from their use as part of our research against the cost to their welfare. This chapter has so far highlighted the benefits to research of transgenesis and of using the mouse as a model of human disease; however, the cost to animals used in this research may include suffering due to the phenotype caused by the genetic modification, embryonic or post-natal death as a result of the effect of the genetic modification, or animal welfare may be comprised during the production of the GM mice *e.g.* by superovulation, vasectomy, embryo collection and transfer.

There are many ways to ensure that a minimal number of animals are used to gain the maximum amount of information. One essential method is to ensure that animal experiments are carried out in combination with *in vitro*, cellular and bioinformatic techniques and are fully informed by the findings of these other research methods. The number of animals used can be reduced through good technical practice. It is also important that a research project is continually assessed in the light of new published data and that animal requirements are regularly reassessed to ensure minimal animal usage to maximise statistical analyses throughout the course of a project.

Transgenesis is a powerful technology which has been and will continue to be successfully used to increase our knowledge of gene function in health and disease. It is therefore incumbent on the researcher to ensure that exemplary practice regarding animal welfare is used to ensure that the cost: benefit ratio is correctly balanced.

5.9 CONCLUSIONS

The generation of mouse models using a variety of transgenesis techniques in order to study gene function has been immensely popular amongst researchers since the introduction of the techniques in the 1980s. This chapter gives an outline of several of the fundamental technologies involved but there are many aspects of transgenesis which have not been addressed here including the use of other model systems from plants, non-vertebrates, other vertebrates and even large mammals—all of which have an important place in research to determine gene function.

Over recent years we have gathered immense amounts of information regarding mammalian gene function; however, we are still a long way from knowing the function of the majority of the genes in the genome.[56] The next major breakthrough in the use of transgenesis to understand mammalian gene function will be the success of large-scale projects which are systematically mutating every one of the 21 000 protein encoding genes in the genome and generating and phenotyping the resulting knockout mouse models.

International efforts under the umbrella of International Knockout Mouse Consortium (IKMC, http://www.knockoutmouse.org) have developed mutant ES cells for protein-coding genes and the International Mouse Phenotyping Consortium (IMPC) is undertaking a 10 year programme to complete the generation and phenotyping of the 21 000 mouse mutant lines by 2021. This massive undertaking will give us a comprehensive view of mammalian gene function on which we can build our understanding of human health and disease.

ACKNOWLEDGEMENTS

We would like to thank Graham Morrissey and Lynnette Knowles for some of the photographs used in this chapter.

REFERENCES

1. J. D. McPherson, M. Marra, L. Hillier, R. H. Waterston, A. Chinwalla, J. Wallis, M. Sekhon, K. Wylie, E. R. Mardis, R. K. Wilson, *et al.*, *Nature*, 2001, **409**, 934.

2. International Human Genome Sequencing Consortium, *Nature*, 2004, **431**, 931.

3. G. R. Abecasis, D. Altshuler, A. Auton, L. D. Brooks, R. M. Durbin, R. A. Gibbs, M. E. Hurles and G. A. McVean, *Nature*, 2010, **467**, 1061.

4. J. W. Gordon, G. A. Scangos, D. J. Plotkin, J. A. Barbosa and F. H. Ruddle, *Proc. Natl. Acad. Sci. U. S. A.*, 1980, **77**, 7380.

5. R. L. Brinster, H. Y. Chen, M. Trumbauer, A. W. Senear, R. Warren and R. D. Palmiter, *Cell*, 1981, **27**, 223.

6. R. D. Palmiter, H. Y. Chen and R. L. Brinster, *Cell*, 1982, **29**, 701.

7. D. Oceandy, E. J. Cartwright, M. Emerson, S. Prehar, F. M. Baudoin, M. Zi, N. Alatwi, K. Schuh, J. C. Williams, A. L. Armesilla and L. Neyses, *Circulation*, 2007, **115**, 483.

8. K. Schuh, T. Quaschning, S. Knauer, K. Hu, S. Kocak, N. Roethlein and L. Neyses, *J. Biol. Chem.*, 2003, **278**, 41246.

9. J. F. Schiltz, A. Rustighi, M. A. Tessari, J. Liu, P. Braghetta, R. Sgarra, M. Stebel, G. M. Bressan, F. Altruda, V. Giancotti, *et al.*, *Biochem. Biophys. Res. Commun.*, 2003, **309**, 718.

10. N. Iguchi, H. Tanaka, S. Yamada, H. Nishimura and Y. Nishimune, *Biol. Reprod.*, 2004, **70**, 1239.

11. A. A. Migchielsen, M. L. Breuer, M. S. Hershfield and D. Valerio, *Hum. Mol. Genet.*, 1996, **5**, 1523.

12. T. Choi, M. Huang, C. Gorman and R. Jaenisch, *Mol. Cell. Biol.*, 1991, **11**, 3070.

13. L. Kjer-Nielsen, K. Holmberg, J. D. Perera and J. McCluskey, *Transgenic Res.*, 1992, **1**, 182.

14. R. L. Brinster, H. Y. Chen, M. E. Trumbauer, M. K. Yagle and R. D. Palmiter, *Proc. Natl. Acad. Sci. U. S. A.*, 1985, **82**, 4438.

15. B. Hogan, R. Beddington, F. Costantini and E. Lacy, *Manipulating the Mouse Embryo: A Laboratory Manual*, Cold Spring Harbor Laboratory Press, New York, 1994.

16. F. Costantini and E. Lacy, *Nature*, 1981, **294**, 92.

17. F. Buchholz, Y. Refaeli, A. Trumpp and J. M. Bishop, *EMBO Rep.*, 2000, **1**, 133.

18. E. A. Lindsay, A. Botta, V. Jurecic, S. Carattini-Rivera, Y. C. Cheah, H. M. Rosenblatt, A. Bradley and A. Baldini, *Nature*, 1999, **401**, 379.

19. J. A. Cearley and P. J. Detloff, *Transgenic Res.*, 2001, **10**, 479.

20. P. Dickinson, W. L. Kimber, F. M. Kilanowski, S. Webb, B. J. Stevenson, D. J. Porteous and J. R. Dorin, *Transgenic Res.*, 2000, **9**, 55.

21. Y. Geng, W. Whoriskey, M. Y. Park, R. T. Bronson, R. H. Medema, T. Li, R. A. Weinberg and P. Sicinski, *Cell*, 1999, **97**, 767.

22. N. Hosen, T. Shirakata, S. Nishida, M. Yanagihara, A. Tsuboi, M. Kawakami, Y. Oji, Y. Oka, M. Okabe, B. Tan, *et al.*, *Leukemia*, 2007, **21**, 1783.
23. M. Ikeya, M. Kawada, Y. Nakazawa, M. Sakuragi, N. Sasai, M. Ueno, H. Kiyonari, K. Nakao and Y. Sasai, *Int. J. Dev. Biol.*, 2005, **49**, 807.
24. F. L. Lin, K. Sperle and N. Sternberg, *Proc. Natl. Acad. Sci. U. S. A.*, 1985, **82**, 1391.
25. O. Smithies, R. G. Gregg, S. S. Boggs, M. A. Koralewski and R. S. Kucherlapati, *Nature*, 1985, **317**, 230.
26. E. Robertson, A. Bradley, M. Kuehn and M. Evans, *Nature*, 1986, **323**, 445.
27. T. Doetschman, R. G. Gregg, N. Maeda, M. L. Hooper, D. W. Melton, S. Thompson and O. Smithies, *Nature*, 1987, **330**, 576.
28. K. R. Thomas and M. R. Capecchi, *Cell*, 1987, **51**, 503.
29. H. te Riele, E. R. Maandag and A. Berns, *Proc. Natl. Acad. Sci. U. S. A.*, 1992, **89**, 5128.
30. A. Ayadi, M. C. Birling, J. Bottomley, J. Bussell, H. Fuchs, M. Fray, V. Gailus-Durner, S. Greenaway, R. Houghton, N. Karp, *et al.*, *Mamm. Genome*, 2012, **23**, 600.
31. H. Gu, J. D. Marth, P. C. Orban, H. Mossmann and K. Rajewsky, *Science*, 1994, **265**, 103.
32. S. M. Dymecki, *Proc. Natl. Acad. Sci. U. S. A.*, 1996, **93**, 6191.
33. F. Buchholz, L. Ringrose, P. O. Angrand, F. Rossi and A. F. Stewart, *Nucleic Acids Res.*, 1996, **24**, 4256.
34. X. Wang, B. Nadarajah, A. C. Robinson, B. W. McColl, J. W. Jin, F. Dajas-Bailador, R. P. Boot-Handford and C. Tournier, *Mol. Cell. Biol.*, 2007, **27**, 7935.
35. C. S. Branda and S. M. Dymecki, *Dev. Cell*, 2004, **6**, 7.
36. R. Kühn, F. Schwenk, M. Aguet and K. Rajewsky, *Science*, 1995, **269**, 1427.
37. R. Feil, J. Brocard, B. Mascrez, M. LeMeur, D. Metzger and P. Chambon, *Proc. Natl. Acad. Sci. U. S. A.*, 1996, **93**, 10887.
38. D. S. Sohal, M. Nghiem, M. A. Crackower, S. A. Witt, T. R. Kimball, K. M. Tymitz, J. M. Penninger and J. D. Molkentin, *Circ. Res.*, 2001, **89**, 20.
39. I. Imayoshi, T. Ohtsuka, D. Metzger, P. Chambon and R. Kageyama, *Genesis*, 2006, **44**, 233.
40. M. Takahashi, L. Altschmied and W. Hillen, *J. Mol. Biol.*, 1986, **187**, 341.
41. A. Harper, *Biochim. Biophys. Acta*, 2010, **1802**, 785.
42. A. Kistner, M. Gossen, F. Zimmermann, J. Jerecic, C. Ullmer, H. Lubbert and H. Bujard, *Proc. Natl. Acad. Sci. U. S. A.*, 1996, **93**, 10933.
43. H. Prosser and S. Rastan, *Trends Biotechnol.*, 2003, **21**, 224.
44. M. Gossen and H. Bujard, *Proc. Natl. Acad. Sci. U. S. A.*, 1992, **89**, 5547.
45. J. Heineke, M. Auger-Messier, J. Xu, T. Oka, M. A. Sargent, A. York, R. Klevitsky, S. Vaikunth, S. A. Duncan, B. J. Aronow, *et al.*, *J. Clin. Invest.*, 2007, **117**, 3198.
46. M. Gossen, S. Freundlieb, G. Bender, G. Muller, W. Hillen and H. Bujard, *Science*, 1995, **268**, 1766.

47. P. A. Furth, L. St Onge, H. Boger, P. Gruss, M. Gossen, A. Kistner, H. Bujard and L. Hennighausen, *Proc. Natl. Acad. Sci. U. S. A.*, 1994, **91**, 9302.

48. *Nat. Methods*, 2012, **9**, 1.

49. T. Gaj, C. A. Gersbach and C. F. Barbas 3rd, *Trends Biotechnol.*, 2013, **31**, 397.

50. H. Wang, H. Yang, C. S. Shivalila, M. M. Dawlaty, A. W. Cheng, F. Zhang and R. Jaenisch, *Cell*, 2013, **153**, 910.

51. J. D. Horton, N. A. Shah, J. A. Warrington, N. N. Anderson, S. W. Park, M. S. Brown and J. L. Goldstein, *Proc. Natl. Acad. Sci. U. S. A.*, 2003, **100**, 12027.

52. X. Wang, A. J. Merritt, J. Seyfried, C. Guo, E. S. Papadakis, K. G. Finegan, M. Kayahara, J. Dixon, R. P. Boot-Handford, E. J. Cartwright, *et al.*, *Mol. Cell. Biol.*, 2005, **25**, 336.

53. H. Fuchs, V. Gailus-Durner, T. Adler, J. A. Aguilar-Pimentel, L. Becker, J. Calzada-Wack, P. Da Silva-Buttkus, F. Neff, A. Gotz, W. Hans, *et al.*, *Methods*, 2011, **53**, 120.

54. D. C. Rogers, E. M. Fisher, S. D. Brown, J. Peters, A. J. Hunter and J. E. Martin, *Mamm. Genome*, 1997, **8**, 711.

55. X. L. Tian, S. L. Yong, X. Wan, L. Wu, M. K. Chung, P. J. Tchou, D. S. Rosenbaum, D. R. Van Wagoner, G. E. Kirsch and Q. Wang, *Cardiovasc. Res.*, 2004, **61**, 256.

56. S. D. Brown and M. W. Moore, *Dis. Model Mech.*, 2012, **5**, 289.

57. W. Liu, M. Zi, R. Naumann, S. Ulm, J. Jin, D. M. Taglieri, S. Prehar, J. Gui, H. Tsui, R. P. Xiao, *et al.*, *Circulation*, 2011, **124**, 2702.

58. K. Schuh, E. J. Cartwright, E. Jankevics, K. Bundschu, J. Liebermann, J. C. Williams, A. L. Armesilla, M. Emerson, D. Oceandy, K. P. Knobeloch and L. Neyses, *J. Biol. Chem.*, 2004, **279**, 28220.

59. R. H. Waterston, K. Lindblad-Toh, E. Birney, J. Rogers, J. F. Abril, P. Agarwal, R. Agarwala, R. Ainscough, M. Alexandersson, P. An, *et al.*, *Nature*, 2002, **420**, 520.

60. J. A. Richardson and D. K. Burns, *Ilar J.*, 2002, **43**, 89.

61. W. H. Colledge, B. S. Abella, K. W. Southern, R. Ratcliff, C. Jiang, S. H. Cheng, L. J. MacVinish, J. R. Anderson, A. W. Cuthbert and M. J. Evans, *Nat. Genet.*, 1995, **10**, 445.

62. S. J. Engle, D. E. Womer, P. M. Davies, G. Boivin, A. Sahota, H. A. Simmonds, P. J. Stambrook and J. A. Tischfield, *Hum. Mol. Genet.*, 1996, **5**, 1607.

63. H. A. Wallace, F. Marques-Kranc, M. Richardson, F. Luna-Crespo, J. A. Sharpe, J. Hughes, W. G. Wood, D. R. Higgs and A. J. Smith, *Cell*, 2007, **128**, 197.

CHAPTER 6

Molecular Engineering of Antibodies

JAMES D. MARKS

Department of Anesthesia, University of California, San Francisco,
Rm 3C-38, San Francisco General Hospital, 1001 Potrero Ave,
San Francisco, CA 94110, USA
Email: marksj@anesthesia.ucsf.edu

6.1 INTRODUCTION

Antibodies are one of the effector molecules of the vertebrate humoral immune system. They are generated *in vivo* in response to the presence of foreign pathogens or molecules (antigens), bind specifically to the antigen, and result in its neutralization and elimination. One of the characteristics of antibodies is that they can bind with high affinity and specificity to only the target antigen and not to any of the tens of thousands of other proteins and potential antigens in the circulation. This specific and high affinity binding led to the appreciation that antibodies could be the so-called 'magic bullets' proposed by Paul Ehrlich at the turn of the century; molecules that could selectively target a disease causing organism and deliver a toxic payload, killing only the organism targeted.

As will be described below, antibodies in the serum of immunized animals were some of the first therapeutics used for infectious diseases, at a time before antibiotics had been discovered. Yet it took more than 100 years from this initial use of serum therapy for antibodies to begin to be approved by the Food and Drug Administration (FDA) for the treatment of human diseases. The first such antibodies entered clinical practice in the 1990's and today there are currently 30 therapeutic antibodies approved by the FDA which had sales of more than $50 billion in 2013

Molecular Biology and Biotechnology, 6th Edition
Edited by Ralph Rapley and David Whitehouse
© The Royal Society of Chemistry 2015
Published by the Royal Society of Chemistry, www.rsc.org

(see http://www.landesbioscience.com/journals/mabs/about/ and http://biotechspain.com/en/article.cfm?iid = market_therapeutic_antibodies). Most of these antibodies have been approved for clinical use within the last 15 years and it is estimated that there are at least a hundred antibodies in the different phases of human clinical trials for a wide range of diseases including cancer, inflammatory diseases and infectious diseases.[1,2]

The era of antibodies as therapeutics became possible with the advent of hybridoma technology in 1975, a technological breakthrough that resulted in the ability to clone single B-cells and the single antibody made by that B-cell (Figure 6.1).[3] Such monoclonal antibodies (mAbs), unlike the hundreds to thousands of antibodies in serum, recognized only a single antigen and could be made in virtually unlimited quantities. Unfortunately, the technology was developed to make mAbs from the B-cells of immunized mice and it has proven technically challenging to apply the technology to generate human mAbs.[4] As it turns out, when murine mAbs are administered to humans, they elicit an immune response called the human anti-mouse antibody response (HAMA) that either results in unacceptable systemic reactions or results in rapid clearance of the mAb from the bloodstream.[5,6] While many mAbs from hybridomas entered clinical trials, few were approved by the FDA due to the limitations described above.

With the advent of molecular cloning and protein engineering technologies in the late 1980's, it has proven possible to engineer murine mAbs to have sequences more similar to human mAbs, so called chimeric[7–9] and humanized antibodies.[6] Such antibodies have proven significantly less immunogenic than murine mAbs and many of these are now approved for clinical use. More recently, it has proven possible to make mAbs that are fully human in sequence using antibody gene diversity libraries and display technologies (see references 1, 10 and 11 for reviews), as well as mice that are transgenic for the human immunoglobulin loci.[12] While transgenic mice (and other species) are an important source of human antibodies, this technology will not be discussed further in this chapter as it is not a true antibody engineering technology.

In the following sections, we will review the increasing importance of antibodies as a therapeutic class and review antibody structure, generation, and function. We will use this information as background to describe the molecular engineering techniques of chimerization and humanization that have yielded the first widely successful therapeutic antibodies. We will then describe how the more recent techniques using diversity libraries and display technologies can be used to generate fully human antibodies and to evolve antibody affinity to values not typically generated by the humoral immune system.

6.2   ANTIBODIES AS THERAPEUTICS

Antibody therapy began approximately a century ago with the discovery that serum from animals immunized with toxins, *e.g.* diphtheria toxin or viruses,

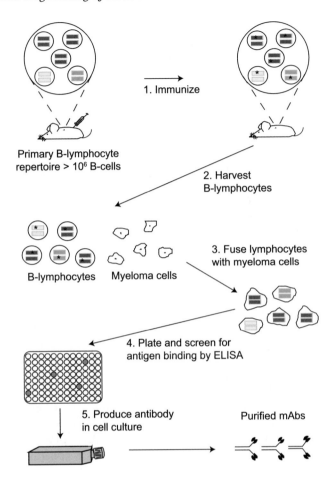

Figure 6.1 Generation of monoclonal antibodies using hybridoma technology. The naïve mouse generates a primary repertoire of more than 10^6 rearranged V_H and V_L genes (colored bars) in B-cells, coding for antibodies that are displayed as membrane-bound molecules. Immunization (step 1) causes antigen-driven proliferation and somatic hypermutation ('stars' within V-genes represent mutations introduced by the somatic hypermutation machinery). To make hybridomas, B-cells are harvested from the spleen (step 2) or marrow of the mouse and fused with immortal myeloma cells (wrinkled edged cells, step 3) to generate immortalized, antibody secreting hybridomas. Hybridomas are plated into microtiter plates and the supernatants containing secreted antibody screened by ELISA for antigen binding (step 4). Hybridomas are expanded into tissue culture flasks and the secreted monoclonal antibodies purified (step 5).

was an effective therapy for the disease caused by the same agent in humans. While capable of potent antigen neutralization, such sera contain hundreds to thousands of different antibodies with only approximately 1% of the antibodies binding to the immunizing antigen. In the 1880s von Behring developed an antitoxin that neutralized the toxin that the Diptheria bacillus released into the body and was awarded the first Nobel Prize in Medicine in 1901 for his role in the discovery and development of serum therapy for diphtheria.

Following the initial successes in the late 1800s, sera from humans or animals containing antibodies were widely used for prophylaxis and therapy of viral and bacterial diseases.[13,14] Serum therapy of most bacterial infections was abandoned in the 1940s, however, after antibiotics became widely available.[13] Polyclonal antibody preparations have continued to be used for some toxin-mediated infectious diseases and venomous bites.[15] Serum immunoglobulin has also continued to be used for viral diseases where there are few treatments available, although immunoglobulin is largely used for pre- or post-exposure prophylaxis.[16,17]

While serum polyclonal antibody preparations have been clinically effective in many cases, they have problems related to toxicity including a significant risk of allergic reactions including anaphylactic shock and serum sickness.[18] Other limitations of polyclonal antibodies include lot to lot variations in potency and side effects and uncertainty in dosing due to these effects.[15] In addition, the active antigen-specific antibodies in a polyclonal preparation typically represent a relatively small portion of the total antibodies (1%); the rest of the antibodies are not only ineffective but could be even toxic or immunogenic. However, until the 1970s it was not possible to produce large amounts of antibodies with the desired specificity from a single antibody producing B-cell.

Development of hybridoma technology made it possible to clone single B-cells and the single antibody made by that B-cell (Figure 6.1).[3] Molecular cloning techniques then made it possible to genetically replace the mouse constant regions of the mouse antibody with human constant regions yielding chimeric antibodies, antibodies that are approximately 90% human in sequence (Figure 6.2).[7-9] Chimeric antibodies are far less immunogenic than murine antibodies. Therapeutic antibodies were ushered into a 'take off' phase by the 1997 launch of the chimeric antibody Rituxan (rituximab) for non-Hodgkin's lymphoma (NHL). Rituxan represented the first mAb product to succeed commercially in a high-revenue/high-growth market (oncology) and to provide significant enhancements in the efficacy of treatment *versus* existing non-mAb therapies. As a result, Rituxan rapidly became established as the gold-standard therapy for NHL and the first launched mAb product which went on to achieve blockbuster status (revenues above $1 billion per year). Rituxan approval by the FDA was followed by approval of Remicade (Infliximab) which binds tumor necrosis factor-alpha (TNF-α) and which is used to treat rheumatoid arthritis and other inflammatory diseases mediated by TNF, such as psoriasis and Crohn's disease.

Further advances in antibody engineering resulted in humanized antibodies, mAbs where the variable region framework regions as well as the constant regions were replaced with human sequences (Figure 6.2).[65] Humanized antibodies, which are greater than 90% human in sequence,

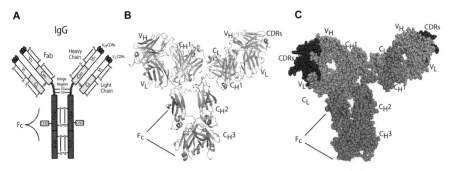

Figure 6.2 Structure of IgG antibody and chimeric and humanized antibodies. (A) IgG Structure. IgG antibody consists of a pair of heavy and light chains. Each chain consists of the antigen binding variable domains (V_H and V_L) and one or more constant domains (C_H1, C_H2, C_H3, and C_L). Each V or C domain contains an intramolecular disulfide bond (S–S). A single glycosylation site (CHO) exists in the C_H2 domain of the heavy chain. The Fab consists of the light chain and the V_H-C_H1 domains; each IgG consists of two Fab arms. The Fv is the minimal antigen binding unit, consisting of the V_H and V_L domains. The V_H and V_L domains contact antigen *via* the amino acids in the complementarity determining regions (CDRs). The Fc region elicits antibody effector functions such as ADCC and CDC. (B) Alpha carbon backbone tracing of a chimeric antibody. The alpha carbon backbone tracing of a chimeric IgG antibody is shown. The chimeric antibody consists of murine V_H and V_L domains (colored cyan and green respectively) and human constant domains (colored shades of gray). (C) Space filling model of a humanized antibody. Murine CDRs in each of the Fab arms are colored red. The framework regions of the V-domains and all of the C-domains are colored grey.

are potentially less immunogenic than chimeric antibodies. A number of blockbuster humanized antibodies have been approved by the FDA including Herceptin (trastuzumab) for treatment of breast cancer, Avastin (bavacizumab) for treatment of colon cancer, and Synagis (palivizumab) for prevention of respiratory syncytial virus (RSV) infection. In 2002, Humira (adalimumab), the first fully human antibody (produced *via* phage display technology), was licensed by the FDA. It too has achieved blockbuster status, with most recent annual sales of greater than \$10 billion/year. Currently, more than 100 mAbs are in various stages of clinical trials to treat a range of human diseases including cancers, inflammatory, and infectious diseases.[1,2,19] These antibodies are primarily humanized and fully human antibodies generated and optimized using antibody engineering technologies.[1,2,19] Moreover, a number of engineering strategies have been deployed to enhance antibody potency and the resulting antibody based drugs have entered clinical trials.[2] These include antibody–drug conjugates,[20] antibody–toxin fusion proteins,[21] and antibodies where the Fc portion of the antibody has been engineered to more effectively elicit antibody dependent cellular cytotoxicity (ADCC) or complement dependent cytotoxicity (CDC).[22] Two antibody drug conjugates, trastuzumab emtansine and brentuximab vedotin, were recently approved by the FDA for treatment of HER2 positive breast cancer and Hodgkin's lymphoma.[23,24] The engineering techniques used to generate therapeutic antibodies will be discussed in the following sections, after a review of antibody structure and function.

6.3 ANTIBODY STRUCTURE AND FUNCTION

Antibodies are Y-shaped dichain glycoproteins with molecular mass of approximately 150 kDa (Figure 6.2). They consist of two heavy chains and two light chains, with each chain consisting of two or more domains that share the immunoglobulin fold. Each domain is a pleated beta sheet with an intramolecular disulfide bond. The light chain consists of two domains, a light chain variable domain (V_L) and a light chain constant domain (C_L). There are two types of light chains, kappa and lambda. In humans, approximately two-thirds of light chains are kappa and one-third are lambda. The heavy chain consists of four to give domains, depending on the antibody isotype, and includes a heavy chain variable domain (V_H), and three to four constant domains (C_H1, C_H2, C_H3, and C_H4). There are a number of different isotypes of antibodies which include IgG, IgD, IgA, IgM, and IgE. The major isotype present in the circulation is IgG, which is also the isotype most frequently used for therapeutic antibodies. The light chains are disulfide linked to the V_H-C_H1 domains to make the antigen-binding fragment (Fab). The two Fabs are linked *via* the hinge region to the Fc domain, which consists of CH2-CH3 or CH2-CH4 domains. The Fc region is involved in the elicitation of antibody effector functions including the ability of antibody to mediate cellular killing by immune effector cells (antibody dependent cellular cytotoxicity, ADCC) or complement (antibody dependent cellular cytotoxicity, CDC).

The smallest antibody fragment capable of binding antigen is the fragment variable or Fv. The Fv consists of the V_H and V_L domains (Figure 6.2). Within each of the V-domains are three complementarity-determining regions (CDRs) which form loops that comprise the antigen-binding region of the V-domains. CDRs were first identified by Kabat as regions with the greatest sequence diversity when different V-domains were compared to each other and are also called the hypervariable regions.[25] The six CDRs contribute the majority of the amino acid side chains that make direct contact with antigen.[26,27] Each CDR is flanked by regions of more conserved sequence that make up the variable region framework regions.[28,29] The frameworks fold into the β-strands that comprise the two β-sheets of each V-domain and which support the CDRs.

6.4 CHIMERIC ANTIBODIES

As described in the therapeutic antibodies section above, the development of chimeric antibodies was a significant leap forward for the therapeutic antibody field.[7,30,31] To reduce the immunogenicity of murine antibodies, the murine variable regions were grafted onto human kappa light chain and gamma heavy chain constant regions (Figure 6.2B). This is relatively straightforward as the V-regions and C-regions are contiguous pieces of DNA and each domain folds independently.

For chimerization, the DNA encoding the antibody heavy (V_H) and light (V_L) chain variable region genes must be cloned from the hybridoma cell. Prior to the development of the polymerase chain reaction, this required cDNA cloning of the variable region genes. With the advent of PCR, 5′ PCR primers were initially designed based on N-terminal protein sequencing of the variable regions from the heavy and light chain.[32] Design of the 3′ primer was more straightforward, as it could be based in the constant regions. A major breakthrough was the development of partially degenerate universal V-region primers based on sequence databases that could be used to amplify and directionally clone most murine variable regions without the need for protein sequencing (Figure 6.3).[33,34] Additional groups have designed sets of universal V-gene primers containing internal or appended restriction sites suitable for amplification of murine[35–40] human,[41–45] chicken,[46] and rabbit[47] V-genes. Once the V_H and V_L genes have been cloned, they can be directly forced cloned into cloning vectors by incorporating restriction sites into the primers.[42,48] Alternatively, PCR can be used to directly combine V_H and V_L genes with linker DNA using splicing by overlap extension to generate antigen binding antibody fragment constructs such as those encoding single chain Fv antibodies (scFv, Figure 6.3, part 3b).[42] By incorporating restriction sites into the primers, the scFv genes can be directionally cloned into bacterial secretion vectors (see below and Figure 6.4) which allow antibody fragment expression and screening for antigen binding.[42,48]

While PCR greatly simplified the cloning of V-genes, mutations introduced by the somatic hypermutation machinery into the regions where the primers anneal may make PCR amplification difficult or impossible, necessitating another amplification approach such as RACE or oligoligation PCR.[49–51] Cloning the correct V_H and V_L can also be complicated by the presence of several immunoglobulin transcripts, some of them arising from the fusion partner.[52]

In the original examples of chimerization, vectors containing murine V-regions fused to human constant regions were constructed and used to transfect mouse myeloma cell lines to produce antibody to confirm antigen binding, affinity, and specificity.[31] This process, which ensures that the correct V_H and V_L genes have been cloned from the hybridoma, takes several months if one constructs stable mammalian cell lines, for example in myeloma cells. An alternative would be to express the chimeric antibody transiently in mammalian cells such as COS-7 cells, cutting the time from cloning to screening for antigen binding to several weeks. Commercially, full length IgG (chimeric, humanized, or human) are typically expressed in Chinese Hamster Ovary cells at yields greater than 1 gram/liter.

A major step forward in antibody engineering occurred with the discovery that antibody fragments could be expressed in bacteria, such as *E. coli* (Figure 6.4). Due to the rapid growth of bacteria, time from V-gene cloning to

1. First strand cDNA synthesis

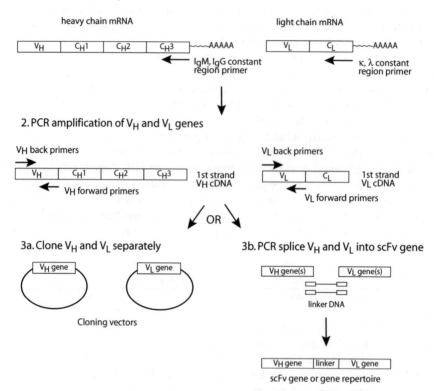

2. PCR amplification of V$_H$ and V$_L$ genes

Figure 6.3   PCR cloning of antibody V-genes. (1) mRNA is isolated from hybridoma cells, peripheral blood lymphocytes, spleen, or bone marrow, and antibody genes are reverse transcribed (by using reverse transcriptase) using IgG, IgM, κ, or λ constant-region specific primers, creating 1st strand cDNA. (2) The V$_H$ and V$_L$ variable region genes are amplified using PCR and universal primer mixtures specific for the 5′ (back primers) and 3′ (forward primers) ends of the heavy and light chain V-genes. (3a) Amplified V$_H$ and V$_L$ gene DNA is purified and cloned into cloning vectors for DNA sequencing. By incorporating restriction sites into the primers, it is possible to directionally clone the V-genes. (3b) Alternatively, the amplified V$_H$ and V$_L$ genes can be combined with a short 'linker DNA' which overlaps the 3′ and 5′ ends of the V$_H$ and V$_L$ genes, respectively, and PCR amplified to yield one continuous DNA fragment. The linker DNA encodes a flexible peptide that links the V$_H$ and V$_L$ gene to generate a single chain Fv (scFV) gene. A final PCR reaction (not shown) adds flanking restriction sites to the assembled scFv gene for cloning into bacterial secretion vectors or display vectors.

screening of antibody fragments for binding can as short as several days. It is now common practice after cloning V-genes from hybridomas to express them as antibody fragments in bacteria to verify antigen binding prior to construction of chimeric IgG or prior to humanization (see below). Initial attempts to express full length antibody in the cytoplasm of *E. coli* resulted in

Secreted Antibody Fragments

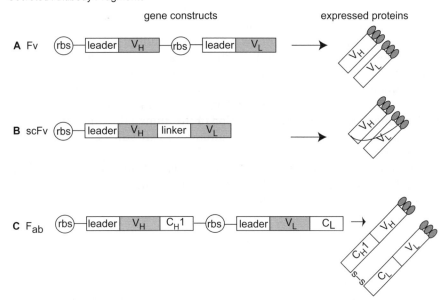

Figure 6.4 Antibody fragments that can be expressed in *E. coli*. Fv, single chain Fv (scFv) and Fab antibody fragments can be expressed in *E. coli* by directing the V-domains into the bacterial periplasm using a leader sequence. rbs: ribosome binding site; leader: bacterial secretion signal directing the expressed protein to the bacterial periplasm; V_H: heavy chain variable domain; V_L: light chain variable domain; linker: flexible scFv linker between V_H and V_L domains; C_H1: heavy chain constant domain 1; C_L: light chain constant domain; S–S indicates the disulfide bond between the C_H1 and the C_L domains.

very low yields of insoluble protein that required solubilization and re-folding.[53,54] A significant breakthrough occurred when it was discovered that antigen-binding fragments of antibodies could be produced in properly folded soluble form if expression of the antibody fragment was directed into the periplasmic space. Secretion into the periplasm was directed by attaching signal sequences such as pelB to the N-terminus of the antibody fragment genes (Figure 6.4). The oxidizing environment of the periplasm results in proper folding and intramolecular disulfide bond formation. Both the Fab and the Fv antigen binding fragments can be expressed in *E. coli* where they can be harvested from the bacterial periplasm.[55,56] Fv's, however, are not particularly stable; the V_H and V_L domains are not covalently linked and they tend to dissociate at typically expressed concentrations.[57,58] Fv's can be stabilized by physically linking the V_H and V_L domains together with a flexible peptide linker to create the single chain Fv antibody fragments (scFv).[59,60] Alternatively, cysteines can be engineered into both the V_H and V_L domains resulting in disulfide bond formation when the V-domains pair, a so-called disulfide linked Fv.[61]

As described in the section on therapeutic antibodies (above), a number of chimeric antibodies have been approved by the FDA and entered clinical practice, including rituximab and infliximab. Chimerization reduces the immunogenicity of murine mAbs and allows multiple and repeated dosing.[62,63] Replacing the murine Fc with a human Fc also results in more efficient effector functions such as ADCC and CDC as the human Fc interacts with human Fc receptors with higher affinity than the murine Fc. Similarly, chimeric antibodies have a longer half-life in humans than murine antibodies, due to more efficient interaction of the human Fc with FcRn. Despite their successes, however, chimeric antibodies can still be immunogenic. A human anti-chimeric antibody (HACA) response is frequently observed and in some cases can be severe and require discontinuation of the antibody or result in its ineffectiveness.[62-64] As a result, techniques to generate more fully human antibodies were developed.

6.5 ANTIBODY HUMANIZATION

As illustrated in Figure 6.2, the V_H and V_L CDRs comprise the antigen-binding site and contain the majority of amino acids that make direct contact with antigen. Jones *et al.* hypothesized that the CDRs from a murine antibody could be 'grafted' onto human frameworks to make a so called 'humanized' antibody.[65] Grafting the CDRs from the V_H domain of a murine antibody to a hapten antigen onto human frameworks resulted in an antibody that bound the hapten with comparable affinity when combined with the murine light chain variable domain.[65] Such CDR grafting has been applied to an anti-lymphocyte mAb and to an antibody to respiratory syncytial virus (RSV) but is not always successful due to the critical role that framework residues may play in both supporting the proper conformation of the CDRs as well as contributing antigen contacting amino acids.[66,67]

With the determination of a large number of antibody X-ray crystallographic structures, many now take a more directed approach to humanization.[68] After the murine V_H and V_L genes are sequenced, human frameworks are selected that are the most homologous to the murine V-genes. This is straightforward given the large number of antibody sequence databases, including those of Kabat (http://www.bioinf.org.uk/abs/seqtest.html) and IMGT (http://imgt.cines.fr/). Modeling the human frameworks and mouse CDRs onto a homologous antibody structure allows identification of murine residues that may either contact antigen or that directly or indirectly influence the conformation of the CDRs. Finally, amino acids in the human frameworks that are 'rare' or 'unusual' are replaced with residues more typical of those positions. Approaches like this tend to retain more murine residues in the framework regions than with CDR grafting, and have been used to humanize antibodies to the interleukin receptor and interferon-gamma.[68,69]

A number of variations have been developed to the above approach. These include replacing murine CDR residues that are outside the antigen binding loops with human residues, as done by Presta and colleagues,[70,71] the use of human germline V-genes as the acceptor frameworks, and grafting the murine CDRs into human frameworks with the most homologous CDRs, as opposed to most homologous V-genes or frameworks (super-humanization).[72] Alternatively, library approaches, as described below, can be used to humanize murine antibodies.[73,74] In these approaches, one of the murine V-domains, for example the V_H domain, is paired with a library of human light chains and a chimeric antibody containing murine V_H and human V_L selected. The murine V_H is then replaced with a library of human V_H's paired to the new human V_L and a fully human antibody is selected. With any of these methodologies, it can be anticipated that the binding constants of the humanized antibodies may be less than those of the parental antibodies. If so, additional constructs can be generated and evaluated or library approaches can be used to return affinity to that of the parental antibody (see below).

More than half of the antibody therapeutics licensed by the FDA are humanized antibodies. These include such 'blockbuster' antibodies as Herceptin (trastuzmab) for the treatment of breast cancer, Avastin (bavacizumab) for treatment of colon cancer, and Synagis (palivizumab) for prevention of respiratory syncytial virus (RSV) infection. Humanized antibodies are clearly less immunogenic than murine antibodies. They also appear to be less immunogenic in some instances than chimeric antibodies.[64] Humanized antibodies are not completely free of immunogenicity, and anti-humanized antibody responses (HAHA) can be detected in some patients. Such responses may be related to the number of non-human amino acids retained in the humanized antibodies, as well as the dose, the immunocompetence of the individual, and the specific target of the antibody.

6.6 ANTIBODIES FROM DIVERSITY LIBRARIES AND DISPLAY TECHNOLOGIES

The pharmaceutical industry discovers small molecule drugs by screening very large compound libraries to get lead 'hits' with affinities in the micromolar range. The tools of medicinal chemistry are then applied to these leads to improve their binding to the drug target resulting in a new drug entity. In the late 1980's advances in molecular biology made it possible to apply the same type of strategy to the discovery of fully human antibodies. Libraries of millions to billions of different antibodies could be generated and methods were developed that allowed the isolation of rare lead antibodies binding to a specific target antigen. As with the medicinal chemists, similar methods could be applied to diversify the structure of lead antibodies and allow selection of antibodies with considerably higher affinities. The methods used to create such libraries are called display technologies and include phage display, yeast display, ribosome display, and a number of other less frequently used technologies. Display technologies have three characteristics in common: 1) A method for generating antibody gene diversity; 2) a method for linking the antibody genotype with the expressed antibody phenotype; and 3) a method for isolating rare antigen binding antibodies (and their genes) from a majority of non-binding antibodies.

6.6.1 Antibody Phage Display

Phage display was the first antibody display technology to be developed and is the one most widely used today (see Figure 6.5 for overview and references 10, 11, 75 and 76 for reviews). Antibody phage display resulted from concurrent progress in prokaryotic expression of antibody fragments, PCR cloning of antibody gene repertoires, and display of peptides and proteins on filamentous bacteriophages. PCR cloning of antibody genes from hybridomas was covered in Section 6.4, above. By applying PCR and V-region specific primers to RNA prepared from human peripheral blood lymphocytes or mouse splenocytes, it is possible to amplify repertoires of V_H and V_L genes (see Figures 6.3 and 6.5).[42,43,48] The V-gene repertoires can then be either separately cloned sequentially or spliced by overlap extension to create human or murine scFv or Fab gene repertoires (see Figures 6.3 and 6.5).[42,48] An alternative to the use of naturally occurring V-gene repertoires is to synthesize V-genes *in vitro* using cloned or synthetic V-gene segments and random oligonucleotides encoding part of the antigen combining site.[75,77] Using this approach it is possible to ensure that the potential V-gene repertoire diversity is large enough so that a library will only contain a single member of each sequence. In contrast, there may be considerable duplication of sequences using naturally occurring V-genes.

The filamentous bacteriophages that infect *E. coli*, such as M13 and fd phage, are the phages used for antibody display. Such phages are single stranded DNA viruses where the phage contains one copy of the viral DNA which is covered by a protein coat. The coat consists of approximately 3000 copies of the major coat protein pVIII, three to five copies of the minor coat protein pIII, and the coat proteins pVII and pIX (Figure 6.6). scFv and Fab antibody fragments have been successfully displayed on pIII, pVII, pVIII, and pIX (Figure 6.6).[78–80] Antibody phage display is accomplished by cloning antibody genes into a vector where they are in-frame with one of the coat proteins. They then are expressed as fusion proteins on the phage surface (Figure 6.6). The phage provides a physical link between the antibody fragment on its surface and the gene that encodes the antibody contained within the phage. The first example of antibody phage display was reported in 1990 by John McCafferty who demonstrated that a scFv fragment of the anti-lysozyme Mab D1.3 could be functionally displayed on pIII in the phage vector fd and that the scFv retained antigen binding (Figure 6.6).[80] Moreover, phage displaying the scFv could be enriched from a mixture with wild-type phage by affinity chromatography on a lysozyme column. Enrichment factors of 10^3 for one round of selection and 10^6 for two rounds of selection were achieved.

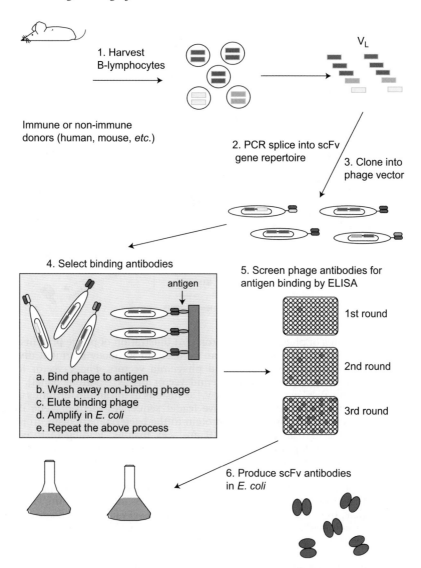

Figure 6.5 For phage display, B-cells are isolated from immunized mice (as in panel A) or naive or immunized humans. Heavy and light chain V-genes (shaded bars) are amplified by PCR and assembled as single-chain Fv antibody genes (scFv). Alternatively, rearranged V-genes can be generated entirely *in vitro* from cloned V-segments and synthetic oligonucleotides. The repertoire of scFv genes are cloned into a phage display vector, where the encoded scFv proteins (colored ovals) are displayed as fusion proteins to one of the phage coat proteins. The phage contains the appropriate scFv gene within. Multiple rounds of selection with immobilized antigen allow isolation of even rare antigen binding phage antibodies, which are identified by ELISA. Native scFv can be expressed from *E. coli* and purified for characterization and use in assays.

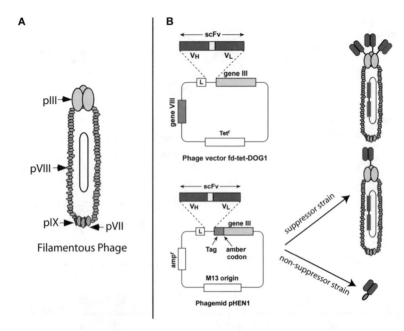

Figure 6.6 Features of phage display vectors. (A) Illustration of phage proteins pIII, pVII, pVIII, and pIX which can be used for antibody fragment display. (B) Representative phage and phagemid vectors. Phage vector fd-tet-DOG1 (top panel) and phagemid vector pHEN1 (bottom panel). Both vectors display scFv (V_H-linker-V_L) as fusions to the amino terminus of the pIII protein. Both vectors have leader sequences (pelB or gene III) to direct the expressed fusion protein to the bacterial periplasm. In phage fd-tet-DOG1, all copies of pIII are scFv fusions, leading to three–five copies displayed per phage particle. With phagemid pHEN1, expression in supressor strains of *E. coli* allows the amber codon following the scFv-tag to be read as a glutamine, causing the scFv to be fused to the pIII protein. In phagemids, both wild-type pIII (from the helper phage) and fusion pIII (from the phagemid) compete for inclusion in the viral particle. In non-suppressers *E. coli* strains, the scFv is expressed as a soluble protein with a myc epitope tag for detection of binding in ELISA.

A large number of vector systems have subsequently been described for the display of antibody fragments (Figure 6.6). The vectors differ primarily with respect to the type of antibody fragment displayed, Fab[79,81,82] or scFv,[42,48,80] the fusion partner pIII[39,82] or pVIII[79,83] and whether the vector is a phage[48,80] or phagemid.[39,42,81,82] The most widely used vectors by far are those that display scFv or Fab antibody fragments as pIII fusions in phagemid vectors (Figure 6.6). pIII display appears to be more robust than display on the other phage proteins; phagemids result in much higher transformation efficiencies, leading to larger antibody library sizes. Use of a phagemid vector results in 'monovalent' antibody fragment display compared to display in true phage vectors like fd (Figure 6.6). This is because phagemid vectors do not contain the genes for phage proteins other than pIII.

The other genes, and regulatory DNA, are provided in trans by infecting *E. coli* with helper phage. Wild-type pIII from the helper phage competes with the antibody fragment-pIII fusion for incorporation into the phage coat.

scFv or Fab gene repertoires can be cloned into phage or phagemid display vectors to create what are called 'libraries' of phage, each with a different antibody fragment on its surface and the gene for the antibody fragment inside (see Figure 6.5). Given the high transformation of *E. coli*, it is possible to create libraries containing millions to billions of different antibodies (Figure 6.5). Rare antibodies binding specific antigens can be isolated from non-binding antibodies by a range of different types of affinity chromatography. For example, antigen can be absorbed to a plastic plate and then phage incubated with immobilized antigen, non-binding phage removed by washing, and bound phage eluted. A single round of selection will result in a 20 to 1000 fold enrichment for binding phage.[80] Eluted phage are used to infect *E. coli*, which produce more phage for the next round of selection. Repetition of the selection process makes it possible to isolate binding phage present at frequencies of less than 1 in a billion.

Adsorption of protein antigens to plastic may lead to partial or complete antigen denaturation. The problem of antigen denaturation can be overcome by selecting on soluble antigen in solution. The antigen can be chemically tagged, for example by biotinylation, and captured along with bound phage using avidin or streptavidin magnetic beads. Alternatively, the antigen can be genetically tagged with a hexahistidine tag with capture on Ni-NTA agarose, or expressed as GST or maltose binding protein fusions with capture on either glutathione or maltose columns. Selections can be performed on more complex mixtures of antigen, including intact cells, as long as measures are taken to prevent enrichment of phage antibodies which bind to non-relevant cell-surface antigens. Cell surface antigen specific antibodies have been isolated by selecting phage antibody libraries on adult erythrocytes,[84] fetal erythrocytes,[85] lymphocytes,[86] melanoma cells,[87] and breast tumor cells.[88] It has also proven possible to directly select phage antibodies that trigger receptor mediated endocytosis.[88] Such antibodies can be used for the targeted delivery of therapeutics to the cytosol.[89,90]

6.6.1.1 Bypassing Hybridoma Technology Using Phage Display. Phage display can be used to bypass hybridoma technology to make scFv or Fab antibodies from immunized mice, rabbits, chickens, or humans, or from humans who have mounted an immune response as a result of infection or a disease process. Immune lymphocytes are harvested from peripheral blood lymphocytes, spleen, or bone marrow and RNA prepared. scFv or Fab gene repertoires are prepared using PCR as described above and used to construct phage antibody libraries from which antigen specific antibodies are isolated using affinity chromatography.

In the first example, a scFv phage antibody library was created from the V-genes of a mouse immunized with the hapten phenyloxazolone.[48] An scFv phage antibody library was created from murine V-genes and binding phage antibodies were isolated by selection on an antigen column. After two rounds of selection, more than twenty unique scFv were isolated. The K_d of the highest affinity phage antibodies (1.0×10^{-8} M) were comparable to the affinities of IgG from hybridomas constructed from mice immunized with the same hapten. Similar panels of scFv have been obtained using phage display and mice immunized with EGF receptor[91] and Botulinum neurotoxin type A.[35] This approach has also been used to produce monoclonal chicken[46] and rabbit[47] antibody fragments using species specific primers.

Antigen specific antibodies can also be isolated from phage antibody libraries constructed from the V-genes of immunized humans or from humans mounting an immune response to an infection or disease process. Since hybridoma technology has not worked well to generate human hybridomas, this approach is especially useful for generation of human antibodies where an immune response can be induced or is mounted. Human monoclonal antibody fragments have been isolated from immunized volunteers or infected patients against Tetanus toxin,[81] botulinum neurotoxin,[92] HIV-1,[93] Hepatitis B,[94] hepatitis C,[95] respiratory syncytial virus,[96] and hemophilus influenza.[97] Autoimmune antibodies have been isolated from patients with SLE[98] and myasthenia gravis,[99] as well as other autoimmune diseases. There has been success in isolating antibodies to self-antigens, including tumor antigens, from libraries constructed from patients with disease[100] or after vaccination with tumor antigen.[87]

6.6.1.2 *Bypassing Immunization Using Phage Display.*

In many instances, it is not feasible or ethical to immunize humans with antigens to which one desires to produce human antibodies. In additions, many antigens are evolutionarily conserved and not immunogenic. Fortunately, phage display technology offers a route to directly produce human antibodies without immunization. Very large and diverse scFv or Fab phage antibody libraries are constructed from which it is theoretically possible to generate panels of antigens to virtually any antigen. In the first example, a human scFv gene repertoire was cloned into the phagemid pHEN-1 to create a non-immune phage antibody library of 3.0×10^7 members.[42] From this single non-immune library, scFv were isolated against more than 20 different antigens, including a hapten, three different polysaccharides, and 16 different proteins. The scFv were highly specific for the antigen used for selection and had affinities typical of the primary immune response with dissociation equilibrium constants ranging from 1 µM to 15 nM. Larger or more diverse phage antibody libraries theoretically provide higher affinity antibodies against a greater number of epitopes on all antigens used for selection.[101] Many published examples of such very large libraries exist, with sizes ranging between 10^9 and 10^{11}. For example, we have constructed a 6.7×10^9 member scFv phage antibody library from the V-genes of healthy humans.[102] An average of nine scFv with K_d as high as 3.7×10^{-10} M were isolated to 10 different protein antigens. Antibodies from non-immune libraries have been used for Western blotting, epitope mapping, cell agglutination assays, cell staining and FACS. A number have also entered human clinical trials for the treatment of a range of different diseases, with phage antibodies adalimumab and belimumab approved by the FDA to treat rheumatoid arthritis and systemic lupus erythematosus, respectively.

Improvements in the performance of non-immune phage antibody libraries have focused on increasing functional library size. The quality (number and affinity) of antibodies from phage antibody libraries is largely dependent on functional library size (the number of unique expressing Abs).[103] Initially libraries were made from natural repertoires extracted from human mRNA.[42,102,104] Advantages of such libraries included the absence of stop codons and V-domain structures shaped by evolution and biased amino acid usage to encode functional domains.[105] Disadvantages include an inability to control redundancy resulting from mRNA amplification[106] and V, D, J diversification mechanisms that produce correlations between adjacent amino acids,[107] and low stability and/or *E. coli* display or expression levels inherent in some of the V-genes.[108,109] More recently, libraries have been built on a small number or even single V_H and V_K domains that express well in bacteria and are stable. Degenerate synthetic DNA, frequently encoding a limited amino acid set, is used to introduce diversity into the antigen binding complementarity determining regions (CDRs).[110–113] As with the natural repertoire much or all of this diversity is introduced into the V_H CDR3 (H3), whose length varies from four to >25 amino acids. Advantages of synthetic repertories include greater genetic diversity reducing redundant sequences and the potential for better Ab stability and expression levels. However the use of a restricted V-domain set limits the canonical structural diversity of V_H CDR1 and 2 and V_L CDR1 perhaps restricting the diversity of antigen specific rAbs. In addition, without natural selection mechanisms to select for functional diversity, synthetic library designs are vulnerable to generating non-functional rAbs, especially when multiple CDRs are diversified.[111,112,114,115] Not surprisingly, it has been observed that biases that occur in natural repertoires are selected for in the functional binding rAbs recovered from synthetic libraries.[102,110] As a result, including known biases from the natural Ab repertoire has been demonstrated to contribute to successful synthetic designs[110,111,116] and has the potential to combine the best of both library types.

The non-immune libraries described above represent the current state of the art in antibody engineering; the ability to generate high affinity human monoclonal scFv or Fab antibody fragments to any antigen within weeks and without immunization.

6.6.2 Alternative Display Technologies

While a number of alternative display technologies have been developed, the two most widely used are yeast display and ribosome display.[117–120] For yeast display, scFv antibody gene repertoires are fused to the Saccharomyces cell surface protein AgaII.[117] Antigen specific scFv are isolated by subjecting the yeast antibody library to fluorescent activated cell sorting (FACS) using flow cytometry. Yeast display has been used to generate lead antibodies from both immune and non-immune antibody libraries.[121,122] Its major use, however, has been to increase the affinity and cross-reactivity of antibodies (antibody affinity maturation, see Section 6.7).[123–126] In the most impressive example of affinity maturation using yeast display, the affinity of an scFv was increased more than 1000 fold to 48 femptomolar.[123] Potential advantages of yeast compared to phage display include the ability to more efficiently select higher affinity antibodies using flow cytometry and the ability to measure antibody fragment affinity with the antibody in the display format. This obviates the need for antibody fragment expression and purification. Potential disadvantages of yeast display include the lower transformation efficiencies of yeast compared to bacteria.

In ribosome display, scFv antibody gene repertoires encoded by mRNA are translated *in vitro* in cell free systems such that the mRNA remaining attached to the ribosome along with the scFv.[119] The ribosome provides the physical link between genotype and phenotype, taking the place of the phage. Antigen binding scFv are selected on antigen, and the mRNA encoding the scFv genes amplified by RT-PCR. Transcription and translation of the scFv DNA provide the display library for the next round of selection. The major advantage of ribosome display is that no cloning is required resulting in libraries much greater in size than those created by cloning. Ribosome display has been successfully applied to generate scFv from immune[127] and non-immune[128] libraries and for affinity maturation of existing antibodies.[128]

6.7 ENGINEERING ANTIBODY AFFINITY

Display technologies have been widely used to increase antibody affinity to values not typically achievable using hybridoma technology.[129,130] Mutations are introduced into the DNA of an antigen-specific antibody fragment either randomly, for example using error prone PCR[131] or specifically using spiked oligonucleotides.[132] The mutated antibody fragment DNA is then used to create a display library either in phage, yeast, or by using ribosome display. Higher affinity antibody fragments are selected from lower affinity ones by a variety of different affinity chromatography approaches, including flow cytometry for yeast-display libraries. For example, we reported increasing the affinity of a HER2 scFv using phage display more than 1000 fold from 16 nM to 13 pM.[132] Others have reported comparable levels of affinity maturation to values down to low pM to femptomolar affinities using either phage or yeast display.[123,133] These *in vitro* affinity maturation approaches have now become routine in the biotechnology field.

6.8 IMPROVING ANTIBODY MANUFACTURING PROPERTIES

An essential therapeutic antibody feature is that can be manufactured at high titer (typically greater than 1 gram/L of cell culture) and that it is soluble and stable once formulated and does not aggregate. Depending on the variable region sequence, antibodies can differ significantly in their 'manufacturability' and stability, features that have been termed 'developability'.[134,135] Identification of a 'poorly behaved' antibody late in the development process can result in significant efforts to correct,[136] costing time and money, or result in the need to identify new lead antibodies. Two parallel approaches have been used to identify and improve developability earlier, designing antibody libraries based on stable variable domains that express well[113,137] and selection or screening technologies[138–140] to identify poorly behaved antibodies early in the discovery process.

6.9 ENHANCING ANTIBODY POTENCY

While antibodies are an effective class of therapeutics, there is certainly room for increasing their potency. This is especially true in the field of cancer therapy. The response rate of antibodies, either as single drugs or in combination with chemotherapy, ranges from 10–50%. Existing antibody oncologics work by interrupting essential signaling pathways, such as the epidermal growth factor receptor signaling family, and/or by eliciting ADCC, CDC, or apoptosis. There are a number of approaches under investigation to enhance antibody potency by either conjugating antibodies to toxic compounds or by engineering the Fc portion of the antibody to increase ADCC or CDC.

The concept of enhancing antibody potency *via* delivery of a toxic payload dates to the time of Paul Ehrlich, at the turn of the century. Ehrlich conceptualized magic bullets, molecules that could selectively target a disease-causing organism and deliver a toxic payload. One of the first examples of such antibody delivery of toxic payloads are immunotoxins, antibodies or antibody fragments fused to toxins.[141] Toxins that have been utilized, such as pseudomonas exotoxin, are highly toxic, with only a few molecules being capable of cell killing. Therapeutic development of immunotoxins has required overcoming a number of obstacles, including immunogenicity of bacterial and plant toxins when administered to humans, the high toxicity of the toxins used for immunotoxin construction, and development of methods to fuse toxins to antibodies or antibody fragments.[142] Discovery of scFv antibody fragments made for natural fusion partners to toxins which could now be encoded in a single gene and expressed as a single polypetide chain.[143] While immunogenicity still remains an issue, immunotoxins are now in clinical trials for cancer, and may be especially efficacious against hematologic malignancies.[144] An alternative approach is to deliver chemotherapeutic drugs to antibodies rather than toxins, either directly conjugated to antibodies (antibody drug conjugates) or *via* antibody-targeted drugs containing nanoparticles, such as immunoliposomes.[89,90] Such anti-body-targeted drug conjugates show dramatic therapeutic affects in pre-clinical models and are currently in human clinical trials as cancer therapies.[145] Recently, two antibody–drug conjugates, trastuzumab emtan-sine and brentuximab vedotin, were approved by the FDA for treatment of HER2 positive breast cancer and Hodgkin's lymphoma.[23,24]

An alternative way to increase the potency of antibodies is to enhance their inherent cell killing activities that occur *via* the Fc portion of the antibody and Fc receptors. This interaction is responsible for antibody dependent cellular cytotoxicity (ADCC) and complement dependent cellular cytotoxicity (CDC).[146,147] The Fc binds to both activating and inhibitory Fc receptors and, depending on the affinity of the specific interactions, cell killing occurs. Recently, it has proven possible to engineer the Fc to more efficiently bind activating Fc receptors compared to inhibitory Fc receptors.[22] Such engin-eered Fc's show enhanced ADCC in preclinical models of cancer and will be evaluated for enhanced efficacy in human clinical trials.

6.10 ANTIBODY ENGINEERING TOOLS ON THE HORIZON

Two promising tools that may significantly enhance the discovery and op-timization of antibodies are next generation high throughput DNA se-quencing and the rationale (computational) design of antibodies. DNA sequencing has become a commodity and it is now possible to sequence antibody libraries and repertoires at a depth of millions of sequences at an affordable price. This approach has already been applied to better define the diversity of natural antibody repertoires and design more functional anti-body libraries.[116] It is also being deployed to identify binding antibodies from immune repertoires and is being explored as a tool to identify rare antibodies after antibody library selection.[148,149] Rational antibody design applies the tools of structural and computational biology to either the *de novo* design of antibodies binding a target antigen or to modulate the binding of antibody to antigen (higher affinity, alerting specificity). While ligand-binding proteins have been designed,[150] this has yet to be accom-plished for antibodies and the importance and role of this tool remains to be demonstrated.

6.11 SUMMARY

The field of antibody engineering effectively began with the engineering of the first chimeric antibodies back in 1984. In the ensuing 30 years, we have seen the development of humanized antibodies as well as library and transgenic approaches to directly generate fully human antibodies to virtu-ally any antigen, including those that are evolutionarily conserved. Library techniques have also developed to the point where it is possible to tune the antibody binding site to virtually any affinity for antigen. At the same time, these bench-side scientific achievements have translated into a novel therapeutic pipeline, biologic antibodies. There are now 30 therapeutic antibodies approved by the FDA and in clinical use, with at least a hundred more in the various stages of clinical trials. While once the purview of bio-technology companies, essentially every major pharmaceutical company now has a significant effort in the field of therapeutic antibodies. Antibodies, as protein therapeutics are capable of interrupting protein–protein inter-actions, something difficult to achieve with traditional small molecule drugs. It is thus reasonable to expect that antibodies will continue to be an important and evolving novel class of therapeutics. We are now seeing the next generation of engineered antibodies entering clinical trials. These antibodies are 'armed' to be more potent by conjugation to toxins or toxic drugs, or by increased effector functions resulting from engineering of the antibody Fc. It is reasonable to expect that these advances, as well as others, will result in antibodies with even greater therapeutic efficacy.

REFERENCES

1. D. S. Dimitrov and J. D. Marks, *Methods Mol. Biol.*, 2009, **525**, 1–27.
2. J. M. Reichert, *mAbs*, 2014, **6**, 5–14.
3. G. Kohler and C. Milstein, *Nature*, 1975, **256**, 495–497.
4. K. James and G. T. Bell, *J. Immunol. Meth.*, 1987, **100**, 5–40.
5. G. J. Jaffers, T. C. Fuller, A. B. Cosimi, P. S. Russel, H. J. Winn and R. B. Colvin, *Transplantation*, 1986, **41**, 572–578.
6. C. H. J. Lamers, J. W. Gratama, S. O. Warnaar, G. Stoter and R. L. H. Bolhuis, *Int. J. Cancer*, 1995, **60**, 450–457.
7. M. S. Neuberger, G. T. Williams, E. B. Mitchell, S. S. Jouhal, J. G. Flanagan and T. H. Rabbitts, *Nature*, 1985, **314**, 268–270.
8. J. Sharon, M. L. Gefter, T. Manser, S. L. Morrison, V. T. Oi and M. Ptashne, *Nature*, 1984, **309**, 364–367.
9. S. Takeda, T. Naito, K. Hama, T. Noma and T. Honjo, *Nature*, 1985, **314**, 452–454.
10. A. R. Bradbury and J. D. Marks, *J. Immunol. Methods*, 2004, **290**, 29–49.
11. J. D. Marks, H. R. Hoogenboom, A. D. Griffiths and G. Winter, *J. Biol. Chem.*, 1992, **267**, 16007–16010.
12. M. J. Mendez, L. L. Green, J. R. Corvalan, X. C. Jia, C. E. Maynard-Currie, X. D. Yang, M. L. Gallo, D. M. Louie, D. V. Lee, K. L. Erickson, J. Luna, C. M. Roy, H. Abderrahim, F. Kirschenbaum, M. Noguchi, D. H. Smith, A. Fukushima, J. F. Hales, S. Klapholz, M. H. Finer, C. G. Davis, K. M. Zsebo and A. Jakobovits, *Nat. Genet.*, 1997, **15**, 146–156.
13. A. Casadevall and M. D. Scharff, *Antimicrob. Agents Chemother.*, 1994, **38**, 1695–1702.
14. A. Casadevall and M. D. Scharff, *Clin. Infect. Dis.*, 1995, **21**, 150–161.
15. A. Casadevall, *Clin. Immunol.*, 1999, **93**, 5–15.
16. J. Bayry, S. Lacroix-Desmazes, M. D. Kazatchkine and S. V. Kaveri, *Trends Pharmacol. Sci.*, 2004, **25**, 306–310.

17. L. A. Sawyer, *Antiviral Res.*, 2000, **47**, 57–77.

18. R. E. Black and R. A. Gunn, *Am. J. Med.*, 1980, **69**, 567–570.

19. J. M. Reichert and V. E. Valge-Archer, *Nat. Rev. Drug Discovery*, 2007, **6**, 349–356.

20. E. L. Sievers and P. D. Senter, *Annu. Rev. Med.*, 2013, **64**, 15–29.

21. I. Pastan, R. Hassan, D. J. Fitzgerald and R. J. Kreitman, *Nat. Rev. Cancer*, 2006, **6**, 559–565.

22. G. A. Lazar, W. Dang, S. Karki, O. Vafa, J. S. Peng, L. Hyun, C. Chan, H. S. Chung, A. Eivazi, S. C. Yoder, J. Vielmetter, D. F. Carmichael, R. J. Hayes and B. I. Dahiyat, *Proc. Natl. Acad. Sci. U. S. A.*, 2006, **103**, 4005–4010.

23. A. Ballantyne and S. Dhillon, *Drugs*, 2013, **73**, 755–765.

24. A. Younes, *Hematol. Oncol. Clin. North Am.*, 2014, **28**, 27–32.

25. E. A. Kabat and T. T. Wu, *Ann. NY Acad. Sci.*, 1971, **190**, 382–393.

26. I. M. Tomlinson, J. P. Cox, E. Gherardi, A. M. Lesk and C. Chothia, *EMBO J.*, 1995, **14**, 4628–4638.

27. I. M. Tomlinson, G. Walter, J. D. Marks, M. B. Llewelyn and G. Winter, *J. Mol. Biol.*, 1992, **227**, 776–798.

28. C. Chothia and A. M. Lesk, *J. Mol. Biol.*, 1987, **196**, 901–917.

29. C. Chothia, A. M. Lesk, E. Gherardi, I. M. Tomlinson, G. Walter, J. D. Marks, M. Llewelyn and G. Winter, *J. Mol. Biol.*, 1992, **227**, 799–817.

30. G. L. Boulianne, N. Hozumi and M. J. Shulman, *Nature*, 1984, **312**, 643–646.

31. S. L. Morrison, M. J. Johnson, L. A. Herzenberg and V. T. Oi, *Proc. Natl. Acad. Sci. U. S. A.*, 1984, **81**, 6851–6855.

32. J. W. Larrick, Y. L. Chiang, R. Sheng-Dong, G. Senck and P. Casali, in *In vitro Immunisation in Hybridoma Technology*, ed. C. A. K. Borrebaeck, Elsevier Science Publishers B.V., Amsterdam, 1988, pp. 231–246.

33. J. W. Larrick, L. Danielsson, C. A. Brenner, M. Abrahamson, K. E. Fry and C. A. Borrebaeck, *Biochem. Biophys. Res. Commun.*, 1989, **160**, 1250–1256.

34. R. Orlandi, D. H. Gussow, P. T. Jones and G. Winter, *Proc. Natl. Acad. Sci. USA*, 1989, **86**, 3833–3837.

35. P. Amersdorfer, C. Wong, S. Chen, T. Smith, S. Desphande, R. Sheridan, R. Finnern and J. D. Marks, *Infect. Immun.*, 1997, **65**, 3743–3752.

36. S. A. Iverson, L. Sastry, W. D. Huse, J. A. Sorge, S. J. Benkovic and R. A. Lerner, *Cold Spring Harbor Symp. Quant. Biol.*, 1989, **1**, 273–281.

37. C. A. Kettleborough, J. Saldanha, K. H. Ansell and M. M. Bendig, *Eur. J. Immunol.*, 1993, **23**, 206–211.

38. R. D. LeBoeuf, F. S. Galin, S. K. Hollinger, S. C. Peiper and J. E. Blalock, *Gene*, 1989, **82**, 371–377.

39. H. Orum, P. S. Andersen, A. Oster, L. K. Johansen, E. Riise, M. Bjornvad, I. Svendsen and J. Enberg, *Nucl. Acids Res.*, 1993, **21**, 4491–4498.

40. L. Sastry, M. M. Alting, W. D. Huse, J. M. Short, J. A. Sorge, B. N. Hay, K. D. Janda, S. J. Benkovic and R. A. Lerner, *Proc. Natl. Acad. Sci. U. S. A.*, 1989, **86**, 5728–5732.

41. D. R. Burton, C. F. Barbas, M. A. A. Persson, S. Koenig, R. M. Chanock and R. A. Lerner, *Proc. Natl. Acad. Sci. U. S. A.*, 1991, **88**, 10134–10137.

42. J. D. Marks, H. R. Hoogenboom, T. P. Bonnert, J. McCafferty, A. D. Griffiths and G. Winter, *J. Mol. Biol.*, 1991, **222**, 581–597.

43. J. D. Marks, M. Tristrem, A. Karpas and G. Winter, *Eur. J. Immunol.*, 1991, **21**, 985–991.

44. M. A. Persson, R. H. Caothien and D. R. Burton, *Proc. Natl. Acad. Sci. U. S. A.*, 1991, **88**, 2432–2436.

45. D. Sblattero and A. Bradbury, *Nature Biotech.*, 2000, **18**, 75–80.

46. E. Davies, J. Smith, C. Birkett, J. Manser, D. Anderson-Dear and J. Young, *J. Immunol. Meth.*, 1995, **186**, 125–135.

47. I. Lang, C. r. Barbas and R. Schleef, *Gene*, 1996, **172**, 295–298.

48. T. Clackson, H. R. Hoogenboom, A. D. Griffiths and G. Winter, *Nature*, 1991, **352**, 624–628.

49. J. B. Edwards, J. Delort and J. Mallet, *Nucl. Acids Res.*, 1991, **19**, 5227–5232.

50. A. Heinrichs, C. Milstein and E. Gherardi, *J. Immunol. Meth.*, 1995, **178**, 241–251.

51. F. Ruberti, A. Cattaneo and A. Bradbury, *J. Immunol. Meth.*, 1994, **173**, 33–39.

52. S. M. Kipriyanov, O. A. Kupriyanova and G. Moldenhauer, *J. Immunol. Meth.*, 1996, **13**, 51–62.

53. M. A. Boss, J. H. Kenten, C. R. Wood and J. S. Emtage, *Nucl. Acids Res.*, 1984, **12**, 3791–3806.

54. S. Cabilly, A. D. Riggs, H. Pande, J. E. Shively, W. E. Holmes, M. Rey, L. J. Perry, R. Wetzel and H. L. Heyneker, *Proc. Natl. Acad. Sci. U. S. A.*, 1984, **81**, 3273–3277.

55. M. Better, C. P. Chang, R. R. Robinson and A. H. Horwitz, *Science*, 1988, **240**, 1041–1043.

56. A. Skerra and A. Pluckthun, *Science*, 1988, **240**, 1038–1041.

57. R. Glockshuber, M. Malia, I. Pfitzinger and A. Pluckthun, *Biochemistry*, 1990, **29**, 1362–1367.

58. C. Horne, M. Klein, I. Polidoulis and K. J. Dorrington, *J. Immunol.*, 1982, **129**, 660–664.

59. R. E. Bird, K. D. Hardman, J. W. Jacobson, S. Johnson, B. M. Kaufman, S. M. Lee, T. Lee, S. H. Pope, G. S. Riordan and M. Whitlow, *Science*, 1988, **242**, 423–426.

60. J. S. Huston, D. Levinson, H. M. Mudgett, M. S. Tai, J. Novotny, M. N. Margolies, R. J. Ridge, R. E. Bruccoleri, E. Haber, R. Crea and H. Oppermann, *Proc. Natl. Acad. Sci. U. S. A.*, 1988, **85**, 5879–5883.

61. Y. Reiter, U. Brinkmann, S. Jung, I. Pastan and B. Lee, *Protein Eng.*, 1995, **8**, 1323–1331.

62. M. B. Khazaeli, R. M. Conry and A. F. LoBuglio, *J. Immunother. Emphasis Tumor Immunol.*, 1994, **15**, 42–52.

63. K. Kuus-Reichel, L. S. Grauer, L. M. Karavodin, C. Knott, M. Krusemeier and N. E. Kay, *Clin. Diagn. Lab. Immunol.*, 1994, **1**, 365–372.

64. W. Y. Hwang and J. Foote, *Methods*, 2005, **36**, 3–10.
65. P. T. Jones, P. H. Dear, J. Foote, M. S. Neuberger and G. Winter, *Nature*, 1986, **321**, 522–525.
66. L. Riechmann, M. Clark, H. Waldmann and G. Winter, *Nature*, 1988, **332**, 323–327.
67. P. R. Tempest, P. Bremner, M. Lambert, G. Taylor, J. M. Furze, F. J. Carr and W. J. Harris, *Biotechnology (N Y)*, 1991, **9**, 266–271.
68. C. Queen, W. P. Schneider, H. E. Selick, P. W. Payne, N. F. Landolfi, J. F. Duncan, N. M. Avdalovic, M. Levitt, R. P. Junghans and T. A. Waldmann, *Proc. Natl. Acad. Sci. U. S. A.*, 1989, **86**, 10029–10033.
69. A. B. Thakur and N. F. Landolfi, *Mol. Immunol.*, 1999, **36**, 1107–1115.
70. P. Carter, L. Presta, C. M. Gorman, J. B. Ridgway, D. Henner, W. L. Wong, A. M. Rowland, C. Kotts, M. E. Carver and H. M. Shepard, *Proc. Natl. Acad. Sci. U. S. A.*, 1992, **89**, 4285–4289.
71. L. G. Presta, H. Chen, S. J. O'Connor, V. Chisholm, Y. G. Meng, L. Krummen, M. Winkler and N. Ferrara, *Cancer Res.*, 1997, **57**, 4593–4599.
72. P. Tan, D. A. Mitchell, T. N. Buss, M. A. Holmes, C. Anasetti and J. Foote, *J. Immunol.*, 2002, **169**, 1119–1125.
73. M. Figini, J. D. Marks, G. Winter and A. D. Griffiths, *J. Mol. Biol.*, 1994, **239**, 68–78.
74. L. S. Jespers, A. Roberts, S. M. Mahler, G. Winter and H. R. Hoogenboom, *Biotechnology (N Y)*, 1994, **12**, 899–903.
75. H. R. Hoogenboom, J. D. Marks, A. D. Griffiths and G. Winter, *Immunol. Rev.*, 1992, **130**, 41–68.
76. C. Marks and J. D. Marks, *N. Engl. J. Med.*, 1996, **335**, 730–733.
77. A. Nissim, H. R. Hoogenboom, I. M. Tomlinson, G. Flynn, C. Midgley, D. Lane and G. Winter, *EMBO J.*, 1994, **13**, 692–698.
78. C. Gao, S. Mao, C. H. Lo, P. Wirsching, R. A. Lerner and K. D. Janda, *Proc. Natl. Acad. Sci. U. S. A.*, 1999, **96**, 6025–6030.
79. A. S. Kang, C. F. Barbas, K. D. Janda, S. J. Benkovic and R. A. Lerner, *Proc. natl. Acad. Sci. U. S. A.*, 1991, **88**, 4363–4366.
80. J. McCafferty, A. D. Griffiths, G. Winter and D. J. Chiswell, *Nature*, 1990, **348**, 552–554.
81. C. F. Barbas, A. S. Kang, R. A. Lerner and S. J. Benkovic, *Proc. Natl. Acad. Sci. U. S. A.*, 1991, **88**, 7978–7982.
82. H. R. Hoogenboom, A. D. Griffiths, K. S. Johnson, D. J. Chiswell, P. Hudson and G. Winter, *Nucl. Acids Res.*, 1991, **19**, 4133–4137.
83. W. Huse, T. Stinchcombe, S. Glaser, L. M. Starr, K. Hellstrom, I. Hellstrom and D. Yelton, *J. Immunol.*, 1992, **149**, 3914–3920.
84. J. D. Marks, W. H. Ouwehand, J. M. Bye, R. Finnern, B. D. Gorick, D. Voak, S. Thorpe, N. C. Hughes-Jones and G. Winter, *Bio/Technology*, 1993, **11**, 1145–1149.
85. M. A. Huie, M. C. Cheung, M. O. Muench, B. Becerril, Y. W. Kan and J. D. Marks, *Proc. Natl. Acad. Sci. U. S. A.*, 2001, **98**, 2682–2687.
86. J. de Kruif, L. Terstappen, E. Boel and T. Logtenberg, *Proc. Natl. Acad. Sci. U. S. A.*, 1995, **92**, 3938–3942.

87. X. Cai and A. Garen, *Proc. Natl. Acad. Sci. U. S. A.*, 1995, **92**, 6537–6541.

88. M. A. Poul, B. Becerril, U. B. Nielsen, P. Morisson and J. D. Marks, *J. Mol. Biol.*, 2000, **301**, 1149–1161.

89. U. B. Nielsen, D. B. Kirpotin, E. M. Pickering, K. Hong, J. W. Park, M. Refaat Shalaby, Y. Shao, C. C. Benz and J. D. Marks, *Biochim. Biophys. Acta*, 2002, **1591**, 109–118.

90. J. W. Park, K. Hong, D. B. Kirpotin, G. Colbern, R. Shalaby, J. Baselga, Y. Shao, U. B. Nielsen, J. D. Marks, D. Moore, D. Papahadjopoulos and C. C. Benz, *Clin. Cancer Res.*, 2002, **8**, 1172–1181.

91. C. Kettleborough, K. Ansell, R. Allen, E. Rosell-Vives, D. Gussow and M. Bendig, *Eur. J. Immunol.*, 1994, **24**, 952–958.

92. P. Amersdorfer, C. Wong, T. Smith, S. Chen, S. Deshpande, R. Sheridan and J. D. Marks, *Vaccine*, 2002, **20**, 1640–1648.

93. C. F. Barbas, T. A. Collet, W. Amberg, P. Roben, J. M. Binley, D. Hoekstra, D. Cababa, T. M. Jones, A. Williamson, G. R. Pilkington, N. L. Haigwood, E. Cabezas, A. C. Satterthwait, I. Sanz and D. R. Burton, *J. Mol. Biol.*, 1993, **230**, 812–823.

94. S. L. Zebedee, C. F. Barbas, Y.-L. Hom, R. H. Cathoien, R. Graff, J. DeGraw, J. Pyatt, R. LaPolla, D. R. Burton and R. A. Lerner *et al.*, *Proc. Natl. Acad. Sci. U. S. A.*, 1992, **89**, 3175–3179.

95. S. Chan, J. Bye, P. Jackson and J. Allain, *J. Gen. Virol.*, 1996, **10**, 2531–2539.

96. C. Barbas, J. Crowe, D. Cababa, T. Jones, S. Zebedee, B. Murphy, R. Chanock and D. Burton, *Proc. Natl. Acad. Sci. U. S. A.*, 1992, **89**, 10164–10168.

97. D. Reason, T. Wagner and A. Lucas, *Infect. Immun.*, 1997, **65**, 261–266.

98. S. Barbas, H. Ditzel, E. Salonen, W. Yang, G. Silverman and D. Burton, *Proc. Natl. Acad. Sci. U. S. A.*, 1995, **92**, 2529–2533.

99. Y. Graus, M. de Baets, P. Parren, S. Berrih-Aknin, J. Wokke, v. Breda, P. Vriesman and D. Burton, *J. Immunol.*, 1997, **158**, 1919–1929.

100. M. A. Clark, N. J. Hawkins, A. Papaioannou, R. J. Fiddes and R. L. Ward, *Clin. Exp. Immunol.*, 1997, **109**, 166–174.

101. A. S. Perelson and G. F. Oster, *J. Theor. Biol.*, 1979, **81**, 645–670.

102. M. D. Sheets, P. Amersdorfer, R. Finnern, P. Sargent, E. Lindquist, R. Schier, G. Hemingsen, C. Wong, J. C. Gerhart and J. D. Marks, *Proc. Natl. Acad. Sci. U. S. A.*, 1998, **95**, 6157–6162.

103. A. D. Griffiths, S. C. Williams, O. Hartley, I. M. Tomlinson, P. Waterhouse, W. L. Crosby, R. E. Kontermann, P. T. Jones, N. M. Low, T. J. Allison, T. D. Prospero, H. R. Hoogenboom, A. Nissim, J. P. L. Cox, J. L. Harrison, M. Zaccolo, E. Gherardi and G. Winter, *EMBO J.*, 1994, **13**, 3245–3260.

104. T. J. Vaughan, A. J. Williams, K. Pritchard, J. K. Osbourn, A. R. Pope, J. C. Earnshaw, J. McCafferty, R. A. Hodits, J. Wilton and K. S. Johnson, *Nature Biotechnol.*, 1996, **14**, 309–314.

105. M. Zemlin, M. Klinger, J. Link, C. Zemlin, K. Bauer, J. A. Engler, H. W. Schroeder, Jr. and P. M. Kirkham, *J. Mol. Biol.*, 2003, **334**, 733–749.

106. N. Jiang, J. A. Weinstein, L. Penland, R. A. White, 3rd, D. S. Fisher and S. R. Quake, *Proc. Natl. Acad. Sci. U. S. A.*, 2011, **108**, 5348–5353.
107. T. Mora, A. M. Walczak, W. Bialek and C. G. Callan, Jr., *Proc. Natl. Acad. Sci. U. S. A.*, 2010, **107**, 5405–5410.
108. S. Ewert, A. Honegger and A. Pluckthun, *Biochemistry*, 2003, **42**, 1517–1528.
109. S. Ewert, T. Huber, A. Honegger and A. Pluckthun, *J. Mol. Biol.*, 2003, **325**, 531–553.
110. F. A. Fellouse, K. Esaki, S. Birtalan, D. Raptis, V. J. Cancasci, A. Koide, P. Jhurani, M. Vasser, C. Wiesmann, A. A. Kossiakoff, S. Koide and S. S. Sidhu, *J. Mol. Biol.*, 2007, **373**, 924–940.
111. B. J. Hackel, M. E. Ackerman, S. W. Howland and K. D. Wittrup, *J. Mol. Biol.*, 2010, **401**, 84–96.
112. A. Knappik, L. Ge, A. Honegger, P. Pack, M. Fischer, G. Wellnhofer, A. Hoess, J. Wolle, A. Pluckthun and B. Virnekas, *J. Mol. Biol.*, 2000, **296**, 57–86.
113. S. S. Sidhu, B. Li, Y. Chen, F. A. Fellouse, C. Eigenbrot and G. Fuh, *J. Mol. Biol.*, 2004, **338**, 299–310.
114. C. Rothe, S. Urlinger, C. Lohning, J. Prassler, Y. Stark, U. Jager, B. Hubner, M. Bardroff, I. Pradel, M. Boss, R. Bittlingmaier, T. Bataa, C. Frisch, B. Brocks, A. Honegger and M. Urban, *J. Mol. Biol.*, 2008, **376**, 1182–1200.
115. S. S. Sidhu and F. A. Fellouse, *Nat. Chem. Biol.*, 2006, **2**, 682–688.
116. W. Zhai, J. Glanville, M. Fuhrmann, L. Mei, I. Ni, P. D. Sundar, T. Van Blarcom, Y. Abdiche, K. Lindquist, R. Strohner, D. Telman, G. Cappuccilli, W. J. Finlay, J. Van den Brulle, D. R. Cox, J. Pons and A. Rajpal, *J. Mol. Biol.*, 2011, **412**, 55–71.
117. E. T. Boder and K. D. Wittrup, *Nat. Biotechnol.*, 1997, **15**, 553–557.
118. P. M. Bowers, R. A. Horlick, M. R. Kehry, T. Y. Neben, G. L. Tomlinson, L. Altobell, X. Zhang, J. L. Macomber, I. P. Krapf, B. F. Wu, A. D. McConnell, B. Chau, A. D. Berkebile, E. Hare, P. Verdino and D. J. King, *Methods*, 2014, **65**, 44–56.
119. C. Schaffitzel, J. Hanes, L. Jermutus and A. Pluckthun, *J. Immunol. Methods*, 1999, **231**, 119–135.
120. E. S. Smith and M. Zauderer, *Curr. Drug Discovery Technol.*, 2014, **11**, 48–55.
121. D. R. Bowley, A. F. Labrijn, M. B. Zwick and D. R. Burton, *Protein Eng. Des. Sel.*, 2007, **20**, 81–90.
122. M. J. Feldhaus, R. W. Siegel, L. K. Opresko, J. R. Coleman, J. M. Feldhaus, Y. A. Yeung, J. R. Cochran, P. Heinzelman, D. Colby, J. Swers, C. Graff, H. S. Wiley and K. D. Wittrup, *Nat. Biotechnol.*, 2003, **21**, 163–170.
123. E. T. Boder, K. S. Midelfort and K. D. Wittrup, *Proc. Natl. Acad. Sci. U. S. A.*, 2000, **97**, 10701–10705.
124. A. Razai, C. Garcia-Rodriguez, J. Lou, I. N. Geren, C. M. Forsyth, Y. Robles, R. Tsai, T. J. Smith, L. A. Smith, R. W. Siegel, M. Feldhaus and J. D. Marks, *J. Mol. Biol.*, 2005, **351**, 158–169.

125. C. Garcia-Rodriguez, I. N. Geren, J. Lou, F. Conrad, C. Forsyth, W. Wen, S. Chakraborti, H. Zao, G. Manzanarez, T. J. Smith, J. Brown, W. H. Tepp, N. Liu, S. Wijesuriya, M. T. Tomic, E. A. Johnson, L. A. Smith and J. D. Marks, *Protein Eng. Des. Sel.*, 2011, **24**, 321–331.

126. C. Garcia-Rodriguez, R. Levy, J. W. Arndt, C. M. Forsyth, A. Razai, J. Lou, I. Geren, R. C. Stevens and J. D. Marks, *Nat. Biotechnol.*, 2007, **25**, 107–116.

127. J. Hanes, L. Jermutus, S. Weber-Bornhauser, H. R. Bosshard and A. Pluckthun, *Proc. Natl. Acad. Sci. U. S. A.*, 1998, **95**, 14130–14135.

128. J. Hanes, C. Schaffitzel, A. Knappik and A. Pluckthun, *Nat. Biotechnol.*, 2000, **18**, 1287–1292.

129. J. Foote and H. N. Eisen, *Proc. Natl. Acad. Sci. U. S. A.*, 1995, **92**, 1254–1256.

130. J. Foote and C. Milstein, *Nature*, 1991, **352**, 530–532.

131. R. E. Hawkins, S. J. Russell and G. Winter, *J. Mol. Biol.*, 1992, **226**, 889–896.

132. R. Schier, A. McCall, G. P. Adams, K. Marshall, M. Yim, H. Merritt, R. S. Crawford, W. L. M. C. Marks and J. D. Marks, *J. Mol. Biol.*, 1996, **263**, 551–567.

133. W.-P. Yang, K. Green, S. Pinz-Sweeney, A. T. Briones, D. R. Burton and C. F. Barbas, *J. Mol. Biol.*, 1995, **254**, 392–403.

134. X. Yang, W. Xu, S. Dukleska, S. Benchaar, S. Mengisen, V. Antochshuk, J. Cheung, L. Mann, Z. Babadjanova, J. Rowand, R. Gunawan, A. McCampbell, M. Beaumont, D. Meininger, D. Richardson and A. Ambrogelly, *mAbs*, 2013, **5**, 787–794.

135. T. M. Lauer, N. J. Agrawal, N. Chennamsetty, K. Egodage, B. Helk and B. L. Trout, *J. Pharm. Sci.*, 2012, **101**, 102–115.

136. N. Chennamsetty, V. Voynov, V. Kayser, B. Helk and B. L. Trout, *Proc. Natl. Acad. Sci. U. S. A.*, 2009, **106**, 11937–11942.

137. J. Prassler, S. Thiel, C. Pracht, A. Polzer, S. Peters, M. Bauer, S. Norenberg, Y. Stark, J. Kolln, A. Popp, S. Urlinger and M. Enzelberger, *J. Mol. Biol.*, 2011, **413**, 261–278.

138. S. A. Jacobs, S. J. Wu, Y. Feng, D. Bethea and K. T. O'Neil, *Pharm. Res.*, 2010, **27**, 65–71.

139. Y. Xu, W. Roach, T. Sun, T. Jain, B. Prinz, T. Y. Yu, J. Torrey, J. Thomas, P. Bobrowicz, M. Vasquez, K. D. Wittrup and E. Krauland, *Protein Eng. Des. Sel.*, 2013, **26**, 663–670.

140. S. V. Sule, J. K. Cheung, V. Antochshuk, A. S. Bhalla, C. Narasimhan, S. Blaisdell, M. Shameem and P. M. Tessier, *Mol. Pharmaceutics*, 2012, **9**, 744–751.

141. I. Pastan, M. C. Willingham and D. J. FitzGerald, *Cell*, 1986, **47**, 641–648.

142. I. Pastan, R. Hassan, D. J. FitzGerald and R. J. Kreitman, *Annu. Rev. Med.*, 2007, **58**, 221–237.

143. V. K. Chaudhary, J. K. Batra, M. G. Gallo, M. C. Willingham, D. J. FitzGerald and I. Pastan, *Proc. Natl. Acad. Sci. U. S. A.*, 1990, **87**, 1066–1070.

144. R. J. Kreitman and I. Pastan, *Curr. Drug Targets*, 2006, **7**, 1301–1311.
145. D. Schrama, R. A. Reisfeld and J. C. Becker, *Nat. Rev. Drug Discovery*, 2006, **5**, 147–159.
146. F. Nimmerjahn and J. V. Ravetch, *Nat. Rev. Immunol.*, 2008, **8**, 34–47.
147. F. Nimmerjahn and J. V. Ravetch, *Curr. Opin. Immunol.*, 2007, **19**, 239–245.
148. U. Haessler and S. T. Reddy, *Methods Mol. Bio.*, 2014, **1131**, 191–203.
149. S. Kodangattil, C. Huard, C. Ross, J. Li, H. Gao, A. Mascioni, S. Hodawadekar, S. Naik, J. Min-Debartolo, A. Visintin and J. C. Almagro, *mAbs*, 2014, **6**, 628–636.
150. C. E. Tinberg, S. D. Khare, J. Dou, L. Doyle, J. W. Nelson, A. Schena, W. Jankowski, C. G. Kalodimos, K. Johnsson, B. L. Stoddard and D. Baker, *Nature*, 2013, **501**, 212–216.

CHAPTER 7

Protein Engineering

JOHN ADAIR*[a] AND DUNCAN McGREGOR[b]

[a] Ithaka Life Sciences Ltd, NETPark Incubator, Thomas Wright Way, Sedgefield, Co. Durham, TS21 3FD, UK; [b] Cyclogenix Ltd, Greenburn Road, Bucksburn, Aberdeen, AB21 9SB, UK
*Email: john.adair@ithaka.co.uk

7.1 INTRODUCTION

Protein engineering is the process of constructing novel protein molecules by design, from first principles or by altering an existing structure. There are two main reasons for wishing to do this. Firstly, there is the desire to understand for its own sake how proteins are assembled and what elements of the primary sequence contribute to folding, stability and function. These features can be probed by altering one or more specific amino acids in a directed manner within a protein and observing the outcome after production of the altered version. Often, related proteins with similar but not identical sequences exist in nature that have slightly different properties and these differing sequences can be used as guides for the alterations.

A second reason for wishing to change a protein is that the protein may be suitable, in principle, for a particular technology purpose but the version found in nature does not have the optimal properties required for the task. For example, an enzyme may be considered as part of an industrial process but a feature of the protein, such as the temperature stability or pH optimum for the catalytic activity, or the need for a co-factor may not be compatible with the process. Amino acid changes can be made that can tailor the enzyme so that it functions better in the new environment. There are many other examples of how proteins can be altered to make them better suited to

Molecular Biology and Biotechnology, 6th Edition
Edited by Ralph Rapley and David Whitehouse
© The Royal Society of Chemistry 2015
Published by the Royal Society of Chemistry, www.rsc.org

commercial and technological activities and some of these are noted in Section 7.3 below.

To engineer a protein implies an understanding of protein structural principles, knowledge of proteins as a material, and an appreciation of the limits of the material so that rational design, or alteration, of the properties can be achieved. In addition the engineer should have to hand the tools to produce and analyse the desired protein. The tools and an increasing understanding of the underlying principles have been developing in a parallel but intertwined manner over the past two decades (see references 1–3 for reviews of early developments in the field). In this chapter we will provide a background to protein engineering and summarize some recent developments.

7.1.1 Protein Structures

Our current knowledge of how proteins look, how they behave and the principles that determine how a primary structure folds derives from two sources. Firstly, studies on the structures of proteins using physical techniques, along with biochemical studies on the properties and physical interactions of the amino acids within proteins and with their environment, have shown that the amino acids of proteins adopt particular secondary structures such as the α-helix, β-strands and β-turns, and these secondary structures in turn are folded into tertiary structure motifs.[4] Small proteins may comprise only one such tertiary structure motif, however larger proteins are often comprised of a number of motifs, which may also be termed domains, that are themselves folded into a particular arrangement with specific interactions between the domains. In some cases separate proteins are organized into larger complexes by inter-domain interactions to form quaternary structures. Within the tertiary structures, particular amino acid groupings combine to provide the function of the protein. For example, certain amino acid side chains located at different points in the primary amino acid sequence may be brought into close association by the tertiary structure to generate a catalytic function or a specific ligand-binding surface.

Secondly, and drawing on the first, theorizing and calculation have provided insights into the ways proteins may assemble and function. For example, early observations on the likelihood of particular amino acids to participate in different secondary structure motifs provided the initial impetus for protein structure prediction algorithms (for example, Chou and Fasman[5]). Many basic questions about the properties of proteins remain unresolved. For example, how proteins fold (and methods of simulation of protein folding) remains a subject of intense enquiry (see for example references 4, 6–9). This may not be unexpected, given that the number of protein sequences that have currently been obtained, and protein structures that have been determined, are still a small fraction of the available repertoire occurring in nature. Fortunately nature, *via* selection, repeats successful structural motifs and this redundancy allows an insight into the way that different amino acid sequences can adopt similar secondary and tertiary structures. However, care needs to be taken in inferring structure from sequence homology as some highly homologous sequences can adopt different structures and mutations within a sequence can lead to change of secondary structures.[10,11] As further protein sequences become available and structures determined, and as methods become more efficient at placing new sequences within existing sequence and structure families (recently summarised in reference 4) then the "who, what and when" of protein structures may become more accessible. The "why" and the "how" still require a lot more effort. Protein engineering seeks to accelerate our understanding of the "why" and the "how".

In addition, and adding a further layer of complexity, correct protein folding *in vivo* in the crowded intracellular environment appears to be assisted by accessory protein chaperone complexes that have developed that allow incorrectly folded proteins to be rescued, or slow folding domains to be assisted in their folding.[6]

In the next section the basic tools that are required for protein engineering are described, followed by a series of examples of successful protein engineering.

7.2 TOOLS OF THE TRADE

7.2.1 Sequence Identification

Sequence identification by protein or gene sequencing is now a relatively straightforward process. Large databases now exist containing many thousands of sequences (for example the Protein Data Bank[12,13]) and genome-sequencing projects are providing new DNA sequences and open reading frames at an ever-increasing rate.[14] The functions of many of these sequences are known, from biochemical or genetic data, or by sequence homology to other known sequences. This gives the protein engineer an approach to rational design.

7.2.2 Structure Determination and Modelling

High-resolution structural information, determined by X-ray crystallography or nuclear magnetic resonance (NMR) techniques, is at the core of understanding of protein biochemistry. There are a large number of proteins for which high-resolution structural information is available but this number remains well below the number of sequences identified by protein or gene sequencing. It has been suggested that a "relatively" small number (around 2000) of unique fold topologies may exist in nature (see for example reference 4, and references therein for a recent summary). Proteins with little primary sequence homology have been observed to adopt similar folds and can be grouped into protein superfamilies. These groupings may reflect ancient evolutionary relationships (but care needs to be taken in inferring functional activities to regions that may be conserved for purely physical reasons).

Structural genomics approaches and international cooperation to determine structures of representative proteins may allow better linkage of sequences to structures (see references 15 and 16) for recent summaries of the Protein Structure Initiative, a coordinated international effort to increase protein structure and sequence information to underpin biology research).

In parallel, and partly in response to the ever-increasing number of sequences for which there are currently no structures, there is an ongoing drive to predict protein folds and structures, by a number of means. Progress has been made in predicting structures by modelling newly obtained sequences onto the known structures of homologous sequences or subsequences, where these are available and homologies can be discerned.[17] Improvements have also been made in *ab initio* calculation/prediction of protein structures as a result of increased understanding of the protein folding process, and of computational approaches to finding energetically stable structures for new sequences, both in methodologies and computing capacity (for example in distributed processing, such as the Folding@home project; see for example references 18–22). A question that recurs is how close does a calculated model approach reality and what confidence can be placed in the detail, before protein engineering experiments can begin? As will be seen below, some of the manipulation methods seek to circumvent this problem.

7.2.3 Sequence Modification

Modification of an existing protein by alteration of the gene sequence, from whole domains down to single amino acids, is now a routine process and this area is now the least challenging part of the engineering process. Procedures for preparing novel DNA sequences have been available since the early days of recombinant DNA technology (for example reference 23). Large sequence segments were initially generated by synthesis of the desired double-strand (ds) DNA sequence from short oligonucleotide segments that are annealed and ligated (for example references 24 and 25). The Polymerase Chain Reaction (PCR[26]) method can also be used to gap-fill and amplify partially overlapping oligonucleotides for gene synthesis (see Figure 7.1). Synthesis of domain sized blocks of DNA for subsequent expression is now often performed by contract houses for less than \$1 per base pair (bp). Further reductions in cost and increases in length of dsDNA segments are able to be achieved by use of DNA microchip technologies for microscale synthesis (see for example references 27 and 28).

Individual nucleotide changes or alterations of short segments of sequence in a pre-existing sequence can be performed by oligonucleotide-directed site-specific mutagenesis or the use of the now ubiquitous PCR (see below).

These methods are restricted to gene-coded amino acids. Sequences may also be generated by chemical protein synthesis[29,30] which allows the possibility of introducing variation into the sequence during synthesis, and also synthesis of "mirror image" proteins based on D-amino acids.[31,32]

GENE ASSEMBLY BY PCR

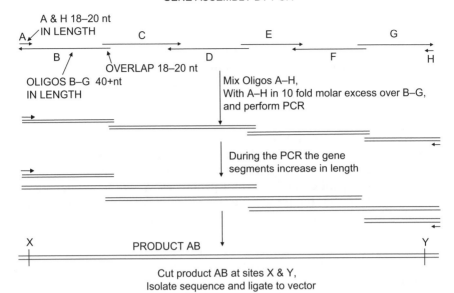

Cut product AB at sites X & Y,
Isolate sequence and ligate to vector

Figure 7.1 Gene assembly PCR. The required amino acid sequence is determined and a suitable nucleotide coding sequence is generated. The sequence may incorporate codon bias to assist with expression in a host cell and will usually also take advantage of the redundancy of the genetic code to include several useful restriction sites for later manipulation of the sequence. The sequence is then marked off into segments as shown in the figure (A–H), in which the sequences used alternate from the top (usually the sense strand) and bottom (antisense) strand. Oligonucleotide A has the sequence of the 5′ end of the top strand of the planned nucleotide sequence and will anneal to the 3′ region of oligonucleotide B. Oligonucleotide H has the sequence of the 5′ end of the bottom strand planned nucleotide sequence and will anneal to the 3′ region of oligonucleotide G. Oligonucleotides A and H are usually short oligonucleotides and will be present in the PCR in molar excess to drive the amplification of the full-length sequence once this has assembled in the early stages of the PCR. Oligonucleotides B to G in the figure cover the whole sequence to be assembled and can be of any length desired; the shorter the length the more that are required. With ever increasing fidelity of long oligonucleotide synthesis, lengths in excess of 100 nucleotides can be used. The 5′ region of oligonucleotide B can anneal to the 5′ region of oligonucleotide C and the 3′ region of oligonucleotide C can anneal to the 3′ region of oligonucleotide D and so on, until the terminal oligonucleotide is reached. Oligonucleotides B to G are mixed with molar excess of A and H and PCR is performed. In the early rounds of the reaction individual pairs of sequence anneal *e.g.* C to D and are extended. Similarly A anneals to B and converts it to a dsDNA sequence. In the next and subsequent rounds the CD product can now anneal to the AB product to give product covering the AD region. The CD product can also anneal to the EF product and so on. Similar annealing and extension events occur during subsequent rounds until a full-length sequence is assembled. The presence of excess A and H then ensures that this product is rapidly amplified. In the nucleotide sequence design the terminal oligonucleotides (B and G) in the figure will encode restriction sites (X and Y in the figure) suitable for insertion into a cloning vector. This figure is adapted from reference 171 with permission of the publisher.

7.2.3.1 Defined Sequence Alterations. In many situations specific alter-ation of one or a small number of amino acids is desired. This can con-veniently be done by site-specific mutagenesis procedures. There are two basic site-specific mutagenesis procedures. Both involve annealing of one or more oligonucleotides to a region of (at least temporarily) single stranded DNA (ssDNA) followed by *in vitro* DNA polymerase-directed exten-sion of the oligonucleotide(s).

7.2.3.1.1 Non-PCR Methods. In the older, non-PCR, procedures the DNA sequence to be modified is linked to a replication origin (for example in a plasmid, bacteriophage ['phage] or phagemid), allowing *in vivo* ampli-fication of the modified and parental genotypes. A synthetic oligonucleo-tide harbouring the required mutation is annealed either to circular ssDNA template, for example the genome of an ssDNA bacteriophage such as M13 or ΦX174 or the ssDNA form of a phagemid (Figure 7.2a) or to a partially single stranded dsDNA template (Figure 7.2b). Partial ssDNA tem-plates can be generated by a variety of enzymic methods.

The annealed oligonucleotide acts as a primer for *in vitro* DNA synthesis using a DNA polymerase, and in the presence of DNA ligase, closed circular double strand DNA (dsDNA) molecules are generated (Figure 7.2c). The dsDNA is introduced into a suitable host cell where replication and hence segregation of parental and mutant daughter strand occurs. The desired modification is then identified by one of a number of screening or selection procedures. The product Betaseron[33] (see Section 7.3.1.1 below) was gener-ated in this manner. Firnberg and Ostermeier[34] have recently described a version of this procedure that can be used to produce libraries of variants at all positions of a protein domain.

7.2.3.1.2 PCR-Based Methods. While the non-PCR methods are well known and have been optimized over a period of many years, and com-mercial kits are available for the various procedures, PCR-based methods have now become more widespread.

In the PCR-based approaches the desired modification is generated and amplified *in vitro* by annealing to denatured target DNA a synthetic oligo-nucleotide harbouring the required mutation along with an oligonucleotide that can act as a primer for replication of the complementary DNA strand (Figure 7.3).

There are a number of variants to the generalized method outlined in Figure 7.3, some of which involve including restriction sites into the muta-genic oligonucleotides. This can reduce the number of reaction involved (for example, if in Figure 7.3 primers A and D have useful restriction sites in the sequences and in D the site is 3′ to the mutagenic sequence, the first re-action products can then be directly cloned). The amplified DNA fragment is then ligated into a bacterial replication origin in a plasmid or 'phage and cloned by introduction into bacteria.

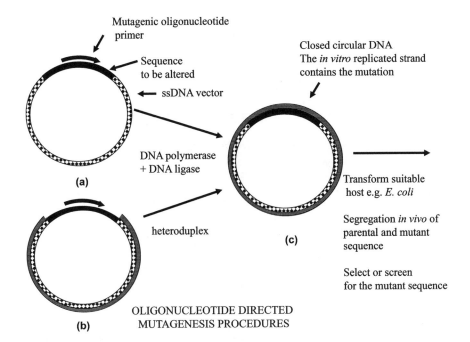

(a)

Mutagenic oligonucleotide primer

Sequence to be altered

ssDNA vector

DNA polymerase + DNA ligase

heteroduplex

(b)

Closed circular DNA
The *in vitro* replicated strand contains the mutation

(c)

Transform suitable host e.g. *E. coli*

Segregation *in vivo* of parental and mutant sequence

Select or screen for the mutant sequence

OLIGONUCLEOTIDE DIRECTED
MUTAGENESIS PROCEDURES

Figure 7.2 Strategy for non-PCR mutagenesis of ssDNA. a) A single strand DNA (ssDNA) template containing the sequence to be mutated is obtained. This is usually achieved by cloning the required sequence into a phage or phagemid vector and generating the ssDNA form. For a single point mutation an oligonucleotide of between 15–20 nucleotides with the proposed mutant sequence located centrally in the sequence is synthesized and is mixed in molar excess with and annealed to the ssDNA template. For more complex mutations it is generally useful to have 15–18 matched nucleotides at either side of the sequence mismatch. A DNA polymerase is added along with DNA ligase and the remaining ssDNA regions are converted to dsDNA and closed by the ligase (Figure 7.2c). T7 DNA polymerase is often used which has efficient $3'$–$5'$ proof-reading activity and good processivity but which lacks the detrimental $5'$–$3'$ exonuclease activity. b) As an alternative to the procedure in Figure 7.2a the vector containing the parental is rendered partially single stranded across the region to be mutated. A very simple way to achieve this is by cloning the required sequence into a phage or phagemid vector and generating the ssDNA form. Then the dsDNA vector without the insert is annealed to the ssDNA form generating a heteroduplex in which the cloned sequence is exposed as ssDNA. The mutation is then generated using a synthetic oligonucleotide, DNA polymerase and DNA ligase as in Figure 7.2a. c) The closed circular DNA that is generated has the mutation coded in the *in vitro* generated strand and the parental sequence in the vector strand. At the site of the mutation there is a DNA mismatch. The mutant and parental sequences are separated by transformation into a suitable host and allowing replication to occur. The mutant sequence can then be identified by one of a number of screening or selection procedures (reviewed in reference 171). In either procedure multiple mutations can be incorporated within one oligonucleotide or by annealing several oligonucleotides to the ssDNA region.
This figure is adapted from reference 171 with permission of the publisher.

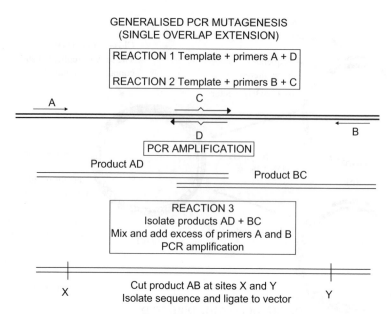

Figure 7.3 PCR mutagenesis by single overlap extension. In this method the mutation to be introduced is coded into two oligonucleotides shown as C and D. These are designed to be complementary to each of the DNA strands at the site to be mutated and are capable of stably annealing. For a single point mutation the oligonucleotides are likely to be around 15–20 nucleotides. This usually allows the selection of an incubation temperature for the PCR that maintains the initial stability of the annealed oligonucleotide. Oligonucleotides A and B are often complementary to sequences within the vector and can be used for DNA sequence confirmation of the mutation when it is made. In a first round of parallel reactions oligonucleotides A and D are mixed and annealed in molar excess over the dsDNA template and similarly for B and C. PCR is performed and new dsDNA fragments AD and BC are produced that each have the mutation. After purification of the fragments an aliquot of each is mixed with molar excess of oligonucleotides A and B and PCR again performed. The result is the fragment AB. This fragment can then be cloned into a suitable vector using sites within the sequence, or quite frequently, sites introduced to the sequence in the oligonucleotides A and B. This method has been termed single overlap extension (SOE) PCR.[172]
This figure is adapted from reference 171 with permission of the publisher.

A variant of the procedure allows the linking together of separate sequences, which may code for domains from different proteins, or allow the reorganization of domains within a protein. In this procedure, outlined in Figure 7.4, the primers C and D are hybrids containing sequences that can anneal to both of the domains in question. The first reactions then generate dsDNA fragments that now overlap in sequence and the desired product can then be generated in a third reaction using the primers A and B. Recently Gibson *et al.*[35] described a new version of this procedure which can be used for assembly large blocks of DNA of up to several hundred kilobases of DNA.

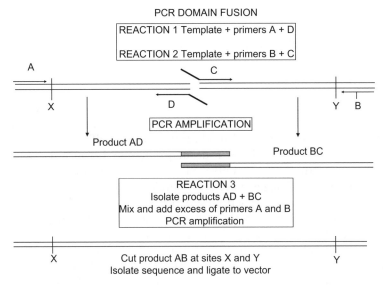

Figure 7.4 PCR domain fusion. The domain fusion PCR is similar to the operation described in Figure 7.3. In this case the design of the central primers is more complex. Oligonucleotide A has the sequence of the sense strand at the N terminal end of the first domain and so can anneal to the anti-sense strand of the coding sequence of the first domain. Oligonucleotide B has the sequence of anti-sense strand at the C terminal end of the second domain. Oligonucleotides A and B are usually 18–20 nucleotides in length and these lengths can be adjusted to ensure that the oligonucleotides remain annealed at the temperature of the PCR extension reaction. Oligonucleotide C is a hybrid of two sequences. At the 5′ end the first 20 (approx.) nucleotides are the same as the coding sequence at the C terminus of the first domain. This part of the oligonucleotide sequence ends at the point where the fusion will occur. The second portion of the oligonucleotide sequence (also often around 20 nucleotides) has the sequence of the sense strand at the N terminal end of the second domain. Therefore oligonucleotide C can anneal to the anti-sense strand of the coding sequence of the second domain, leaving a non-annealed "tail". Oligonucleotide D is a similar hybrid sequence but can anneal to the sense strand of the first domain, also leaving a non-annealed "tail". Two PCR reactions are done. In the first oligonucleotides A and D are in molar excess over the coding sequence for the first domain. In the second reaction oligonucleotides B and C are in molar excess over the coding sequence for the second domain. At the end of the reactions products AD and BC are formed. Product AD codes for the first domain and also codes for a number of amino acids from the N terminal region of the second domain, with the point of fusion exactly as dictated by the sequence in oligonucleotide D. In product BC, the DNA sequence codes for the C terminal amino acids of the first domain and the amino acids of the second domain. Importantly the products AD and BC now have a significant nucleotide sequence homology. Products AD and BC are purified and an aliquot of each is mixed with molar excess of oligonucleotides A and B in a third PCR reaction. The resultant product AB now has the coding sequence of both domains in-frame and joined exactly at the desired point. This fragment can then be cloned into a suitable vector using sites near the ends of the sequence, or quite frequently, using sites introduced to the sequence in the oligonucleotides A and B.
This figure is adapted from reference 171 with permission of the publisher.

In all of these cases the design of the oligonucleotide sequences and the reaction conditions need to be considered carefully, along with the correct choice of thermostable DNA polymerase. However, these procedures are very rapid and efficient.

7.2.3.2 Molecular Evolution. A recurring difficulty with protein engineering is, and may continue to be, knowing what to alter in a particular circumstance. Sometimes directed modification of a specific sequence is not a suitable procedure for obtaining a desired outcome, because it may not be clear where the true target amino acid(s) reside(s) and to what to alter it/them. A number of strategies have been developed to produce and test members of large libraries or repertoires of variants of a particular sequence. Methods of this kind have become the workhorse for the protein engineer and are termed directed evolution (see for example references 36–38).

The approach relies on three main features: firstly, that a method exists to generate the large number of variants. Secondly, that the nucleic acid that codes for the protein sequence of interest remains physically associated with the protein during the selection process. Thirdly, a screening or selection strategy is available that will enrich from among the collection of variants those novel protein sequences that have the phenotype of interest.

7.2.3.2.1 Generation of Variants. Molecular evolution methods require that a number of variants of a parental sequence be generated. The collection of variants is called a library. There are a variety of ways of generating this library of variants. For example if selection among all possible amino acids at one location is required then the codon at that point can be replaced by the triplet NNN where N means each of the four bases. To avoid codon bias, more usually, NNK codons $(K = G/T)$ are employed, to reduce the appearance of termination codons, or "trinucleotide" encoded codons can also be used where each amino acid is represented by a single defined trinucleotide, instead of random mononucleotides.[39] This can be achieved either by direct replacement of a double strand section with NNN in both strands, or by PCR using a priming oligonucleotide with NNN at the desired position. This strategy can be extended for a number of amino acid locations or to make single point mutations across an entire protein (see for example reference 34). Alternatively, larger scale variation can be achieved in certain circumstances by a technique known generically as DNA shuffling.[40,41] In the basic procedure the nucleotide sequences of a number of closely related proteins are used. Breaks are introduced randomly in the sequences and the pooled DNA is then denatured and annealed. Gaps are filled by PCR and then the resultant new sequences are cloned. The procedure provides a means for crossover between multiple original sequences and generates new sequences in which the sequence differences are recombined (see for example reference 42 for a summary of DNA shuffling methods that have emerged since the original description and reference 43 for a recent paper describing optimization of the procedure).

Even with such procedures only small segments of a protein can be altered at one time because of the number of possible combinations that are involved. Strategies that require the presentation and amplification in a cellular system are limited by the ability of the host to take up individual members, and libraries of 10^9–10^{10} members are considered large, *in vitro* libraries can exceed these numbers so that libraries of 10^{12}–10^{14} can be formed. However the possible combinations of all residues in even small protein domains far exceeds this.

7.2.3.2.2 Linking Coding Sequence to Product. The physical linkage of populations of peptides or proteins to their encoding nucleic acids to create large, diverse display libraries has provided a rich source of ligands to a wide range of target molecules. Enrichment of ligands from these libraries is achieved by the co-selection of target-binding peptides along with their associated encoding nucleic acids, which allows the subsequent identification of the selected peptide sequences. This approach has been most widely exemplified using the 'phage display system. In this system, a range of options exists that provide for retaining a link between the coding sequence and the expressed product.[44,45]

In the original procedure the coding sequence of interest is inserted into the gene sequence for one of the coat proteins of a 'phage, (usually the filamentous 'phages, as used for the site directed mutagenesis methods noted above). When expressed during the course of infection the fusion is assembled into the virus coat. The hybrid 'phage can then be selected for by binding to a specific ligand, for example an antibody. The hybrid protein remains associated with the coding sequence inside the 'phage. The 'phage genome can then be rescued and amplified and further rounds of binding can be performed. This procedure provides the basis for a means of selection among variants.

In the original peptide system, libraries of peptide were screened for binding to antibody to identify binding epitopes. However, the same principle applies to protein–protein interactions.

As an alternative to a 'phage, a phagemid can be used. Here, the 'phage coat protein that is used to generate the presenting fusion has been inserted into a plasmid. The host bacterium, containing the phagemid vector, must then be co-infected with an autonomously replicating bacteriophage, termed helper 'phage, to provide the full complement of proteins necessary to produce mature phage particles. Today this technique provides the basis for most library screening methods.

A variant of this procedure involves generating a fusion of the protein or peptide of interest with a DNA binding protein and arranging for the recognition motif to be included in the 'phage genome along with the coding sequence for the fusion. When the fusion is expressed it binds to its cognate DNA sequence on the 'phage genome and becomes packaged. The protein of interest protrudes through the 'phage coat and is available for ligand binding and selection. The subsequent steps of selection and amplification are the same as for 'phage display.[46] In addition to presentation on the surface of a 'phage, proteins and peptides can also be presented on the surface of bacteria,[47,48] baculovirus,[49] yeast,[50,51] or mammalian[52–54] cells.

Another *in vivo* approach is exemplified by the "peptides on plasmids" approach where peptides are fused to the *lacI* DNA-binding protein and expressed in the cytoplasm of bacteria, resulting in a protein–DNA complex library that can be selected against targets after bacterial lysis to release the complexes.[55]

Alternatively, *in vitro* procedures have been devised. The challenge here is to be able to maintain the specific coding sequence–product link and to separate out the individual products. The *in vivo* procedures described above, require the insertion of the library DNA into bacterial cells and the efficiency of bacterial transformation restricts library sizes to the 10^9–10^{10} range. Similarly, the display of peptides on bacterial or yeast surfaces or the display of peptides directly on their encoding plasmids all require a transformation step which imposes a limit to the library sizes that can be constructed.

It is generally accepted that there is a correlation between library size and the affinity of ligands that can be isolated from them.[56] This has prompted the development of display technologies in which the size-limiting transformation step is unnecessary, allowing ever-larger display libraries to be constructed. These new technologies enable higher affinity ligands to be obtained through the sampling of an increased structural repertoire which is made possible through the generation of libraries which are up to four orders of magnitude larger than those that can be constructed for phage display. These can be separated into RNA and DNA-based procedures.

RNA-based procedures were developed first. The most well-known of these are known as "ribosome display"[57] and "mRNA display"[58] (recently reviewed by Plückthun[59]). In both of these systems protein-coding mRNA is transcribed and translated *in vitro*. In ribosome display, the translated protein folds while emerging from the ribosome and remains linked to the mRNA during selection for function, after which the coding mRNA sequence is reverse transcribed to DNA, amplified and transcribed to make more mRNA for further rounds of selection. This is achieved through linking the library protein to a spacer protein (a number of different spacer proteins can be used) to allow space for the library protein to fold correctly unhindered by proximity to the ribosome, and by deleting the stop codon from the coding sequence. This latter change causes ribosome stalling, and hence linkage between the protein and its mRNA molecule. Both prokaryotic and eukaryotic cell free translation systems can be used, although the prokaryotic methods are more prevalent. The key is to ensure the folded protein does not disengage from the ribosome while the selection for function is occurring. This requires that selections are performed at low temperature, and in the presence of Magnesium ions.

In mRNA display, mRNA molecules encoding protein or peptide libraries are tagged at the 3′-end with the antibiotic puromycin. During translation, when the ribosome reaches the puromycin molecule, the antibiotic is covalently transferred to the elongating peptide chain (while still attached to the mRNA), thereby linking the mRNA to the encoded protein. While this is more robust than ribosome display, there is no method for ensuring attachment between the correct elongating peptide chain and its encoding mRNA molecule. As both methods are mRNA-based, there are stability issues that must be considered when using these procedures.

DNA-based methods can be divided into non-covalent (*CIS* display[60]), covalent[61,62] and emulsion-based systems.[63]

CIS display exploits the high-fidelity *cis*-activity that is exhibited by a group of bacterial plasmid DNA-replication initiation proteins typified by RepA of the R1 plasmid. In this context, *cis*-activity refers to the property of the RepA family of proteins to bind exclusively to the template DNA from which they have been expressed. By genetically fusing peptide libraries to the N-terminus of the RepA protein, Odegrip *et al.*[60] demonstrated that it is possible to achieve a direct linkage of peptides to the DNA molecules that encode them *in vitro*. The rationale of the system is that *in vivo*, R1 plasmid replication is initiated through the binding of RepA to the plasmid origin of replication (*ori*). Ori is separated from the RepA-coding sequence by a DNA element termed *CIS*. The consensus model for *cis*-activity is that the *CIS* element, which contains a rho dependent transcriptional terminator, causes the host RNA polymerase to stall. This delay allows a nascent RepA polypeptide emerging from a translating ribosome to bind transiently to *CIS*, which in turn directs the protein to bind to the adjacent *ori* site, thus, linking genotype to phenotype.

DNA-based Covalent Display approaches are essentially similar to *CIS* display, but require a covalent interaction between protein and DNA to retain the linkage of genotype to phenotype, through the *cis* action of the crosslinking protein. Two requirements are needed for successful use of this technique. Firstly, proteins are required which interact *in vitro* with the DNA sequence which encodes them (*cis* action), and secondly, said proteins must establish a covalent linkage to their own DNA template. This method suffers from the fact that the DNA is chemically modified which can prevent the recovery and identification of the binding peptide of interest. Despite this, Reiersen *et al.*[62] demonstrated the isolation of scFv antibody fragments from an immunized library using a covalent display approach based on the *cis*-acting P2 A DNA binding protein.

Tawfik and Griffiths[63] demonstrated that directed evolution procedures could be undertaken in cell free emulsions in which transcription and translation occur in water microdroplets in an oil emulsion. This allows the physical co-localization of coding sequence and product and has particular uses for evolution of enzymatic function where substrate can be encapsulated and product can be linked to genotype. More recently the procedure has been extended using liposomes as the compartmentalizing agent[64] and by whole cell encapsulation techniques.[65]

7.2.3.2.3 Identification of the Required Phenotype. Thirdly, these methods require a screening or selection strategy that will enrich from among the library those novel protein sequences that have the phenotype of interest. Screens which involve binding to a ligand can easily be devised by immobilizing the target incubating the library members with the target, washing off unbound members and then recovering the bound fraction which can then be amplified *in vitro* and the binding cycle repeated as desired. By having a soluble target in the incubation mix selection for higher affinity binders can be achieved from among variant members of the library. Recently high throughput "deep" sequencing technologies have been applied to the display process so that instead of only analysing representatives of an enriched population, or driving the selection process by multiple cycles so that a dominant genotype remains, large numbers of bound library members can be sequenced and therefore a greater understanding of the relationship between sequence and function can be developed (see for example references 66 and 67).

7.2.4 Production

Once a novel sequence has been identified and designed it must be expressed to determine that the function is as required. A range of expression systems is available, from microbial systems (*Escherichia coli* being the most frequently used[68]) yeasts (for example *Saccharomyces cerevisiae* and *Pichia pastoris*), insect cells, filamentous fungi and mammalian cell cultures through to transgenic animals and plants. Cell free systems for the expression of relatively small quantities of product for initial analysis are also available. These can now produce sufficient quantities of product for early stage analysis.[69–71] For protein engineering purposes there is also the opportunity to extend beyond the normally found 20 amino acids that are gene coded. Non-natural amino acids can be incorporated during chemical peptide synthesis, but can also be included *via* cellular mechanisms where use of engineered transfer RNAs can lead to incorporation of novel amino acids.[72–74] The scale of production required also varies. In the early stages of development of a novel protein, for example where a protein has been identified among a number from a display library, only small quantities (μgs) of the protein may be needed to confirm a biological property. In the second stage larger quantities (10's to 100's of mgs) of purer material are required, typically to obtain structural information and to perform more demanding *in vitro* and *in vivo* bioassays. In some situations larger quantities (grams to multi-kg or tonne-scale) of material purified by rigorous (and usually costly) procedures may be required if the protein is required for commercial purposes (for example for industrial enzymes or for therapeutic purposes).

Often a number of approaches may need to be tried to find a suitable expression host, as it cannot safely be assumed that an altered protein will express in the same manner as the parental sequence, and in the case of novel sequences no precedent exists. These factors are becoming of increasing importance as more strategies rely on library display systems, where members of a library may be lost or underrepresented because of poor expression or folding properties.

7.2.5 Analysis

Methods must exist for the analysis of the properties of the modified proteins. Where a function is being altered (introduced, modified or removed) a biological assay may already be available to measure the function and activity of the parental protein or can usually be devised to measure the ability of the newly produced protein (this assay may well be the same as the screening assay). The issue is often of ensuring that the assay can function with very small amounts of materials, such as that obtained from library amplification procedures. Methods that include physical adsorption or binding may be better suited to this than those that rely on a detection of a catalytic product. In the latter, diffusible product must remain close to the point of production and so water:oil emulsion compartmentalization or single well techniques are better suited here.

In many cases acquiring and demonstrating the new function may be all that is required. However, in addition to functional assays, interpretation of the results may require a structural understanding of the alteration, particularly to gain further insights into protein folding. Assuming a reasonable quantity of the novel protein can be prepared, gross properties can readily be detected by spectroscopic or other large-scale measurements (for example circular dichroism, turbidity, enthalpic, sedimentation or chromatographic properties). However, detailed structural information about the end product is still the rate-limiting step in rational protein engineering and design. Casimiro *et al.*[75] showed that PCR-based gene synthesis, expression in *E. coli* of milligram quantities of isotopically labelled protein and NMR spectroscopy can be achieved in a reasonably short space of time, with first NMR spectra available for a novel sequence as little as 2 months after the initiation of the gene synthesis steps. More recently, there has been interest in developing methods for rapid determination, at small scale, of the structural information about coded sequences identified in genomic projects, which can be applied to engineered proteins (for example, references 76 and 77).

7.3 APPLICATIONS

Applications of protein engineering are everywhere in biology, as a means of producing novel molecules and as a means of understanding the basic properties of proteins. Therefore to attempt a comprehensive survey is fruitless. However it is useful to note some specific examples of the different types of molecules that can be achieved, from the simplest point mutations and domain rearrangements to more demanding multi-site alterations and *de novo* designs. The examples given in the next sections reflect our personal bias towards biopharmaceuticals, and antibodies in particular, but hopefully will prove of general interest. The examples demonstrate that protein engineering procedures have been used to develop beneficial medicines for nearly 30 years now. Walsh[78] has recently tabulated approved recombinant protein medicines, several of which are the results of protein engineering experiments while Carton and Strohl have provided a useful survey of protein therapeutics, which also tabulates approved protein biotherapeutics.[79]

7.3.1 Point Mutations

Individual point mutations in proteins can be readily achieved once the gene sequence is available, using the techniques noted earlier. One opportunity that arises is to determine the role of a particular side chain by removing the side chain back to the Cβ atom in the peptide by changing the residue to an alanine using the mutagenesis methods outlined earlier. This can be applied to make changes to residues over the protein surface (so called "alanine scanning") to examine, for example, protein–protein interactions (for example references 80 and 81).

Numerous examples have been described where one or a small number of amino acids have been changed in a protein to examine or improve its function. Many of the early proteins considered, and approved, as pharmaceuticals have now been modified in various ways and modified forms are now available as drug products. Some key examples are given below.

7.3.1.1 Betaseron/Betaferon (r-Interferon β-1b). One of the earliest approved biopharmaceutical products which was a result of protein engineering was the production of interferon β-1b. This novel protein was generated by the substitution of a Cysteine (Cys) for Serine (Ser) at residue 17 of the 154 amino acid interferon β effectively making a one atom change from a sulfur atom to an oxygen atom.[33] This substitution reduces the possibility of incorrect disulfide bridge formation during synthesis *in E. coli*, and also removes a possible site for post-translational oxidation. The resultant protein is expressed in *E. coli* at a specific activity approaching that of native fibroblast-derived interferon β. The molecule has been licensed since 1993 for use for the reduction of frequency and degree of severity of relapses in ambulatory patients with relapsing remitting Multiple Sclerosis. In 2012 this product had world-wide sales of $1.6bn.[82]

7.3.1.2 Novel Insulins. Insulin is a pancreatic hormone that acts to regulate blood glucose levels. Insulin is used in the treatment of diabetes to help regulate blood glucose levels. Recombinant insulin (Humulin) was the first biotechnology therapeutic product in 1982. Since the early approval of recombinant insulin a number of altered forms have been developed that seek to improve the effectiveness of the delivered insulin. The first of these, Humalog® (insulin lispro), was approved in 1996.[83] Humalog® is an engineered, fast acting, analogue of insulin. In Humalog® the two residues at the C terminus of the B chain, Pro28 and Lys29, have been reversed in their order. The C terminal alterations were designed based on structural and sequence homology to insulin-like growth factor 1 (IGF-1) and the sequence reversal reduces the dimerization of the B subunit. This reduced self-association means that the monomer is more readily available to function after food uptake and therefore can be administered very shortly before meals. Humalog sales in 2012 were nearly double, at $2.4bn, those of Humulin ($1.24bn).

Since 1996 a number of recombinant forms have been licensed (see for example, Walsh[78]).

7.3.1.3   Aranesp® (darbopoietin α) Modified Erythropoietin. Erythropoietin (EPO) is a hormone that stimulates the formation of red blood cells.[84] EPO is approved for treatment of anaemia. Following its approval in 1989, EPO became a blockbuster biotechnology drug with multi-billion $ sales. In 2001 a modified version, Aranesp®, was approved. Aranesp® is a hyper-glycosylated form of EPO in which the amino acid sequence has been altered at two locations so as to introduce 2 new N-linked carbohydrate attachment motifs. Aranesp® therefore bears five N-linked carbohydrate structures compared to the three in the native form. This results in a longer *in vivo* half-life for the product and allows for less frequent dosing of the patient.[85] In 2012 Aranesp® had worldwide sales of $2bn.

7.3.1.4   Antibody Point Mutations to Alter Structure and Function. Recombinant protein products based on the antibody molecule are now a major biopharmaceutical product class,[86,87] with combined sales now in excess of $60bn per year. Strohl and Strohl[88] have recently provided a very comprehensive and timely summary of therapeutic antibody engineering spanning methods, achievements and current challenges. Antibodies are large multi-domain proteins; the IgG antibody is a Y or T shaped structure comprising two identical 50 kiloDaltons (kDa) "heavy" chains and two identical 25 kDa "light" chains. The heavy chain comprises an N terminal "variable" domain (VH) and three "constant" domains (CH1, CH2, CH3) with a short flexible hinge region between CH1 and CH2. The light chain comprises an N terminal "variable" domain, VL, and a "constant" domain CL. The antigen-binding region, known as the Fab region, comprises the VH and CH1 domain of one heavy chain and one complete light chain. The C terminal CH2 and CH3 domains of the two heavy chains associate together to form the Fc region. The Fc region is involved with interactions with various cells of the immune system and with complement, and is also important in determining the serum half-life of antibodies, at least of the IgG type. This modular structure of the protein is mirrored at the genetic level with each domain being coded by separate exons and provides many opportunities to develop novel molecules.

A wide range of interesting modifications have been described over the years since the beginnings of protein engineering where single point mutations in an antibody have, for example: altered the affinity of binding for target antigen (for example references 89 and 90 among many); improved the structural stability *in vivo* of the IgG molecule;[91] provided specific locations for attaching other molecules, such as drugs or radioisotopes, that can be used to kill cancer cells;[92,93] provided a means of bringing together the binding specificities of two separate antibodies to make a "bispecific" antibody by making complementary changes in the CH3 domain so that heterodimerization is preferred over homodimerization of the IgG.[94,95] Point mutations that can alter the ability of IgG to interact with other cells of the immune system, or with complement or to modify serum half-life have been described by a number of groups since the late 1980's and some of these entities are now in clinical development.[80,81,96–99]

More recently attention has been turned to how individual residues within domains can affect product stability (resistance to unfolding) and reduce protein aggregation, which are concerns for manufacturability, utility in industrial processes, and also development of immune responses when biopharmaceuticals are administered to patients.[100–102] As well as defined single changes, as noted earlier,[34] library methods can be used to make point mutations across a protein domain such that, in most instances, only one alteration is made over the domain, but at each amino acid sequence all amino acids can potentially be substituted, so that sequence–function relationships can be developed (see also references 66 and 67). Firnberg and Ostermeier[34] made changes across the β-lactamase gene. In another example, Wodak[103] described a similar mutagenesis and deep sequencing approach to identify variants with higher binding affinity of small helical proteins that recognize a conserved epitope on influenza H1 haemagglutinin.

Library methods can be used to make multiple mutations across a protein–protein interaction surface; Genentech scientists have shown that the binding site for an antibody can be altered so that the antibody now recognizes two distinct proteins through the same binding site.[104,105]

7.3.2 Domain Shuffling (Linking, Swapping and Deleting)

Most large proteins are composed of smaller independently folding domains, generally corresponding to the protein folds mentioned earlier. These domains are often linked together by short peptide sequences and in many, but not all, cases the domains are identifiable as separate exons in the gene sequence. Using standard molecular biology techniques it is possible to add, remove or swap domains from one protein to another to rebuild proteins.

7.3.2.1 *Linking Domains*

7.3.2.1.1 Domain Fusions for Cell Targeting. Amongst the very earliest examples of protein engineering, the binding site region of an antibody was genetically linked to an enzyme. The antibody–enzyme gene fusions were

constructed by taking the DNA sequence that codes for the heavy chain component of the Fab region of the antibody and linking this either to staphylococcal nuclease[106] or to *E. coli* DNA polymerase coding sequences.[107] Introduction of these fusion genes in a cell that produces the antibody light chain, and expression of the fusion gene, reconstituted both antigen binding and enzymic activity.

These early examples provided the foundation for many other domain fusion examples where a binding function (such as an antibody binding domain, a peptide, cytokine, growth factor or extracellular ligand binding domain [ECD] of a receptor) is linked to an effector function (such as a protein toxin,[108,109] enzyme,[110] or cytokine[111,112]). Also, a non-antibody binding domain can be joined to the Fc region of an antibody to take advantage of the long serum half-life of antibodies to improve the pharmacokinetic properties of a designed molecule.[113,114]

Some of these domain fusions are now licensed for human use. For example, Enbrel® (etanercept) comprises the extracellular domains of the receptor for the cytokine tumour necrosis factor α (TNFα) genetically fused to the Fc regions of IgG. Enbrel® is used to block the activity of TNFα[115] and is currently licensed for treatment of a number of autoimmune conditions including: rheumatoid arthritis, polyarticular juvenile idiopathic arthritis, psoriatic arthritis, ankylosing spondylitis and plaque psoriasis. It had sales in 2012 of $8.4bn. Other more recently approved fusions with similar architecture include Orencia® and Eylea®, each with 2012 sales approaching $1bn.

Ontak® is another early example of a genetic fusion in which the cytokine interleukin-2 (IL-2) is used to replace the receptor-binding domain of diphtheria toxin.[116] Ontak® is used to target the Diphtheria A and B toxin domains to leukaemic cells which over-express the receptor for IL-2. Ontak® was approved in 1999 to treat certain patients with advanced or recurrent cutaneous T-cell lymphoma (CTCL) when other treatments have not worked.

7.3.2.1.2 Fusions to Stabilize Dimeric Proteins. In other situations it is useful to genetically fuse together domains which are normally associated by non-covalent interactions or disulfide bridges. This allows more convenient production and ensures that the domains remain in proximity even at the low concentrations that would be observed during *in vivo* dosing. Examples include joining the VH and VL domains of antibodies together with a linker peptide to form molecules called single chain Fv's (scFv's). The scFv molecule is a veteran of protein engineering, having first been disclosed during the 1980s.[117,118] Another example is the linking of domains of heterodimeric cytokines (for example Interleukin 12 (IL12) to produce Flex-12[119] and more recently IL27 to make scIL27[120]).

7.3.2.2 Swapping Protein Domains. Another simple procedure used in protein engineering is whole domain swapping, in which analogous domains from different sources are swapped into a multi-domain protein to provide a novel functionality.

7.3.2.2.1 Chimeric Mouse–Human Antibodies.   Here the antigen binding domains from a mouse monoclonal antibody are linked to the constant regions (which as noted earlier provide immune effector functions and dictate biological half-life) from a human antibody.[121] This switching can markedly reduce unwanted immunogenicity when used in man compared to the original mouse monoclonal antibody.[122] A number of these chimeric antibody products have been licensed as pharmaceuticals, with the lead example, the anti-TNFα antibody Remicade, having sales in 2012 of $7.47bn.

7.3.2.2.2 Polyketide Synthases (PKSs).   Polyketides are a class of chemical compounds which include many pharmaceutical compounds, including antibiotics, antifungals and immunosuppressants, and which together account for many £billions of sales per annum. These chemicals are produced in various soil microorganisms and fungi. Their synthesis involves the regulated action of a sequence of enzymes, the polyketide synthases (PKSs). The PKSs occur either as small proteins that have a small number of distinct non-repeating catalytic domains (iterative PKSs), or as large multi-domain polypeptides. In the latter, while each domain has a specific enzyme function, the same function may be present a number of times (modular PKSs). In both cases the product of one catalytic domain forms the substrate for the reaction of a neighbouring domain, the modular polyketides producing the more complex molecules. By selectively adding, deleting and rearranging the order of the catalytic domains to generate novel PKSs, a novel order of catalytic reactions is established and hence new products can be generated (see for example references 123–125 and references therein).

These multidomain megasynthases provide lead examples of efforts to generate complex biological molecules by manipulation of multienzyme complexes. These also provide an entry into the emerging discipline of synthetic biology to develop microbial factories for complex chemical syntheses, including polyketides and the non-ribosomally synthesized peptides, the latter also providing further opportunities to develop novel peptide antibiotics (for example references 126 and 127).

7.3.2.3 Deleting Domains.   In some situations only a subset of the domains in a naturally occurring protein are required for development of a functional therapeutic. Tissue plasminogen activator (tPA) is a serine protease secreted by endothelial cells. Following binding to fibrin tPA activates plasminogen to plasmin which then initiates local thrombolysis. Reteplase is a variant form in which three of the five domains of tPA have been deleted. One of the domains that confers fibrin selectivity and the catalytic domain are retained (summarized by Nordt and Bode[128]). Reteplase is licensed as Retavase for the treatment of acute myocardial infarction to improve blood flow in the heart.

7.3.3 Whole Protein Shuffling

Many proteins exist as multi-member families in which the homologues display slightly different biological activities determined by the sequence variation. In the early days of protein engineering hybrid genes were generated by swapping DNA sequence stretches using convenient common restriction sites (for example reference [129]) or by *in vivo* recombination between homologous genes in a more random manner.[130] These methods produced small numbers of novel genes that could be examined individually. As noted earlier, more recently DNA shuffling techniques allows the generation of libraries of novel variants from among protein homologues. New variants, such as novel enzyme variants, can then be identified by selection or screening procedures (for example reference 43).

7.3.4 Protein–Ligand Interactions

We noted earlier some of the examples where antibodies have been altered to improve their ability to interact with their target antigens or with cells or other protein components of the immune system. Another significant class is the modification of enzymes at their active sites to improve their function.

7.3.4.1 Enzyme Modifications. Many examples of protein engineering involve changes to enzymes to examine and modify enzyme–substrate interactions. These experiments can be traced back to the earliest attempts at protein engineering, using tyrosyl-tRNA synthetase (see for example references 131 and 132 reviewed in reference 133). Many of the methods described in the preceding sections are applied to enzyme modifications.[134]

The types of changes include: enhancing catalytic activity; modification of substrate specificity, including the *de novo* generation of novel catalytic functions; alterations to pH profiles so that an enzyme can function in non-physiological conditions; improving oxidation resistance by replacing oxidation sensitive amino acids such as Cys, Tryptophan (Trp) or Methionine (Met) by sterically similar non-oxidizable amino acids such as Ser, Phenylalanine (Phe) or Glutamate (Glu), respectively; improving stability to heavy metals by replacing Cys and Met residues and surface carboxyl groups; removing, modifying or redesigning protease cleavage motifs; removing sites at which catalytic product might otherwise bind to induce allosteric feedback inhibition. Li *et al.*[135] and Craik *et al.*[136] have recently summarized progress with protein engineering of proteases, the largest enzyme class, for industrial and therapeutic uses.

7.3.4.2 Substitution of Binding Specificities. A number of examples are available where the specificity for a ligand has been transferred from one protein background to another. Some of the examples require small changes, some require large-scale transfer of many residues. An early example described the change in specificity of the hormone prolactin by substitution of eight amino acids at the receptor binding surface so that the modified hormone now binds to the receptor for growth hormone.[137] Protein domains often have loop regions that span between other secondary structure motifs (α-helix, β-strand). In some cases the functionality of the protein resides in these loops and they can be the target for relatively simple substitution experiments. For example, the specificity of basic fibroblast growth factor was altered to that of acidic fibroblast growth factor by the substitution of one particular loop region.[138] On a more adventurous scale, antibody humanization involves substitution from a non-human antibody to a human antibody of up to six loop regions that form the antigen binding surface and which extend from the β-sheet frameworks of the antigen binding domains. Since the first demonstration of this technology in 1986[25] it has now progressed to the point where more than a dozen humanized antibodies are approved for human use, with combined 2012 worldwide sales around \$23bn.

7.3.5 Towards *De Novo* Design

7.3.5.1 Development of Novel Binding Molecules. An intermediate step between experiments of the kind described above and "true" *de novo* design is to take a particular protein scaffold (for example α-helical bundle or β-barrel) and modify it so as to introduce new functions, or retain function on a radically altered scaffold. This is often done by taking a small compact domain of known structure and using phage display procedures to modify part of the surface of the protein and to screen or select for the introduction of a new function. Early examples included examination of the structural plasticity of a protein fold by repositioning the N and C termini within the sequence of the IL-4 cytokine, circularly permuting the sequence without affecting the overall fold.[139] In another early example, Martin *et al.*[140] described how a "minibody" β scaffold derived from an immunoglobulin binding domain and retaining only two loops and six β-strands was used to obtain a novel antagonist for IL-6.

Over the last several years there has been a growing examination of naturally occurring, small, stably-folding domains as the basis for developing novel proteins with specific functions, usually as binding domains[141–147] (summarized recently by, for example, references 148, 149). The desired specificity and affinity features can be introduced using the molecular evolution procedures described earlier and significant progress has been made in preparing novel molecules, some of which are now in clinical development. These newer molecules aim to provide high affinity binding and to avoid some of the practical issues surrounding the antibody molecules (for example engineering and production and not least the crowded intellectual property arena). A long-term aim of examining such molecules is to bring protein or peptide biopharmaceutical closer to the "normal" pharmaceutical paradigm of an orally available, synthetic, low molecular weight compound.

The plant microprotein family of knottins or cyclotides are a case in point: The cyclotide family (28–40 amino acids) have a unique topology, which combines a circular peptide backbone and a tightly knotted disulfide network that forms a cyclic cysteine knot (CCK) motif and makes the more than 350 known cyclotides exceptionally stable.[150–153] The cyclotides are resistant to thermal unfolding, chemical denaturants and proteolytic degradation. These remarkable chemical and biological properties make cyclotides ideal candidates as display library scaffolds and as templates for the grafting of biologically active peptide epitopes. Certain naturally-occurring cyclotides are known to have orally active biological activities, and this has been exploited to orally deliver biologically active bradykinin antagonist cyclotides.[154] Additionally cyclotides have the potential to act as scaffolds with enhanced blood–brain barrier penetration and even intracellular delivery may be possible.[155,156]

Beyond this there is the possibility of true *de novo* or *ab initio* design.

7.3.5.2 De Novo Design. Discussion of, and attempts to, design protein domains from first principles have been underway for some time (see for example references 157–159). *Ab initio* design brings different challenges. In principle, for any protein of n residues there are 2×10^n different possible sequences. As noted earlier however, known structure space is populated by a relatively small number of stable fold types. The structures and sequences databases demonstrate that a range of sequences can be accommodated into similar protein folds, and a range of sequences can fulfil similar functions. One approach to designing a functional protein is to decide what fold may be appropriate for a particular circumstance and then identify what sequence(s) would be needed to generate the desired fold and function. This can be approached by computation or by designed library methods.[160–163] The next decade should bring significant developments in this area.

7.4 CONCLUSIONS AND FUTURE DIRECTIONS

Protein engineering, as a methodology, is now a mature technology. Through the pragmatic approach allowed by molecular evolution methods, novel proteins are being produced for research and commercial applications at an ever-increasing rate. In the near future there will be an increasing effort to exploit the technology on areas that have been less tractable to date: in better understanding the structure and function of membrane proteins such as G-protein coupled receptors (GPCRs) that are the focus for many pharmaceutical drug development strategies (see for example, references 65, 164 and 165), and in the developing field of nano-technology,[166,167] including biosensors[168] and bioelectronics.

Rational, *ab initio*, protein design is still at an early stage and remains the domain of experts (see for example reference 169 for a discussion of current progress with protein design). Much will need to be done to achieve the goal of routine *ab initio* design and it might be argued that evolutionary methods, which provide means for finding the desired needle from among a stack of needles, are the way forward. It is likely each route will find a niche (see for example reference 170). There seems little doubt however that protein engineering will provide the basis for enormous opportunities in healthcare and the materials sciences.

ACKNOWLEDGEMENTS

We would like to acknowledge the helpful suggestions for recent literature and methods from, in particular, Roy Jefferis, Stephen Kent, Andy Popplewell, Joe Sheridan and Bill Strohl.

REFERENCES

1. R. Wetzel, *Protein Eng.*, 1986, **1**, 3.
2. W. W. Shaw, *Biochem. J.*, 1987, **246**, 1.
3. J. A. Brannigan and A. J. Wilkinson, *Nat. Rev. Mol. Cell Biol.*, 2002, **3**, 964.
4. R. D. Schaeffer and V. Daggett, *Protein Eng. Des. Sel.*, 2011, **24**, 11.
5. P. Y. Chou and G. D. Fasman, *Biochemistry*, 1974, **13**, 222.
6. F. U. Hartl and M. Hayer-Hartl, *Nat. Struct. Mol. Biol.*, 2009, **16**, 574.
7. A. I. Bartlett and S. E. Radford, *Nat. Struct. Mol. Biol.*, 2009, **16**, 582.
8. R. B. Best, *Curr. Opin. Struct. Biol.*, 2012, **22**, 52.
9. Z. Zhang and H. S. Chan, *Proc. Natl. Acad. Sci. U. S. A.*, 2012, **109**, 20919.
10. P. A. Alexander, Y. He, Y. Chen, J. Orban and P. N. Bryan, *Proc. Natl. Acad. Sci. U. S. A.*, 2007, **104**, 11963.
11. Y. He, Y. Chen, P. A. Alexander, P. N. Bryan and J. Orban, *Structure*, 2012, **20**, 283.
12. H. M. Berman, J. Westbrook, Z. Feng, G. Gilliland, T. N. Bhat, H. Weissig, I. N. Shindyalov and P. E. Bourn, *Nucleic Acids Res.*, 2000, **28**, 235.
13. S. K. Burley, *Biopolymers*, 2013, **9**, 165.
14. 1000 Human Genomes Consortium, *Nature*, 2012, **491**, 56.
15. D. Lee, T. A. P. de Beer, R. A. Laskowski, J. M. Thornton and C. A. Orengo, *BMC Struct. Biol.*, 2011, **11**, 2.
16. G. T. Montelione, *F1000 Biology Reports*, 2012, **4**, 7.
17. E. di Luccio and P. Koehl, *BMC Bioinformatics*, 2011, **12**, 48.
18. Y. Zhang, *Curr. Opin. Struct. Biol.*, 2009, **18**, 145.
19. Y. Zhang, *Curr. Opin. Struct. Biol.*, 2009, **19**, 342.

20. J. Lee S. Wu and Y. Zhang, in *From Protein Structure to Function with Bioinformatics*, ed. D. J. Rigden, Springer, Netherlands, 2009, p. 3.
21. A. Jothi, *Protein Pept. Lett.*, 2012, **19**, 1194.
22. D. S. Marks, T. A. Hopf and C. Sander, *Nat. Biotechnol.*, 2012, **30**, 1072.
23. K. Itakura, T. Hirose, R. Crea, A. D. Riggs, H. L. Heyneker, F. Bolivar and H. W. Boyer, *Science*, 1977, **198**, 1056.
24. M. D. Edge, A. R. Green, G. R. Heathcliffe, P. A. Meacock, W. Schuch, D. B. Scanlon, T. C. Atkinson, C. R. Newton and A. F. Markham, *Nature*, 1981, **292**, 756.
25. P. T. Jones, P. H. Dear, J. Foote, M. S. Neuberger and G. Winter, *Nature*, 1986, **321**, 522.
26. R. K. Saiki, S. Scharf, F. Faloona, K. B. Mullis, G. T. Horn, H. A. Erlich and N. Arnheim, *Science*, 1985, **230**, 1350.
27. S. Kosuri, N. Eroshenko, E. M. Leproust, M. Super, J. Way, J. B. Li and G. M. Church, *Nat. Biotechnol.*, 2010, **28**, 1295.
28. H. Kim, H. Han, J. Ahn, J. Lee, N. Cho, H. Jang, H. Kim, S. Kwon and D. Bang, *Nucleic Acids Res.*, 2012, **40**, e140.
29. B. L. Bray, *Nat. Rev. Drug Discov.*, 2003, **2**, 587.
30. T. Durek, V. Y. Torbeev and S. B. Kent, *Proc. Natl. Acad. Sci. U. S. A.*, 2007, **104**, 4846.
31. S. Kent, Y. Sohma, S. Liu, D. Bang, B. Pentelute and K. Mandal, *J. Pept. Sci.*, 2012, **18**, 428.
32. K. Mandal, M. Uppalapati, D. Ault-Riché, J. Kenney, J. Lowitz, S. S. Sidhu and S. B. Kent, *Proc. Natl. Acad. Sci. U. S. A.*, 2012, **109**, 14779.
33. D. F. Mark, S. D. Lu, A. A. Creasey, R. Yamamoto and L. S. Lin, *Proc. Natl. Acad. Sci. U. S. A.*, 1984, **81**, 5662.
34. E. Firnberg and M. Ostermeier, *PLoS ONE*, 2012, **12**, e52031.
35. D. G. Gibson, L. Young, R. Y. Chuang, J. C. Venter, C. A. Hutchison 3rd and H. O. Smith, *Nat. Methods*, 2009, **6**, 343.
36. M. J. Dougherty and F. H. Arnold, *Curr. Opin. Biotechnol.*, 2009, **20**, 486.
37. P. A. Romero and F. H. Arnold, *Nat. Rev. Mol. Cell Biol.*, 2009, **10**, 866.
38. S. Lutz, *Curr. Opin. Biotechnol.*, 2010, **21**, 734.
39. B. Virnekäs, L. Ge, A. Plückthun, K. C. Schneider, G. Wellnhofer and S. E. Moroney, *Nucleic Acids Res.*, 1994, **22**, 5600.
40. W. P. Stemmer, *Nature*, 1994, **370**, 389.
41. W. P. Stemmer, *Proc. Natl. Acad. Sci. U. S. A.*, 1994, **91**, 10747.
42. W. L. Tang and H. Zhao, *Biotechnol. J.*, 2009, **4**, 1725.
43. L. He, A. M. Friedman and C. Bailey-Kellogg, *BMC Bioinformatics*, 2012, **13**(Suppl 3), S3.
44. G. P. Smith, *Science*, 1985, **228**, 1315.
45. J. K. Scott and G. P. Smith, *Science*, 1990, **249**, 386.
46. D. McGregor and S. Robins, *Anal. Biochem.*, 2001, **294**, 108.
47. G. Georgiou, H. L. Poetschke, C. Stathopoulos and J. A. Francisco, *Trends Biotechnol.*, 1993, **11**, 6.
48. S. A. Kenrick and P. S. Daugherty, *Protein Eng. Des. Sel.*, 2010, **23**, 9.
49. Y. Boublik, P. DiBonito and I. M. Jones, *Bio/Technology*, 1995, **13**, 1079.

50. E. T. Boder and K. D. Wittrup, *Nat. Biotechnol.*, 1997, **15**, 553.
51. T. A. Whitehead, A. Chevalier, Y. Song, C. Dreyfus, S. J. Fleishman, C. De Mattos, C. A. Myers, H. Kamisetty, P. Blair, I. A. Wilson and D. Baker, *Nat. Biotechnol.*, 2012, **30**, 543.
52. G. C. Rice, D. V. Goeddel, G. Cachianes, J. Woronicz, E. Y. Chen, S. R. Williams and D. W. Leung, *Proc. Natl. Acad. Sci. U. S. A.*, 1992, **89**, 5467.
53. P. M. Bowers, A. Horlick, T. Y. Neben, R. M. Toobian, G. L. Tomlinson, J. L. Dalton, H. A. Jones, A. Chen, L. Altobell III, X. Zhang, J. L. Macomber, I. P. Krapf, B. F. Wu, A. McConnell, B. Chau, T. Holland, A. D. Berkebile, S. S. Neben, W. J. Boyle and D. J. King, *Proc. Natl. Acad. Sci. U. S. A.*, 2011, **108**, 20455.
54. C. Zhou, F. W. Jacobsen, L. Cai, Q. Chen and W. D. Shen, *mAbs*, 2010, **2**, 508.
55. M. G. Cull, J. F. Miller and P. J. Schatz, *Proc. Natl. Acad. Sci. U. S. A.*, 1992, **89**, 1865.
56. A. S. Perelson and G. F. Oster, *J. Theor. Biol.*, 1979, **81**, 645.
57. L. C. Mattheakis, R. R. Bhatt and W. J. Dower, *Proc. Natl. Acad. Sci. U. S. A.*, 1994, **91**, 9022.
58. M. He and M. J. Taussig, *Nucleic Acids Res.*, 1997, **25**, 5132.
59. A. Plückthun, *Methods Mol. Biol.*, 2012, **805**, 3.
60. R. Odegrip, D. Coomber, B. Eldridge, R. Hederer, P. A. Kuhlman, C. Ullman, K. FitzGerald and D. McGregor, *Proc. Natl. Acad. Sci. U. S. A.*, 2004, **101**, 2806.
61. J. Bertschinger and D. Neri, *Protein Eng. Des. Sel.*, 2004, **17**, 699.
62. H. Reiersen, I. Lobersli, G. A. Løset, E. Hvattum, B. Simonsen, J. E. Stacy, D. McGregor, K. Fitzgerald, M. Welschof, O. H. Brekke and O. J. Marvik, *Nucleic Acids Res.*, 2005, **33**, e10.
63. D. S. Tawfik and A. D. Griffiths, *Nat. Biotechnol.*, 1998, **16**, 652.
64. T. Nishikawa, T. Sunami, T. Matsuura and T. Yomo, *J. Nucleic Acids*, 2012, **2012**, 923214.
65. D. J. Scott and A. Plückthun, *J. Mol. Biol.*, 2013, **425**, 662.
66. D. M. Fowler, C. L. Araya, S. J. Fleishman, E. H. Kellogg, J. J. Stephany, D. Baker and S. Fields, *Nat. Methods*, 2010, **7**, 741.
67. C. L. Araya and D. M. Fowler, *Trends Biotechnol.*, 2011, **29**, 435.
68. T. Makino, G. Skretas, T. H. Kang and G. Georgiou, *Metab. Eng.*, 2011, **13**, 241.
69. M. G. Casteleijn, A. Urtti and S. Sarkhel, *Int. J. Pharm.*, 2013, **440**, 39.
70. J. F. Zawada, G. Yin, A. R. Steiner, J. Yang, A. Naresh, S. M. Roy, D. S. Gold, H. G. Heinsohn and C. J. Murray, *Biotechnol. Bioeng.*, 2011, **108**, 1570.
71. G. Yin, E. D. Garces, J. Yang, J. Zhang, C. Tran, A. R. Steiner, C. Roos, S. Bajad, S. Hudak, K. Penta, J. Zawada, S. Pollitt and C. J. Murray, *mAbs*, 2012, **4**, 217.
72. Q. Wang, A. R. Parrish and L. Wang, *Chem. Biol.*, 2009, **16**, 323.
73. J. Gubbens, S. J. Kim, Z. Yang, A. E. Johnson and W. R. Skach, *RNA*, 2010, **16**, 1660.

74. H. Cho, T. Daniel, Y. J. Buechler, D. C. Litzinger, Z. Maio, A. M. Putnam, V. S. Kraynov, B. C. Sim, S. Bussell, T. Javahishvili, S. Kaphle, G. Viramontes, M. Ong, S. Chu, G. C. Becky, R. Lieu, N. Knudsen, P. Castiglioni, T. C. Norman, D. W. Axelrod, A. R. Hoffman, P. G. Schultz, R. D. DiMarchi and B. E. Kimmel, *Proc. Natl. Acad. Sci. U. S. A.*, 2011, **108**, 9060.

75. D. R. Casimiro, P. E. Wright and H. J. Dyson, *Structure*, 1997, **5**, 1407.

76. W. Peti, R. Page, K. Moy, M. O'Neil-Johnson, I. A. Wilson, R. C. Stevens and K. Wüthrich, *J. Struct. Funct. Genomics*, 2005, **6**, 259.

77. D. R. Jensen, C. Woytovich, M. Li, P. Duvnjak, M. S. Cassidy, R. O. Frederick, L. F. Bergeman, F. C. Peterson and B. F. Volkman, *Protein Sci.*, 2010, **19**, 570.

78. G. Walsh, *Nat. Biotechnol.*, 2010, **28**, 917.

79. J. M. Carton and W. R. Strohl, *Introduction to Biological and Small Molecule Drug Research and Development: Theory and Case Studies*, ed. C. R. Ganellin, R. Jefferis and S. M. Roberts, Elsevier, Oxford, UK, 2013, pp. 127–159.

80. A. R. Duncan, J. M. Woof, L. J. Partridge, D. R. Burton and G. Winter, *Nature*, 1988, **332**, 563.

81. A. R. Duncan and G. Winter, *Nature*, 1988, **332**, 738.

82. All product sales figures are from the www.pipelinereview.com 2012 summary of biological sales: Blockbuster Biologics 2012, published May 2013.

83. J. Uy, L. Fogelfeld and Y. Guerra, *Diabetes Metab. Syndr. Obes.*, 2012, **5**, 1.

84. M. Joyeux-Faure, *J. Pharmacol. Exp. Ther.*, 2007, **323**, 759.

85. A. M. Sinclair and S. Elliott, *J. Pharm. Sci.*, 2005, **94**, 1626.

86. J. M. Reichert, *mAbs*, 2012, **4**, 413.

87. J. M. Reichert, *mAbs*, 2013, **5**, 1.

88. W. R. Strohl and L. M. Strohl, *Therapeutic Antibody Engineering*, Woodhead Publishing, Cambridge, UK, 2012.

89. S. Roberts, J. C. Cheetham and A. R. Rees, *Nature*, 1987, **328**, 731.

90. D. Kuroda, H. Shirai, M. P. Jacobson and H. Nakamura, *Protein Eng. Des. Sel.*, 2012, **25**, 507.

91. S. Angal, D. J. King, M. W. Bodmer, A. Turner, A. D. Lawson, G. Roberts, B. Pedley and J. R. Adair, *Mol. Immunol.*, 1993, **30**, 105.

92. A. Lyons, D. J. King, R. J. Owens, G. T. Yarranton, A. Millican, N. R. Whittle and J. R. Adair, *Protein Eng.*, 1990, **3**, 703.

93. J. R. Junutula, H. Raab, S. Clark, S. Bhakta, D. D. Leipold, S. Weir, Y. Chen, M. Simpson, S. P. Tsai, M. S. Dennis, Y. Lu, Y. G. Meng, C. Ng, J. Yang, C. C. Lee, E. Duenas, J. Gorrell, V. Katta, A. Kim, K. McDorman, K. Flagella, R. Venook, S. Ross, S. D. Spencer, W. Lee Wong, H. B. Lowman, R. Vandlen, M. X. Sliwkowski, R. H. Scheller, P. Polakis and W. Mallet, *Nat. Biotechnol.*, 2008, **26**, 925.

94. J. B. Ridgway, L. G. Presta and P. Carter, *Protein Eng.*, 1996, **9**, 617.

95. S. Atwell, J. B. Ridgway, J. A. Wells and P. Carter, *J. Mol. Biol.*, 1997, **270**, 26.

96. G. L. Moore, H. Chen, S. Karki and G. A. Lazar, *mAbs*, 2010, **2**, 181.
97. J. Zalevsky, A. K. Chamberlain, H. M. Horton, S. Karki, I. W. Leung, T. J. Sproule, G. A. Lazar, D. C. Roopenian and J. R. Desjarlais, *Nat. Biotechnol.*, 2010, **28**, 157.
98. T. T. Kuo and V. G. Aveson, *mAbs*, 2011, **3**, 422.
99. R. Jefferis, in *Antibody-Fc: Linking Adaptive and Innate Immunity*, ed. M. E. Ackermann and F. Nimmerhahn, Elsevier, Oxford, UK, 2013, in press.
100. N. Chennamsetty, V. Voynov, V. Kayser, B. Helk and B. L. Trout, *Proc. Natl. Acad. Sci. U. S. A.*, 2009, **106**, 11937.
101. S. J. Wu, J. Luo, K. T. O'Neil, J. Kang, E. R. Lacy, G. Canziani, A. Baker, M. Huang, Q. M. Tang, T. S. Raju, S. A. Jacobs, A. Teplyakov, G. L. Gilliland and Y. Feng, *Protein Eng. Des. Sel.*, 2010, **23**, 643.
102. K. Dudgeon, R. Rouet, I. Kokmeijer, P. Schofield, J. Stolp, D. Langley, D. Stock and D. Christ, *Proc. Natl. Acad. Sci. U. S. A.*, 2012, **109**, 10879.
103. S. J. Wodak, *Nat. Biotechnol.*, 2012, **30**, 502.
104. J. Bostrom, S. F. Yu, D. Kan, B. A. Appleton, C. V. Lee, K. Billeci, W. Man, F. Peale, S. Ross, C. Wiesmann and G. Fuh, *Science*, 2009, **323**, 1610.
105. J. Bostrom, L. Haber, P. Koenig, R. F. Kelley and G. Fuh, *PLoS ONE*, 2011, **6**, e17887.
106. M. S. Neuberger, G. T. Williams, E. B. Mitchell, S. S. Jouhal, J. G. Flanagan and T. H. Rabbitts, *Nature*, 1985, **314**, 268.
107. G. T. Williams and M. S. Neuberger, *Gene*, 1986, **43**, 319.
108. W. Liu, M. Onda, B. Lee, R. J. Kreitman, R. Hassan, L. Xiang and I. Pastan, *Proc. Natl. Acad. Sci. U. S. A.*, 2012, **109**, 11782.
109. N. Becker and I. Benhar, *Antibodies*, 2012, **1**, 39.
110. M. Borriello, P. Laccetti, G. Terrazzano, G. D'Alessio and C. De Lorenzo, *Br. J. Cancer*, 2011, **104**, 1716.
111. S. D. Gillies, Y. Lan, T. Hettmann, B. Brunkhorst, Y. Sun, S. O. Mueller and K. M. Lo, *Clin. Cancer Res.*, 2011, **17**, 3673.
112. K. L. Gutbrodt and D. Neri, *Antibodies*, 2012, **1**, 70.
113. A. Beck and J. M. Reichert, *mAbs*, 2011, **3**, 415.
114. D. M. Czajkowsky, J. Hu, Z. Shao and R. J. Pleass, *EMBO Mol. Med.*, **4**, 1015.
115. K. M. Mohler, D. S. Torrance, C. A. Smith, R. G. Goodwin, K. E. Stremler, V. P. Fung, H. Madani and M. B. Widmer, *J. Immunol.*, 1993, **151**, 1548.
116. D. P. Williams, K. Parker, P. Bacha, W. Bishai, M. Borowski, F. Genbauffe, T. B. Strom and J. R. Murphy, *Protein Eng.*, 1987, **1**, 493.
117. R. E. Bird, K. D. Hardman, J. W. Jacobson, S. Johnson, B. M. Kaufman, S. M. Lee, T. Lee, S. H. Pope, G. S. Riordan and M. Whitlow, *Science*, 1988, **242**, 423.
118. J. S. Huston, D. Levinson, M. Mudgett-Hunter, M. S. Tai, J. Novotný, M. N. Margolies, R. J. Ridge, R. E. Bruccoleri, E. Haber, R. Crea and H. Oppermann, *Proc. Natl. Acad. Sci. U. S. A.*, 1988, **85**, 5879.

119. G. J. Lieschke, P. K. Rao, M. K. Gately and R. C. Mulligan, *Nat. Biotechnol.*, 1997, **15**, 35.
120. T. Sasaoka, M. Ito, J. Yamashita, K. Nakajima, I. Tanaka, M. Narita, Y. Hara, K. Hada, M. Takahashi, Y. Ohno, T. Matsuo, Y. Kaneshiro, H. Tanaka and K. Kaneko, *Am. J. Physiol. Gastrointest. Liver Physiol.*, 2011, **300**, G568.
121. S. L. Morrison, M. J. Johnson, L. A. Herzenberg and V. T. Oi, *Proc. Natl. Acad. Sci. U. S. A.*, 1984, **81**, 6851.
122. A. F. LoBuglio, R. H. Wheeler, J. Trang, A. Haynes, K. Rogers, E. B. Harvey, L. Sun, J. Ghrayeb and M. B. Khazaeli, *Proc. Natl. Acad. Sci. U. S. A.*, 1989, **86**, 4220.
123. W. Zhang, Y. Li and Y. Tang, *Proc. Natl. Acad. Sci. U. S. A.*, 2008, **105**, 20683.
124. J. M. Crawford and C. A. Townsend, *Nat. Rev. Microbiol.*, 2010, **8**, 879.
125. X. Gao, P. Wang and Y. Tang, *Appl. Microbiol. Biotechnol.*, 2010, **88**, 1233.
126. J. Du, Z. Shao and H. Zhao, *J. Ind. Microbiol. Biotechnol.*, 2011, **38**, 873.
127. J. M. Winter and Y. Tang, *Curr. Opin. Biotechnol.*, 2012, **23**, 736.
128. T. K. Nordt and C. Bode, *Heart*, 2003, **89**, 1358.
129. A. G. Porter, L. D. Bell, J. Adair, G. H. Catlin, J. Clarke, J. A. Davies, K. Dawson, R. Derbyshire, S. M. Doel, L. Dunthorne, M. Finlay, J. Hall, M. Houghton, C. Hynes, I. Lindley, M. Nugent, G. J. O'Neil, J. C. Smith, A. Stewart, W. Tacon, J. Viney, N. Warburton, P. G. Boseley and K. G. McCullagh, *DNA*, 1986, **5**, 137.
130. H. Weber and C. Weissmann, *Nucleic Acids Res.*, 1983, **11**, 5661.
131. G. Winter, A. R. Fersht, A. J. Wilkinson, M. Zoller and M. Smith, *Nature*, 1982, **299**, 756.
132. A. J. Wilkinson, A. R. Fersht, D. M. Blow, P. Carter and G. Winter, *Nature*, 1984, **307**, 187.
133. A. Fersht and G. Winter, *Trends Biochem. Sci.*, 1992, **17**, 292.
134. M. B. Quin and C. Schmidt-Dannert, *ACS Catal.*, 2011, **1**, 1017.
135. Q. Li, P. Marek and B. L. Iverson, *FEBS Lett.*, 2013, **587**, 1155.
136. C. S. Craik, M. J. Page and E. L. Madison, *Biochem. J.*, 2011, **435**, 1.
137. B. C. Cunningham, D. J. Henner and J. A. Wells, *Science*, 1990, **247**, 1461.
138. A. P. Seddon, D. Aviezer, L.-Y. Li, P. Bohlen and A. Yayon, *Biochemistry*, 1995, **34**, 73.
139. R. J. Kreitman, R. K. Puri and I. Pastan, *Proc. Natl. Acad. Sci. U. S. A.*, 1994, **91**, 6889.
140. F. Martin, C. Toniatti, A. L. Salvati, G. Ciliberto, R. Cortese and M. Sollazzo, *J. Mol. Biol.*, 1996, **255**, 86.
141. H. K. Binz, P. Amstutz and A. Pluckthun, *Nat. Biotechnol.*, 2005, **23**, 1257.
142. M. Gebauer and A. Skerra, *Curr. Opin. Chem. Biol.*, 2009, **13**, 245.
143. D. Lipovsek, *Protein Eng. Des. Sel.*, 2011, **24**, 3.

144. A. Koide, J. Wojcik, R. N. Gilbreth, R. J. Hoey and S. Koide, *J. Mol. Biol.*, 2012, **415**, 393.

145. R. Tamaskovic, M. Simon, N. Stefan, M. Schwill and A. Plückthun, *Methods Enzymol.*, 2012, **503**, 101.

146. K. Gehlsen, R. Gong, D. Bramhill, D. Wiersma, S. Kirkpatrick, Y. Wang, Y. Feng and D. S. Dimitrov, *mAbs*, 2012, **4**, 466.

147. S. Jacobs and K. O'Neil, in *Protein Engineering*, ed. P. Kaumaya, Intech, Rijeka, Croatia, 2012, 7, p. 145.

148. F. Zoller, U. Haberkorn and W. Mier, *Molecules*, 2011, **16**, 2467.

149. C. S. Mintz and R. Crea, *BioProcess Intl.*, 2013, **11**, 40.

150. C. K. Wang, Q. Kaas, L. Chiche and D. J. Craik, *Nucleic Acids Res.*, 2008, **36**, D206.

151. D. J. Craik, *Toxicon.*, 2010, **56**, 1092.

152. L. Cascales and D. J. Craik, *Org. Biomol. Chem.*, 2010, **8**, 5035.

153. K. Jagadish and J. A. Camarero, *Biopolymers*, 2010, **94**, 611.

154. C. T. Wong, D. K. Rowlands, C. H. Wong, T. W. Lo, G. K. Nguyen, H. Y. Li and J. P. Tam, *Angew Chem. Int. Ed. Engl.*, 2012, **51**, 5620.

155. J. Contreras, A. Y. Elnagar, S. F. Hamm-Alvarez and J. A. Camarero, *J. Control Release*, 2011, **155**, 134.

156. K. P. Greenwood, N. L. Daly, D. L. Brown, J. L. Stow and D. J. Craik, *Int. J. Biochem. Cell Biol.*, 2007, **39**, 2252.

157. K. E. Drexler, *Proc. Natl. Acad. Sci. U. S. A.*, 1981, **78**, 5275.

158. C. Pabo, *Nature*, 1983, **301**, 200.

159. L. Regan and W. F. DeGrado, *Science*, 1988, **241**, 976.

160. M. H. Hecht, A. Das, A. Go, L. H. Bradley and Y. Wei, *Protein Sci.*, 2004, **13**, 1711.

161. A. Das, Y. Wei, I. Pelczer and M. H. Hecht, *Protein Sci.*, 2011, **20**, 702.

162. L. Dai, Y. Yang, H. R. Kim and Y. Zhou, *Proteins*, 2012, **78**, 2338.

163. C. A. Smith and T. Kortemme, *PLoS ONE*, 2011, **6**, e20451.

164. T. Nugent and D. T. Jones, *Proc. Natl. Acad. Sci. U. S. A.*, 201, **109**, E1540.

165. M. W. Lluis, J. I. Godfroy 3rd and H. Yin, *Protein Eng. Des. Sel.*, 2013, **26**, 91.

166. M. Di Marco, S. Shamsuddin, K. A. Razak, A. A. Aziz, C. Devaux, E. Borghi, L. Levy and C. Sadun, *Int. J. Nanomedicine*, 2010, **5**, 37.

167. D. C. Julien, S. Behnke, G. Wang, G. K. Murdoch and R. A. Hill, *mAbs*, 2011, **3**, 467.

168. M. M. Stratton and S. N. Loh, *Protein Sci*, 2011, **20**, 19.

169. Z. Li, Y. Yang, J. Zhan, L. Dai and Y. Zhou, *Annu. Rev. Biophys.*, 2013, **42**, 1.

170. A. Barrozo, R. Borstner, G. Marloie and S. C. L. Kamerlin, *Int. J. Mol. Sci.*, 2012, **13**, 12428.

171. J. R. Adair and T. P. Wallace, in *Molecular Biomethods Handbook*, ed. R. Rapley and J. Walker, Humana Press Inc, Totowa, NJ, USA, 1998, pp. 347–360.

172. S. N. Ho, H. D. Hunt, R. M. Horton, J. K. Pullen and L. R. Pease, *Gene*, 1989, 77, 51.

CHAPTER 8

Biosensors

MARTIN FRANK CHAPLIN

London South Bank University, Borough Road, London SE1 0AA, UK
Email: martin.chaplin@lsbu.ac.uk

8.1 INTRODUCTION

Biosensors are analytical devices that convert biological actions into electrical signals in order to quantify them.[1-3] In this chapter, only biosensors that make use of the specificity of biological processes are described, that is, the recognition of enzymes for their substrates or other ligands, antibodies for their antigens, lectins for carbohydrates and nucleic acids or peptide nucleic acids for their complementary sequences. Biosensor science is interdisciplinary, bringing together chemistry, physics, biology, electronics and engineering to solve real-world analytical problems. The parts of a typical biosensor are shown in Figure 8.1.

The primary advantage of using biologically active molecules as part of a biosensor is their high specificity and hence, high discriminatory power. Thus, they are generally able to detect particular molecular species from within complex mixtures of other materials with similar structure that may be present at comparable or substantially higher concentrations. Often, samples can be analysed with little or no prior clean-up. In this aspect they show distinct advantages over most 'traditional' analytical methods; for example, colorimetric assays like the Lowry assay for proteins.

Biosensors may serve a number of analytical purposes. In some applications, for example in clinical diagnosis, it is sometimes important only to determine whether the analyte is above or below some pre-determined threshold, whereas in process control there often needs to be a continual

Molecular Biology and Biotechnology, 6th Edition
Edited by Ralph Rapley and David Whitehouse
© The Royal Society of Chemistry 2015
Published by the Royal Society of Chemistry, www.rsc.org

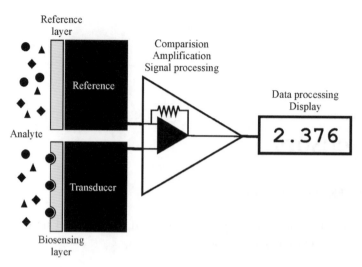

Reference
layer

Comparision
Amplification
Signal processing

Data processing
Display

Reference

2.376

Analyte

Transducer

Biosensing
layer

Figure 8.1   The functional units of a biosensor. The biological reaction usually takes place in close contact with the electrical or optical transducer, here shown as a 'black-box'. This intimate arrangement ensures that most of the biological reaction is detected. The resultant electrical or optical signal is compared with a reference signal that is usually produced by a similar system without the biologically active material. The difference between these two signals, which optimally is proportional to the material being analysed (*i.e.* the analyte), is amplified, processed and displayed or recorded. The reference electrode stabilises the output, so reducing signal drift and improving reproducibility.

accurate but imprecise feed-back of the level of analyte present. In the former case, the biosensor must be designed to give the minimum number of false positives and, more importantly, false negatives. In the latter case, it is generally the rapid response to changes in the analyte that is more important than its absolute precision. Minimising false negatives is often more important in clinical analyses than minimising false positives, as in the former case a diagnosis may be missed and therapy delayed whereas in the latter case further tests may show the error. Other biosensor analytical applications may require accuracy and precision over a wide analytical range.

If they are to be acceptable, biosensors must show advantages over the use of the free active biomolecules, which possess at least equal specificity and discriminatory power. Their main advantages usually involve rapid response, the reduced need for sample pre-treatment, their ease of use and often their reusability and portability. Repetitive re-use of the same biologically active sensing material generally ensures that similar samples give similar responses as this avoids any need for accurate aliquoting of such biological materials to contain precisely similar activity. This circumvents the possibility of introducing errors by inaccurate pipetting or dilution and is a necessary prerequisite for automated and on-line monitoring. Repetitive and reagentless methodology offers considerable savings in terms of reagent

Table 8.1 The properties required of a successful biosensor.

Required property	Achievable with ease?
Specificity	Yes
Discrimination	Yes, except for very low concentrations in complex biological fluids
Repeatability	Yes
Precision	Yes
Safe	Yes
Biocompatibility	Problematic, but only required for *in vivo* sensors
Accuracy	Yes, as easily calibrated
Appropriate sensitivity	Yes, except in trace analysis
Fast response	Yes, usually
Miniaturizable	Yes
Small sample volumes	Yes
Temperature independence	Yes, may be electronically compensated
Low production costs	Yes, but only if mass-produced
Reliability	Yes, if to be useful
Marketable	Difficult, if there is pre-existing competing methodology
Drift free	Difficult but possible; may be compensated
Continuous use	Yes, for short periods (days)
Robust	Getting there, but generally need careful handling
Stability	No, except on storage (months) or in the short term (weeks)
Sterilizable	No, except on initial storage
Autoclavable	Not currently achievable

costs, so reducing the cost per assay. In addition, the increased operator time per assay and the associated higher skill required for 'traditional' assay methods also involves a cost penalty which is often greater than that due to the reagents. These advantages must be sufficient to encourage the high investment necessary for the development of a biosensor and the purchase price to the end-user. Table 8.1 lists a number of important attributes that a successful biosensor may be expected to possess. In any particular case only some of these may be achievable.

The different types of biosensors have their own advantages and disadvantages that are summarised in Table 8.2. Apart from the important related area of colorimetric test strips, the most important commercial biosensors are electrochemical, but there are exciting new developments in optical and piezoelectric biosensors. Amperometric biosensors are available for glucose, lactate, glycerol, ethanol, lactose, L-amino acids, cholesterol, fish freshness and micro-organisms. Potentiometric devices have been marketed for glucose, other low molecular weight carbohydrates and alcohols. Receptor-based and immunosensors have been marketed utilising surface plasmon resonance devices and piezocrystals.

Most of the market share for biosensor applications is for glucose analysis for use in the monitoring and control of diabetes.[4] Only a few years ago, market research projections for the growth of the biosensor industry

Table 8.2 The properties of some important biosensor configurations. Constant development reduces the more negative features when large-scale production is planned.

Biosensor	Amperometric	Potentiometric	Piezoelectric	Optical
Cost	Low	Low	Medium	High
Reliability	High	Medium	Improving	Medium
Complexity	Medium	Low	Medium	High, becoming less
Selectivity	High	High	Medium	High
Sensitivity	High	Medium	Low but improving	Medium
Speed of response	Medium	Slow	Fast	Sometimes fast
Applicability	High	Medium	Medium	Medium
Present usage	Highest	Medium	Medium but growing	Medium but growing
Future prospects	High	Medium	High	High

predicted expansion at an impressively fast pace of almost 50% per year. These were over-optimistic, with current growth about 20%. However, industry sales as a whole were over \$12 000 million in 2012. Although for many years there have been over a thousand scientific biosensors papers published each year, in the past biosensor science has not progressed as significantly as other areas of modern biotechnology.[5] A vast number of biosensor systems have been published, some being very imaginative, but with just a few reaching the marketplace and even these often utilise well-established ideas. It is now accepted that there are substantial and investment-intensive difficulties involved in producing such robust and reliable commercial analytical devices that are able to operate under authentic real-life conditions, even where a novel and highly promising prototype device has been produced. The cost of the biosensor is often of importance and considerable effort has been expended in the production of disposable devices using cheap integrated chip technology; disposable technology offering opportunities for continuing and increased revenue. Such silicon-based methodology[6] has opened up the possibility for having a number of different sensors on one device, allowing multi-parameter assays. A realistic target for the future is a density of a million sensors per square centimetre. However, there must be a demand for very large numbers of such devices in order for their production to be economically viable. Relatively unsophisticated processes, involving screen-printing or ink-jet technologies are currently preferred, due to their lower start-up costs, whereas photolithography can be used for the production of sub-micron structures with well-defined geometry.

By far the largest biosensor application area is in clinical diagnostics.[7] This includes monitoring of critical metabolites during surgery. The major target markets are concerned with use for self-testing at home, within the physician's office for screening and within casualty and other hospital departments for point-of-care diagnosis. These application areas are potentially very wide. The use of rapid biosensor techniques in doctors' surgeries and in hospitals avoids the need for expensive and, most importantly, time-consuming testing at central clinical laboratories. Thus diagnosis and treatment may start during the first visit of a patient. This removes the need to wait for a return visit after the clinical tests have been completed elsewhere and allowing time, perhaps, for the patient's condition to deteriorate somewhat. Also, there is less likelihood of the sample being mishandled or contaminated. As an example, testing for diabetes using glucose biosensors is now routine even if a patient has an apparently unrelated complaint. Centralised clinical analytical facilities remain a necessity due to the need, in many cases, for multiple different analyses on the same sample and difficulties such as regulatory compliance. Legislation and the increasing possibility of medical malpractice claims often impose stringent quality assurance and control standards on clinical analyses. This makes it much more expensive to bring novel clinical biosensors to the market place today than in the past but generally allows developments in those biosensors that are already established, giving them a distinct competitive edge.

Home diagnosis and testing is an area that is being opened up by, for example, pregnancy and ovulation test kits, blood cholesterol and diabetes control.[8] Clearly, there are risks and problems involved with their more widespread use, but many people prefer to use them as indications for whether a trip to the doctor's surgery is really necessary or not. As counselling may be necessary with some potential home diagnosis applications (for example, cancer and AIDS), controversy exists over their development.

One of the major potential uses for biosensors is for *in vivo* applications such as for the close control of type 1 diabetes.[9] The purpose here is to monitor continuously the levels of metabolites such that corrective action can be employed immediately when necessary. Clearly such biosensors must be biocompatible and miniaturised so that they are implantable. In addition they should be reagentless, the reaction being controlled only by the presence of the metabolite and the stabilised bioreagent. The signal generated must be drift-free over the period of interest. At the present time, such biosensors have a relatively short lifespan of a few days at most, due mainly to problems which arise from the body's response.

Industrial analyses involve food, cosmetics and fermentation process control and quality control and monitoring. The defence industry is interested in detectors for explosives, nerve gases, viruses, and microbial spores and toxins. Environmental uses of biosensors are mainly in areas of water quality and pollution control. A typical application might be to detect parts per million of particular molecular species such as an industrial toxin within the highly complex mixtures produced as process effluent.

8.2 THE BIOLOGICAL REACTION

An important factor in most biosensor configurations is the sensing surface.[10] This normally consists of a thin layer of biologically active material in intimate contact with the electronic or optical transducer. In some cases the biological material may be covalently or non-covalently attached to the surface but often in electrochemical biosensors it forms part of a thin membrane covering the sensing surface. Generally the conversion of the biological process into an electronic or optical signal is most efficient where there is minimal distance between where the biological reaction or binding takes place and where the electronic transduction takes place. In addition, it is important for the retention of biological activity that the biological material is not lost into analyte solutions. The immobilisation technology for holding the biocatalyst in place is extensive, with examples of immobilisation methods summarised in Table 8.3. Much current research is directed at stabilising enzymes[11] and increasing their sensitivity[12] for use in biosensors. At suitable pH, polyelectrolytes such as diethylaminoethyldextran wrap around enzymes so restricting their movement and reducing their tendency to denature. Polyelectrolytes in combination with polyalcohols, such as sorbitol and lactitol, have been shown to considerably stabilise enzymes against thermal inactivation for use in biosensors. More recently, enzymes have been specifically engineered for their use in biosensors.[12]

Often the sensitivity is an important sensor parameter, particularly when looking for disease, pathogen or allergen markers. This may be assessed

Table 8.3   Examples of biosensor immobilisation methods.

Physical entrapment
Biologically active material held next to the sensing surface by a semi-permeable membrane that prevents it from escaping to the bulk phase but allows passage of the analyte. Often this membrane is synthesised *in situ* and sometimes it is made such that it increases the specificity of the sensing or reduces unwanted side-reactions. This is usually a simple and inexpensive method.

Non-covalent binding
Adsorption of proteins to a porous carbon or silicon electrode. This may suffer from a gradual leaching of the enzyme to the bulk phase.

Covalent binding
Treatment of the biosensor surface with 3-aminopropyltriethoxysilane followed by coupling of biologically active material to the reactive amino groups remaining on the cross-linked siloxane surface. Proteins may be attached by use of carbodiimides, which form amide links between amines and carboxylic acids. Such methods permanently attach the biological material but are difficult to reproduce exactly and sometimes cause a large reduction in activity. Recent developments make use of high surface area nanoparticles.

Membrane entrapment
Crosslink proteins with glutaraldehyde within a cellulose or nylon supporting net. Due to limiting diffusion, this method can only be used for low molecular weight analytes.

from the analyte level when both false negatives and false positives arise and will depend on their relative importance. Often the sensitivity is recognised as the level when, due to the signal noise, there is less than a 5% chance of a positive value in the absence of analyte.

8.3   THEORY

In the absence of diffusion effects (see later), most biological reactions and binding processes can be described in terms of saturation kinetics:

$$\text{biological material} + \text{analyte} \rightleftharpoons \text{bound analyte} \qquad (8.1)$$

The bound analyte is then detected optically or by weight, or gives rise to a biological action, so generating the electronic response. This electronic or optical response varies with the extent of binding or biological response that, in turn, varies with the concentration of the bound analyte. Apart from the logarithmic relationship in potentiometric biosensors, the optical, biological and electronic responses are often proportional.

bound analyte $\rightarrow$ biological or optical response $\rightarrow$ electronic response

$$\text{electronic response} = \frac{(\text{maximum electronic response possible}) \times (\text{analyte concentration})}{(\text{half-saturation constant}) + (\text{analyte concentration})} \qquad (8.2)$$

where the half-saturation constant is equal to the analyte concentration which gives rise to half the maximum electronic response possible (Figure 8.2). The response is linear, to within 95%, at analyte concentrations

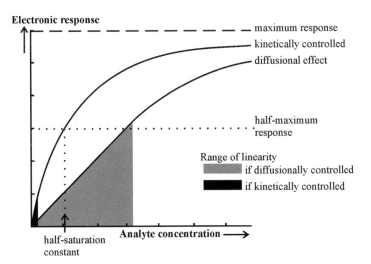

Figure 8.2   The range of response of a biosensor under biocatalytic kinetic and diffusional control.

up to about one-twentieth of the half-saturation constant. A biosensor may be used over a wider, non-linear range if it has compensatory electronics.

This relationship holds for both reacting (*e.g.* enzymes) and non-reacting (*e.g.* immunosorption) processes. Where the analyte reacts as part of the biological response (*e.g.* during a biocatalytic reaction utilising enzyme(s) and/or microbial cells), an additional factor is the diffusion of the analyte from the bulk of the solution to the reactive surface. If this rate of diffusion is less than the rate at which the analyte would otherwise react, there follows a reduction in the local concentration of analyte undergoing reaction. The rate of diffusion increases as the concentration gradient increases.

$$\text{rate of diffusion} = (\text{diffusivity constant}) \times (\text{analyte concentration gradient}) \quad (8.3)$$

where the analyte concentration gradient is given by the difference between the bulk analyte concentration and the local (microenvironmental) analyte concentration on the sensing surface of the biosensor divided by the distance through which the analyte must diffuse.

As most biocatalytic biosensor configurations utilise a membrane-entrapped biocatalyst this concentration gradient depends not only on the analyte concentration in the bulk and within the membrane but also on the membrane's thickness. The thicker the membrane, the greater the diffusive distance from the bulk of the solution to the distal sensing surface of the biosensor and the greater the amount of biocatalyst encountered. Both effects increase the likelihood that the overall reaction will be controlled by diffusion. Hence, such biosensors can be designed to be under diffusional or kinetic control by varying the membrane thickness. When the rate of analyte diffusion is slower than the rate at which the biocatalyst can react, the electronic response decreases due to the lower level of analyte available for reaction. A steady state is rapidly established when the rate of arrival of the analyte equals its rate of reaction. This steady state condition may be determined wholly by the rate of diffusion (diffusional control) or wholly by the rate of reaction (kinetic control) or by an intermediate dependence. Where the reaction rate depends solely on the rate of diffusive flux of the analyte, this determines the electronic response.

$$\text{electronic response} \propto \text{rate of diffusive flux} \quad (8.4)$$

As the rate of diffusion depends on the bulk concentration of the analyte, this electronic response is linearly related to the bulk analyte concentration and, most importantly and intriguingly, is independent of the properties of the enzyme. Thus, the biosensor is linear over a much wider range of substrate concentrations (see Figure 8.2) and relatively independent of changes in the pH and temperature of the biocatalytic membrane, so long as the system remains diffusion controlled. It should be noticed, however, that under these conditions the response is reduced relative to a system containing the same amount of biocatalyst but not diffusionally limited.

Maximum sensitivity to analyte concentration would be accomplished by the utilisation of thin membranes containing a high biocatalyst activity and a well-stirred analyte solution. The overall kinetics of most biosensor configurations are difficult to predict. They depend on the diffusivities in the bulk phase and within the biocatalytic volume, the nature, porosity and physical properties of any membrane, the intrinsic biocatalytic kinetics, the electronic transduction process and kinetics, the way in which the analyte is presented, and on other non-specific factors. Generally, such overall kinetics are determined experimentally using the complete bio-sensor and, hence, it is very important that the biosensor configuration is reproducible.

In biosensors utilising binding only, such as immunosensors, the major problem encountered is non-specific absorption that blocks the binding sites. There is need to minimise this and maximise the specific binding. As binding is an equilibrium process, high sensitivity necessitates a very high affinity between the analyte and the sensor surface.

8.4 ELECTROCHEMICAL METHODS

Electrochemical biosensors are generally fairly simple devices.[13] There are three types utilising electrical current, potential or resistive changes:

1. Amperometric biosensors, which determine the electric current associated with the electrons involved in redox processes;
2. Potentiometric biosensors, which use ion selective electrodes to determine changes in the concentration of chosen ions (*e.g.* hydrogen ions); and
3. Conductimetric biosensors, which determine conductance changes associated with changes in the ionic environment.

There has been much progress in miniaturising these devices using microfabrication technologies developed by the electronics industry. These include the use of screen-printing and the deposition of nanolitre volumes of enzymes using advanced ink-jet printing and conducting inks.

8.4.1 Amperometric Biosensors

Enzyme catalysed redox reactions can form the basis of a major class of biosensors if the flux of redox electrons can be determined. Normally a constant potential is applied between two electrodes and the current, due to the electrode reaction is determined. The potential is held relative to that of a reference electrode and is chosen such that small variations do not affect the rate of the electrode reaction. Direct electron transfer between the electrode and the redox site associated with the enzyme is kinetically hindered due to the distance between them or the presentation of an unfavourable pathway for the electrons. This can usually be addressed by the use of mediators and/or special electrode materials.

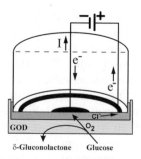

Figure 8.3 Amperometric glucose biosensor based on the oxygen electrode utilising glucose oxidase (GOD).

The first and simplest biosensor was based on this principle. It was for the determination of glucose and made use of the Clark oxygen electrode.[14] Figure 8.3 shows a section through such a simple amperometric biosensor. A potential of -0.6 V is applied between the central platinum cathode and the surrounding silver/silver chloride reference electrode (the anode). Dissolved molecular oxygen at the platinum cathode is reduced and the circuit is completed by means of the saturated KCl solution. Only oxygen can be reduced at the cathode due to its covering by a thin Teflon or polypropylene membrane through which the oxygen can diffuse but which acts as a barrier to other electroactive species.

$$\text{Pt cathode reaction}\,(-0.6\text{ V):}\quad O_2 + 4H^+ + 4e^- \rightarrow 2H_2O \qquad (8.5)$$

$$\text{Anode reaction:}\quad 4Ag^\circ + 4Cl^- \rightarrow 4AgCl + 4e^- \qquad (8.6)$$

The biocatalyst is retained next to the electrode by means of a membrane, which is permeable only to low molecular weight molecules including the reactants and products.

Glucose may be determined by the reduction in the dissolved oxygen concentration when the redox reaction, catalysed by glucose oxidase, occurs (Figure 8.4a).

$$\text{glucose} + O_2 \xrightarrow{\text{glucose oxidase}} \delta\text{-gluconolactone} + H_2O_2 \qquad (8.7)$$

It is fortunate that this useful enzyme is also one of the most stable oxidoreductases found.

Conditions can be chosen such that the rate at which oxygen is lost from the biocatalyst-containing compartment is proportional to the bulk glucose concentration. Other oxidases can be used in this biosensor configuration and may be immobilised as part of a membrane by treatment of the dissolved enzyme(s), together with a diluent protein, with glutaraldehyde on a cellulose or nylon support. An alternative method of determining the rate of

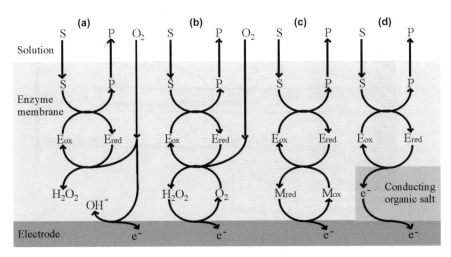

Figure 8.4 The redox mechanisms for various amperometric biosensor configurations.[18]

reaction is to detect the hydrogen peroxide produced directly by reversing the polarity of the electrodes (Figure 8.4b). This is the principle used in YSI (Yellow Springs) analysers. The use of a covering, such as the highly anionic Nafion membrane, prevents electroactive anions such as ascorbate reaching the electrode without restricting glucose. Internally, a cellulose acetate membrane replaces the Teflon membrane to allow passage of the hydrogen peroxide. This arrangement has a higher sensitivity than that utilising an oxygen electrode but, in the absence of highly selective membranes, is more prone to interference at the electrode surface.

$$\text{Cathode reaction: } 2AgCl + 2e^- \rightarrow 2Ag^\circ + 2Cl^- \tag{8.8}$$

$$\text{Pt anode reaction } (+0.6\text{ V}): H_2O_2 \rightarrow O_2 + 2H^+ + 2e^- \tag{8.9}$$

These electrodes can be developed further for the determination of substrates for which no direct oxidase enzyme exists. Thus sucrose can be determined by placing an invertase layer over the top of the glucose oxidase membrane in order to produce glucose (from the sucrose), which can then be determined. Interference from glucose in the sample can be minimised by including a thin anti-interference layer of glucose oxidase and peroxidase over the top of both layers, which removes the glucose without significantly reducing the oxygen diffusion to the electrode. A clever alternative approach to assay sucrose and glucose together[15] makes use of the lag period in the response due to the necessary inversion of the sucrose that delays its response relative to the glucose (Figure 8.5). This arrangement shows the typical craft required for using existing biosensors to analyse alternative analytes.

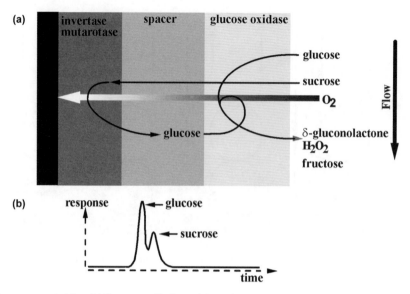

Figure 8.5 A kinetically controlled anti-interference membrane for sucrose. The sample is presented as a rapid pulse of material in the flowing stream. Use of phosphate buffer, which catalyses mutarotation, removes the need for the mutarotase enzyme.

Many biochemicals may be analysed by similar biosensors. One interesting example is the biosensor for determining the artificial sweetener aspartame in soft drinks where three enzymes are necessary in the sensing membrane in order to produce the H_2O_2 for the electrode reaction.

$$\text{aspartame} + H_2O \xrightarrow{\text{peptidase}} \text{L-aspartate} \atop + \text{L-phenylalanine methyl ester} \tag{8.10}$$

$$\text{L-aspartate} + \alpha\text{-ketoglutarate} \xrightarrow{\text{aspartate aminotransferase}} \text{L-glutamate} \atop + \text{oxaloacetate} \tag{8.11}$$

$$\text{L-glutamate} + O_2 + H_2O \xrightarrow{\text{glutamate oxidase}} \alpha\text{-ketoglutarate} + NH_4^+ + H_2O_2 \tag{8.12}$$

Fish freshness can be determined using a similar concept; the nucleotides in fish changing due to a series of reactions after death. Fish freshness can be quantified in terms of its **K** value where

$$K = \frac{(\text{HxR} + \text{Hx}) \times 100}{(\text{ATP} + \text{ADP} + \text{AMP} + \text{IMP} + \text{HxR} + \text{Hx})} \tag{8.13}$$

where HxR, IMP and Hx represent inosine, inosine-5'-monophosphate and hypoxanthine, respectively. After fish die, their ATP undergoes catabolic degradation through a series of reactions outlined below:

$$ATP \rightarrow ADP \rightarrow AMP \rightarrow IMP \rightarrow HxR \rightarrow Hx \rightarrow Xanthine \rightarrow Uric\ acid \quad (8.14)$$

The accumulation of the intermediates inosine and hypoxanthine relative to the nucleotides is an indicator of how long the fish has been dead and its handling and storage conditions and hence its freshness. A commercialised fish freshness biosensor has been devised which utilises a triacylcellulose membrane containing immobilised nucleoside phosphorylase and xanthine oxidase over an oxygen electrode.

$$inosine + phosphate \xrightarrow{\text{nucleoside phosphorylase}} \begin{array}{l} hypoxanthine \\ + ribose - phosphate \end{array} \quad (8.15)$$

$$hypoxanthine + O_2 \xrightarrow{\text{xanthine oxidase}} xanthine + H_2O_2 \quad (8.16)$$

$$xanthine + O_2 \xrightarrow{\text{xanthine oxidase}} uric\ acid + H_2O_2 \quad (8.17)$$

The electrode may be used to determine the reduction in oxygen[16] or the increase in hydrogen peroxide. The inosine content may be determined after the hypoxanthine content by the addition of the necessary phosphate. The nucleotides can be determined using the same electrode and sample, subsequent to the addition of nucleotidase and adenosine deaminase. Typically, K values below 20 show the fish is very fresh and may be eaten raw. Fish with a K value between 20 and 40 must be cooked but those with a K value above 40 are not fit for human consumption. Critical K values vary amongst species but can be a reliable indicator applicable to fish that is frozen, smoked or stored under modified atmospheres. Clearly, a relatively simple probe that accurately and reproducibly determines fish freshness has significant economic importance to the fish industry. In its absence freshness is determined completely subjectively by inspection.

Although such biosensors are easy to produce they do suffer from some significant drawbacks. The reaction is dependent on the concentration of molecular oxygen, which precludes its use in oxygen deprived environments such as *in vivo*. Also the potential used is sufficient to cause other redox processes to occur, such as ascorbate oxidation/reduction, which may interfere with the analyses. Much research has been undertaken on the development of substances that can replace oxygen in these reactions.[17] Generally, oxidases are far more specific for the oxidisable reactant than they are for molecular oxygen itself, as the oxidant. Many other materials can act as the oxidant. The optimal properties of such materials include fast electron transfer rates, the ability to be easily regenerated by an electrode reaction and retention within the biocatalytic membrane. In addition, they should not react with other molecules, including molecular oxygen that may

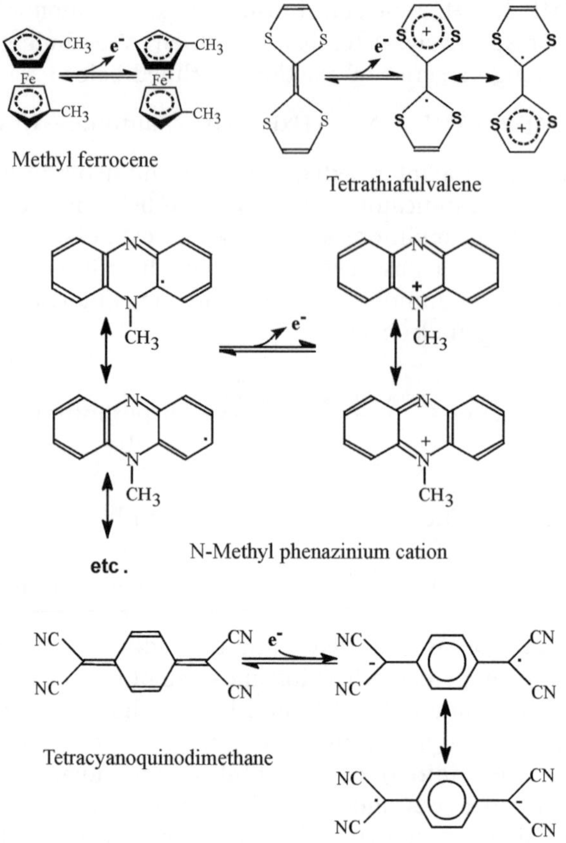

Figure 8.6 The redox reactions of amperometric biosensor mediators. Tetra-cyanoquinodimethane acts as a partial electron acceptor whereas ferro-cene, tetrathiafulvalene, and *N*-methylphenazinium can all act as partial electron donors.

be present. Many such oxidants, now called mediators, have been developed. Their redox reactions, which together transfer the electrons from the substrate to the electrode so producing the electrical signal, are summarised in Figure 8.6. As they allow reduced working potentials, interference from natural redox materials is reduced.

The mediated biosensor reaction consists of three redox processes:

$$\text{substrate}_{(\text{reduced})} + \text{enzyme}_{(\text{oxidised})} \xrightarrow{\text{enzyme reaction}} \text{product}_{(\text{oxidised})} + \text{enzyme}_{(\text{reduced})} \quad (8.18)$$

$$\text{enzyme}_{(\text{reduced})} + \text{mediator}_{(\text{oxidised})} \xrightarrow{\text{enzyme reaction}} \text{enzyme}_{(\text{oxidised})} + \text{mediator}_{(\text{reduced})} \quad (8.19)$$

$$\text{mediator}_{(\text{reduced})} \xrightarrow{\text{electrode reaction}} \text{mediator}_{(\text{oxidised})} + e^- \quad (8.20)$$

When a steady state response has been obtained, the rates of all of these processes and the rate of the diffusive flux in must be equal (see Figure 8.4c). Any of these, or a combination, may be the controlling factor. Although the level of response is reproducible, it is therefore difficult to predict its magnitude.

Several blood-glucose biosensors, for the control of diabetes, have been built and marketed based on this mediated system (Figure 8.7).[18] The sensing area is a single-use disposable electrode, typically produced by screen-printing onto a plastic strip, and consisting of an Ag/AgCl reference electrode and a carbon working electrode containing glucose oxidase and a derivatized ferrocene mediator. Both electrodes are covered with wide mesh hydrophilic gauze to enable even spreading of a blood drop sample and to prevent localized cooling effects due to uneven evaporation. The electrodes are kept dry until use and have a shelf life of six months when sealed in aluminium foil. They can detect glucose concentrations of 2–25 mM in single tiny drops of blood (1–4 µl) and accurately display the result within 5–40 seconds. The pricing is such that the biosensor is sold cheaply with the major profit arising from the sale of the necessary disposable electrodes. Such biosensors are currently used by millions of people with diabetes in more than 50 countries worldwide. The profits from this area, although by far the greatest in the biosensor industry, are levelling off as, while the world diabetic population is stands at over 200 million and is steadily expanding, diabetic patients use their biosensors less often when their diabetes appears under control. Although the subject of much research, the use of continuous subcutaneous *in vivo* glucose sensing, using electrode-based biosensors has not overcome the difficulties caused by the body's response and poor patient acceptability. New non-biosensor technologies are being developed,[19] involving near-infra-red light which can penetrate tissue and detect glucose in the blood, albeit at low sensitivity and in the face of almost overwhelming interference.

When an oxidase is unable to react rapidly enough with available mediators, horseradish peroxidase, which rapidly reacts with ferrocene mediators, can be included with the enzyme. This catalyses the reduction of the hydrogen peroxide produced by the oxidase and consequent oxidation of the mediator. In this case the mediator is acting as an electron donor rather than acceptor. The oxidised mediator then can be rapidly reduced at the electrode at moderate redox potential.

$$\text{mediator}_{(reduced)} + 1/2 H_2O_2 + H^+ \xrightarrow{\text{peroxidase}} \text{mediator}_{(oxidised)} + H_2O \quad (8.21)$$

$$\text{mediator}_{(oxidised)} + e^- \xrightarrow{\text{electrode reaction}} \text{mediator}_{(reduced)} \quad (8.22)$$

A major advance in the development of microamperometric biosensors came with the discovery that pyrrole can undergo electrochemical oxidative polymerization (Figure 8.8) under conditions mild enough to entrap

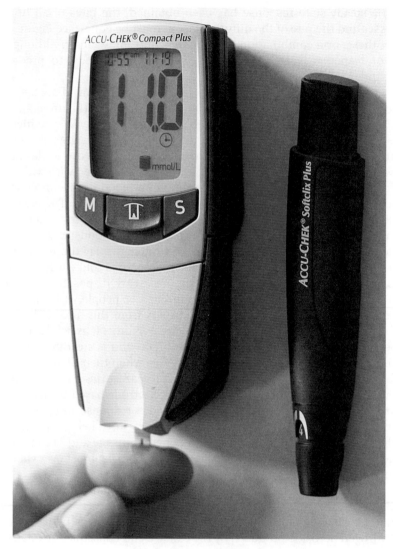

Figure 8.7 A biosensor for glucose used in diabetes monitoring. A drop of blood is applied to the disposable electrode at the bottom with its glucose concentration being produced on the read-out a few seconds later. On the right is a system for the controlled pricking of a finger.

enzymes and mediators at the electrode surface without denaturation.[20] A membrane, entrapping the biocatalyst and mediator, can be formed at the surface of even extremely small electrodes by polymerising pyrrole in the present of biocatalyst. The polypyrrole tightly adheres to platinum, gold or carbon electrodes. This allows silicon chip microfabrication methods to be used and for many different sensors to be laid down on the same chip (Figure 8.9). Similar membranes may be prepared less wastefully *in situ* from

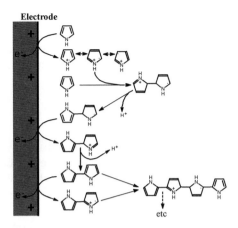

Figure 8.8 The mechanism for the electrochemical oxidative polymerisation of pyrrole.

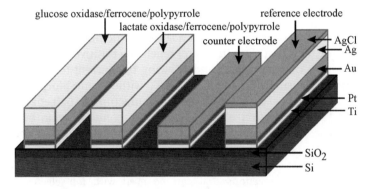

Figure 8.9 A combined microelectrode for glucose and lactate.

the oxidation of pyrrole by chloroauric acid to give sensing surfaces with excellent electron transfer properties formed from polypyrrole-immobilized enzymes and gold nanoparticles.

Another advance has been the use of conducting organic salts on the electrode. These allow the direct transfer of electrons from the reduced enzyme to the electrode without the use of any (other) mediator (Figure 8.4d). Conducting organic salts consist of a mixture of two types of planar aromatic molecules, electron donors and electron acceptors (see Figure 8.6), which partially exchange their electrons. These molecules form segregated stacks, containing either the donor or acceptor molecules, with some of the electrons from the donors being transferred to the acceptors. The partially transferred electrons are mobile up and down the stacks giving the organic crystals a high conductivity. There must not be a total electron transfer between the donor and acceptor molecules or the crystal becomes an insulator through lack of electron mobility. These electrodes give the somewhat misleading appearance of direct electron transfer to the electrode. As both the components of the organic salts, in the appropriate redox state, are able to mediate the reaction, it is highly probable that these electrodes are behaving as a highly insoluble mediator prevented from large-scale leakage by electrostatic effects.

8.4.2 Potentiometric Biosensors

Changes in ionic concentrations are easily determined by use of ion-selective electrodes. This forms the basis for potentiometric biosensors.[21] Many biocatalysed reactions involve charged species, each of which will absorb or release hydrogen ions according to their pK_a and the pH of the environment. This allows a relatively simple electronic transduction using the commonest ion-selective electrode, the pH electrode. Table 8.4 shows some biocatalytic reactions that can be utilised in potentiometric biosensors. Potentiometric biosensors can be miniaturised by the use of field effect transistors (FETs).

Ion-selective field effect transistors (ISFETs) are low cost devices that are in mass production. Figure 8.10 shows a diagrammatic cross-section through an npn hydrogen ion-responsive ISFET with an approximately 0.025 mm^2 biocatalytic membrane covering the ion selective membrane. The build-up of positive charge on this surface (the gate) repels the positive holes in the p-type silicon causing a depletion layer and allowing the current to flow. The reference electrode is usually an identical ISFET without any biocatalytic membrane. A major practical problem with the manufacture of such enzyme-linked FETs (ENFETs) is protection of the silicon from contamination by the solution, hence the covering of waterproof encapsulant. Because of their small

Table 8.4 Biocatalytic reactions that can be used with ion-selective electrode biosensors.

Electrode	Reactions
Hydrogen ion	
Penicillin	penicillin $\xrightarrow{\text{penicillinase}}$ penicilloic acid $+ H^+$
Lipid	triacylglycerol $\xrightarrow{\text{lipase}}$ glycerol $+$ fatty acids $+ 3H^+$
Urea	$H_2NCONH_2 + H_2O + 2H^+ \xrightarrow{\text{urease (pH 6)}} 2NH_4^+ + CO_2$
Ammonia	
L-Phenylalanine	L-phenylalanine $\xrightarrow{\text{phenylalanine ammonia-lyase}} NH_4^+ +$ trans-cinnamate
L-Asparagine	L-asparagine $+ H_2O \xrightarrow{\text{asparaginase}} NH_4^+ + $ L-aspartate
Adenosine	adenosine $+ H_2O + H^+ \xrightarrow{\text{adenosine deaminase}} NH_4^+ +$ inosine
Creatinine	creatinine $+ H_2O \xrightarrow{\text{creatininase}}$ creatine $\longrightarrow$
Creatine	creatine $+ H_2O \xrightarrow{\text{creatinase}} H_2NCONH_2 +$ sarcosine $\longrightarrow$
Urea	$H_2NCONH_2 + 3H_2O \xrightarrow{\text{urease (pH 7)}} 2NH_4^+ + HCO_3^- + OH^-$
Iodide	
Peroxide	$H_2O_2 + 2I^- + 2H^+ \xrightarrow{\text{peroxidase}} 2H_2O + I_2$

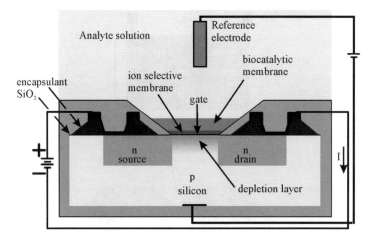

Figure 8.10 An FET-based potentiometric biosensor.

size, they only require minute amounts of biological material and can be produced in a form whereby they can determine several analytes simultaneously. A further advantage is that they have a more rapid response rate when compared with the larger sluggish ion-selective electrode devices. The enzyme may be immobilised to the silicon nitride gate using polyvinyl butyral deposited by solvent evaporation and cross-linked with glutaraldehyde. Such devices still present fabrication problems such as reproducibility, drift, sensitivity to light and the need for on-chip temperature compensation.

Use of membranes selective for ions other than hydrogen ions, such as ammonium, allows many related biosensors to be constructed. Potentiometric biosensors for DNA have been developed which use anti-DNA monoclonal antibodies conjugated with urease. DNA is bound to a membrane, placed on the electrode, and quantified by the change in pH on addition of urea (see Table 8.4).

8.4.3 Conductimetric Biosensors

Many biological processes involve changes in the concentrations of ionic species. Such changes can be utilised by biosensors that detect changes in electrical conductivity. A typical example of such a biosensor is the urea sensor, utilising immobilised urease,[22] and used as a monitor during renal surgery and dialysis (Figure 8.11). The reaction gives rise to a large change in ionic concentration at pH 7.0, making this type of biosensor particularly attractive for monitoring urea concentrations.

$$NH_2CONH_2 + 3H_2O \xrightarrow{\text{electrode reaction}} 2NH_4^+ + HCO_3^- + OH^- \qquad (8.23)$$

An alternating field between the two electrodes allows the conductivity changes to be determined whilst minimising undesirable electrochemical

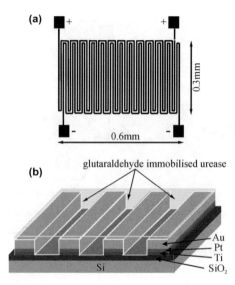

Figure 8.11 Parts of a conductimetric biosensor electrode arrangement. (a) Top view, (b) cross-sectional view. The tracks are about 5 μm wide and the thickness of the various layers are approximately; SiO$_2$ 0.55 μm, Ti 0.1 μm, Pt 0.1 μm, Au 2 μm.

processes. The electrodes are interdigitated to give a relatively long track length (∼1 cm) within a small sensing area (0.2 mm^2). A steady state response can be achieved in a few seconds allowing urea to be determined within the range 0.1–10 mM. The output is corrected for non-specific changes in pH and other factors by comparison with the output of a non-enzymic reference electrode pair on the same chip. The method can easily be extended to use other enzymes and enzyme combinations that produce ionic species, for example amidases, decarboxylases, esterases, phosphatases and nucleases. As molecular iodine molecules change the conductivity of iodine-sensitive phthalocyanine films, they may be used in peroxidase-linked immunoassays where the peroxidase converts iodide ions to iodine. Conductimetric biosensors have also be devised to determine DNA hybridization.[23]

8.5 PIEZOELECTRIC BIOSENSORS

The piezoelectric effect is due to some crystals containing positive and negative charges that separate when the crystal is subjected to a stress, causing the establishment of an electric field. As a consequence, if this crystal is subjected to an electric field it will deform. An oscillating electric field of a resonant frequency will cause the crystal to vibrate with a characteristic frequency dependent on its composition and thickness as well as the way it has been cut. As this resonant frequency varies when molecules absorb to the crystal surface, a piezoelectric crystal may form the basis of a biosensor. Even small changes in resonant frequencies are easy to determine with precision and accuracy using straightforward electronics. Differences in mass, even as small as an attogram, can be measured when adsorbed to the sensing surface. Changes in frequency are generally determined relative to a similarly treated reference crystal but without the active biological material. As an example, a biosensor for cocaine in the gas phase may be made by attaching cocaine antibodies to the surface of a piezoelectric crystal. This biosensor changes frequency by about 50 Hz for a 1 ppb atmospheric cocaine sample and can be reused on flushing for a few seconds with clean air. The relative humidity of the air is important as if it is too low, the response is less sensitive and if it is too high the piezoelectric effect may disappear altogether. Cocaine in solution can be determined after drying such biosensors.[24]

Enzymes with gaseous substrates or inhibitors can also be attached to such crystals, as has been proven by the production of biosensors for formaldehyde incorporating formaldehyde dehydrogenase and for organophosphorus insecticides incorporating acetylcholinesterase.

One of the drawbacks that initially prevented the more widespread use of piezoelectric biosensors was the difficulty in using them to determine analytes in solution. The frequency of a piezoelectric crystal depends on the liquid's viscosity, density and specific conductivity. Under unfavourable conditions, the crystal may cease to oscillate completely. There is also a marked effect of temperature due to its effect on viscosity. The binding of material to the crystal surface may be masked by other intermolecular effects at the surface and bulk viscosity changes consequent upon even small concentration differences. There is also the strong possibility of interference due to non-specific binding.

Piezoelectricity is also utilised in surface acoustic wave (SAW) devices[25] where a set of interdigitated electrodes is microfabricated at each end of a rectangular quartz plate (Figure 8.12). Binding of molecules to the surface affects the propagating wave, generated at one end, such that its frequency is reduced before reception at the other. The sensitivity of these devices is proportional to the square of their frequency and dependent on minimizing the interference from non-specific binding.

New developments include the use of piezoelectric polyvinylidene fluoride polymer thin films ($\sim$30 $\mu\mu$ thick) where the frequency change for an additional mass load is greater in water than in air, although at a lower base resonant frequency. Key to the optimal functioning of such devices is the introduction of stress into the piezoelectric membrane by the presence of a vacuum chamber on its non-sensing back surface.

Another development that overcomes some of these drawbacks is the use of atomic force microbalances where microscale vibrating 'diving boards' can be made very sensitive to the amount of material bound (Figure 8.13).[26] These devices have been developed for determining the presence of single viral particles.[27]

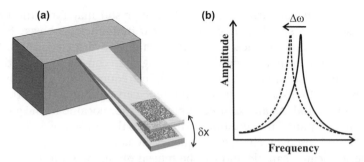

Figure 8.12   Atomic force microscopy. (a) The use of a thin resonating micron-sized 'diving board' allows sensitive measurement of changes in the mass bound. (b) The frequency change $\Delta\omega$ of the resonance is caused by addition of mass to the cantilever and such changes in the deflection (δx) may be shown by means of changes in the capacitance between the cantilever and a static electrode or from reflected laser light.

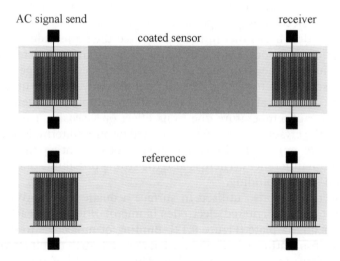

Figure 8.13   A surface acoustic wave (SAW) biosensor.

8.6   OPTICAL BIOSENSORS

Optical biosensors (also called optodes) have generated considerable interest, particularly with respect to the use of fibre optics and optoelectronic transducers.[28] These allow the safe non-electrical remote sensing of materials in hazardous or sensitive (*i.e. in vivo*) environments. An advantage of optical biosensors is that no reference sensor is needed; a comparative signal is generally easily generated by splitting the light source used by the sampling sensor. A simple example of an optical biosensor is the fibre optic lactate sensor (Figure 8.14) which senses changes in molecular oxygen concentrations by determining its quenching of a fluorescent dye.

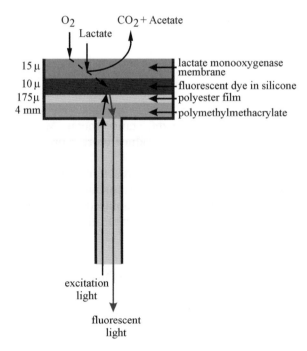

Figure 8.14 A fibre optic lactate biosensor.

$$O_2 + lactate \xrightarrow{\text{lactate monooxygenase}} CO_2 + acetate + H_2O \qquad (8.24)$$

The presence of oxygen quenches (reduces) the amount of fluorescence generated by the dyed film. An increase in lactate concentration reduces the oxygen concentration reaching the dyed film so alleviating the quenching and consequentially causing an increase in the fluorescence output.

Simple colorimetric changes can be monitored in some biosensor configurations. A lecithin biosensor has been developed containing phospholipase D, choline oxidase and bromothymol blue. The change in pH, due to the formation of the acid betaine from the release choline, causes a change in the bromothymol blue absorbance at 622 nm.[29] Gas phase reactions can also be monitored.[30] For example, alcohol vapour can be detected by the colour change of a dry dispersion of alcohol oxidase and peroxidase plus the redox dye 2,6-dichloroindophenol.

One of the most widely established biosensor technologies is the low-technology single-use colorimetric assay based on a paper pad impregnated with reagents. This industry revolves mainly round blood and urine analysis with test strips costing only a few cents. A particularly important use for these colorimetric test strips is the detection of glucose. In this case the strips contain glucose oxidase and horseradish peroxidase together with a chromogen (*e.g. o*-toluidine) which changes colour when oxidised by the peroxidise-catalysed reaction with the hydrogen peroxide produced by the aerobic oxidation of glucose.

$$\text{(reduced) chromogen (2H)} + H_2O_2 \xrightarrow{\text{peroxidase}} \text{(oxidised) dye} + 2H_2O \quad (8.25)$$

The colour produced can be evaluated by visual comparison to a test chart or by the use of a portable reflectance meter. Many test strips incorporate anti-interference layers to produce more reproducible and accurate results.

It is possible to link up luminescent reactions to biosensors, as light output is a relatively easy phenomenon to transduce to an electronic output. As an example, the reaction involving immobilised (or free) luciferase can be used to detect the ATP released by the lysis of micro-organisms.[31]

$$\text{luciferin} + \text{ATP} + O_2 \xrightarrow{\text{luciferase}} \text{oxyluciferin} + CO_2 + \text{AMP} \\ + \text{pyrophosphate} + \text{light} \quad (8.26)$$

This allows the rapid detection of urinary infections by detecting the microbial content of urine samples.

8.6.1   Evanescent Wave Biosensors

A light beam will be totally reflected when it strikes an interface between two transparent media, from the side with the higher refractive index, at angles of incidence (θ) greater than the critical angle (Figure 8.15a). This is the principle that allows transparent fibres to be used as optical waveguides. At the point of reflection an electromagnetic field is induced which penetrates into the medium with the lower refractive index; usually air or water. This field is called the evanescent wave and it rapidly decays exponentially with the penetration distance and generally has effectively disappeared within a few hundred nanometres. The exact depth of penetration depends on the refractive indices and the wavelength of the light and can be controlled by the angle of incidence. The evanescent wave may interact with the medium and the resultant electromagnetic field may be coupled back into the higher refractive index medium (usually glass) by essentially the reverse process. This gives rise to changes in the light emitted down the waveguide. Thus, it can be used to detect changes occurring in the liquid medium. The necessary surface interactions impose a limitation on the sensitivity of such devices at about 10 pg mm^{-2} and the requirement to limit non-specific absorption.

Various effects, dependent on the biological sensing processes, can be determined including changes in absorption, optical activity, fluorescence and luminescence. Because of the small degree of penetration, this system is particularly sensitive to biological processes in the immediate vicinity of the surface and independent of any bulk processes or changes. Due to the small

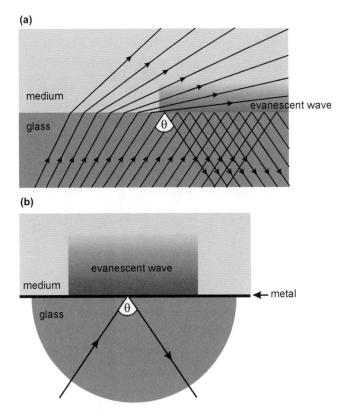

Figure 8.15 Production of (a) an evanescent wave (b) surface plasmon resonance. At acute enough angles of incidence the light is totally internally reflected at the glass surface. In (a) an evanescent wave extends from this surface into the air or water medium. This process is amplified in (b) by the presence of the thin metal film.

path-length through the solution, it can even be used for the continuous monitoring of apparently opaque solutions.

This biosensor configuration is particularly suitable for immunoassays as there is no need to separate bulk components since the wave only penetrates as far as the antibody-antigen complex. Surface-bound fluorophores may be excited by the evanescent wave and the excited light output detected after it is coupled back into the fibre (Figure 8.16). Sensors can be fabricated which measure oxidase substrates using the principle of quenching of fluorescence by molecular oxygen as described earlier. Another advantage of only sensing a surface reaction less than 1 μm thick is that the volume of analyte needed may be very small indeed; that is, less than 1 nL using suitable fluid transfer microfluidics.

Protein A, an important immunoglobulin-binding protein from *Staphylococcus aureus*, has been determined by this method using a plastic optical fibre coated with its antibody. Detection was by the fluorescence of a

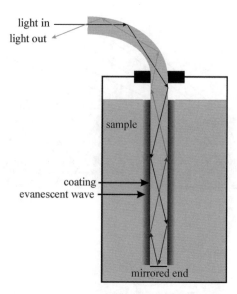

Figure 8.16 The principle behind evanescent wave immunosensor. The light output
is reduced by absorption within the evanescent wave.

fluorescein-bound anti-Protein A immunoglobulin which was subsequently
bound, sandwiching the Protein A.[32]

8.6.2   Surface Plasmon Resonance

The evanescent field generated by the total internal reflection of mono-
chromatic plane-polarised light within a fibre optic or prism may be utilised
in a different type of optical biosensor by means of the phenomena of surface plasmon resonance (SPR).[33] If the surface of the glass is covered with a
very thin layer of metal (usually pure gold, silver or palladium just a nano-
metre or so thick) then the electrons at the surface may be caused to oscillate
in resonance with the photons (as surface plasmon polaritons). This generates a surface plasmon wave and amplifies the evanescent field on the far
side of the metal (Figure 8.15b). If the metal layer is thin enough to allow
penetration of the evanescent field to the opposite surface, the effect is
critically dependent on the 100 nm or so of medium that is adjacent to the
metal. This effect occurs only when the light is at a specific angle of incidence dependent on its frequency, the thickness of the metal layer, and the
refractive index of the medium immediately adjacent the metal surface
within the evanescent field.[34] The generation of this surface plasmon resonance adsorbs some of the energy of the light so reducing the intensity of
the internally reflected light (Figure 8.17). Changes occurring in the medium
caused by biological interactions may be followed using the consequential
changes in the intensity of the reflected light or the resonance angle.
Figure 8.17 shows the change in the resonance angle of a human chorionic

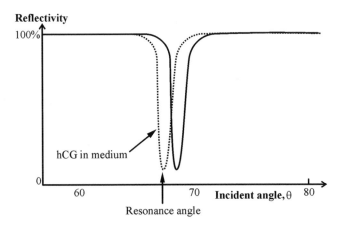

Figure 8.17 The change in absorption due to surface plasmon resonance.

gonadotrophin (hCG) biosensor on binding hCG to surface bound hCG antibody.[35] The sensitivity in such devices is limited by the degree of uniformity of the surface and the bound layer and the more-sensitive devices minimise light scattering. Under optimal conditions just 20 or 30 protein molecules bound to each square micrometre of surface may be detected. As with other immunosensors, the main problem occurring in such devices is non-specific absorption.

The biological sensing can be achieved by attaching the bioactive molecule to the solution side of the metal film. Physical adsorption may be used but, because this may lead to undesired denaturation and weak binding, covalent binding is often preferred. Gold films can be coated with a monolayer of long-chain 1, ω-hydroxyalkylthiols, which are copolymerized to a flexible un-crosslinked carboxymethylated dextran gel enabling the subsequent binding of bioactive molecules. This flat plate system, marketed as Biacore,[36] allows the detection of parts per million of protein antigen where the appropriate antibody is bound to the gel. Similarly, biosensors for DNA detection can be constructed by attaching a DNA or RNA probe to the metal surface when as little as a few femtograms of complementary DNA or RNA can be detected and, as a bonus, the rate of hybridization may be determined. Such biosensors retain the advantages of the use of evanescent fields as described earlier. They can also be used to investigate the kinetics of the binding and dissociation processes.[37] In spite of the relatively low costs possible for producing biosensing surfaces, the present high cost of the instrumentation is restricting the developments in this area. However, developments in the use of low-cost light-emitting diodes and lens arrays together with the use of porous surfaces and nanoparticles to increase surface area, microfluidics and integrated devices with in-built waveguided interferometry of reference beams are allowing the field to progress rapidly.

Table 8.5 Whole cell biosensors.

Analyte	Organism	Biosensor
Ammonia	*Nitrosomonas* sp.	Amperometric (O_2)
Biological oxygen demand (BOD)	Many	Amperometric (O_2/mediated), polarographic (O_2) or potentiometric (FET/H_2)
Cysteine	*Proteus morganii*	Potentiometric (H_2S)
Glutamate	*Escherichia coli*	Potentiometric (CO_2)
Glutamine	*Sarcina flava*	Potentiometric (NH_3)
Herbicides	Cyanobacteria	Amperometric (mediated)
Nicotinic acid	*Lactobacillus arabinosus*	Potentiometric (H^+)
Sulphate	*Desulfovibrio desulfuricans*	Potentiometric (SO_3^-)
Thiamine	*Lactobacillus fermenti*	Amperometric (mediated)

8.7 WHOLE CELL BIOSENSORS

As biocatalysts, whole microbial cells can offer some advantages over pure enzymes when used in biosensors.[38] Generally, microbial cells are cheaper, have longer active lifetimes and are less sensitive to inhibition, pH and temperature variations than the isolated enzymes. Against these advantages, such devices usually offer longer response and recovery times, a lower selectivity and they are prone to biocatalytic leakage. They are particularly useful where multistep or coenzyme-requiring reactions are necessary. The microbial cells may be viable or dead. The advantage of using viable cells is that the sensor may possess a self-repair capability but this must be balanced against the milder conditions necessary for use and problems that might occur due to membrane permeability and cellular outgrowth. Different types of whole cell biosensors are shown in Table 8.5.

Biochemical oxygen demand (BOD) biosensors most often use a single selected microbial species. Although rapid, linear and reproducible, they give a different result to conventional testing, which involve incubation over five days and reflect the varying metabolism of a mixed microbial population. Thermophilic organisms have been used in such a biosensor for use in hot wastewater.

8.8 RECEPTOR-BASED SENSORS

Receptor-based sensors (Table 8.6) include immunosensors,[39] lectin sensors and nucleic acid sensors using DNA, RNA and peptide nucleic acids.[9] This is a rapidly expanding field and some of these biosensors have been mentioned earlier. Most biosensor configurations may be used, with Figure 8.18 showing some of those possible. Direct binding of the antigen to immobilised antibody (Figure 8.18a) or antigen-antibody sandwich (Figure 8.18b) may be detected using piezoelectric or SPR devices,[40] as can antibody release due to free antigen (Figure 8.18c). Binding of enzyme-linked antigen (Figure 8.18d) or antibody can form the basis of all types of immunosensors but has proved particularly useful in amperometric devices. The amount of enzyme activity

Table 8.6 A selection of receptor-based sensors.

Analyte	Sensing method	Biosensor
Anthrax spores (warfare)	Antibody	Surface acoustic wave
Cholera toxin (warfare)	Antibody	Potentiometric
Cocaine	Antibody/Protein A	Piezoelectric
Dengue fever virus	Antibody	Atomic force microscopy
Human chorionic gonadotrophin	Antibody/catalase	Amperometric (O_2)
	Antibody	Surface plasmon resonance
Hepatitis B surface antigen	Antibody/peroxidase	Potentiometric (I^-)
Insulin	Antibody/catalase	Amperometric (O_2)
Ricin (warfare)	Antibody	Evanescent wave
Trinitrotoluene (TNT)	Antibody/labelled antigen	Fluorescence
Verotoxin-producing *E. coli*	Peptide nucleic acid	Surface plasmon resonance

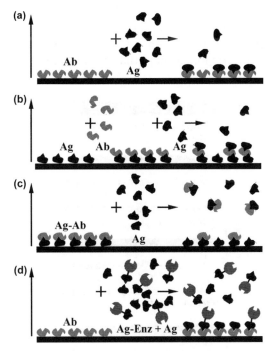

Figure 8.18 Different configurations for biosensor immunoassays: (a) antigen binding to immobilised antibody; (b) immobilised antigen binding antibody which binds free second antigen; (c) antibody bound to immobilised antigen partially released by competing free antigen; (d) immobilised antibody biding free antigen and enzyme-labelled antigen in competition.

bound in these receptor-based sensors is dependent on the relative concentrations of the competing labelled and unlabelled ligands and so it can be used to determine the concentration of unknown antigen concentrations.

The main problems involved in developing receptor-based sensors centre on non-specific binding and incomplete reversibility of the binding process both of which reduce the active area, and hence sensitivity, on repetitive assay. Single use biosensing membranes are a way round this, but they necessitate strict quality control during production.

Biosensors involving DNA hybridization, between a single stranded DNA target and its complementary single stranded probe, form a rapidly expanding area of sensor development.[38] They may be sensors for particular DNA strands (genosensors) or as microarrays (gene chips) to investigate a large range of target structures. Many detection mechanisms may be used which involve either labeled or unlabeled DNA, where labeling enable more sensitive determination but at greater cost. DNA chips have been produced that contain tens of thousands of different but known single stranded polynucleotides immobilised on a silicon chip. Thus, fluorescence labelled single-stranded DNA can be probed for structure due to the specificity of DNA–DNA binding.[41] Introduction of redox materials that can intercalate between double-stranded DNA, but not bind to single-stranded DNA allows the development of electrochemical versions of these chips without the need for fluorescent probes or fluorescence readers (Figure 8.19).[42] A simple example involves ferrocene-linked naphthalene diimide.[43]

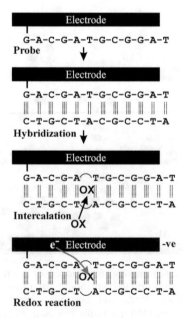

Figure 8.19   DNA biosensor. Binding of the complementary strand allows certain redox molecules to intercalate into the DNA fragment to allow detection *via* amperometry.

Binding of DNA to complementary strands attached to surfaces may change the surface electrical or optical properties, so allowing detection. Even unlabelled DNA strands cause changes in impedance causing easily detectable variation in the electrical response.[44] Variations on this theme allows the detection of single base mismatches in DNA for use in determining single nucleotide polymorphisms.[45]

Proteins, such as antibodies, may also be attached to DNA chips by means of covalently attached complementary probes, so allowing thousands of differently binding proteins to be specifically attached to single chips. The high negative charge on the complementary linkers almost totally removes the chance of non-specific binding. This area is likely to expand considerably over the next few years.

8.9 CONCLUSION

Biosensors form an interesting and varied part of biotechnology. They have been applied to solve a number of analytical problems and some have achieved notable commercial success. They have been slow to evolve from research prototype to the marketplace and are only just reaching towards their full potential. Many more commercial products are expected over the next few years, particularly in medical diagnostics and the war against terrorism.[46]

REFERENCES

1. J. D. Newman and S. J. Setford, *Mol. Biotechnol.*, 2006, **32**, 249.
2. S. Kumar, N. Dilbaghi, M. Barnela, G. Bhanjana and R. Kumar, *Bionanoscience*, 2012, **2**, 196.
3. B. Van Dorst, J. Mehta, K. Bekaert, E. Rouah-Martin, W. De Coen, P. Dubruel, R. Blust and J. Robbens, *Biosens. Bioelectron.*, 2010, **26**, 1178.
4. J. D. Newman and A. P. F. Turner, *Biosens. Bioelectron.*, 2005, **20**, 2435.
5. P. T. Kissinger, *Biosens. Bioelectron.*, 2005, **20**, 2512.
6. Z. Zhu, J. Zhang and J. Zhu, *Sens. Lett.*, 2005, **3**, 71.
7. C. I. L. Justino, T. A. Rocha-Santos and A. C. Duarte, *Trends Anal. Chem.*, 2010, **29**, 1172.
8. J. Wang, *Electroanalysis*, 2001, **13**, 983.
9. L. Ricotti, T. Assaf, P. Dario and A. Menciassi, *J. Artif. Organs*, 2013, **16**, 9–22.
10. J. P. Chambers, B. P. Arulanandam, L. L. Matta, A. Weis and J. J. Valdes, *Curr. Issues Mol. Biol.*, 2008, **10**, 1.
11. T. D. Gibson, *Analusis*, 1999, **27**, 632.
12. M. Campàs, B. Prieto-Simón and J.-L. Marty, *Seminars Cell Dev. Biol.*, 2009, **20**, 3.
13. M. Pohanka and P. Skládal, *J. Appl. Biomed.*, 2008, **6**, 57.
14. L. C. Clark and C. H. Lyons, *Ann. New York Acad. Sci.*, 1962, **102**, 29.
15. E. Watanabe, M. Takagi, S. Takei, M. Hoshi and C. Shu-gui, *Biotechnol. Bioeng.*, 1991, **38**, 99.
16. S. S. Hu and C. C. Liu, *Electroanalysis*, 1997, **9**, 1174.

17. F. W. Scheller, F. Schubert, B. Neumann, D. Pfeiffer, R. Hintsche, I. Dransfeld, U. Wollenberger, R. Renneberg, A. Warsinke, G. Johansson, M. Skoog, X. Yang, V. Bogdanovskaya, A. Bückmann and S. Y. Zaitsev, *Biosens. Bioelectron.*, 1991, **6**, 245.
18. J. Wang, *Chem. Rev.*, 2008, **108**, 814.
19. H. E. Koschwanez and W. M. Reichert, *Biomaterials*, 2007, **28**, 3687.
20. M. Singh, P. K. Kathuroju and N. Jampana, *Sens. Actuat. B*, 2009, **143**, 430.
21. R. Koncki, *Anal. Chim. Acta*, 2007, **599**, 7.
22. N. F. Sheppard, D. J. Mears and A. Guiseppielie, *Biosens. Bioelectron.*, 1996, **11**, 967.
23. Y. Q. Fu, J. K. Luo, X. Y. Du, A. J. Flewitt, Y. Li, G. H. Markx, A. J. Walton and W. I. Milne, *Sens. Actuat. B*, 2010, **143**, 606.
24. S.-J. Park, A. Taton and C. A. Mirkin, *Science*, 2002, **295**, 1503.
25. B. S. Attili and A. A. Suleiman, *Microchem. J.*, 1996, **54**, 174.
26. J. Lee, R. Chunara, W. Shen, K. Payer, K. Babcock, T. P. Burgm and S. R. Manalis, *Lab Chip*, 2011, **11**, 645.
27. R. L. Caygill, G. E. Blair and P. A. Millner, *Anal. Chim. Acta*, 2010, **681**, 8.
28. X. Fan, I. M. White, S. I. Shopova, H. Zhu, J. D. Suter and Y. Sun, *Anal. Chim. Acta*, 2008, **620**, 8.
29. V. P. Kotsira and Y. D. Clonis, *J. Agric. Food Chem*, 1998, **46**, 3389.
30. S. Lamare and M. D. Legoy, *Trends Biotechnol.*, 1993, **11**, 413.
31. M. F. Chaplin, in *Physical Methods for Microorganisms Detection*, ed. W. M. Nelson, CRC Press, Inc., Boca Raton, USA, 1991, p. 81.
32. Y. H. Chang, T. C. Chang, E. F. Kao and C. Chou, *Biosci., Biotechnol., Biochem.*, 1996, **60**, 1571.
33. X. D. Hoa, A. G. Kirk and M. Tabrizian, *Biosens. Bioelectron.*, 2007, **23**, 151.
34. J. Homola, *Anal. Bioanal. Chem.*, 2003, **377**, 528.
35. J. W. Attridge, P. B. Daniels, J. K. Deacon, G. A. Robinson and G. P. Davidson, *Biosens. Bioelectron.*, 1991, **6**, 201.
36. R. L. Rich and D. G. Myszka, *Drug Discovery Today: Technol.*, 2004, **1**, 301.
37. M. Fivash, E. M. Towler and R. J. Fisher, *Curr. Opin. Biotechnol.*, 1998, **9**, 370.
38. Y. I. Korpan and A. V. Elskaya, *Biochemistry (Moscow)*, 1995, **60**, 1517.
39. O. Lazcka, F. J. Del Campo and F. X. Muñoz, *Biosens. Bioelectron.*, 2007, **22**, 1205.
40. A. Sassolas, B. D. Leca-Bouvier and L. J. Blum, *Chem. Rev.*, 2008, **108**, 109.
41. R. C. Anderson, G. McGall and R. J. Lipshutz, *Top. Curr. Chem.*, 1997, **194**, 117.
42. A. L. Ghindilis, M. W. Smith, K. R. Schwarzkopf, K. M. Roth, K. Peyvan, S. B. Munro, M. J. Lodes, A. G. Stöver, K. Bernards, K. Dill and A. McShea, *Biosens. Bioelectron.*, 2007, **22**, 1853.
43. J. Wang, *Anal. Chim. Acta*, 2002, **469**, 63.
44. A. Bonanni and M. del Valle, *Anal. Chim. Acta*, 2010, **678**, 7.
45. L. E. Ahangar and M. A. Mehrgardi, *Biosens. Bioelectron.*, 2012, **38**, 252.
46. J. J. Gooding, *Anal. Chim. Acta*, 2006, **559**, 137.

CHAPTER 9

Human and Animal Cell Culture

JOHN DAVIS

School of Life and Medical Sciences, University of Hertfordshire,
College Lane, Hatfield, AL10 9AB, UK
Email: j.m.davis@herts.ac.uk

9.1 INTRODUCTION

In modern biotechnology, the *in vitro* growth, maintenance, manipulation, differentiation and use of human and other animal cells is of huge and ever-increasing importance. Over the last few decades, the application of such cell cultures has expanded vastly, from being a useful research tool (which it still is) to being a central testing and production platform in industry. To give three examples: all animal testing of cosmetics is now banned in the EU, and most of these former *in vivo* tests have been replaced by *in vitro* tests on cultured cells; the production of influenza vaccines is increasingly being performed in cultured animal cells, in order to overcome some of the many disadvantages of the older production methods that employed hens' eggs; and the production of all but one of the 30 + licensed pharmaceutical monoclonal antibodies is performed in large-scale animal cell culture, in fermenters of up to 20 000 litres. Not only that, but the hugely promising stem cell and regenerative medicine fields are intimately dependent on the culture of human and other cells. Thus the importance of human and animal cell culture to the future of medical care is difficult to overemphasise.

The basic scientific and technical details of cell culture are far too extensive to address in the space available in this chapter, and such information

Molecular Biology and Biotechnology, 6th Edition
Edited by Ralph Rapley and David Whitehouse
© The Royal Society of Chemistry 2015
Published by the Royal Society of Chemistry, www.rsc.org

can be found in reference books on the subject.[1,2] Instead, this chapter will focus on the general principles that must be applied to all cell culture work, and particularly that where the intended application (directly or indirectly) is the improvement of human or animal health.

Six crucial principles that need to be followed in order to ensure the quality (and thus usefulness) of any cell culture work were identified in the *Guidance on Good Cell Culture Practice.*[3] In essence, these were:

- As far as possible, understand and standardize all the *in vitro* (and, where relevant, *in vivo*) systems that you are employing.
- Check/control the quality of all the materials you are using.
- Make sure you obtain or generate (as appropriate) all the documentation you might need now or in the future.
- Work safely.
- Make certain that all work is carried out in compliance with current legal and ethical requirements.
- Ensure that the education and training of everyone involved (directly or peripherally) in the work is *and remains* suitable for the purpose and up-to-date.

These principles are examined in more detail in the sections below.

9.2   UNDERSTANDING AND STANDARDIZING *IN VITRO* CULTURE SYSTEMS

It is essential that, in order to use *in vitro* cell culture systems in a meaningful way, their behaviours, and the factors that affect them, are well understood. Yet our scientific understanding of such systems is imperfect, and their standardization is difficult and depends on multiple factors. Consequently, it is essential to work using a system that controls all those factors that are controllable, and where the impact of any uncontrolled variables is minimized. This requires—as a minimum—close attention to the following aspects of cell culture systems:

9.2.1   The Cells

As soon as cells are removed from the host organism they are inevitably subjected to conditions they would not normally encounter *in vivo*. Many cell types are extremely sensitive to environmental conditions, and small changes can alter their growth, differentiation, gene expression and many other characteristics. Often, as soon as cells are put into *in vitro* culture, characteristic properties, and indeed cell populations, are lost. This is usually due to technological constraints combined with limitations in our current knowledge of relevant aspects of the cells' biology. Clearly, then, it is essential to adequately characterize the cell lines as soon as possible after establishment in culture. This applies irrespective of the source (an organism, a cell bank or another laboratory) from which the cells came. However, the nature of the intended culture process may limit the degree of characterization that is appropriate; most notably, in the case of primary cells that are not intended for subculture *in vitro*, characterization may be limited to the identification of cell markers. Characterization is discussed further in Sections 9.3.1 and 9.3.2.

Characterization at this stage is performed for the purposes of:

a) *Ensuring the culture contains only the type of cells you think it should, and that they display the expected properties.*
   In vitro culture systems can often support the growth of cell types other than those required and (particularly if the culture is freshly derived from the parental organism) such undesired cells could contaminate or overgrow the cells of interest. Cell lines obtained from a cell bank or another laboratory may (for reasons ranging from mis-labelling to inadvertent contamination) not be what they are purported to be, or may have been subjected to conditions—particularly during transit—that have affected their properties. Also, when you initiate culture of these cells in your laboratory, the conditions will almost inevitably not be precisely identical to those under which they grew previously, and this may exert selective pressures on the cells, potentially leading to loss of the phenotype of interest.

   Other organisms can also contaminate the cells of interest; in particular, cell culture media is very good at supporting the growth of a wide variety of microorganisms. Bacteria, fungi and yeast, if present, will usually overgrow and destroy the culture, although in some cases their effects may be subtle and their presence difficult to detect. Mycoplasma, though, are almost invariably difficult to detect and seldom prevent the growth of mammalian cells. Despite this, they can adversely affect almost every aspect of the behaviour of cultured cells. The presence of endogenous or exogenous viruses also has the potential to cause very real problems, and the ability of any cell line to harbour and support the propagation of pathogenic prions is worthy of consideration.[4]

b) *Comparing the cells* in vitro *with the corresponding cells* in vivo.
   It is essential, if using cells as an *in vitro* model intended to reflect the characteristics of the corresponding cells *in vivo*, to compare the cultured cells to the 'normal' cells in the organism from which they originated. The degree and nature of the characterization will depend on factors such as the type of studies envisioned, and the presence of relevant features that can be compared. Unfortunately, in far too many cases this type of comparison has been neglected, resulting in time and money being wasted, misleading papers being published, and scientific reputations being damaged because studies were scientifically invalid—see, for example, reference 5.

Table 9.1 Some of the changes that may be observed with increasing population doubling number in cultured non-immortalized fibroblasts.[14]

Shortening of telomeres
Changes to the cell cycle
Decreased rate of DNA synthesis
Increased population doubling time
Changes in plasma membrane organisation and structure
Decreases in protein turnover and amino acid transport
Increased aneuploidy

---

c) *Establishing a baseline against which to compare future cultures.*
The properties of cells grown *in vitro* tend to change with time.[6–12] These changes may include:

- Genetic changes—mutations will tend to accumulate with time in culture and the number of divisions the cells go through.
- Epigenetic changes—again, these tend to accrue with time.
- Stochastic changes—these can arise rapidly within a cultured cell population,[13] and if certain cells gain a selective advantage they may very quickly take over the whole culture.
- Cellular ageing—*in vitro*, cells that are not immortalized display progressive alterations in numerous characteristics as the number of population doublings through which they have gone increases (Table 9.1). Indeed, such non-immortalized cells can only undergo a limited number of population doublings (generally dependent, amongst other things, on the species and age of the donor organism) before they become senescent and stop multiplying.

Thus it is essential, as soon as possible after starting to culture a cell line *in vitro*, to establish a baseline profile of important characteristics. Subsequent cultures can then be compared against them, and they can also be used to define the maximum number of population doublings that the cells can safely be allowed to go through without significant loss of important characteristics. Experimentation should then be limited to cells within this range. This naturally means that a continual record must be kept of the population doubling level of each culture.

9.2.2 Culture Conditions

Culture conditions are artificial and, as already mentioned, will inevitably differ from those the cell would encounter *in vivo*. In the organism, the conditions within particular niches are often very closely controlled, and changes in these conditions may cause the cells to undergo changes that may be difficult to predict. Thus it is essential to control conditions within the culture closely. The parameters that will need controlling include:

- *Incubation temperature*—For optimal growth and expression of function, most cells must be maintained within a limited temperature range. This is dependent on a number of factors including the species and tissue of origin. As a generalisation, most cells will tolerate temperatures below their optimum better than those above. However, even modest drops can lead to changes in metabolism and gene expression.[15–17] Most cells cultured *in vitro* are subjected to temperature drops every time the incubator door is opened or the culture is removed from the incubator for examination or manipulation. Established cultures will tolerate this non-physiological situation because they have always been treated in this manner, *i.e.* they have been selected to be resistant to this treatment. However, it should always be borne in mind that selection for one characteristic may be selection against another.
- *pH*—There is a very limited pH range which is optimal for cells, yet during culture they may be subjected to more alkaline or more acidic conditions.
  - More *alkaline* conditions can occur early during culture, or when cell population densities are low, *i.e.* before the cells have secreted significant quantities of acidic metabolites into the medium. As most media use a HCO_3^-/CO_2 buffering system, optimal pH is only maintained while the culture is in contact with an atmosphere containing the relevant concentration of carbon dioxide. In open/vented culture vessels, this only occurs when they are in the incubator. Once removed from the incubator for examination or manipulation, and particularly if the culture volume is small and/or has a high ratio of surface area to volume (as in a Petri dish), the pH will rise rapidly. This can clearly be seen if the medium contains Phenol Red as an indicator.
  - More *acidic* conditions will tend to be encountered as culture progresses and cell metabolism causes cells to excrete acidic products (most importantly lactic acid and CO_2) into their surroundings. Although in the organism (and some specialized culture systems) these acidic products are removed by perfusion, most *in vitro* culture takes place in closed systems where such products will act to decrease the pH of the culture medium. Whilst all media contain a buffering system, the HCO_3^-/CO_2 system most commonly used has limited buffering capacity, and can be overwhelmed as culture progresses, particularly in high population density cultures. It may be advantageous to gain extra buffering capacity by adding a non-CO_2-dependent buffer—such as HEPES—to the medium. However, such chemicals can sometimes contain impurities that are toxic to cells, so each batch should be tested before use.

    Generally, cells are more tolerant of acidic rather than alkaline conditions, and this should be taken into account when growing sensitive cultures, notably those at extremely low population densities such as used when performing single-cell cloning. It may be advantageous in such circumstances to use a higher percentage of CO_2 in the incubator's atmosphere than that normally recommended for the medium in use, just to ensure that the pH does not err on the alkaline side, but this should be checked out on a case-by-case basis.

- *Dissolved oxygen tension*—Historically, most cell cultures were comprised of a monolayer of cells under a thin (2–3 mm) layer of medium overlaid by a gas phase with an oxygen content close to that of air. Thus oxygen saturation was close to 100%, but as larger culture systems were employed oxygen transfer rates often became limiting and cells were cultured at much lower oxygen tensions. This often had little adverse effect on the cells, which is perhaps not surprising as the oxygen tensions in cellular environments within an organism are frequently very much lower than atmospheric. Indeed certain stem cell types differentiate spontaneously under normoxic conditions, and can be more successfully propagated under low oxygen tensions.[18,19] Thus this parameter too should be examined for any effect, and where necessary must be closely controlled. The requirement for some cells to be handled—at every stage of culture—under sub-atmospheric oxygen concentrations, has led to the development of special workstations and these are now commercially available.

- *Concentrations of nutrients, co-factors, trace elements, growth/differentiation factors, inhibitors, and other chemicals in the medium*—At one time animal or human sera were added to almost all basal media as a rich source of many (at that time undefined) factors necessary for the maintenance and proliferation of cells in culture. However there was great batch-to-batch variation, and serum-free formulations have been devised in order to better define and control cell growth and differentiation. Depending on their constituents, they may also reduce the potential for introducing adventitious contaminants (microbes, viruses, *etc.*) into the culture. The logical conclusion of this has been to formulate completely chemically-defined media.[20] There are clearly huge potential benefits to such media in terms of controlling medium constituents and understanding the nutritional, metabolic and hormonal requirements of cells. However, it may take a huge investment of time, money and energy to define even an adequate chemically-defined medium, let alone an optimal one, for any particular cell line, and it may be inadequate for the culture of even closely related lines. Consequently, the potential benefits of such media must be balanced against the effort required to define and formulate them. For more details on this subject, see reference 20.

- *Feeder cells*—Mitotically inhibited feeder cells are often used to support the growth of other cells, particularly under conditions of extremely low population density. This has similar drawbacks to those already mentioned with regard to serum supplementation of medium, *i.e.* lack of definition and reproducibility, and the potential for introducing adventitious contaminants. In addition, in certain critical applications for human therapy, mouse feeder cells (and/or other animal-derived culture components) have been shown to have the potential to contaminate human stem cells with non-human antigens.[21,22] Thus, like serum, the use of feeder cells must be critically assessed and avoided where possible.

9.2.3 Handling and Maintenance

It essential that when cells are first handled in the laboratory they are held under quarantine conditions. Until tested and characterized it is not possible to be certain that the cells are what you think they are, and they may contain contaminants that could potentially spread to other cultures within the laboratory. A dedicated quarantine laboratory is ideal for the purpose, but few laboratories possess these and other approaches can be taken. Cells in quarantine can, for example, be handled in the normal cell culture laboratory, but at the end of the day once work on other cultures has been completed. However, wherever possible incubation should be in a dedicated incubator. Cells should only be released from quarantine once an initial stock has been frozen in liquid nitrogen, successfully thawed, subjected to microbial and other quality control testing and yielded satisfactory results.

Unless justifiable on a scientific basis, cells should always be cultured under antibiotic-free conditions. In far too many laboratories, antibiotics are employed in an attempt to cover up sloppy technique. Yet antibiotics themselves can affect many aspects of a cell's metabolism and behaviour (see for example references 23–25). In addition, they frequently suppress but do not fully eliminate the growth of bacteria, and this promotes the outgrowth of antibiotic-resistant strains. These are then very difficult to eliminate from the laboratory.

It is important that all handling procedures are clearly defined, well controlled, and consistently applied. Wherever possible, *i.e.* with techniques that are not experimental but rather are finalised and repeatedly used in the same manner, work should be performed according to a Standard Operating Procedure (SOP) that clearly defines the materials to be used and how the method is to be applied.

Subculturing of cells is a prime example of a technique carried out repeatedly in the cell culture laboratory and for which there should be an SOP. Details of the technique will often vary between cell lines, in which case each line will need its own SOP. Subculturing is a very non-physiological process which places cells under particular stress and selective pressure, and it is essential that the process is well defined and controlled.

Most cells in culture grow attached to a substrate (*e.g.* the plastic of a flask), and in order to subculture them they will need to be treated with a proteolytic enzyme and usually other chemicals such as EDTA, in order to release them from the substrate. This treatment will tend to remove or degrade cell surface proteins, and too much exposure to such treatment will lead to cell death. Also, it will take time, usually hours but in some cases days, for viable cells to re-synthesize surface components and regain their normal cell surface phenotype and function. Use of a well developed SOP will help ensure that the cells are efficiently removed from the substrate with the minimum amount of damage.

Amongst other things, an SOP for subculturing should include criteria for:

- *Deciding if cells appear healthy.* Always examine cells under a microscope before use (and preferably at regular intervals during growth). Do they have the expected morphology?
- *Deciding if they are at a suitable population density for subculturing.* Some cell types change their properties if allowed to become confluent (*i.e.* cover all the substrate area); Caco-2 cells, for example, differentiate and may lose their ability to proliferate.[11]
- *Ensuring that treatment is adequate for cell removal whilst limiting damage.* The vast majority of the cells should be released from the substrate with minimum exposure to those substances that can damage cells.

It is essential that once the proteolytic enzyme has released the cells from the substrate, an enzyme inhibitor is added immediately to stop proteolysis. Serum contains numerous anti-proteases, as well as proteins such as albumin which will act as alternative substrates for any residual protease activity. When culturing under serum- or protein-free conditions, however, it will be necessary to add a specific inhibitor; for example soybean trypsin inhibitor can be used with serum-free (but not protein-free) cultures.

Recently, a plastic substrate (UpCell™) has been developed by Nunc that allows mammalian cells to detach simply by dropping the incubation temperature to 32 °C. This makes chemical treatment redundant but, as mentioned above, temperature changes themselves have effects on cells.

Subculturing serves as an illustration that most if not all *in vitro* procedures subject the cells to conditions they would not normally encounter *in vivo*. These conditions may be potentially damaging and/or subject the cells to selective pressures, and consequently must be carefully considered and controlled in order to give reproducible results.

9.2.4 Cryopreservation

Cells kept in culture continuously can be expected to undergo ageing and/or other changes (some of which have been described above), and the chance of losing a culture to microbial contamination increases with every manipulation. Thus it is crucial that a standard, well characterized stock of cells is continually available to which one can turn in order to avoid or minimize the impact of such factors. With mammalian cells this can currently only be achieved by cryopreserving banks of cells at temperatures of −130 °C or lower.

With any important cell line, the best approach is to use a tiered cell banking system. This ensures that stocks of cells are always available, and works as follows:

As soon as possible after isolation or receipt, an initial stock of cells is cryopreserved as a *token freeze*, frequently of just 3–5 vials. Because initial attempts at freezing the cells may be unsuccessful, culture must continue at least until the token freeze has been prepared, frozen, test thawed and characterized. Characterization at this stage should be fairly extensive, and give assurance that microbial contamination is absent and the desired cell characteristics have been retained. Once this has been completed, there can be reasonable confidence that the cells are fit for purpose.

A *Master Cell Bank* (MCB) is then prepared. The number of ampoules required in the MCB will depend on the level and frequency of usage envisioned. MCBs containing 200 ampoules are not unusual in the biopharmaceutical industry, but for research purposes, banks will usually be much smaller. The testing and characterisation of the MCB will be extensive, and will depend in part on the use to which the cells are to be put and must anticipate the demands of any relevant regulatory bodies. In particular, tests for adventitious contaminants will need to be more extensive than those carried out on the token freeze.[26–29]

Once the MCB has been tested and characterised, a *Working Cell Bank* (WCB) is prepared from an ampoule of the MCB. The number of ampoules in the WCB will again depend on the projected rate of usage of the cells. Characterization and testing of the WCB will be required, but depending on the circumstances it may be possible to justify less testing than was performed on the MCB.

As its name implies, the Working Cell Bank is used as the source of cells for experiments or production runs. When the WCB is nearly depleted, a fresh WCB is created from another ampoule of the MCB. When the MCB is nearly depleted, a new MCB must be prepared, either from the token freeze or more usually from an ampoule of the existing MCB.

This approach ensures an essentially inexhaustible supply of important cells. To illustrate this, imagine that a biopharmaceutical company made an MCB of 100 ampoules and a WCB containing 200 ampoules, and used two ampoules from the WCB every week. In theory it would be 100 weeks (nearly 2 years) before a new WCB was required, and nearly 200 years before a new MCB was needed.

Further information on cryopreservation and cell banking can be found in references 30–32.

9.3 QUALITY ASSURANCE

In the context of cell culture, quality assurance means confirmation of the identity, purity, consistency, traceability and reproducibility of all the important materials and systems used. In practice this is achieved by having in place appropriate procedures and specifications, along with corresponding checks and tests to identify real or potential problems. Ideally there should be SOPs for all routine culture and related operations along with associated checks/tests. Two examples of such operations are:

- *Cell accessioning* (obtaining cells from external sources and introducing them into the laboratory). SOPs would incorporate amongst other things criteria for accepting the cells (*e.g.* checks on the paperwork supplied with the cells), the procedures for quarantining the cells and producing the token freeze, and acceptance tests that need to be carried out and passed (and the parameters for acceptance or failure) prior to cells being released from quarantine.
- *Subculturing.* SOPs might cover not only the materials and methods to be used, but also preliminary checks on the cells (*e.g.* is the morphology as expected, is the cell population density/degree of confluence within the parameters set), checks on release of the cells from the substrate, checks on the quantity and quality of the cells (*e.g.* cell yield and viability) and parameters for the re-establishment of culture (*e.g.* medium, seeding population density, incubation conditions).

As the name implies, each component within quality assurance helps assure the quality of the materials or procedures employed, and hence the scientific validity of the work performed.

9.3.1 Cell Identity Testing and Other Forms of Characterization

Contamination of cell cultures with other (unwanted) cells is an extremely important issue, which, despite having been identified as a problem over 50 years ago, still plagues mammalian cell culture. Cross-contamination with cells from other species was first reported in 1962 by Coriell,[33] and this was followed a few years later by the identification of intraspecies contamination.[34,35] In the intervening period, cross-contamination of one cell type with another has been reported repeatedly.[36-46] The German Cell Bank (DSMZ) reported in 1999 that 18% of all 'new' cell lines submitted to it were actually other established cell lines.[43] Scandalously, other than in biopharmaceutical production (where the issue has been rigorously addressed) the problem shows little sign of going away.[5,47-50] due largely to appalling levels of complacency and ignorance. The resultant waste of time and research funds, not to mention the damage to scientific reputations and the number of misleading papers that have been published, is almost beyond belief. There have been repeated calls for action over the last 14 years,[51-53] and these seem at last to be bearing fruit, with increasing numbers of journals now refusing to publish papers on cell lines that have not been identity tested.

All cell lines, then, MUST be authenticated at an early stage. The easiest initial check can be performed as soon as you contemplate using the cells; simply use an internet search engine to check that the cell line has not already been shown to be misidentified. Better still, check the list of misidentified cell lines maintained by Amanda Capes-Davis and Ian Freshney and posted on the websites of Cell Bank Australia (www.cellbankaustralia.com/Misidentified-Cell-Line-List/default.aspx) and other major cell banks. At the time of writing, the latest version (dated August 2013) lists some 438 cell lines for which there is no known un-contaminated stock that corresponds to the original claims for the cell line.

If the cell line you wish to use passes this initial check, or has been recently isolated, actual cross-contamination tests on the cells must be performed at the first opportunity, ideally on the token freeze. Subsequently it will also be necessary to repeat these tests at carefully chosen points (*e.g.* on the WCB).

Nowadays, tests for inter- and intra-species contamination and, in humans at least (but also in certain select species of animal), identification of the individual cell line and/or donor can be performed rapidly in suitably equipped laboratories. Such testing is commercially available and relatively cheap. For identity testing on human cells, short tandem repeat (STR) pro-filing is now the method of choice and the American National Standards Institute in collaboration with the American Type Culture Collection has published a standard on the subject (ASN-0002-2011, available from www.ansi.org). *COI* DNA sequencing is currently the best method for use with non-human cells. Nonetheless, older techniques such as iso-enzyme analysis (Figure 9.1) and DNA fingerprinting may still have their place for certain applications. For more details on this topic, see reference 54.

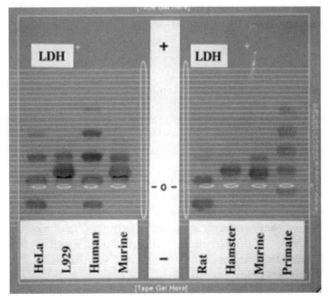

Figure 9.1 Isoenzyme analysis of six cell lines from various species, compared to the two standards HeLa (human) and L929 (mouse). The test enzyme is lactate dehydrogenase.

Table 9.2   Some phenotypic characteristics that can be used for the characterization of cell lines.

Morphology
Growth characteristics
Response to specific growth factors
Cell surface and intracellular molecular markers
Secreted products
Gene expression profiles
Differentiation potential

Whilst identity testing is essential, this is only part of the story as cells with identical genomes can have different phenotypes (think of the different cell types in your own body). Thus phenotypic characterization is also important. Appropriate tests will have to be chosen on a case-by-case basis, dependent on the anticipated phenotype, but should be capable of differentiating between cells of this phenotype and closely related cells or cells that might reasonably have a chance of being present, *e.g.* other types of cell from the organ of origin. Some typical characteristics that might be used for this purpose are given in Table 9.2.

As mentioned previously, in the case of a cell line it is also important to characterize its stability, *i.e.* identify the number of doublings it can go through without significant phenotypic change, and to check that the cellular characteristics have not changed after cryopreservation. This applies to all cell lines, not just those that are non-immortalized; whilst immortalized or transformed cells may not senesce, they can still change their characteristics.

9.3.2   Microbial Contamination Testing

Any cell culture can be or become contaminated with microorganisms. While some contaminations—particularly those due to bacteria, yeast or fungi—may be overt, others may be cryptic and require specialized tests to detect them.

In the cell culture laboratory it is particularly important to test for mycoplasma, as contamination is seldom overt and the organisms do not overgrow cultures in the way that bacteria, for example, do. Rather, they are cellular parasites that can adversely affect many aspects of a cell's behaviour, including the rate of proliferation,[55] morphology,[56] amino acid metabolism[57] and macromolecular synthesis,[58] and can induce chromosomal aberrations and cell transformation.[59]

All cell lines must be tested for microbial contamination as soon as possible after introduction to the laboratory—*i.e.* whilst still under quarantine—and at regular intervals thereafter as well as whenever there is cause for concern. Most importantly, every culture should be tested at the end of an

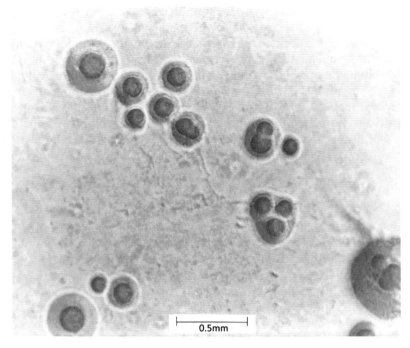

Figure 9.2 Mycoplasma colonies growing on agar, showing the characteristic "fried egg" morphology.

experiment or period of continuous culture; only if no contamination can be detected at this stage is it reasonable to assume that the culture has remained contamination-free throughout the period of use.

Testing for bacteria, fungi and yeasts is generally simple, and entails the culture of samples on or in suitable nutrient media under conditions where microorganisms might be expected to grow. Mycoplasma can also be tested for in a similar way, and grow into colonies having a characteristic "fried egg" morphology (Figure 9.2). However, certain strains have long been known not to grow under these conditions, and other tests have been (and continue to be) devised. These include Hoechst 33258 staining, usually after inoculation into an indicator cell line such as Vero, as well as tests based on the polymerase chain reaction (PCR) or on the detection of specific mycoplasma enzymes.

The presence of viruses is more difficult (and expensive) to test for, but both highly specific tests and tests for general categories of viruses are available.[54] The range of viruses it is appropriate to test for should be considered on a case-by-case basis, and will depend on the species of the cells being used, their history in culture (including any other cells or animal products with which they have come into contact) and their intended use.

Whilst the meaning of a positive test may be clear, a negative test result only implies a failure to *detect* any contamination; it does not guarantee there are no contaminating organisms present. All such tests have limitations with regard to specificity (not all organisms, or strains of organisms, may be detected by a particular test) and sensitivity (there will be a minimum number of organisms that will be needed to give a positive result in the test and within its timeframe, and these will need to be present in the volume of sample being tested). As discussed previously, the use of antibiotics in culture media may suppress but not eliminate the growth of microorganisms, and will also have an adverse effect on cultivation-based tests for microbes. Thus all cultures to be tested for the presence of microbes must be grown for at least two passages on antibiotic-free medium before testing. Further details of microbial contamination testing can be found in reference 54.

9.3.3 Quality Assurance of Reagents and Other Materials

The quality control of media, supplements, other additives, and cultureware (flasks, pipettes *etc.*) is highly specialized, complex and time-consuming, and as far as possible should be performed by the supplier working to appropriate industry standards. The supplier should supply relevant documentation (for example a Certificate of Analysis [CoA]) at the time of delivery, and maintain their original quality records in safe storage for many years thereafter. This does not entirely relieve the personnel in the receiving laboratory of responsibility, however, and in addition to checking that the supplied paperwork is complete, correct, and satisfies the laboratory's criteria for acceptance, other checks should be performed. These may range from simple visual checks to tests, for example, that the actual pH and osmolality of medium as supplied agrees with the values stated on the CoA. Very occasionally such tests may consistently reveal differences between the CoA and measurements carried out in the receiving laboratory, in which case investigations must be instigated to find the root cause. If no faults can be found with the tests at the receiving laboratory, then it will be necessary to enter into discussions with the manufacturer and may require an audit of their manufacturing facility.

Additionally, it is important to check that materials are suitable for their intended use, and that they are correctly stored. The properties of materials of biological origin (such as sera) can vary considerably from batch to batch, and checks should be carried out to monitor this. Indeed, critical reagents may need to be batch-tested prior to use, particularly where crucial properties of the reagent are not (or cannot) be tested by the supplier. For example, serum to be used in the production of a virus would need to be batch-tested before use to ensure the absence of virus-neutralizing antibodies.

Various other features of the materials used in and contributing to cell culture systems will need checking and controlling, including (but not limited to):

- Cleanliness (of equipment and the laboratory environment).
- Sterility (of everything coming into direct contact with cell cultures, media, *etc.*).
- Lack of toxicity (of anything that contributes to the culture system).

9.3.4 Quality Assurance of Equipment and Methods

It is essential that appropriate procedures are put in place for (as appropriate) the selection, purchase, installation, calibration, monitoring, maintenance and use of all critical equipment used in the cell culture laboratory. These include:

- Microbiological safety cabinets and other clean workstations.
- Incubators.
- Pressurized systems (including fermenters and pressurized gases).
- Refrigerators, freezers, and cryogenic storage and transport vessels.
- Autoclaves and sterilizing ovens.
- Reusable (Gilson- or Eppendorf-type) pipettes, and automatic dispensers.
- Analytical and production equipment.

It will usually be best to qualify or validate equipment prior to use, and a programme of regular preventive maintenance and calibration may also be required in order to ensure continued safe and accurate operation. In some cases these procedures may be a legal or regulatory requirement. For an introduction to the topic of validation, see reference 60.

In a similar fashion, important methods—governed by an SOP—will need to be validated to ensure that they are suitable for use, and give reliable, accurate and reproducible results. The nature and degree of validation that is appropriate will need to be assessed individually for each method. Some guidance on the validation of analytical techniques can be found in references 61 and 62, but it is important to note that these are regulatory documents aimed at the pharmaceutical industry; in other environments, and particularly in most research laboratories, the sort of extensive validation described will not be necessary (and indeed may not be financially feasible).

9.4 DOCUMENTATION

In any practical scientific work, adequate documentation of the experimental design and its rationale, the methods and analyses employed, the observations and results obtained, and the conclusions reached is essential in order to permit traceability, interpretation, and repetition of the work. This applies in full to cell culture work, but in addition the recording and retention of certain information can become crucial where the eventual aim of the work is to see it applied to the improvement of human or animal health. Consequently, it is essential right from the initiation of any work with cells to maintain comprehensive records, which should be signed and dated by the person generating them. A non-exhaustive list of some of the information that should be recorded is given in Table 9.3.

It is essential that all documentation be retrievable for an indefinite period, certainly for many years, and thus the low-tech method of recording

Table 9.3   Some of the essential information to record when performing cell culture work (note that this list is by no means exhaustive).

- The objective of the work and the rationale behind it.
- The source or donor of the cells:
  ○ Species
  ○ Sex
  ○ Age
  ○ Identity
  ○ Health
  ○ Organ
  ○ Method of isolation
  ○ Other important details (it may be worth taking and placing in long-term, possibly cryogenic, storage other samples from a cell donor (*e.g.* a blood sample) so that testing can be performed in the future when questions arise and/or when new technology becomes available)
- All the characterization and other tests performed on the cells.
- All the materials used during culture, including the medium (supplier, type, constituents, product code and batch number, testing performed, expiry date), and similar details of serum or serum substitutes, other additives/reagents, the cultureware, and any other substrate such as an extracellular matrix product or chemical used to enhance attachment
- The cell preservation and storage procedures used, including:
  ○ Identity of the cells
  ○ Passage number at freezing
  ○ Cryoprotectant used and its concentration
  ○ Number and concentration of cells per vial
  ○ Freezing, storage and resuscitation procedures
  ○ Details of tests performed to confirm successful resuscitation of the cells
  ○ Methods for the traceability of all stored samples
- Full details of all the methods employed (including SOPs where appropriate) for:
  ○ Media preparation
  ○ Cell culture
  ○ Analytical procedures
  ○ Equipment operation
  ○ Experimental or production procedures
  ○ Waste disposal
- The raw data obtained, with details of how they were processed/analysed, and the results derived.
- Records of all quality assurance and control procedures and results, and other relevant records such as those of waste disposal or laboratory environmental testing.

as much of this information as possible in ink on paper (for example in a high-quality, hardback laboratory book) has much to commend it. Data stored on computers can, even after a few years, be difficult to retrieve due to rapid changes in hardware and software; for example how many computers these days can retrieve information from a $5\frac{1}{4}$-inch floppy disk? Thus careful and detailed consideration should be given to how computer data should be stored, not only in terms of retrieval but also—and particularly if storage "in the cloud" is considered—security and retention of ownership.

Some types of study, notably those addressing environmental issues or the safety of pharmaceuticals, will need to be carried out in compliance with Good Laboratory Practice (GLP). Systems for documentation are central to GLP, and Orton[63] has supplied a useful introduction to GLP specifically aimed at cell culture scientists. Whilst workers outside the GLP environment will neither want (because of the burden of expense and time) nor need to comply with GLP requirements, Orton's chapter[63] is well worth reading and will provide useful food for thought to anyone engaged in cell culture work.

9.5 SAFETY

There are national and local laws (in most countries) that cover aspects of safety in the workplace, and clearly the current versions of these laws must always be understood, implemented and adhered to. Particular procedures, such as genetic modification, may also be regulated by law. In addition, many laboratories have their own rules. Numerous guidance documents are available covering various aspects of laboratory safety, and some of these specifically address issues pertinent to the cell culture laboratory.[64-67] What follows aims merely to supplement these, stressing and illustrating some of the particular hazards that may be encountered during cell culture.

In comparison to other laboratories, the cell culture laboratory that uses well characterized sub-primate cells not known to contain pathogens is a relatively safe place to work. Nonetheless, many of the same safety principles apply as in other laboratories.

9.5.1 Risk Assessment

Maintaining safety—in any situation—is achieved by assessing the risks involved and promptly instituting actions to avoid or minimise those risks. Consequently, in the laboratory a preliminary Risk Assessment must be made *and documented* before any new procedure is started.[68] Should the worst happen, this documentation is the first thing that any investigation will look for. In addition, it allows the risks and necessary safety procedures to be communicated to all staff involved, and facilitates training. All risk assessments should be reviewed:

- At least every 2 years.
- At any time when equipment, circumstances or the related environment changes.
- Whenever new technologies are introduced.
- When advances are made in relevant scientific knowledge.
- When new regulations or laws are introduced.

In addition, a preliminary Risk Assessment will need updating once a procedure has been carried out a number of times and the *real* (as opposed to perceived) hazards have been identified. It is, of course, essential that all

relevant staff know *and implement* all the necessary control measures identified in the Risk Assessment.

There are six main hazards in the cell culture laboratory, as follows:

9.5.2 Potential Pathogens

Cell cultures can contain pathogens, and the closer to humans the cells are
phylogenetically, the greater the potential risk. Human cells can of course
harbour human pathogens, but other primate cells can contain dangerous
viruses too, and deaths have been reported of workers handling monkey cell
lines.[69] Rodent cell lines can also contain viruses potentially pathogenic to
humans.[54,68] Risk assessment of cell lines, and the procedures in which they
are used, must take this into account, and (amongst other things) identify:

- Appropriate containment measures.[70–74]
- Any necessary testing or monitoring of the cells or of the individuals
  using them.
- Any other steps that can be taken to minimise risks (for example,
  workers using cell lines known to produce certain viruses might be
  required to be immunised against the relevant virus before starting to
  work with the cell line).

Containment measures will include the use of a laboratory designed to
handle organisms of the relevant risk category, use of the relevant type and
class of MSC, and other measures. In some cases, particularly where extended
use is to be made of a particular cell line, it may be appropriate to test the cells
for the presence of particular viruses or classes of viruses,[54,68] although this is
liable to be expensive and the cost/benefit ratio will need assessing. In addition it should be noted, as mentioned previously, that a negative result in a
test does not necessarily mean that the relevant organism is not actually
present. Furthermore, with time more and more viruses are being identified,
and as-yet-unknown viruses could be present in cultures. Thus, it is important
that the principle of universal caution should always be employed when using
cells in culture, *i.e.* treat every culture as if it *might* contain a pathogen.

9.5.3 The Cells Themselves

Theoretically, the closer cells are phylogenetically to humans the greater the
risk they pose. Yet, perhaps surprisingly, only one case has been reported of
accidental transplantation (of a human adenocarinoma *via* needlestick injury) with subsequent growth in the recipient. The resultant superficial lump
was excised and there was no evidence of subsequent recurrence.[75]

Cells can harbour oncogenes that could potentially be dangerous, and
these may have long latency periods. Genetically modified cells could
also pose a risk, depending on the nature of the cells and the
modifications made.

9.5.4 Pressurized Gases

A number of compressed gases are commonly used in the cell culture laboratory (carbon dioxide, carbon dioxide/air mixtures, nitrogen, oxygen, compressed air) and others may be used in ancillary equipment, so measures must be taken to minimize the risks they pose. Gases other than compressed air can be poisonous, asphyxiant (*e.g.* nitrogen, carbon dioxide), encourage combustion (oxygen) or themselves be combustible (*e.g.* hydrogen). All pressurized gases can become dangerous in the event of leaks, explosive release of pressure, or misuse. The commonest cause of such problems is the use of inappropriate, *i.e.* non-pressure-rated, tubing. Note that, in the UK, many pressurized systems will fall under the remit of the Pressure Systems Safety Regulations 2000, and will need checking regularly by suitably qualified individuals. Some components may also come under The Carriage of Dangerous Goods and Use of Transportable Pressure Equipment Regulations 2004. Similar regulations may apply in other countries.

9.5.5 Liquid Nitrogen

Arguably, in cell culture laboratories where neither human nor primate cells nor known pathogens are being used, liquid nitrogen—widely used for the cryopreservation and storage of cells—poses the greatest hazard to staff. It can kill by asphyxiation,[76] cause frostbite due to its low temperature (-196 °C), or if used with ill-designed or poorly maintained vessels can cause explosions.[77] Also, if incompletely sealed, vials stored under liquid nitrogen (LN) can fill up with LN during storage and explode on thawing.

Safety advice can be sought from LN suppliers, but some useful tips are:

- Make sure all staff are trained in the correct handling of LN.
- Handle all LN, and vessels containing it, in a well-ventilated area.
- Immediately before handling significant volumes, make sure someone not involved knows what you are doing, and where, and inform them when you have finished.
- Always wear the correct personal protective equipment:
  - Laboratory coat with long sleeves (ideally without pockets)
  - Visor
  - Insulated gloves (ideally loose fitting but with elasticated wrists)
  - Closed-toed shoes (*i.e.* not sandals or similar)
  - Use of a personal oxygen depletion monitor is also worth consideration.
- Take special care when filling warm vessels. Large volumes of nitrogen vapour can be generated extremely rapidly, and this can spray LN significant distances.

Further safety advice on the use of LN can be found in Appendix 1 of reference 3.

9.5.6   Other Chemicals

In many countries the use of substances hazardous to health falls under national regulations such as the COSHH Regulations in the UK, and such substances must be treated accordingly. It is worth noting that certain potentially hazardous chemicals are commonly encountered in the cell culture laboratory. Foremost among these are:

- Dimethyl sulphoxide (frequently used as a cryoprotectant in cell cryo-preservation). This passes easily through the skin, and has the potential to carry other, possibly hazardous, substances with it.
- Methotrexate (used in the culture of certain recombinant cell lines). This is teratogenic, and care must be taken not to expose women of child-bearing age to significant quantities.
- Propidium iodide (used in certain methods to stain non-viable cells) is mutagenic.

9.5.7   Breakage/Malfunction of Equipment

There are many conceivable risks that could fall into this category, but three are particularly likely to occur in the cell culture laboratory:

- *Needlestick injuries*. Avoid using hypodermic needles wherever possible. If unavoidable, devise careful handling techniques and methods that do not require the re-sheathing of needles, and dispose of sharps correctly.
- *The explosion of vials stored under liquid nitrogen*. Avoid the use of glass ampoules. Thoroughly check that ampoules are properly sealed before storage. Store in the vapour phase of liquid nitrogen where appropriate.
- *The breakage of glass pipettes*. This can occur during insertion of such pipettes into a pipetting aid. The cause is the pipette is not being held close to the point of insertion, resulting in laceration of the hand or wrist with the broken glass of the pipette. So hold pipettes close to the point of insertion, or simply use plastic pipettes.

9.6   LEGAL, ETHICAL AND REGULATORY COMPLIANCE

The importance of this really needs no explanation, and each individual worker has a duty to ensure that he or she, as well as managers and co-workers in the laboratory, is aware of relevant laws, requirements and guidelines covering the work being undertaken. In some cases it may not even be necessary actually to be performing work in order to come under the provisions of a specific regulation; in the UK, for example, it is only necessary to store a genetically modified organism to come under the provisions of the Biological Agents and Genetically Modified Organisms (Contained Use) Regulations 2010.

9.7 EDUCATION AND TRAINING

Cell culture science, techniques and technology are continually evolving, and it is essential that all workers involved with it have an appropriate education, that they are well trained at the outset of their work, and that continuing education and training is provided. It should be backed up by adequate supervision in the laboratory by experienced staff who are capable of spotting bad practice, identifying training needs and giving practical support. Even experienced staff, when entering a new laboratory, will need training in local procedures and the use of equipment with which they may be unfamiliar. In order to track staff training and professional development it is useful for each individual to maintain a formal training record, and where their work may involve or lead to the development or production of material for clinical use, the maintenance of such training records is a regulatory requirement. Thus it is recommended that such formal training records, with evidence of continuing professional development, are maintained for all staff.

More detailed suggestions for training can be found in reference 3.

9.8 CONCLUSION

In this chapter it has only been possible to look briefly at the issues that surround *in vitro* cell culture. For greater depth of coverage and further guidance on principles, readers should refer to the Guidance on Good Cell Culture Practice,[3] the Guidelines for the Use of Cell lines in Biomedical Research[78] and some of the many good books and review articles on the subject (such as references 1, 2 and 79) and the many references therein.

REFERENCES

1. J. Davis, *Animal Cell Culture: Essential Methods*, Wiley-Blackwell, Chichester, UK, 2011.
2. R. I. Freshney, *Culture of Animal Cells: A Manual of Basic Technique and Specialized Applications*, 6th edition, John Wiley and Sons, Hoboken, NJ, USA, 2010.
3. S. Coecke, M. Balls, G. Bowe, J. Davis, G. Gstraunthaler, T. Hartung, R. Hay, O.-W. Merten, A. Price, L. Schechtman, G. Stacey and W. Stokes, *Altern. Lab. Anim.*, 2005, **33**, 261–287.
4. I. Vorberg, A. Raines, B. Story and S. A. Priola, *J. Infect. Dis.*, 2004, **189**, 431–439.
5. J. J. Boonstra, R. van Marion, D. G. Beer, L. Lin, P. Chaves, C. Ribeiro, A. D. Pereira, L. Roque, S. J. Darnton and N. K. Altorki, *J. Natl Cancer Inst.*, 2010, **102**, 271–274.
6. E. G. Langeler, C. J. van Uffelen, M. A. Blankenstein, G. J. van Steenbrugge and E. Mulder, *Prostate*, 1993, **23**, 213–223.

7. M. J. Briske-Anderson, J. W. Finley and S. M. Newman, *Proc. Soc. Exp. Biol. Med.*, 1997, **214**, 248–257.
8. C. M. Chang-Liu and G. E. Woloschak, *Cancer Lett.*, 1997, **113**, 77–86.
9. M. Esquenet, J. V. Swinnen, W. Heyns and G. Verhoeven, *J. Steroid Biochem. Mol. Biol.*, 1997, **62**, 391–399.
10. H. Yu, T. J. Cook and P. J. Sinko, *Pharm. Res.*, 1997, **14**, 757–762.
11. Y. Sambuy, I. De Angelis, G. Ranaldi, M. L. Scarino, A. Stammati and F. Zucco, *Cell. Biol. Toxicol.*, 2005, **21**, 1–26.
12. X. Y. Li, Q. Jia, K. Q. Di, S. M. Gao, X. H. Wen, R. Y. Zhou, W. Wei and L. Z. Wang, *Cell Tissue Res.*, 2007, **327**, 607–614.
13. J. R. Smith and R. G. Whitney, *Science*, 1980, **207**, 82–84.
14. J. Davis, in *Animal Cell Culture: Essential Methods*, ed. J. Davis, Wiley-Blackwell, Chichester, UK, 2011, pp. 91–151.
15. W. Slikker, V. G. Desai, H. Duhart, R. Feuers and S. Z. Imam, *Free Radic. Biol. Med.*, 2001, **31**, 405–411.
16. T. C. Kou, L. Fan, Y. Zhou, Z. Y. Ye, X. P. Liu, L. Zhao and W. S. Tan, *J. Biosci. Bioeng.*, 2011, **111**, 365–369.
17. Y. Sumitomo, H. Higashitsuji, H. Higashitsuji, Y. Liu, T. Fujita, T. Sakurai, M. M. Candeias, K. Itoh, T. Chiba and J. Fujita, *BMC Biotechnol.*, 2012; DOI: 10.1186/1472-6750-12-72.
18. S. J. Morrison, M. Csete, A. K. Groves, W. Melega, B. Wold and D. J. Anderson, *J. Neuroscience*, 2000, **20**, 7370–7376.
19. T. Ezashi, P. Dash and R. M. Roberts, *Proc. Nat. Acad. Sci. U.S.A.*, 2005, **102**, 4783–4788.
20. S. F. Gorfien and D. W. Jayme, in *Animal Cell Culture: Essential Methods*, ed. J. Davis, Wiley-Blackwell, Chichester, UK, 2011, pp. 153–184.
21. M. J. Martin, A. Muotri, F. Gage and A. Varki, *Nature Medicine*, 2005, **11**, 228–232.
22. I. Kubikova, H. Konecna, O. Sedo, Z. Zdrahal, P. Rehulka, H. Hribkova, H. Rehulkova, A. Hampl, J. Chmelik and P. Dvorak, *Cytotherapy*, 2009, **11**, 330–340.
23. P. Tulkens and F. Van Hoof, *Toxicology*, 1980, **17**, 195–199.
24. R. Le Gall, C. Marchand and J. F. Rees, *Eur. J. Dermatol.*, 2005, **15**, 146–151.
25. S. Cohen, A. Samadikuchaksaraei, J. M. Polak and A. E. Bishop, *Tissue Eng.*, 2006, **12**, 2025–2030.
26. ICH, *Derivation and Characterisation of Cell Substrates Used for the Production of Biotechnological/Biological Products Q5D*. International Conference on Harmonisation of Technical Requirements for Registration of Pharmaceuticals for Human Use, Geneva, Switzerland, 1997. This can be downloaded at www.ich.org.
27. ICH, *Viral Safety Evaluation of Biotechnology Products Derived from Cell Lines of Human or Animal Origin Q5A (R1)*. International Conference on Harmonisation of Technical Requirements for Registration of Pharmaceuticals for Human Use, Geneva, Switzerland, 1999. This can be downloaded at www.ich.org.

28. World Health Organisation, *Requirements for the use of animal cells as in vitro substrates for the production of biologicals. In: WHO expert committee on biological standarcization. Forty-seventh report.* (WHO Technical Report Series, No. 878, annex 1) WHO, Geneva, Switzerland, 1998.

29. I. Knezevic, G. Stacey and J. Petricciani, *Biologicals*, 2008, **36**, 203–211.

30. R. Fleck and B. Fuller, in *Medicines from Animal Cell Culture*, ed. G. N. Stacey and J. M. Davis, John Wiley and Sons, Ltd, Chichester, UK, 2007, pp. 417–432.

31. International Stem Cell Banking Initiative, *Stem Cell Rev. and Rep.*, 2009, **5**, 301–314.

32. G. N. Stacey, J. R. Hawkins and R. A. Fleck, in *Animal Cell Culture: Essential Methods*, ed. J. Davis, Wiley-Blackwell, Chichester, UK, 2011, pp. 185–203.

33. L. L. Coriell, *Natl. Cancer Inst. Monogr.*, 1962, **7**, 33–53.

34. S. M. Gartler, *Natl. Cancer Inst. Monogr.*, 1967, **26**, 167–195.

35. S. M. Gartler, *Nature*, 1968, **217**, 750–751.

36. W. A. Nelson-Rees, R. R. Flandermeyer and P. K. Hawthorne, *Science*, 1974, **184**, 1093–1096.

37. W. A. Nelson-Rees and R. R. Flandermeyer, *Science*, 1976, **191**, 96–98.

38. W. A. Nelson-Rees, D. W. Daniels and R. R. Flandermeyer, *Science*, 1981, **212**, 446–452.

39. O. Markovic and N. Markovic, *In Vitro Cell. Dev. Biol. Anim.*, 1998, **34**, 1–8.

40. H. G. Drexler, W. G. Dirks and R. A. MacLeod, *Leukemia*, 1999, **13**, 1601–1607.

41. H. G. Drexler, W. G. Dirks, Y. Matsuo and R. A. MacLeod, *Leukemia*, 2003, **17**, 416–426.

42. H. G. Drexler, Y. Matsuo and R. A. MacLeod, *Hum. Cell*, 2003, **16**, 101–105.

43. R. A. MacLeod, W. G. Dirks, Y. Matsuo, M. Kaufman, H. Milch and H. G. Drexler, *Cancer*, 1999, **83**, 555–563.

44. J. R. Masters, *Cytotechnology*, 2002, **39**, 69–74.

45. J. R. Masters, *In Vitro Cell. Dev. Biol. Anim.*, 2004, **40**, 10.

46. G. C. Buehring, E. A. Eby and M. J. Eby, *In Vitro Cell Dev. Biol. Anim.*, 2004, **40**, 211–215.

47. R. Chattergee, *Science*, 2007, **315**, 928–931.

48. P. Hughes, D. Marshall, Y. Reid, H. Parkes and C. Gelber, *BioTechniques*, 2007, **43**(5), 575–583.

49. Y. A. Reid, *BioProcessing Journal*, 2009, **Spring 2009**, 12–17.

50. A. Capes-Davis, G. Theodosopoulos, I. Atkin, H. G. Drexler, A. Kohara, R. A. Macleod, J. R. Masters, Y. Nakamura, Y. A. Reid and R. R. Reddel, *Int. J. Cancer*, 2010, **127**, 1–8.

51. G. N. Stacey, *Nature*, 2000, **403**, 356.

52. R. A. MacLeod and H. G. Drexler, *Lancet Oncol.*, 2001, **2**, 467–468.

53. R. M. Nardone, *Cell Biol. Toxicol.*, 2007, **23**, 367–372.

54. P. Thraves and C. Rowe, in *Animal Cell Culture: Essential Methods*, ed. J. Davis, Wiley-Blackwell, Chichester, UK, 2011, pp. 255–296.

55. M. H. Claesson, T. Tscherning, M. H. Nissen and K. Lind, *Scand. J. Immunol.*, 1990, **32**, 623–630.

56. M. E. Pollock, P. E. Treadwell and G. E. Kenny, *Exp. Cell Res.*, 1963, **31**, 321–328.

57. D. M. Powelson, *J. Bacteriol.*, 1961, **82**, 288–297.

58. D. C. Krause and Y. Y. Chen, *Immun.*, 1988, **56**, 2054–2059.

59. K. Namiki, S. Goodison, S. Provasnik, R. W. Allan, K. A. Iczkowski, C. Urbanek, L. Reyes, N. Sakamoto and C. J. Rosser, *PLOS ONE*, 2009, **4**, e6872.

60. N. Chesterton, in *Medicines from Animal Cell Culture*, ed. G. N. Stacey and J. M. Davis, John Wiley and Sons, Ltd, Chichester, UK, 2007, pp. 285–301.

61. ICH, *Validation of Analytical Procedures: Text and Methodology Q2 (R1)*. International Conference on Harmonisation of Technical Requirements for Registration of Pharmaceuticals for Human Use, Geneva, Switzerland, 2005. This can be downloaded at www.ich.org.

62. USP, Biological Assay Validation. In *U.S. Pharmacopeia*, Chapter 1033, 2010. Accessed at http://www.ipqpubs.com/wp-content/uploads/2010/06/USP_1033.pdf in July 2013.

63. B. Orton, in *Animal Cell Culture: Essential Methods*, ed. J. Davis, Wiley-Blackwell, Chichester, UK, 2011, pp. 323–338.

64. W. E. Barkley, in *Methods in Enzymology*, ed. W. B. Jakoby and I. H. Pastan, Academic Press, New York, vol. 58, 1979, pp. 36–44.

65. J. L. Caputo, *J. Tiss. Cult. Methods*, 1988, **11**, 223–227.

66. World Health Organisation, *Laboratory Biosafety Manual*, 3rd edn, WHO, Geneva, Switzerland, 2004. This can be viewed/downloaded at http://www.who.int/csr/resources/publications/biosafety/WHO_CDS_CSR_LYO_2004_11/en/. Accessed July 2013.

67. *Biosafety in Microbiological and Biomedical Laboratories*, ed. L. C. Chosewood and D. E. Wilson, Department of Health and Human Services, Washington, DC, USA, 5th edn, 2009. This can be viewed/downloaded at http://www.cdc.gov/biosafety/publications/bmbl5/BMBL.pdf. Accessed July 2013.

68. G. N. Stacey, in *Medicines from Animal Cell Culture*, ed. G. N. Stacey and J. M. Davis, John Wiley and Sons, Ltd, Chichester, UK, 2007, pp. 569–588.

69. K. Hummeller, W. L. Davidson, W. Henle, A. C. Laboccetta and H. G. Ruch, *N. Engl. J. Med.*, 1959, **261**, 64–68.

70. R. A. Weiss, *Natl. Cancer Inst. Monogr.*, 1978, **48**, 183–189.

71. W. E. Grizzle and S. S. Polt, *J. Tiss. Cult. Methods*, 1988, **11**, 191–199.

72. Health and Safety Executive, *Biological agents: Managing the risks in laboratory and healthcare premises*, 2005, pp. 68–70. This can be viewed/downloaded at http://www.hse.gov.uk/biosafety/biologagents.pdf. Accessed July 2013.

73. ed. L. C. Chosewood and D. E. Wilson, *Appendix A - Primary Containment for Biohazards: Selection, Installation and Use of Biological Safety Cabinets, in Biosafety in Microbiological and Biomedical Laboratories*, 5th edn, Department of Health and Human Services, Washington, DC, USA,

2009. This can be viewed/downloaded at http://www.cdc.gov/biosafety/publications/bmbl5/BMBL.pdf. Accessed July 2013.

74. Advisory Committee on Dangerous Pathogens, *The Approved List of Biological Agents*, 3rd Edition, HMSO, Norwich, UK, 2013. This can be viewed/downloaded at http://www.hse.gov.uk/pubns/misc208.pdf. Accessed July 2013.

75. E. A. Gugel and M. E. Sanders, *New Engl. J. Med.*, 1986, **315**, 1487.

76. BBC, *Safety problems led to lab death*, 2000. This can be viewed/downloaded at http://news.bbc.co.uk/1/hi/scotland/798925.stm. Accessed July 2013.

77. Health and Safety Executive, *Rupture of a Liquid Nitrogen Storage Tank, Japan*, 28th August 1992. This can be viewed/downloaded at http://www.hse.gov.uk/comah/sragtech/caseliqnitro92.htm. Accessed July 2013.

78. R. Geraghty, A. Capes-Davis, J. M. Davis, J. Downward, R. I. Freshney, I. Knezevic, R. Lovell-Badge, J. R. W. Masters, J. Meredith and G. N. Stacey *et al.*, *Br. J. Cancer*, 2014, **111**, 1021–1046.

79. J. M. Davis and K. L. Shade, in *The Encyclopedia of Industrial Biotechnology: Bioprocess, Bioseparation and Cell Technology*, ed. M. C. Flickinger, John Wiley and Sons, New York, USA, Volume 1, 2010, pp. 396–415.

CHAPTER 10

The Biotechnology and Molecular Biology of Yeast

BRENDAN P. G. CURRAN*[a] AND VIRGINIA C. BUGEJA[b]

[a] School of Biological Sciences, Queen Mary College, University of London, Mile End Road, London, E1 4NS, UK; [b] School of Life & Medical Sciences, University of Hertfordshire, College Lane, Hatfield, Herts., AL10 9AB, UK
*Email: b.curran@qmul.ac.uk

10.1 INTRODUCTION

The yeast *Saccharomyces cerevisiae* plays a central role both in Biotechnology, the profitable exploitation of biological systems by man, and in Molecular Biology, the study and manipulation of biological systems at the molecular level. Its contribution to Biotechnology extends back over two thousand years; its contribution to Molecular Biology, although more recent, is equally as impressive because it is currently the most molecularly characterised eukaryotic organism on the planet.

Today the ancient biotechnological exploitation of yeast for the production of beer, wine and bread is complemented by yeast biotechnologies which manipulate DNA to direct the biological functions of yeast cells in ways that are only limited by scientific ingenuity. Therefore, in addition to its roles in traditional biotechnology, the baker's yeast *Saccharomyces cerevisiae*, and its non-fermentative cousin *Pichia pastoris*, play increasingly important roles as hosts in the production of non-native (heterologous) proteins, many of which are molecular medicines (Table 10.1), and as hosts in which sophisticated recombinant cloning technologies are employed to engineer

Molecular Biology and Biotechnology, 6th Edition
Edited by Ralph Rapley and David Whitehouse
© The Royal Society of Chemistry 2015
Published by the Royal Society of Chemistry, www.rsc.org

Table 10.1 Heterologous protein production in the yeasts *Saccharomyces cerevisiae* and *Pichia Pastoris.*

Produced in *Saccharomyces cerevisiae*[69]

Recombinant protein	Therapeutic use
Hirudan	Anticoagulant
Insulin	Diabetes mellitus
Somatropin	Growth disturbance
Glucagon	Hypoglycemia
Platelet-derived growth factor (PDGF)	Lower-extremity diabetic neuropathic ulcers
Hepatitis B surface antigen (HBsAg)	Hepatitis B vaccination
Major capsid protein from four Human papillomovirus (HPV) types	Vaccination against diseases caused by HPV
Urate oxidase	Hyperuricemia
Granulocyte-macrophage colony stimulating factor (GM-CSF)	Chemotherapy-induced neutropenia
Liraglutide	Type 2 Diabetes

Products under development in *Pichia pastoris*[70,71]

Recombinant protein	Therapeutic use
Angiostatin	Antiangiogenic factor
Endostatin	Antiangiogenic factor
Epidermal growth factor (EGF) analog	Diabetes
Elastase inhibitor	Cystic fibrosis
Human serum albumin	Blood volume stabiliser in burns treatment
Insulin-like growth factor-1	Insulin-like growth factor-1 deficiency

novel biochemical pathways to provide complex glycosylation patterns on heterologous proteins, and valuable novel (heterologous) secondary metabolites. Furthermore, this genetic tractability has also facilitated the development of massively parallel analysis techniques for DNA, RNA, protein and cellular metabolites. These post-genomic technologies, aimed at understanding the detailed molecular biology of *S. cerevisiae*, provide a deep insight into how this simple cell metabolises, grows and divides, which in turn provides a paradigm for the biology of eukaryotic cells—the basic building blocks of every multi-cellular organism on earth. These molecular insights are currently giving rise to an entirely new era for the rational biotechnological exploitation of yeast and other living organisms. This new approach, requiring interdisciplinary studies involving computer scientists, biologists, chemists, physicists, mathematicians and engineers, has produced two complementary, frequently overlapping fields, referred to as Systems Biology and Synthetic Biology. Together these aim to uncover the fundamental principles underlying biological phenomena, and to extrapolate beyond naturally evolved molecular mechanisms to rationally designed modular and more predictable alternatives.

Thus, the Biotechnology and Molecular Biology of yeast includes: the production of heterologous proteins; the use of sophisticated recombinant cloning technologies for the large-scale manipulation of DNA; the application of massively parallel post-genomic analyses technologies; and the rational design of yeast biology.

10.2   THE PRODUCTION OF HETEROLOGOUS PROTEINS BY YEAST

Heterologous, or recombinant, proteins are produced when recombinant DNA technology is used to ensure the expression of a gene product in an organism in which it would not normally be made. It is very challenging to assemble all of the genetic components required to ensure that a specific DNA sequence is transcribed into an mRNA molecule, which can subsequently be translated into a functional protein by a cell. Moreover, due to the bewildering array of molecular processes (*e.g.* glycosylation, acetylation, myristelation, *etc.*) that impact on protein processing, the final protein structure is dependent upon the type of host cell in which the expressed protein is synthesized. Thus, although gene expression vector technology was originally developed in *E. coli*, the production of recombinant proteins, especially if they are for *human use*, often requires an appropriately engineered DNA molecule to be transformed into a *eukaryotic* host cell. Therefore, shortly after the first heterologous proteins were expressed in *E. coli* in 1977,[1] the yeast *S. cerevisiae* was developed as the first eukaryote host system. Many other eukaryotes have been developed as host systems since then, and one of these *Pichia pastoris* is another yeast with considerable biotechnological value (for a review of heterologous protein expression in fungi see reference 2).

10.2.1   The Yeast Hosts

Yeasts are the simplest eukaryotes and they share many of the attributes of bacteria: they are unicellular, grow rapidly, can be transformed with DNA, and can form colonies on an agar plate.

  S. cerevisiae was developed as the first eukaryotic host cell to express heterologous proteins not only because it shared a number of useful attributes with *E. coli* but also because it had a long safe history of use in commercial fermentation processes. This made it particularly suitable for approval by regulatory bodies charged with the responsibility of ensuring the safe production of medically important heterologous proteins. Moreover it also has its own autonomously replicating plasmid, can carry out post-translational modifications of expressed proteins and it secretes a small number of proteins into the growth medium. This, as we shall see, can be exploited both to produce glycosylated proteins and to simplify the purification of heterologous proteins. It has a very primitive glycoslyation pathway however, which makes it unsuitable for many human proteins. That is one of the main reasons why its cousin *P. pastoris* was developed as an alternative yeast host cell. In addition to having a glycosylation pathway which produces more authentic patterns for human heterologous protein products than can *S. cerevisiae*, *P. pastoris* is a methylotroph, one of a small number of yeast species that share a biochemical pathway allowing the cells to utilise methanol as a sole carbon, and can therefore grow to much higher cell densities in fermenters due to the absence of toxic levels of ethanol.

10.2.2 Assembling and Transforming DNA Constructs into the Yeast Hosts

Yeast cells require appropriate yeast expression vectors in order to express foreign proteins. These are generally shuttle vectors, which consist of DNA sequences that ensure appropriate protein expression in the yeast host cells, and a section of bacterial DNA that allows the molecule to be assembled and engineered in *E. coli*. Typical vectors can be seen in Figure 10.1. They consist of a backbone of bacterial DNA with an origin of replication for *E. coli*, a selectable marker such as Ampicillin resistance, and a number of appropriate restriction sites into which the various yeast-specific DNA sequences can be inserted. These include: a yeast selectable marker, a strong yeast promoter and terminator to control mRNA production, appropriate DNA sequences to direct translation when the mRNA has been generated, and a polycloning site for insertion of the DNA sequence encoding the heterologous protein. The heterologous DNA, which is almost always a cDNA, is inserted between the promoter and terminator sequences (Figure 10.1C). This is because yeast cells cannot normally recognise regulatory sequences in DNA, or introns in expressed RNA, from higher eukaryotes such as humans.

There are two main types of expression vectors used in *S. cerevisiae* biotechnology. YEp (Yeast Episomal Plasmid) vectors are based on the ARS (Autonomously Replicating Sequence) sequence from the endogenous yeast 2μ plasmid which contains genetic information for its own replication and segregation. They are capable of autonomous replication, are present at 20–200 copies per cell and under selective conditions are found in 60% to 95% of the cell population (Figure 10.1A). The addition of a centromeric sequence converts a YEp into a Yeast Centromeric plasmid (YCp) (Figure 10.1B). YCps are normally present at one copy per cell, can replicate without integration into a chromosome and are stably maintained during cell division even in the absence of selection. An extremely versatile family of expression vectors providing alternative selectable markers, a range of copy numbers and differing promoter strengths (see Figure 10.1D) has been produced by Mumberg *et al.*[3]

S. cerevisiae vectors are usually transformed as circular molecules into the host cells; linearised expression constructs inserted into the chromosomes are less frequently used. On the other hand *P. pastoris* expression vectors are designed to be cut and linearised before transformation. This allows the constructs to integrate into the host chromosome and, although this limits the copy number, thereafter there is no need for further selection to maintain the construct in the host cells. *S. cerevisiae* and *P. pastoris* expression vectors are freely available from yeast research laboratories and there is increasing availability from commercial sources.

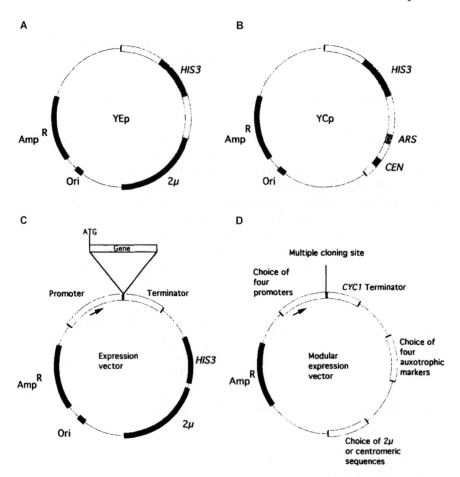

Figure 10.1 Schematic diagrams of yeast cloning vectors. A yeast episomal (YEp) vector (A) consists of a prokaryotic gene for resistance to ampicillin (Amp[R]), a prokaryotic replication origin (ori), a yeast auxotrophic marker (HIS3) and a yeast 2μ DNA sequence. A yeast centromeric (YCp) vector (B) contains a yeast centromere (CEN) and an autonomous replication sequence (ARS) instead of the 2μ DNA sequence. The addition of yeast promoter and terminator sequences generates a yeast expression vector (C); transcription initiation of heterologous genes cloned into a unique cloning site is indicated by the arrow. A extremely versatile series of modular expression vectors (D) provides a choice of promoters (CYC1, ADH, TEF, and GPD); a choice of selectable marker genes (HIS3, LEU2, TRP1 and URA3); a choice of copy number (centromeric or 2μ plasmid) and contains a multiple cloning site between the promoter and terminator sequences (see reference 3).

Regardless of the origin of the vector, transformation of the host by DNA is usually achieved either by treating whole cells with alkali cations (frequently lithium), in a procedure analogous to *E. coli* transformation, or by electroporation, which involves using a brief voltage pulse to facilitate entry of DNA

Table 10.2   Selectable markers for yeast transformation.

Auxotrophic markers	
Gene	*Chromsomal mutation*
*HIS*3	*his3-Δ1*
*LEU*2	*leu2-3,leu2-112*
*TRP*1	*trp1-289*
*URA*3	*ura3-52*

Dominant markers	
Gene	*Selection*
*CUP*1	Copper resistance
G418[R]	G418 resistance (Kanomycin phosphotransferase)
TUN[R]	Tunicamycin resistance

molecules into the cells. Yeast sphaeroplasts exposed to DNA in the presence of calcium ions and polyethylene glycol may also be used. The Saccharomyces Genome database is an excellent resource for strains, protocols and methods.[4]

Transformants are identified by selection. Unlike the dominant antibiotic resistance markers used in *E. coli* transformations, many yeast selectable markers are genes which complement a specific auxotrophy (*e.g.* Leu, His, Trp, *etc.*) and thus require the host cell to contain a recessive, non-reverting mutation. The most widely used selectable markers and their chromosomal counterparts and the dominant selectable marker systems are listed in Table 10.2.

10.2.3   Ensuring Optimal Expression of the Desired Protein

10.2.3.1   Ensuring High Levels of mRNA.   The first step in expressing heterologous proteins is the production of mRNA from the DNA construct. The overall level of heterologous mRNA in the cell is a balance between the production of mRNA and its stability in the cytoplasm. The former is determined by the copy number of the expression vector and the strength of the promoter; the latter by the specific mRNA sequence.

In the case of *S. cerevisiae*, expression vectors based on YEp technology have a high copy number but require selective conditions to ensure their stable inheritance. Plasmid instability can be prevented by introducing a centromere into the vector but at the cost of reducing the plasmid copy number to 1–2 copies per cell.

High level mRNA production is also dependant on the type of promoter chosen to drive expression. The major prerequisite for an expression vector promoter is that it is a strong one. Those most frequently encountered are based on promoters from genes encoding glycolytic enzymes *e.g.* phosphoglycerate kinase (*PGK*), alcohol dehydrogenase 1 (*ADH*1) and glyceraldehyde-3-phosphate dehydrogenase (*GAPDH*), all of which facilitate high level constitutive mRNA production. Constitutive expression can be disadvantageous when the foreign protein has a toxic effect on the cells but this can be circumvented by using a regulatable promoter to induce heterologous gene expression after cells have grown to maximum biomass. The most commonly used one, based on the promoter of the galactokinase gene (*GAL1*), is induced when glucose is replaced by galactose in the medium, but a number of others are also available. Transcript stability is also vitally important in maintaining high levels of mRNA in the cell. Despite the fact that "instability elements" have been identified[5] mRNA half-lives cannot be accurately predicted from primary structural information so they must be empirically determined; unstable transcripts curtail high level expression. *S. cerevisiae* often fails to recognise heterologous transcription termination signals because its own genes lack typical eukaryotic terminator elements. This can result in the production of abnormally long mRNA molecules which are often unstable. As this can result in a dramatic drop in heterologous protein yield, expression vectors frequently contain the 3' terminator region from a yeast gene (*e.g. CYC1, PGK* or *ADH1*) to ensure efficient mRNA termination (Figure 10.1D).

In the case of *P. pastoris* the cells can be grown on methanol using an alternative alcohol oxidase locus, in which case the heterologous protein is continuously expressed, or on glucose, in which case the heterologous gene is repressed until induced by methanol. The tight level of regulation allows for extremely precise control of the expression of the heterologous gene. This easily regulated promoter has practical advantages over the more cumbersome galactose-inducible ones used to regulate heterologous expression in *S. cerevisiae*. Recently, however, a set of novel inducible[6] and repressible[7] promoters have been identified for high level expression of recombinant proteins in *P. pastoris* on different carbon sources.

10.2.3.2 Ensuring High Levels of Protein. A high level of stable heterologous mRNA does not necessarily guarantee a high level of protein production. The protein level depends on the efficiency with which the mRNA is translated and the stability of the protein after it has been produced (see reference 2).

The site of translation initiation in 95% of yeast mRNA molecules corresponds to the first AUG codon at the 5' end of the message. It is advisable to eliminate regions of dyad symmetry and upstream AUG triplets in the heterologous mRNA leader sequence to ensure efficient initiation of translation. The overall context of the sequence on either side of the AUG codon (with the exception of an A nucleotide at the -3 position) and the leader length do not appear to affect the level of translation.[8]

Achieving high level transcription and translation of a heterologous gene in any expression system does not necessarily guarantee the recovery of large amounts of heterologous gene product: some proteins are degraded during cell breakage and subsequent purification; others are rapidly turned over in the cell. The powerful tools provided by a detailed knowledge of yeast molecular biology can be exploited to minimise this. Genes for proteolytic enzymes can be inactivated either by mutation or the presence of protease inhibitors during extraction. A more elegant route is to exploit the yeast secretory pathway to smuggle heterologous proteins out of the cell into the culture medium where protease levels are low.

This not only minimises the exposure of heterologous proteins to intracellular protease activity but also facilitates their recovery and purification due to the very low levels of native yeast proteins normally present in culture media.

Entry into the secretory pathway is determined by the presence of a short hydrophobic 'signal' sequence on the N-terminal end of secreted proteins. The 'signal' sequences from *S. cerevisiae*'s four major secretion products have been attached (by gene manipulation) to the N-terminus of heterologous proteins and used with varying degrees of success to direct their secretion: invertase[9] and acid phosphatase[10] signal sequences target proteins to the periplasmic space whereas α factor[11] and killer toxin[12] signals target the proteins to the culture medium. A typical secretion vector is shown in Figure 10.2. Secretion can also be used to produce proteins that have an amino acid other than methionine at their N-terminus. If a secretory signal is spliced onto the heterologous gene at the appropriate amino acid (normally the penultimate one) then the N-terminal methionine which is obligatory for translation initiation will be on the secretory signal. Proteolytic cleavage of this signal from the heterologous protein in the endoplasmic reticulum (ER) will generate an authentic N-terminal amino acid (Figure 10.3).

Heterologous proteins can also be secreted from *P. pastoris* with the most widely used secretion signal sequences being the *S. cerevisiae* α factor prepro sequence and the signal sequence from *Pichia's* own acid phosphatase gene.

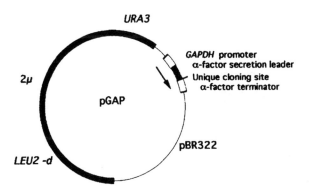

Figure 10.2 Schematic representation of the yeast secretion vector pGAP.[72] The vector contains LEU2-d and URA3 yeast selectable marker genes, pBR322 sequences for amplification in *E. Coli* and 2μ sequences for autonomous replication in yeast. The expression 'cassette' contains a unique cloning site flanked by GAPDH promoter, α-factor secretion leader and α-factor terminator sequences. Transcription iniation is indicated by the arrow.

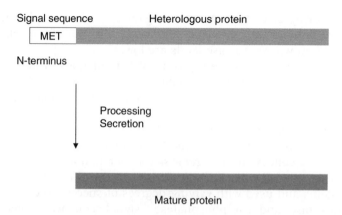

Figure 10.3   Schematic diagram showing the secretion of a heterologous protein
using a signal sequence. Cleavage of the secretory signal in the endoplasmic reticulum removes the N-terminal methionine, thereby generating a heterologous protein with an authentic N-terminal amino acid.

10.2.3.3   Obtaining the Appropriate Protein Structure and Function.   The
objective of heterologous gene expression is the high level production of
biologically active, authentic protein molecules. It is therefore important to
consider the nature of the final product when choosing the expression
system. The protein size, hydrophobicity, normal cellular location, need for
post-translational modification(s) and ultimate use must be assessed before
an appropriate expression system is chosen.

The secretory pathway is often chosen for heterologous protein production
because apart from the fact that it enhances protein stability and can generate proteins lacking an N-terminal methionine, secretion facilitates the
accurate folding of large proteins and contains the machinery for post-
translational modification (for review see reference 13). In a seminal experiment a direct comparison between the intracellular production and extra
cellular secretion of prochymosin and human serum albumin resulted in the
recovery of small quantities of mostly insoluble, inactive protein when they
were produced intracellularly but the recovery of soluble, correctly folded,
fully active protein when they were secreted.[14,15]

The biological activity and/or stability of heterologous proteins can also be
affected by the post-translational addition of carbohydrate molecules to
specific amino acid residues. Glycosylation in yeast is of both the N-linked
(*via* an asparagine amide) and O-linked (*via* a serine or threonine hydroxyl)
types, occurring at the sequences Asn-X-Ser/Thr and Thr/Ser respectively.
Inner core N-linked glycosylation occurs in the ER and outer core glycosylation in the Golgi apparatus. However, it is important to note that the
number and type of outer core carbohydrates attached to glycosylated proteins in yeast are different to those found on mammalian proteins. In many
cases, these differences can be tolerated but if the protein is being produced

for therapeutic purposes they may cause unacceptable immunogenicity problems. One approach to overcoming this problem is to remove the glycosylation recognition site by site-directed mutagenesis. This strategy was successfully used to produce urokinase type plasminogen activator.[16] Alternatively, the entire pathway can be re-engineered (see Section 10.3.1.2). The secretion pathway is extremely complex and, although there are overall principles, the expression of each protein is unique because of the wide variety of variables which must be borne in mind including: the effect of the secretion leader sequence, the type of protein, and the promoter used (for an excellent review see reference 17).

Despite the advantages secretion offers for the production of heterologous proteins in yeast, higher overall levels of protein production are often possible using intracellular expression. Some proteins form insoluble complexes when expressed intracellularly in *S. cerevisiae* but many others do not. Human superoxide dismutase was recovered as a soluble active protein after expression in yeast. It was also efficiently acetylated at the amino terminus to produce a protein identical to that found in human tissue.[18] Other proteins can be produced as denatured, intracellular complexes that can be disaggregated and renatured after harvesting. The first recombinant DNA product to reach the market was a hepatitis B vaccine originally produced in this way.[19]

P. pastoris is regarded as a more efficient and more faithful glycosylator of secreted proteins. Many proteins of therapeutic importance have been successfully made using both intracellular and extracellular production in *Pichia*. These include: proteins involved in the prevention and treatment of clots; peptide hormones and cytokines and protein vaccines including a lucrative hepatitis B vaccine.

Many proteins of biotechnological interest require complex post-translational modification but yeast, as a primitive eukaryote, is unable to effect those molecular changes. However, as we shall see in section 10.3.1, molecular biologists are currently developing a series of yeast strains that have been genetically manipulated so that they have modified and/or completely new biochemical pathways,[20] some of which can now circumvent this limitation.[21,22]

10.3 SOPHISTICATED RECOMBINANT CLONING TECHNOLOGIES FOR THE LARGE-SCALE MANIPULATION OF DNA

The contribution of yeast to Biotechnology and Molecular Biology extends far beyond the production of heterologous proteins. The ease with which yeast cells can be genetically manipulated, and in particular their proficiency at homologous recombination, makes yeast ideally suited as a host in which to develop novel and extremely powerful cloning technologies. These include *metabolic pathway engineering* and the assembly of *artificial chromosomes*. The former is used for the production of novel heterologous secondary metabolites, and the addition of new biological functionalities such as the production of 'humanised' glycoslyation heterologous proteins, to the cell. The latter has been exploited in genome analyses, re-engineering mammalian genomes, and even the assembly of synthetic life forms.

10.3.1 Engineering New Metabolic Pathways in Yeasts

Homologous recombination allows the targeted insertion of chosen DNA sequences into yeast genomes. These can be additional genes or the insertion of selectable markers aimed at deleting existing genome sequences. Gene deletions coupled with the addition of genes from other organisms can thus be used to dramatically modify existing biochemical pathways or introduce completely new ones. This combination of host cell gene deletion and foreign gene insertion has been used to produce secondary metabolites normally found in plants on one hand, and glycoproteins normally produced in mammalian cells on the other. In essence, biochemical pathway engineering projects allow biotechnologists to convert yeast cells into surrogate plant or animal cells.

10.3.1.1 Genetically Engineering the Production of Secondary Metabolites. Many organisms have biochemical pathways which, although useful, are non-essential to cell survival. These secondary metabolic pathways produce a huge array of products many of which are commercially valuable medicines, fragrances and flavours. Commonly isolated from plants, microbes and marine organisms these compounds are normally produced in rather limited amounts and purifying them is often difficult. Moreover their intrinsic complexity makes them very difficult to chemically synthesise. An alternative is to transfer the genes encoding the enzymes required for the biosynthesis of the compounds into microbial cells from whence the products can be easily extracted. The yeast *S. cerevisiae* is increasingly being used in this endeavour (for reviews see references 23 and 24). There are many examples of yeast cells being re-engineered into surrogate plant cells by transforming plant genes into the yeast such that the central metabolic pathway is hijacked to generate the basic building blocks required for the synthesis of the complex molecules in question. Two very exciting examples are presented in Figure 10.4.

Here the central biochemical pathway in yeast is the pathway leading from the products of glycolysis, through isopentenyl pyrophosphate (IPP) and its isomer dimethylallyl pyrophosphate (DMAPP), geranyl diphosphate (GPP) and farensyl diphosphate (FPP) to squalene and thence to ergosterol the central membrane steroid. This pathway can he hijacked into secondary metabolism by expressing the appropriate plant enzymes. Aritemisinin, a highly effective product in the treatment of the malaria causing parasite *Plasmodium falciparum*, can be extracted from the plant *Artemisia annua*. It is a chemically complex molecule and is in short supply. Workers cloned the amorphadiene synthase gene from that plant and expressed it in yeast.[25] This hijacks the central biosynthetic pathway at FPP (see Figure 10.4) and

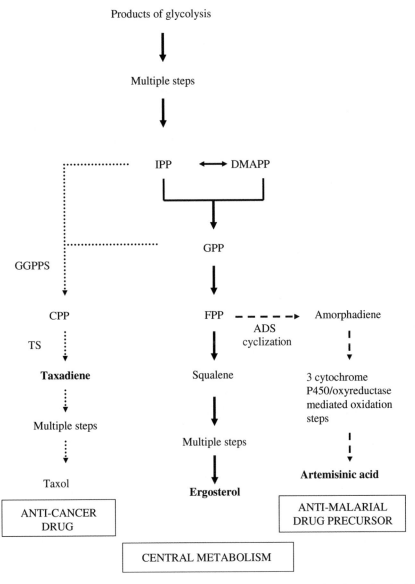

Figure 10.4 Producing plant cell metabolites, used in the production of valuable drugs, by hijacking the central metabolism of yeast. Expression of a plant amorphadiene synthase enzyme converts the yeast metabolite FPP into Amorphadiene. The further coexpression of a cognate cytochrome P450 hydroxylase and cytochrome P450 reductase generates Artemisinic, a precursor to the anti-malarial drug Artemisinin. Alternatively the same pathway can be by 'hijacked' at an earlier point by expressing Genranylgenranyl diphosphate synthase (GGPPS) and Taxadiene synthase (TS) from yew trees. These convert the yeast central metabolites IPP and GPP *via* CPP, into a plant cell metabolite called Taxadiene. This is a precursor in the production of the anticancer drug Taxol.

converts FPP it into the cyclical molecule Amorphadiene. The further cloning and expression of coexpression of a cognate cytochrome P450 hydroxylase and cytochrome P450 reductase then mediates three oxidation steps to generate Artemisinic acid which can then be extracted and treated chemically to generate the desired product. Similarly the expression of GGPPS and TS in a different project hijacked the cell into making a precursor for the anticancer drug Taxol.[26,27] Although these are the most well established examples of this type of pathway engineering for the production of secondary metabolites in yeast many more are in progress (for reviews see references 20 and 28).

10.3.1.2   Humanising the Glycosylation of Heterologous Proteins in P. pastoris *and* S. cerevisiae.   Just as recombinant cloning technology and homologous recombination were used to convert yeast cells into surrogate plant cells they can equally be harnessed to convert yeast cells into surrogate mammalian cells. Yeasts cells share many of the same basic sugar residues and the first steps in protein glycosylation, *core glycosylation*, as mammalian cells. In *core* glycosylation the sugars that are added to proteins are first enzymatically 'activated' as sugar nucleotides in the cytoplasm. Then two UDP-*N*-acetyl glucosamine followed by one GDP-mannose are sequentially added to the phosphate group of dolichol phosphate (a long lipid molecule in the ER with a terminal phosphate group in the cytoplasm). A further four GDP-mannose units are added generating a small oligosaccharide (2 GlcNac 5 Man) tethered to dolichol by its phosphate group. This entire structure is then flipped into the lumen of the ER, where enzyme-bound enzymes add further oligosaccharides and the entire structure transferred onto a growing polypeptide chain *via* specific asparagine residues within specific recognition amino acid sequences. This oligosaccharide is further processed before being exported into the Golgi complex, where *terminal glycoslyation* occurs. In the Golgi complex, N-linked oligosaccharides are further processed, and the addition of sugars to appropriate -OH residues of amino acids (O-linked glycoslyation) also commences.

It is here in the Golgi that yeast and higher organisms diverge in their biochemistry: yeast simply add further mannose residues; higher organisms trim the arriving oligosaccharide and then build up a complex array of sugar moieties onto the exposed sugars. In particular the cytoplasmic biosynthetic pathway for the synthesis of CMP-Sialic Acid, Golgi transporters for UDP-galactose and UDP-GlcNAc and CMP-sialic Acid, and the enzymes required to transfer these critically important saccharides onto the growing oligosaccharide are completely absent from yeast cells (see Figure 10.5).

Thus in order to transform a yeast cell into a surrogate animal cell it was necessary to re-engineer the cells such that:

1. Extra mannose residues were no longer added to the core oligosaccharide structures on glycoproteins after they entered the Golgi complex.

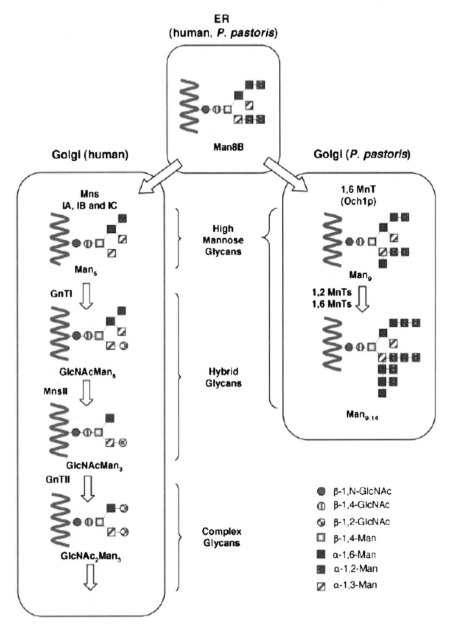

Figure 10.5 N-Linked glycosylation pathway in humans and in *P. pastoris*. This comparison highlights the differences between human and Pichia glycosylation pathways. From ref. 21. Reprinted with permission from AAAS. Yeast simply add further mannose residues; higher organisms trim the arriving oligosaccharide and then build up a complex array of sugar moieties onto the exposed sugars. The production of High Mannose Glyucans was prevented in by deleting genes from the Pichia host. The expression of a series of non-yeast enzymes then converted the host cell's glycosylation pathway into a mammalian surrogate.

2. Transporters for the nucleotide-sugars UDP-GlcNac and CMP-sialic Acid (absent in yeast but are required for human-type oligosaccharide assembly) were expressed and targeted to the Golgi complex.

3. A selection of glycosidase and glycotransferase genes were expressed and targeted to the Golgi in order to assemble the correct type and sequence of sugar moieties.

4. They expressed the biochemical pathway needed to generate CMP-sialic acid—a sugar which is an absolute requirement for humanised therapeutic proteins.

In a series of genetic engineering steps of increasing sophistication and complexity (for review see reference 29) each of these biological obstacles was overcome. In brief:

1. The addition of mannose in the Golgi complex was prevented by using gene knockout technology to delete *OCH1*, the enzyme responsible for transferring mannose sugars onto the core oligosaccharide in the Golgi complex.[30]

2. Genes for the appropriate transporter proteins were isolated from a range of organisms transformed into *P. pastoris* and successfully targeted into the Golgi membrane.[21,30,31]

3. A ground breaking procedure was developed in which one of a range of N-terminal ER (or early Golgi) localization signals from either *S. cerevisiae* or *P. pastoris* was fused in frame to one of a range of mannosidase or transferase catalytic domains from a whole range of organisms. In one such paper more than 600 different combinations were tested in order to identify an mannosidase capable of efficient high level trimming of mannose residues.[30] In the same study a transfer enzyme (GnT1) capable of adding UDP-GlcNac to oligosaccharides in the Golgi complex was isolated by screening a series of 67 different protein fusions engineered between fungal Golgi localization signal peptides and GnT1 catalytic domains from a range of higher organisms.[30] The same group later developed a more complex glycosylation pathway by adding a further two eukaryotic enzymes using the same strategy of creating chimeric proteins consisting of fungal leader sequences fused to catalytic domains.[21]

4. Finally, the same group expressed the genes for four human enzymes which modify nucleotide-sugars UDP-GlcNac into the human specific CMP-sialic acid in the cytoplasm, and then transport it into the Golgi complex. They finally ensured the addition of the terminal sialic acid residues onto the mature glycoprotein by successfully targeted a sialyltransferase/yeast-leader chimeric protein into the Golgi.[31]

This generated a yeast cell which has been genetically engineered by the deletion of native genes and the addition of fourteen new genes, with the latter being generated using DNA sequences from nine different organisms including three yeast species, the fruit fly, mouse, rat, and human in order to generate a fully humanised glycosylation pattern of heterologous human proteins. It has already been used to express EPO which has been shown to have the same biological effects as EPO of human origin.[31] This yeast has also been used to express human antibodies with extremely accurate and reproducible glycosylation patterns[32]—much more consistent in fact than mammalian cell cultures, which produce natural variability in the glycans on expressed proteins due to multiple enzymes competing for the same transient glycan structure. Not only do these yeast cells outshine mammalian cells in terms of expressing human proteins with reproducible glycosylation patterns, but because different yeast strains have been engineered with different combinations of glycosylation enzymes, work can now begin on how these sugar moieties, which are major determinants in the effector function of these complex molecules, mediate a whole range of immune responses.

Finally, following the successful assembly of this complex humanised N-linked glycosylation pathway, the less well characterised O-linked glycosylation pathway is currently being re-engineered in *Pichia*[33] and much progress has also been made humanising heterologous proteins in *S. cerevisiae*.[22]

10.3.2 Engineering Artificial Chromosomes

Our detailed knowledge of yeast molecular biology, coupled with the versatility of its genetic manipulation, led to the development of novel and extremely powerful cloning vectors referred to as Yeast Artificial Chromosomes (YACs), which are specialised vectors capable of accommodating extremely large fragments of DNA (100 Kb–1000 Kb).[34]

10.3.2.1 Genome Analysis using YACS. Schematic diagrams of a YAC and its use as a cloning system are shown in Figure 10.6. YACs contain a centromere, an autonomously replicating sequence, two telomeres and two yeast selectable markers separated by a unique restriction site. They also contain sequences for replication and selection in *E. coli*. YACs are linear molecules when propagated in yeast but must be circularised by a short DNA sequence between the tips of the telomeres for propagation in bacteria. When used as a cloning vehicle, the YAC is cleaved with restriction enzymes to generate two telomeric arms carrying different yeast selectable markers. These arms are then ligated to suitably digested DNA fragments, transformed into a yeast host and maintained as a mini chromosome.

YACs have become indispensable tools for mapping complex genomes such as the human genome[35] because they accommodate much larger fragments of DNA than bacteriophage or cosmid cloning systems thus simplifying the ordering of the human genome library. The complete library can be contained in approximately 10 000 clones, cutting by a factor of five the number of clones required by other vector systems. However, given the huge capacity and

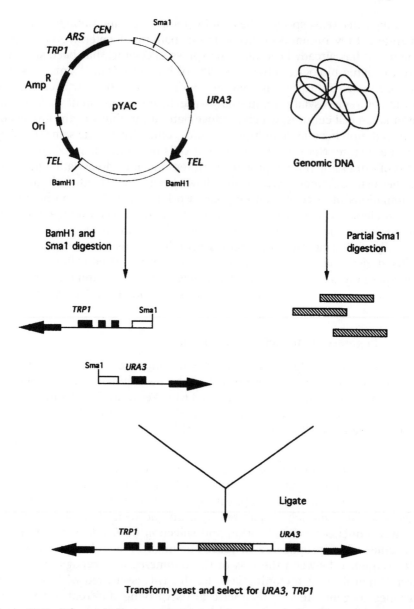

Figure 10.6 Schematic diagram of a YAC cloning vector, indicating prokaryotic gene for resistance to amplicillin (Amp^R), prokaryotic replication origin (ori), yeast auxotrophic markers (URA3, TRP1), autonomous replication sequence (ARS), yeast centromere (CEN) and telomeres (TEL).

stability of these powerful vectors, YACs have been further developed to address much more than genome analysis. YACs are also used as a chassis in which to engineer sections of heterologous genomes in yeast cells by exploiting the extremely accurate homologous recombination system.

10.3.2.2 Manipulating Mammals. Once a YAC has been successfully transformed into a yeast cell the highly efficient homologous recombination system of *S. cerevisiae* can be exploited *in vivo* to extensively manipulate both YAC vector sequences (retrofitting) and their inserts.[36] For example, homologous recombination can be used to retrofit mammalian selectable markers into the vector arms and/or introduce specific mutations into any genomic sequence carried in a YAC thus generating artificial chromosomes that can be used in the production of transgenic mice (see Figure 10.7). A linear DNA fragment consisting of neomycin resistance and *LYS2* genes sandwiched between the 5′ and 3′ ends of the *URA3* gene can be targeted into the *URA3* locus on the right arm of the YAC. Homologous recombination at this locus, which can be selected for by selecting for transformants on a lysine deficient medium, generates a useful uracil minus phenotype and introduces a mammalian selectable marker into the construct.

Homologous recombination and the selectable/counter selectable nature of the uracil phenotype can then be used to introduce desired mutations into the heterologous DNA carried in the YAC (see Figure 10.8): A *URA3* gene is inserted close to a suitably mutated site in a sub-clone of the relevant section of the gene of interest. This is then cut from the plasmid and used to transform yeast. Homologous recombination of the mutated version of the gene into the DNA inserted into the YAC insert is selected for by simply growing the yeast cells on medium lacking uracil. Removal of the *URA3* gene by a second homologous recombination event, selected for by growth on media containing 5-fluoroorotic acid (FOA),[37] generates a YAC vector containing a specifically mutated version of the gene of interest. A simplified diagram of this procedure is shown in Figure 10.8 and precise details can be found in reference 36. The YAC containing the manipulated genes can then be used to create transgenic mice[36] and rats[38] thus enabling the analysis of large genes or multigenic loci *in vivo*. Many such murine and human genes have been introduced into mice and display correct stage and tissue specific expression.[36,39]

10.3.2.3 Building Synthetic Organisms. The homologous recombination mechanism of yeast can also be used to generate an artificial chromosome that can be maintained in yeast but is very different to the types of YAC described above. This process of Transformation-Associated Recombination (TAR) exploits the fact that short specific non-yeast sequences at the ends of a linear DNA will recombine exclusively with the correct sequence in DNA molecules within the cell. In Figure 10.9 the vector has been linearised such that it carries specific human DNA sequences on both ends of the fragment. These will specifically recombine with and circularize with the target sequence. The vector also has a centromere and a selectable marker. However it lacks an autonomous replication sequence (ARS) and so, even if could circularize onto itself, will not survive because it cannot replicate in yeast.

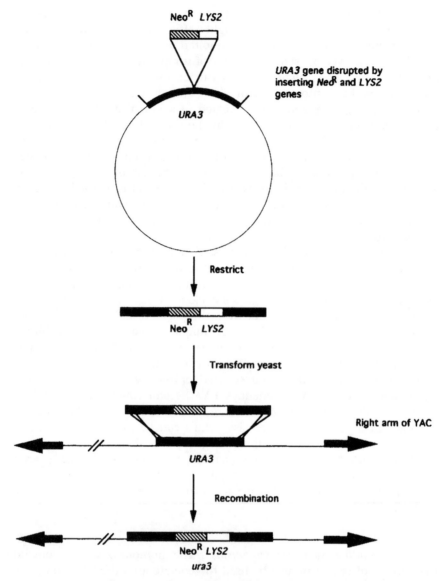

Figure 10.7   Schematic diagram showing the integration of linear yeast DNA into a homologous region of chromosomal DNA carried in a YAC. A linear fragment of DNA carrying the neomycin resistance gene (Neo^R) and the LYS2 gene flanked by URA3 gene sequences is isolated from plasmid DNA and transformed into the YAC carrying yeast strain. Homologous recombination events which result in the replacement of the wild-type (URA3) gene on the right arm of the YAC with this linear fragment are selected for by growth on lysine deficient medium.

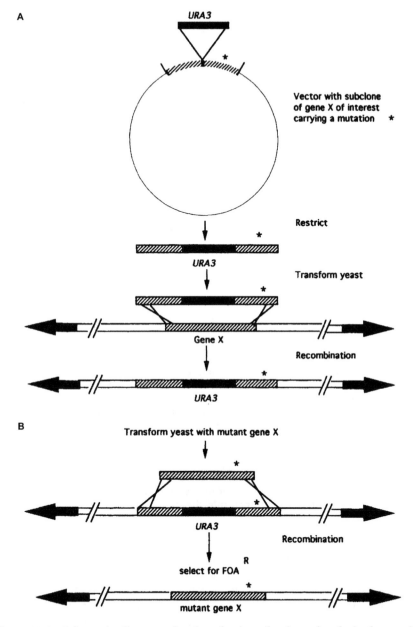

Figure 10.8   Schematic diagram showing the introduction of a desired mutation into a heterologous gene X carried in a YAC. (A) The URA3 gene is inserted into a suitably mutated subclone of gene X (the asterisk indicates a mutation in the gene DNA sequence). Homologous recombination and co-selection with URA3 is then used to introduce this mutation into the YAC insert. (B) Homologous recombination using the mutated subclone without the URA3 gene and counterselecting on FOA medium is then used to generate a specifically mutated version of gene X.

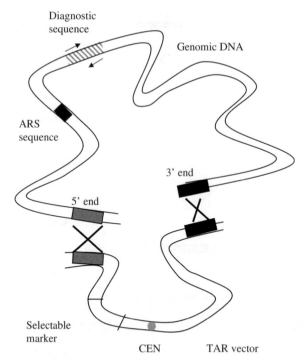

Figure 10.9   Isolation of a specific fragment of geonomic DNA by exploiting Transformation-Associated Recombination (TAR). Random segments of
genomic DNA are co-transformed into yeast with a linearised vector
which lacks an autonomous replication sequence (ARS). The linearised
vector carries a centromere, selectable marker and two specific targets
sequences. Site specific recombination creates an artificial chromosome consisting of the yeast vector and the targeted fragment of
genomic DNA, which supplies an ARS thereby allowing re-circularised
vectors carrying an insert to replicate in the yeast cell. A simple PCR
reaction is then used to confirm that the appropriate segment has been
successfully recombined into the linear vector.

However, human ARS are recognised in yeast cells and therefore vectors
carrying one will survive. However, this is only possible if the vector can
undergo homologous recombination with the specific human target sequences on the chromosomal DNA. Thus rather than generate an entire library of genomic sequences in order to isolate a specific chromosomal
segment, TAR cloning allows the direct isolation of specific targets.[40,41]
A PCR of part of the desired segment confirms if it has been successfully
isolated. There have been a number of extremely useful developments of
this type of TAR technology to join and isolate all types of DNA targets.
These include isolation of a functional copy of the human *BRCA1* gene.[42]
   This technology was also a key player in the synthesis of an entire
genome from laboratory synthesised DNA fragments. Craig Venter and his

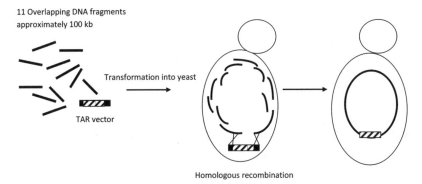

11 Overlapping DNA fragments
approximately 100 kb

Transformation into yeast

TAR vector

Homologous recombination

Figure 10.10 Schematic diagram showing how TAR cloning and homologous re-combination produced the assembly of the eleven ~ 100-kb synthetic DNA fragments into the final synthetic genome of *Mycoplasma mycoides*.

co-workers made extensive use of homologous recombination and TAR cloning in yeast to assemble a synthetic genome. They took one hundred and nine 10 000-base pair sequences and recombined them in sets of ten in yeast to produce eleven ~ 100 kb fragments (see Figure 10.10). These huge fragments were then simultaneously transformed into yeast cells where they were accurately recombined into the complete genome. The synthetic genome, which contained the genome sequence of the mycoplasm *Mycoplasma mycoides*, and some other diagnostic DNA sequences,[43] was successfully purified from the yeast cells and used to transform a related mycoplasm, *M. capricolum*, such that the synthetic genome replaced the host genome. The resulting transformant displayed only the biology encoded by the synthetic genome. Thus YAC technology was used to synthesise the first cell driven by an artificially synthesised genome. This extremely powerful yeast-based technology allows large DNA molecules to be assembled much more rapidly from synthetic or naturally occurring sub-fragments than with any other system described to date,[44] and is therefore set to accelerate many ongoing synthetic biology research projects.

10.4 THE DEVELOPMENT AND APPLICATION OF POST-GENOMIC ANALYSIS TECHNOLOGIES

The contribution yeast has made to both ancient and contemporary biotechnology is matched only by the key role it has played, and is playing, as a model organism in the development of eukaryotic molecular biology. *S. cerevisiae*'s well characterised biology and tractable genetics ensured that it became the first eukaryotic organism to have its entire genome sequenced (for review see reference 45), and ever since then a series of technological developments have maintained *S.cerevisiae* at the forefront of eukaryotic cellular biology.[46] Shortly after the genome was sequenced yeast geneticists set about using the exquisitely precise homologous recombination to knockout each individual reading frame to identify what each gene did. *S. cerevisiae* was thereafter used as the model eukaryote in which to develop methods of globally monitoring cellular mRNAs, proteins, metabolites and the myriad interactions effected by these cellular molecules (see reference 47). Currently this information is being combined with decades of insight gleaned from the literature to generate systems biology, a holistic counterweight to post-genomic reductionism, and synthetic biology, a newly emerging field that seeks to extrapolate from naturally evolved molecules and mechanisms to a much more rationally designed biology.

10.4.1 From Genome to Phenotypes

When the complete genome of *S. cerevisiae* was first revealed in 1996 one of the most intriguing observations was that despite decades of intensive research, biologists had no idea what functions were encoded by most of the genes. Even today, despite almost 20 years of intensive analysis many are still quite obscure. Nevertheless the powerful homologous recombination system in yeast has been used to produce a bar-coded set of yeast strains, each one carrying a precisely deleted single open reading frame. A PCR-based gene disrupted procedure[48] was used to disrupt each Open Reading Frames (ORFs) affording analysis of any resultant physiological effects (Figure 10.11). Long PCR primers, homologous to a section of the DNA sequence under investigation at the 5′ end and homologous to a selectable marker (frequently the *Kan*^{R} gene) at their 3′ terminus end, are used to generate a PCR fragment consisting of two short target gene segments on either side of a selectable marker. The target gene was then disrupted by homologous recombination and the event selected for by growth on appropriate selective medium. By incorporating one shared 18mer and one uniquely identifiable 20mer sequence into the 5′ primer between the ORF and the *Kan*^{R} sequences (see Figure 10.11)[49] each of the 6000 ORFs in the yeast genome has been successfully deleted. Not only does this allow insight into the effect of each deletion, but PCR can be used to monitor the population dynamics of these differently tagged yeast strains growing together under competitive growth conditions. These gene knockouts have provided unparalleled insights into basic eukaryotic cellular biology with perhaps the most dramatic revelation being that many gene deletions could be well tolerated in the haploid yeast cell revealing the robustness of biological systems.

However, this project, which casts light onto how yeast gene sequences mapped onto phenotypes, also threw light onto the genetics of other organisms including humans. Once the yeast genome sequence became available it rapidly became apparent that many human disease genes have homologues in this simple eukaryotic cell. Comparative sequence analysis between *S. cerevisiae* and human genes has already provided valuable insight into human cellular metabolism. A variety of examples are given in

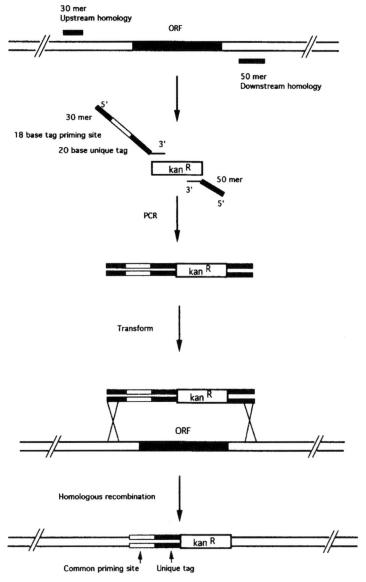

Figure 10.11 PCR gene deletion strategy. The Kanamycin resistance gene (Kan^R) is amplified using an 86 mer forward primer that contains 30 bases of upstream homology to the yeast gene of interest, an 18 base tag priming site, a uniquely indentifiable 20 base sequence tag and 18 bases of homology to the Kan^R gene. The reverse primer is a 68 mer that contains 50 bases of downstream homology to the yeast gene of interest and 18 bases of homology to the Kan^R gene. The PCR products are transformed into a haploid yeast strain and selected for on G-418 containing medium. Homologous recombination replaces the targeted yeast ORF with a common 18 mer priming site, a unique 20 base tag and the Kan^R gene.

Table 10.3 Yeast genes homologous to positionally cloned human genes.

Human disease	Human gene	Yeast gene	Yeast gene function	Yeast phenotype
Hereditary non-polyposis cancer	*MSH2*	*MSH2*	DNA repair protein	Increased mutation frequency
Cystic fibrosis protein	*CFTR*	*YCFl*	Membrane transport	Cadmium sensitivity
Wilson's disease	*WND*	*CCC2*	Copper transport atpase	Iron uptake deficiency
Glycerol kinase deficiency	*GK*	*GUT1*	Defective glycerol kinase	Defective glycerol utilization
Rhizomelic chondrodysplasia punctata	*PEX.5, PEX7*	*PEX.5, PEX7*	Peroxisome targeting mutants	Peroxisome dysfunction
Ataxia telangiectasia	*ATM*	*TEL1*	Phosphoinositol3-kinase	Short telomeres

Table 10.3 of which perhaps the most dramatic to date is the analysis of an autosomal recessively inherited disease Rhizomelic chondrodysplasia punctate, which presents with symptoms of severe growth and mental retardation. Patients with this condition were found to carry mutations leading to defects in peroxisome biogenesis that are functionally equivalent to the yeast peroxisome targeting mutants (*pex5* and *pex7*).[50,51] Yeast molecular biology also serves as an excellent model for many other human diseases including: cancer, neurodegenerative diseases[52] like Alzheimer's, Parkinson's and CJD and various metabolic disorders.

10.4.2 Transcriptomes, Proteomes, and Metabolomes

The DNA sequence of *S. cerevisiae* was just the starting point for large-scale molecular analysis of this eukaryotic cell. In addition to the deletion analysis mentioned above, this extremely tractable model organism rapidly yielded a whole series of molecular secrets on a global scale over the next few years: global mRNA profiling (transcriptomics) used to examine the oldest biotechnology in the world[53] revealed that a staggering 28% of yeast genes underwent a significant alteration in gene expression level as a result of the metabolic shift from fermentation to respiration; global analyses of protein localization[54] and of protein–protein interactions[55,56] (proteomics) found that many members of protein clusters associated with central roles (*e.g.* cell cycle) also had interactions with proteins from many other cluster classes revealing the great interconnectedness of the yeast proteome,[57,58] a viewpoint confirmed by the global characterization of cellular metabolites (metabolomics) which revealed that the average path length to get from any metabolite to any other metabolite is approximately three.[59] In short, this simple eukaryote became the key to post-genomic research (for review see reference 47).

Together these global studies made it possible to ascribe possible functions to previously unknown proteins and to acquire new insights on how cellular systems and networks underpin the biology of *S. cerevisiae*. The extensive research literature that had been accumulated about yeast, coupled with information from the post-genomic analyses of other model organisms, provided a deep insight into biology revealing a great unity: the basic building blocks, their interactions and processes are conserved in millions of organisms and across aeons of evolutionary time. Given these powerful technologies and this insight, biotechnologists can now monitor global changes in yeast cells in response to a variety of cellular insults and novel therapeutic agents in the knowledge that more often than not cells in multi-celled organism such as man will respond in the same way. This is the basis of a whole range of yeast-based biotechnologies for the identification of drug targets[60,61] for use in humans.

10.5 THE RATIONAL DESIGN OF YEAST BIOLOGY

The extraordinary research efforts that produced the current post-genomic era have generated mind-numbing amounts of data about individual cellular constituents. However, cells are dynamic entities and the next challenge is to understand the dynamic fluxes within the cell with a view to producing accurate mathematical and computer models that describe the biology of individual cells, and eventually multicellular organisms. These goals are a long way off but the complementary and partially overlapping fields of systems and synthetic biology, which require interdisciplinary teams of scientists, mathematicians and engineers, are making rapid progress, and once again yeast is leading the way.

10.5.1 Systems Biology

Systems biology is currently being applied to many biological systems and consists of the same core principles regardless of the organism in question: a computer model of the system is designed using all of the data available; the model is then used to predict how cells (or groups of cells) will respond based on how the computer model responds to altering a parameter *in silico*; if the model fails to predict the outcome in the laboratory then any discrepancy creates new information, which can then be used to inform and improve the model. Thus, the current objectives of systems biology are to develop biologically meaningful quantitative descriptions of living cells and to generate new computational tools to facilitate this. Once again, yeast researchers have been pioneers in this area. *S. cerevisiae* has been intensively studied for decades and, with the additional data from post-genomic high throughput analyses, provides an ideal model organism in which to undertake the reiterative process of model building, prediction, perturbation, analysis, new model. Indeed the first yeast paper on systems biology studied the relatively simple and comprehensively characterised galactose regulon[62] and it was here that the types of challenges that lie ahead were revealed: despite 40 years of publications in this area and all of the tools available to test the system, the best model was still incorrect. This gives a foretaste of just how much information will be needed in order to generate biologically meaningful models of even the simplest modules in a cell. However, the same paper[61] revealed the power and exciting possibilities offered by of this type of approach: the inconsistency between the prediction and biological reality revealed a regulatory parameter that had heretofore been unknown. Systems biology may well be in its infancy in yeast but it promises to provide unparalleled insight into basic cell functions[63] and of course how cells respond to changes in substrates, genetic changes and environmental insults.

10.5.2 Synthetic Biology

The term 'synthetic biology' has been used to describe a wide variety of biological endeavours over the past 20–30 years including enzyme engineering, metabolic manipulation and more recently the assembly and use of synthetic genomes. Currently synthetic biology is understood as a discipline which, like systems biology, uses computer modelling and experimental analysis of biological devices and systems but, in addition, uses the vocabulary/mentality of engineers to design and synthesise novel, modular, biological entities (for a review see reference 64). It is different from recombinant DNA technology in that it seeks to embrace principles from engineering such as modularity, standardisation and rigorously predictive models as it seeks a bottom-up study of biology from first principles. Traditional biologists take a reductionist approach aimed at untangling the molecular circuitry by which interacting cellular molecules give rise to observed biological phenomena; synthetic biologists seek to use well-understood cellular components (currently DNA sequences, transcription factors and signal transduction molecules) and, using as simple a design as possible, synthesise biological devices that effect a cellular outcome in a predictable fashion. In addition, such devices are frequently built using non-native parts to satisfy three essential characteristics of synthetic networks: orthogonality (independence from existing cellular networks), modularity (can be moved as a unit from one cell to another) and inducibility (an external stimulus triggers the newly programmed response).

A simple genetic network that acts as a molecular 'memory' device (where 'memory' is defined as a protracted cellular response to a transient stimulus)[65] is depicted in Figure 10.12. It was built using a bacterial DNA binding domain, a viral activation domain and a bacterial promoter sequence to ensure that it is independent of yeast genetic circuits. Fusion protein 'A' protein, is a transcription factor consisting of a reporter Red Fluorescent Protein (RFP) domain fused to Lex A, a bacterial DNA binding domain, and a viral activation domain. This fusion protein is induced in response to the presence of galactose. Under normal circumstances the removal of galactose

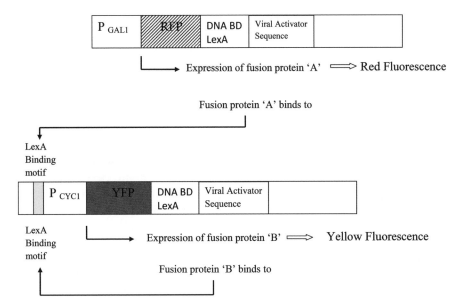

Figure 10.12 This molecular memory device consists of two gene constructs encoding fusion proteins that can act as reporter genes and as transcription factors. The top element in the circuit is induced whenever galactose activates the Gal 1 promoter. The resulting transcript encodes a fusion protein consisting of a Red Fluorescent Protein (RFP) domain fused to Lex A bacterial DNA binding domain (BD), and a viral transcription activation domain. Expression can be monitored by red fluorescence. The LexA domain then binds to the Lex A binding motif in the promoter of the lower element in the circuit thereby inducing fusion protein 'B' which consists of a Yellow Fluorescent Protein (YFP) domain fused to Lex A bacterial DNA binding domain, and a viral transcription activation domain. Expression form this element can be monitored by yellow fluorescence. This protein can then bind to the Lex A binding motif setting up an auto-activation cycle.

would result in the cessation of induction *i.e.* the response would be transient. However, the DNA binding domain of protein 'A' binds to the bacterial promoter element in the second construct inducing the expression of transcription factor 'B'. 'B' is also a fusion protein consisting of a reporter Yellow Fluorescent Protein (YFP) fused to the Lex A DNA binding, and viral activation, domains. However, unlike 'A', fusion protein 'B' has been designed to bind to its own promoter and activate it in an auto-activating cycle. Once triggered, protein 'B' will continue to activate its own expression even in the absence of the original stimulus, resulting in memory as defined above. The cellular 'outputs' can be monitored by the expression of the red and yellow florescent domains in proteins 'A' and 'B' respectively. This simple positive feedback network acts as a molecular memory, other more complex circuits using repressors act as switches and timers whilst other synthetic biologists engineer signal transduction devices using yeast cell receptors, thereby opening the door for using yeast cells to build multicellular synthetic networks.

Synthetic biology is in its infancy but it is already creating devices with potential biotechnological importance. Ellis *et al.*[66] used two mutually repressible bacterial repressors and libraries of promoter sequences to produce a range of timer circuits that controlled the expression of the *FLO1* gene. Using an external stimulus these synthetic devices triggered flocculation and yeast sedimentation on time-scales ranging from 60 to 168 hours depending on the synthetic timer chosen. A very different strategy was used by Babiskin *et al.*[67] to modulate flux in a predictable fashion through the ergosterol biosynthetic pathway. They designed and inserted a library of DNA sequences encoding artificial variants of a naturally occurring regulatory RNA stem-loop structures into the 3′ untranslated (3′ UTR) region of a key gene in the ergosterol pathway. The differential sensitivity of these stem-loops to RNase digestion resulted in predictably altered mRNA levels, thereby providing a new strategy to modulate flux through this biotechnologically important pathway (see Section 10.3.1.1). However, currently the most ambitious synthetic biology project using *S. cerevisiae* is the Sc2.0 project which aims to replace all of the yeast genome with laboratory synthesised DNA and in doing so introduce an inducible evolution system, SCRaMbLE (synthetic chromosome rearrangement and modification by loxP-mediated evolution),[68] which will be used to generate chromosome rearrangements generating significant genetic diversity with the ultimate aim of evolving a minimal yeast genome. This project will also replace all TAG stop codons with TAA stop codons. Thereafter TAG codons will be available as a codon for inserting unnatural amino with new chemistries into expressed proteins. Such a chassis cell will be of enormous value to biotechnology and molecular biology alike.

10.6   FUTURE PROSPECTS

Yeast biotechnology and molecular biology have had an extremely successful and synergistic history. Its central importance in alcohol production made yeast an organism of research value—the more one knew the better one could control it. This basic research in turn made yeast the organism of choice when a eukaryotic cell was needed in which to express heterologous proteins. Subsequently its huge biotechnological value coupled with its genetic tractability and well-characterized biochemistry ensured that it became the first eukaryote to have its genome sequenced. The post-genomic technologies that arose subsequently provided even more insight into its molecular biology such that we know more about it than any other eukaryotic cell and as the model organism without parallel it is set to lead us into a new era of rational biotechnology. The near future will see the development of glycosylation specific strains that will enable the production of a whole new host of heterologous proteins and allow glycobiologists to probe the exciting and elusive world of glycan structure and function in complex organisms. It will also see further developments in synthetic biology facilitating the optimisation of heterologous secondary metabolite production and the development of an ever-increasing number of well-characterized interchangeable biological devices. Likewise, TAR cloning will facilitate the study of the organization and evolution of complex prokaryotic and yeast genomes, thus facilitating the synthesis of novel life-forms with the ability to address global problems of energy limitations and climate change.

In short, yeast biotechnology and molecular biology have just come of age and the best has yet to come.

REFERENCES

1. K. Itakura, T. Hirose, R. Crea, A. D. Riggs, H. L. Heyneker, F. Bolivar and W. H. Boyer, *Science*, 1977, **198**, 1056.
2. B. Curran and V. Bugeja, *Fungi Biology and Applications*, ed. K. Kavanagh, Wiley-Blackwell, Chichester, West Sussex, UK, 2nd edition, 2011, Chapter 8, p. 205.
3. D. Mumberg, R. Muller and M. Funk, *Gene*, 1995, **156**, 119.
4. http://www.yeastgenome.org/.
5. R. Parker and A. Jacobson, *Proc. Natl. Acad. Sci. U. S. A.*, 1990, **87**, 2780.
6. R. Prielhofer, M. Maurer, J. Klein, J. Wenger, C. Kiziak, B. Gasser and D. Mattanovich, *Microb. Cell Fact.*, 2013, **12**, 5.
7. M. Delic, D. Mattanovich and B. Gasser, *Microb. Cell Fact.*, 2013, **12**, 6.
8. T. F. Donahue and A. M. Cigan, *Methods Enzymol.*, 1990, **185**, 366.
9. D. T. Moir and D. R. Dumais, *Gene*, 1987, **56**, 209.
10. A. Hinnen, B. Meyhack and R. Tsapis, in *Gene Expression in Yeast,* ed. M. Kornola, E. Vaisanen, Kauppakirjapaino, Helsinki, 1983, p. 157.
11. A. J. Brake, J. P. Merryweather, D. G. Coit, U. A. Heberlein, F. R. Masiarz, G. T. Mullenbach, M. S. Urdea, P. Valenzuela and P. J. Barr, *Proc. Natl. Acad. Sci. U. S. A.*, 1984, **81**, 4642.
12. N. Skiper, M. Sutherland, R. W. Davies, D. Kilburn, R. C. Miller, A. Warren and R. Wong, *Science*, 1985, **230**, 958.
13. M. Delic, M. Valli, A. B. Graf, M. Pfeffer, D. Mattanovich and B. Gasser, *FEMS Microbiol Rev.*, 2013, **37**, 872.
14. R. A. Smith, M. J. Duncan and D. T. Moir, *Science*, 1985, **229**, 1219.
15. T. Etcheverry, W. Forrester and R. Hitzeman, *Bio/Technology*, 1986, **4**, 726.
16. L. M. Melnick, B. G. Turner, P. Puma, B. Price-Tillotson, K. A. Salvato, D. R. Dumais, D. T. Moir, R. J. Broeze and G. C. Avgerinos, *J. Biol. Chem.*, 1990, **265**, 801.
17. Z. Liu, K. Tyo, J. L. Martínez, D. Petranovic and J. Nielsen, *Biotechnol. Bioeng.*, 2012, **109**, 1259.
18. R. A. Hallewell, R. Mills, P. Tekamp-Olsen, R. Blacker, S. Rosenberg, F. Otting, F. R. Masiarz and C. J. Scandella, *Bio/Technology*, 1987, **5**, 363.

19. P. Valenzuela, A. Medina, W. J. Rutter, G. Ammerer and B. D. Hall, *Nature*, 1982, **298**.
20. L. Liu, H. Redden and H. Alper, *Curr. Opin. Biotechnol.*, 2013, **24**, 1.
21. S. R. Hamilton, P. Bobrowicz, B. Bobrowicz, R. C. Davidson, H. Li, T. Mitchell, J. H. Nett, S. Rausch, T. A. Stadheim, H. Wischnewski, S. Wildt and T. U. Gerngross, *Science*, 2003, **301**, 1244.
22. F. P. Nasab, M. Aebi, G. Bernhard and A. D. Freya, *Appl. Environ. Microbiol.*, 2013, **79**, 997.
23. B. Huang, J. Guo, B. Yi, X. Yu, L. Sun and W. Chen, *Biotechnol. Lett.*, 2008, **30**, 1121.
24. J. A. Chemler, Y. Yan and M. A. G. Koffas, *Microb. Cell Fact.*, 2006, **5**, 20.
25. D.-K. Ro, E. M. Paradise, M. Ouellet, K. J. Fisher, K. L. Newman, J. M. Ndungu, K. A. Ho, R. A. Eachus, T. S. Ham, J. Kirby, M. C. Y. Chang, S. T. Withers, Y. Shiba, R. Sarpong and J. D. Keasling, *Nature*, 2006, **440**, 940.
26. J. M. DeJong, Y. Liu, A. P. Bollon, R. M. Long, S. Jennewein, D. Williams and R. B. Croteau, *Biotechnol. Bioeng.*, 2006, **93**, 212.
27. M. Jiang, G. Stephanopoulos and B. A. Pfeifer, *Appl. Microbiol. Biotechnol.*, 2012, **94**, 841.
28. M. S. Siddiqui, K. Thodey, I. Trenchard and C. D. Smolke, *FEMS Yeast Res.*, 2012, **12**, 144.
29. S. R. Hamilton and T. U. Gerngross, *Curr. Opin. Biotechnol.*, 2007, **18**, 387.
30. B. K. Choi, P. Bobrowicz, R. C. Davidson, S. R. Hamilton, D. H. Kung, H. Li, R. G. Miele, J. H. Nett, S. Wildt and T. U. Gerngross, *Proc. Natl. Acad. Sci. U. S. A.*, 2003, **100**, 5022.
31. S. R. Hamilton, R. C. Davidson, N. Sethuraman, J. H. Nett, Y. Jiang, S. Rios, P. Bobrowicz, T. A. Stadheim, H. Li, B. K. Choi, D. Hopkins, H. Wischnewski, J. Roser, T. Mitchell, R. R. Strawbridge, J. Hoopes, S. Wildt and T. U. Gerngross, *Science*, 2006, **313**, 1441.
32. H. Li, N. Sethuraman, T. A. Stadheim, D. Zha, B. Prinz, N. Ballew, P. Bobrowicz, B. K. Choi, W. J. Cook and M. Cukan, *Nat. Biotechnol.*, 2006, **24**, 210.
33. J. H. Nett, W. J. Cook, M.-T. Chen, R. C. Davidson, P. Bobrowicz, W. Kett, E. Brevnova, T. I. Potgieter, M. T. Mellon, B. Prinz, B.-K. Choi, D. Zha, I. Burnina, J. T. Bukowski, M. Du, S. Wildt and S. R. Hamilton, *PLOS ONE*, 2013, **8**, 68325.
34. D. T. Burke, G. F. Carle and M. V. Olsen, *Science*, 1987, **236**, 806.
35. P. Sudbery, in *Human molecular genetics*, Addison Wesley Longman, England, 1998, p. 209.
36. K. R. Peterson, C. H. Clegg, Q. Li and G. Stamatoyannopoulos, *TIGS*, 1997, **13**, 61.
37. J. D. Boeke, F. LaCroute and G. R. Fink, *Mol. Gen. Genet.*, 1984, **197**, 345.
38. R. Takahashi and M. Ueda, in *Rat Genomics: Methods and Protocols*, ed. I. Anegon, Humana Press, New Jersey USA, **597**, 2010, p. 93.

39. N. S. Rane, A. K. Sandhu, V. S. Zhawar, G. Kaur, N. C. Popescu, R. P. Kandpal, M. Jhanwar-Uniyal and R. S. Athwal, *Cancer Genomics Proteomics*, 2011, **8**, 227.

40. N. Kouprina and L. Vladimir, *FEMS Micro. Rev.*, 2003, **27**, 629.

41. N. Kouprina and L. Vladimir, *Nat. Protoc.*, 2008, **3**, 371.

42. L. A. Annab, N. Kouprina, G. Solomon, P. L. Cable, D. E. Hill, J. C. Barrett, V. Larionov and C. A. Afshari, *Gene*, 2000, **250**, 201.

43. D. G. Gibson, J. I. Glass, C. Lartigue, V. N. Noskov, R. Chuang, M. A. Algire, G. A. Benders, M. G. Montague, L. Ma, M. M. Moodie, C. Merryman, S. Vashee, R. Krishnakumar, N. Assad-Garcia, C. Andrews-Pfannkoch, E. A. Denisova, L. Young, Z. Qi, T. H. Segall-Shapiro, C. H. Calvey, P. P. Parmar, C. A. Hutchison III, H. O. Smith and J. C. Venter, *Science*, 2010, **329**, 52.

44. G. A. Benders, V. N. Noskov, E. A. Denisova, C. Lartigue, D. G. Gibson, N. Assad-Garcia, R. Chuang, W. Carrera, M. Moodie, M. A. Algire, Q. Phan, N. Alperovich, S. Vashee, C. Merryman, J. C. Venter, H. O. Smith, J. I. Glass and C. A. Hutchison III, *Nucleic Acids Res.*, 2010, **38**, 2558.

45. A. Goffeau, B. G. Barrell and H. Bussey *et al.*, *Science*, 1996, **274**, 546.

46. D. Botstein and G. R. Fink, *Genetics*, 2011, **189**, 695.

47. B. Curran and V. Bugeja, in *Fungi Biology and Applications*, ed. K. Kavanagh, Wiley-Blackwell, Chichester, West Sussex, UK, 2nd edition, 2011, Chapter 4, p. 95.

48. A. Baudin, O. Ozier-Kalogeropoulos, A. Denouel, F. Lacroute and C. Cullin, *Nucleic Acids Res.*, 1993, **21**, 3329.

49. D. D. Shoemaker, D. A. Lashkari, D. Morris, M. Mittmann and R. W. Davis, *Nat. Genet.*, 1996, **14**, 450.

50. P. E. Purdue, J. W. Zhang, M. Skoneczny and P. B. Lazarow, *Nat. Genet.*, 1997, **15**, 381.

51. A. M. Motley, E. H. Hettema, E. M. Hogenhout, P. Brites, A. L.M.A. ten Asbroek, F. A. Wijburg, F. Baas, H. S. Heijmans, H. F. Tabak, R. J.A. Wanders and B. Distel, *Nat. Genet.*, 1997, **15**, 377.

52. V. Khurana and S. Lindquist, *Nat. Rev. Neurosci.*, 2010, **11**, 436.

53. J. L. DeRisi, V. R. Iyer and P. O. Brown, *Science*, 1997, **278**, 680.

54. W. K. Huh, J. V. Falvo, L. C. Gerke, A. S. Carroll, R. W. Howson, J. S. Weissman and E. K. O'Shea, *Nature*, 2003, **425**, 686.

55. S. Fields and O. Song, *Nature*, 1989, **340**, 245.

56. P. Uetz, L. Giot, G. Cagney, T. A. Mansfield, R. S. Judson, J. R. Knight, D. Lockshon, V. Narayan, M. Srinivasan and P. Pochart, *et al.*, *Nature*, 2000, **403**, 623.

57. B. Schwikowski, P. Uetz and S. Fields, *Nat. Biotechnol*, 2000, **18**, 1257.

58. A. H. Y. Tong, G. Lesage and G. D. Bader, *et al.*, *Science*, 2004, **303**, 808.

59. M. C. Jewett, G. Hofmann and J. Nielsen, *Curr. Opin. Biotechnol.*, 2006, **17**, 191.

60. N. Bharucha and A. Kumar, *Comb. Chem. High Throughput Screening*, 2007, **10**, 618.

61. A. B. Parsons, R. Geyer, T. R. Hughes and C. Boone, *Prog. Cell Cycle Res.*, 2003, **5**, 159.
62. T. Ideker, V. Thorsson, J. A. Ranish, R. Christmas, J. Buhler, J. K. Eng, R. Bumgarner, D. R. Goodlett, R. Aebersold and L. Hood, *Science*, 2001, **292**, 929.
63. R. Mustacchi, S. Hohmann and J. Nielsen, *Yeast*, 2006, **23**, 227.
64. B. A. Blount, T. Weenink and T. Ellis, *FEBS Letters*, 2012, **586**, 2112.
65. C. M. Ajo-Franklin, D. A. Drubin, J. A. Eskin, E. P. S. Gee, D. Landgraf, I. Phillips and P. A. Silver, *Genes Dev.*, 2007, **21**, 2271.
66. T. Ellis1, X. Wang and J. J. Collins, *Nat. Biotechnol.*, 2009, **27**, 465.
67. A. H Babiskin and C. D Smolke, *Mol. Syst. Biol.*, 2011, **7**, 471.
68. J. Dymond and J. Boeke, *Bioeng. Bugs.*, 2012, **3**, 168.
69. G. Walsh, *Nat. Biotechnol.*, 2010, **28**, 917.
70. T. U. Gerngross, *Nat. Biotechnol.*, 2004, **22**, 1409.
71. J. L Martínez, L. Liu, D. Petranovic and J. Nielsen, *Curr. Opin. Biotechnol.*, 2012, **23**, 965.
72. J. Travis, M. Owen, P. George, R. Carrell, S. Rosenberg, R. A. Hallewell and P. J. Barr, *J. Biol. Chem.*, 1985, **260**, 4384.

CHAPTER 11

Molecular Diagnostics

FRANCINE DE ABREU[a] AND GREGORY J. TSONGALIS*[b]

[a] Department of Pathology, Geisel School of Medicine at Dartmouth, Hanover NH, USA; [b] Dartmouth Hitchcock Medical Center and Norris Cotton Cancer Center, Lebanon, NH, USA
*Email: gregory.j.tsongalis@hitchcock.org

11.1 INTRODUCTION

Recently the use of diagnostic testing to understand the molecular mechanisms of innumerous diseases and for identification of biological biomarkers involved in multiple signaling pathways has been essential for improving general clinical practice. Applications of modern genomic technologies have allowed clinical laboratories the ability to support clinical management decisions for early detection, diagnosis, prognosis, detection of recurrence after therapy, risk assessment, identification of therapeutic targets, prediction of response to therapy, monitoring clinical outcomes, and imaging disease processes).[1]

Currently, molecular technologies are available to identify DNA variation and changes in gene expression associated with infectious diseases, cancer and inherited diseases. The detection of benign and pathogenic alterations in DNA and/or RNA samples facilitates diagnosis, prognosis, and monitoring responses to therapy. The discovery of genes and gene-based markers associated with diseases has allowed physicians to assess disease predisposition and to design and implement improved diagnostic algorithms. Advances in molecular diagnostic testing have been used to improve the sensitivity, specificity and turn-around time for actionable test results in the clinical laboratory.

Molecular Biology and Biotechnology, 6th Edition
Edited by Ralph Rapley and David Whitehouse
© The Royal Society of Chemistry 2015
Published by the Royal Society of Chemistry, www.rsc.org

For this reason, an exceptional demand has been placed on the clinical laboratory to provide increased diagnostic testing with rapid and accurate identification and interrogation of genomic targets. It is evident that molecular diagnostic methods play an increasing and critical role in patient management.

11.2 TECHNOLOGIES

Molecular technologies first entered the clinical laboratories in the early 1980s as manual, labor-intensive procedures that required a working knowledge of chemistry and molecular biology as well as an exceptional skill set. However, recently, these technologies have moved very quickly away from manual and labor intense procedures to cheaper, faster, more semi-automated, and increasingly accurate procedures.[2]

Since the mid-1980s, Southern blot was the method of choice for a variety of clinical applications, such as detection of gene deletions in Duchenne and Becker muscular dystrophy, repeat expansions in Fragile X syndrome, linkage analysis for cystic fibrosis, and other genetic conditions (Figure 11.1). In the early 1990s, the transition of the polymerase chain reaction (PCR) from research to the clinical laboratory began a new era for molecular diagnostics (Figure 11.2). Because PCR is a target-based amplification technology that allows for the detection of specific gene or pathogen sequences with higher sensitivity and specificity it rapidly became the method of choice for diagnostic molecular laboratories. Since then many modifications to the PCR have been introduced, but none as significant as the real time capability.[3–6] The elimination of post-PCR detection systems and the ability to perform the

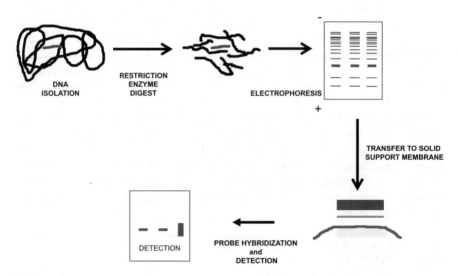

Figure 11.1   Schematic diagram of the Southern blot transfer procedure.

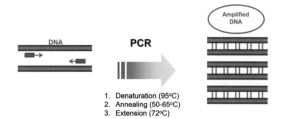

Figure 11.2   Schematic diagram of the polymerase chain reaction (PCR).

entire assay in a closed vessel had significant advantages for the clinical laboratory.

Real time polymerase chain reaction (PCR) allows for the simultaneous amplification and detection of amplified nucleic acid targets. It also increases the specificity of the assay since it can be used with a fluorescent probe that hybridizes to the specific nucleic acid sequence. In routine clinical practice, the main advantages of real-time PCR are: (1) high specificity; (2) the speed with which samples can be analyzed, as there are no post-PCR processing steps required; and (3) the "closed-tube" nature of the technology to eliminate potential sources of contamination. The analysis of results *via* amplification curve and melt curve analysis is very simple and more amenable to routine clinical laboratory quality control procedures.

The rapid development of more complex molecular technologies, such as microarrays, next generation sequencing (NGS) and nanotechnology, has allowed testing of numerous biomarkers and a more detailed classification of diseases, contributing to a personalized prognostic and predictive approach to management.[7] Microarrays can be divided into three main applications: (1) to detect structural alterations (*i.e.* array CGH); (2) to identify phenotype-specific SNPs that are known to drive specific phenotypes, such as panels for retinal degeneration; and (3) to detect genome-wide SNPs that assess risk of multiple common genetic disorders, such as certain cancers and, ophthalmologic, cardiac, renal and neurological disorders.[8] NGS uses massively parallel sequencing chemistries to sequence many genes of interest, the whole exome or the whole genome for variants in a broad range of complex disorders. Nanobiotechnology, using a nanometer length scale, can extend the limits of molecular diagnostics to unprecedented levels on the nanoscale as assays and technologies with better performance characteristics are developed.

Clinical applications of molecular technologies continue to grow and expand into non-traditional diagnostic arenas. Recently, clinical laboratories are applying molecular technologies to newborn screening for highly penetrant disorders, diagnostic and carrier testing for inherited disorders, predictive and pre-symptomatic testing for adult-onset and complex disorders. In addition, detection of infectious disease pathogens, pharmacogenetic

Table 11.1 Current technologies being used in the Molecular Pathology Laboratory at the Dartmouth Hitchcock Medical Center.

Technology	Platform/Instrument
DNA/RNA extraction	1. Manual
	2. Spin column
	3. Qiagen EZ1 robot
	4. Roche Ampliprep
Fluorescence *in situ* hybridization (FISH)	1. Thermobrite elite
	2. Molecular imaging system
Real time PCR	1. Cepheid GeneXpert
	2. AB 7500 Real-time PCR System
	3. Cepheid Smartcycler
	4. Roche Taq48
Capillary electrophoresis	1. AB 3500 Genetic Analyzer
Traditional PCR	1. MJ Research
	2. DNA Engines
Microarray	1. Affymetrix
Next Generation Sequencing	1. Ion Torrent Personal Genome Machine (PGM)
	2. Illumina MiSeq

testing to guide individual drug dosage, selection and response are being performed routinely.[8] Today, molecular diagnostic laboratories are armed with numerous instruments and technologies to address the increasing demand for clinical test results (Table 11.1).

11.3 INFECTIOUS DISEASES

For many years, clinical laboratories used traditional methods of microbial identification and differentiation that rely on phenotypic characteristics (*i.e.* protein, bacteriophage, and chromatographic profiles), biotyping and susceptibility testing. Recently, with the development and application of molecular diagnostic techniques, a new era in microbial identification and characterization has begun.

The diagnosis of an infectious disease has typically depended upon isolation of the pathogen by a culture technique. Even though this approach is adequate for the identification of the majority of common infections, the approach is less than desirable for the detection of organisms that were difficult to grow *in vitro*, including long incubation times. These issues along with performance characteristics of sensitivity and specificity brought about a need for alternative techniques that would allow for direct detection of

A. Cepheid Smartcycler II   **B. Amplification curves**

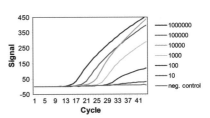

Figure 11.3   (A) The Cepheid Smartcycler II real time PCR instrument. (B) Real time
PCR amplification curves showing cycle threshold differences for various quantities of target sequence.

infectious agents in clinical samples while maintaining a rapid turn-around
time and high performance.

Molecular diagnostic technologies, such as non-amplified nucleic acid
probe and nucleic acid amplification, were rapidly introduced into the
clinical microbiology laboratory. A non-amplified nucleic acid probe could
be bound to the target from direct specimens, blood culture bottles and
colony isolates allowing the identification of infectious agents.[9] Nucleic acid
amplification methods (*i.e.* traditional PCR, reverse transcriptase PCR, nested PCR, multiplex PCR, and real-time PCR) have the ability to selectively
amplify specific targets present in low concentration that could then be
identified by various post-PCR methods or probes incorporated into a real
time PCR assay (Figure 11.3).

Clinical applications of molecular methods for infectious diseases include: (1) virology where it has been applied to resistance testing, genotyping
and viral quantification, and viral detection; (2) bacteriology where it has
been applied to resistance testing, detection of infection due to fastidious
bacteria, and detection of bacterial infection after antibiotics have been
administered; and (3) parasitology and mycology where it has been applied
to difficult fungal infections.[10] Many molecular infectious disease tests that
use PCR-based techniques are approved or cleared by the U.S. Food and
Drug Administration (FDA) and are being performed routinely in clinical
laboratories (Table 11.2).

The next generation of these technologies, such as nucleic acid arrays and
whole genome sequencing, will provide a wealth of information for personalized medicine in the area of infectious diseases. Nucleic acid arrays
with thousands of sequence-specific oligonucleotide probes were developed
to identify, and genotype, influenza viruses, respiratory syncytial virus, herpes viruses, and HPV. Bacterial, parasitic, and fungal organisms can also be
detected, genotyped and evaluated for resistance by this method.[9] Whole
genome sequencing can provide megabases of sequence information from
the whole organism that can help with identification and resistance testing
for specific infectious diseases.

Table 11.2   Molecular infectious disease tests that are currently FDA-cleared
and being performed in clinical laboratories.

Bacteria	Viruses
Bacillus anthracis	Avian flu
Candida albicans and *Candida* spp.	*Cytomegalovirus*
Candida albicans and *Candida glabrata*	Enterovirus
Chlamydia trachomatis	HBV (quantitative)
Clostridium difficile	Hepatitis C virus (qualitative and quantitative)
Enterococcus faecalis	HIV drug resistance
Francisella tularensis	HIV (quantitative)
Gardnerella, *Trichomonas* and *Candida* spp.	HBV/HCV/HIV blood screening assay
Group A Streptococci	Human Papillomavirus
Group B Streptococci	Respiratory viral panel
Legionella pneumophila	West Nile virus
Methicillin resistant *Staphylococcus aureus*	
Mycobacterium tuberculosis	
Mycobacterium spp.	
Staphylococcus aureus	

11.4   GENETICS

The landscape of molecular genetics and genomic testing has rapidly
changed due to advances in DNA analysis technologies and expanded
knowledge in the molecular basis of rare and common disorders. A genetic
disorder can be defined as a condition due to an alteration of DNA, either
inherited or acquired. It can be categorized into three major groups which
include: (1) chromosomal disorders; (2) monogenic or single-gene disorders;
and (3) polygenic or multifactorial disorders.

Chromosomal disorders are due to the loss, gain or abnormal arrangement of one or more chromosomes which results in the presence of excessive or deficient amounts of genetic material. These types of alterations
usually involve large segments of DNA containing numerous genes and can
be organized into two groups: numerical or structural chromosome alterations. Numerical alteration is characterized by an abnormal number of
chromosomes, also known as aneuploidy, which occurs when there is a loss
or excess of one or more chromosomes (monosomy, disomy, trisomy, tet-
rasomy, *etc.*). Structural alteration is characterized by a modification in the
chromosome structure and can be classified into six groups:

(1) Deletion: breakage and/or loss of a portion of a chromosome.
(2) Duplication: a portion of a chromosome is duplicated, resulting in
    extra genetic material.
(3) Translocation: breakage of two chromosomes with transfer of broken
    parts to the opposite chromosome.
(4) Inversion: a portion of the chromosome has broken off, turned upside
    down and reattached to the chromosome.

(5) Insertion: breakage and loss of a portion of one chromosome and insertion of this part to another chromosome.

(6) Isochromosome: splitting at the centromere during mitosis so that one arm is lost and the other duplicated to form one chromosome with identical arms.

Monogenic disorders are the result of a single mutant gene and display traditional Mendelian inheritance patterns including autosomal dominant or recessive and X-linked types. The overall population frequency of monogenic disorders is thought to be approximately 10 per 1000 live births. Polygenic or multifactorial disorders, however, consist of chronic diseases of adulthood, congenital malformations, and dysmorphic syndromes. These disorders result from multiple genetic and/or epigenetic factors which may not conform to traditional Mendelian inheritance patterns.

Recently, the ability to screen individuals for many types of genetic alterations in a clinical setting is expanding at an enormous rate. Many clinical laboratories offer routine testing for common genetic diseases such as the monogenic disorders Cystic fibrosis and Fragile X Syndrome and polygenic disorders such as human cancers (Table 11.3).

Cystic fibrosis (CF) is a monogenic autosomal recessive disease caused by multiple mutations in the *CFTR* (cystic fibrosis transmembrane regulator) gene.

Table 11.3 Genetic and oncologic diseases commonly tested for using molecular diagnostic methods.

Alpha1-antitrypsin deficiency
Angelman Syndrome
ApoE
Colon cancer: KRAS, EGFR
Cystic fibrosis
Drug Metabolizing Enzymes
Duchenne/Becker muscular dystrophy
Factor II (prothrombin)
Factor V Leiden
Fanconi's anemia
Fragile X Syndrome
Gaucher disease
Glioma and Glioblastoma: IDH1, IDH2, SMO
Hemochromatosis
Huntington's disease
HER-2 Status
Melanoma: BRAF, KIT
Methylene tetrahydrofolate reductase (MTHFR)
Non-small cell lung cancer (NSCLC): EGFR, KRAS
Non-small cell lung cancer (NSCLC): ALK fusion
Sex Mismatched Bone Marrow Transplantation
Prader-Willi Syndrome
Tay-Sachs disease

Table 11.4 Panel of 23 recommended mutations used in screening for cystic fibrosis.

ΔF508	N1303K	A455E	621 + 1G > T	2789 + 5G > A
ΔI507	R553X	R560T	711 + 1G > T	3120 + 1G > A
G542X	R347P	R1162X	1898 + 1G > A	2184delA
G551D	R117H	G85E	1717 − 1G > A	3659delC
W1282X	R334W	G85E	849 + 10kbC > T	

According to the CFTR mutation database (http://www.genet.sickkids.on.ca/cftr/StatisticsPage.html), 1951 heterogeneous mutations have been documented in different ethnic groups. Over the past few years, the American College of Medical Genetics (ACMG) and the American College of Obstetrics and Gynecology (ACOG) approved guidelines for a national CF screening panel as a standard of care for CF carrier testing in the general population. This panel consists of 23 mutations in the *CFTR* gene with an allele frequency of at least 0.1% in the US general population and a detection rate ranging from 94%–49% in Ashkenazi Jews and Asian Americans, respectively (Table 11.4).[11,12] Recently, a CF carrier screening panel using next generation sequencing was FDA approved for use with the Illumina MiSeq platform. This panel consists of 156 genomic positions (153 single nucleotide variant and small insertion/deletion sites, and two large deletions and the polyTG/polyT region), which corresponds to 162 unique mutations that provides a more comprehensive detection rate through different ethnic groups.

Fragile X Syndrome (FraX) is a monogenic X-linked neuro-developmental disorder with a non-Mendelian inheritance pattern. It is caused by an expansion of a trinucleotide (CGG) repeat in the *FMR-1* (Fragile X mental retardation 1) gene. The repeat is up to 55 trinucleotides long in the normal population, but it can exceed 200 in patients with FraX, resulting in hypermethylation of the promoter region and silencing of the *FMR-1* gene. This disease is a result of lack of expression of the FMR protein resulting in intellectual disability, hyperactivity, and autistic-like behavior.[13] Since this mutation is X-linked, males are more severely affected than females. Affected females have less severe mental retardation and can pass an expanded allele on to their offspring. The original molecular diagnostic assay for FraX was based on Southern blot transfer analysis and now more commonly, PCR with capillary electrophoresis is used to more accurately size individual alleles (Figure 11.4). The screening for CGG repeats is used as diagnostic tool and as a carrier screening assay whose results are used in genetic counseling of family members.

11.5 HEMATOLOGY

Precise diagnosis and classification of hematologic malignancies are important since treatment options and prognosis vary considerably.

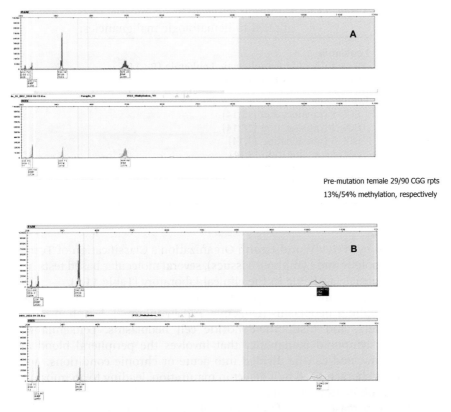

Pre-mutation female 29/90 CGG rpts
13%/54% methylation, respectively

Figure 11.4 Fragile X Syndrome Assessment by PCR and capillary electrophoresis. (A) Premutation female with peaks at 29 and 90 CGG repeats; (B) Full mutation female with peaks at 40 and >200 CGG repeats.

Historically, the diagnosis of myeloid and lymphoid neoplasms was based solely on the histological evaluation of H&E stained slides. With the development of immunophenotyping methods, diagnose and classification of leukemia or lymphoma according to the cell lineage (myeloid, B, T, NK) and stage of maturation became possible. Subsequent availability of cytogenetic and molecular genetic methods to identify unique and specific biomarkers increased the ability to diagnose and classify leukemia and lymphoma providing a newly characterized spectrum of hematologic malignancies according to gene mutations, gene deletions or amplifications, gene or miRNA expression profiles, and chromosomal translocations.[14–17]

Today, the clinical workup of leukemia and lymphoma requires the correlation of traditional and advanced diagnostic techniques (clinical features, morphology, immunophenotyping with cytogenetic and molecular genetic data) to accurately diagnose and classify the disease. Since the current diagnosis and classification method for hematologic malignancies was

Table 11.5   Common molecular diagnostic tests
applied to hematologic malignancies.

Leukemia
  B-Cell Chronic Lymphocytic Leukemia (B-CLL)
  Chromosome 8 Enumeration (CML, AML, MPD, MDS)

Lymphoma
  BCL-1 translocation (11;14)
  BCL-2 translocation (18;14)
  BCL-6 translocation (3;14)
  c-Myc translocation (8;14)
  Immunoglobulin heavy and light chain rearrangements
  NPM-ALK translocation (2;5)
  T-cell receptor gene rearrangements

established by WHO (World Health Organization's Classification of Tumors of Hematopoietic and Lymphoid Tissues), several molecular based tests have become routinely available in the clinical laboratory (Table 11.5).

Hematologic malignancies are divided into myeloid neoplasms (leukemia), lymphoid neoplasms (leukemia and lymphoma), lymphoproliferative disorders, and histiocytic/dendritic cell neoplasms. Leukemia is a myeloid or lymphoid malignancy that involves the peripheral blood and bone marrow, and can be divided into acute or chronic conditions. Acute leukemia is characterized by a defect in maturation, leading to an imbalance between proliferation and maturation, and chronic leukemia is characterized by an increase of proliferation of cells that maintain their capacity to differentiate to end cells. Lymphoma is a lymphoid malignancy that involves lymph nodes and/or other extramedullary sites, and are classified into the two main categories of Hodgkin or non-Hodgkin.

The myeloid neoplasms are characterized by acute myeloid leukemia (AML), myeloproliferative neoplasms (MPN) (*i.e.* chronic myeloid leukemia or CML) and myelodysplastic syndrome (MDS). Acute myeloid leukemia is an aggressive clonal neoplasm that lacks maturation of myeloid cells which then rapidly proliferates causing accumulation of myeloblasts in the bone marrow and/or blood. It is a heterogeneous disease regarding clinical, morphological, immunophenotypic, karyotypic and genetic features (*i.e.* point mutations and chromosomal translocations). The genetic abnormalities involve multiple pathways and have been grouped into the "Class I" oncogenes that confer proliferative advantage to the leukemic cells and "Class II" oncogenes that contribute to myeloid maturation arrest. The association between point mutations and/or translocations of both Class I and Class II oncogenes appears to be important to the development of AML.[18,19] Class I mutations are characterized by point mutations in *FLT3, KIT, RAS* and *JAK2* genes. Class II are characterized by point mutations in *NPM1* and *CEBPA* genes, as well as, chromosomal translocations (*i.e. RUNX1-RNNX1T1, CBFB-MYH11, PML-RARA,* and *MLL* gene rearrangement from t(8;21)(q22;q22), inv(16)(p13.1q22), t(15;17)(q22;q12) and 11q23 transloca-tions, respectively).[20]

Chronic myeloid leukemia (CML) is defined as a clonal hematopoietic stem cell disorder defined by the presence of monocytosis in the blood and the presence of myelodysplastic and myeloproliferative features in the bone marrow.[21] CML is characterized by the Philadelphia chromosome or the *BCR-ABL1* fusion gene, which is a result of a translocation between chromosomes 9 and 22 [t(9;22)(q34;q11)]. However, CML also shows kar-yotypic and genetic abnormalities (*i.e.* point mutations). Some of the karyotype abnormalities include isochromosome 17 [i(17q)], trisomy 8, monosomy 9, and point mutations that involve genes associated with sig-naling pathways and proliferation, such as *RAS*, *JAK2* and *CBL*.[22]

The lymphoid neoplasms are characterized by acute lymphoblastic leu-kemia (ALL), chronic lymphocytic leukemia (CLL) and lymphoma (*i.e.* Hodgkin and non-Hodgkin). Acute lymphoblastic leukemia (ALL) is char-acterized by a group of structural rearrangements, submicroscopic DNA copy number alterations, and sequence mutations. Approximately 20% of B-ALL cases have chromosomal rearrangements of ETV6-RUNX1, TCF3-PBX1, BCR-ABL1, rearrangements of MLL, and mutations of *PAX5*, *JAK1*, *JAK2*, *IKZF1*, *CRLF2*, *IL7R*, *CREBBP*, and *TP53*.[23,24] T-lineage ALL cases are characterized by activating mutations of *NOTCH1* and rearrangements of transcription fac-tors TLX1, TLX3, LYL1, TAL1, and MLL.[23,24] Approximately 75% of the chromosomal alterations are detectable by karyotyping, FISH, or other molecular techniques.

The understanding of the biology of chronic lymphocytic leukemia (CLL) B cells provided the identification and characterization of several genetic ab-normalities (*i.e.* point mutations and chromosomal translocations) that are prognostic indicators of disease progression and survival. The chromosomal abnormalities commonly identified are deletion at 13q14, trisomy 12, de-letion at 11q22-23, and deletion at 17q. CLL patients with 17q deletion have aggressive clinical disease, most likely because of loss of *TP53* gene.[25] Re-cently, next generation sequencing expanded the knowledge of the genomic alterations in CLL, providing the first comprehensive view of somatic mutation in this disease. *NOTCH1* and *SF3B1* mutations were identified in CLL patients and can be considered the new drivers of aggressive forms of the disease.

Non-Hodgkin lymphomas (NHL) are a heterogeneous group of solid tumors of lymphoid cell origin. They are divided into two subtypes, lymphoma derived from B-cells, which correspond to 85% of the NHL cases, or lymphoma derived from T-cells.

For B-cell neoplasms, clonality can often be determined immunopheno-typically by demonstrating the presence of monoclonal surface immuno-globulins or by detecting gene rearrangements in the immunoglobin gene family. In contrast, for T-cell malignancies, there is no immunophenotpyic equivalent to monoclonal surface immunoglobulin. Thus, molecular genetic approaches for the determination of clonality in T-cell lymphoma by detecting gene rearrangements in the T-cell receptor family are especially important.[26] Other molecular genetic applications to the assessment of lymphoid malignancies include detection of chromosomal translocations in the subclassification of non-Hodgkin lymphoma. For example, in a lymph node suspicious for follicular lymphoma, the detection of a translocation, t(14;18), involving the *BCL*-2 proto-oncogene, would confirm this diagnosis. Similarly, the detection of a translocation, t(11;14), involving the *BCL*-1 proto-oncogene, would confirm a diagnosis of mantle cell lymphoma.[26,27]

11.6   ONCOLOGY

Very rapid advances in molecular technologies for identifying somatic and germline mutations in oncology have revolutionized our knowledge base of tumor cell biology. The identification of genetic alterations such as point mutations, copy number alterations, translocations, and/or chromosomal rearrangements, contributed to the characterization of novel biomarkers, leading to a new paradigm for molecular oncology testing. The rapidly growing field of clinical biomarkers, with either a prognostic or predictive importance has contributed to a personalized or precision medicine approach to managing the oncology patient.

Although numerous genes and chromosomal alterations have been identified in many human cancer types, few have had an impact and made it to the clinical laboratory. However, this increased medical and scientific knowledge base associated with the need for early detection and most appropriate treatment selection for the cancer patient has resulted in the rapid uptake of somatic mutation testing for all cancer types. The following examples highlight the utility of molecular diagnostic tools in the assessment of breast, bladder, melanoma, colorectal and lung cancers.

In breast cancer patients, higher resolution imaging, targeted therapies and improved surgical techniques have all contributed to improved patient outcomes by detecting, diagnosing and treating this disease more effectively. Currently, there are two molecular diagnostic assays to identify the presence of metastasis or recurrence in breast cancer patients such as MamaPrint® (Agendia, Amsterdam, The Netherlands) and Oncotype DX (Genomic Health, Inc. Redwood City, CA). MamaPrint®, a 70-gene signature, is a microarray gene expression profiling test that identifies the risk of distant recurrence following surgery. It analyzes 70 critical genes that contains a definitive gene expression signature and stratifies patients into two distinct groups, low risk or high risk of distant recurrence. MamaPrint® was developed by a group of researchers who identified a prognostic signature using a microarray platform in node-negative breast cancer patients under the age of 55.[28] This signature consisted of genes involved in cell cycle, invasion, metastasis, angiogenesis, and signal transduction. The 70-gene prognostic signature was validated in node-negative and node-positive tumors, as well as treated and untreated patients, and proved to be a robust predictor for distant metastatic-free survival, independent of adjuvant treatment, tumor size, histological grade, and age.[29] A second validation was performed in node-negative T1-2 breast tumors not treated with chemotherapy and compared to a traditional clinical factors included in the Adjuvant Online software.

Oncotype DX® is a 21-gene expression assay to provide a recurrence score (RS) as a prognostic indicator. The RS predicts the probability of distant recurrence in node-negative patients treated with tamoxifen, and ER + breast cancer.[30] The assay identifies expression of 21 genes, five reference genes and 16 genes associated with breast cancer, selected from a set of 250 genes previously studied by the NSABP (National Surgical Adjuvant Breast and Bowel Project) clinical studies. Real time, reverse transcriptase (qRT-PCR) in FFPE samples is performed to quantify the expression of the 21 genes and calculate the RS (recurrence score), which classifies patients in three groups: high, intermediate and low risk. The RS was shown to correlate with distant recurrence, relapse-free interval, and overall survival, independent of age and tumor size.

The Genomic Grade Index (GGI) signature was developed to reclassify patients with histologic grade 2 tumors, which is informative for clinical decision making. Sotiriou *et al.*[31] analyzed microarray data from 189 invasive breast cancers and identified 97 genes associated with histologic grade, most of them involved in cell cycle regulation and proliferation. These genes were differently expressed between high-grade and low-grade breast tumors. The intermediate grade tumors showed expression pattern similar to either high-grade or low-grade cases. The GGI may increase the accuracy of tumor grading and improve treatment decisions.

Another example of a molecular biomarker for breast cancer is the HER-2 (human epidermal growth factor receptor 2) gene. It regulates cell proliferation, survival, and other processes important for carcinogenesis. The activation of HER2 occurs through gene amplification which was initially used as a poor prognostic indicator. The occurrence of this abnormality in up to 35% of all breast cancers made it a robust target for new therapies. Currently, HER2 gene amplification is used as a targeted pharmacogenomics assay to identify those breast cancer patients who may benefit from treatment with trastuzumab (Herceptin) (Figure 11.5).[32]

Her2 FISH pharmDx™ Kit (Dako Denmark A/S Glostrup, Denmark), PathVysion® HER-2 DNA Probe kit (Abbott Molecular, Inc., Des Palines, IL),

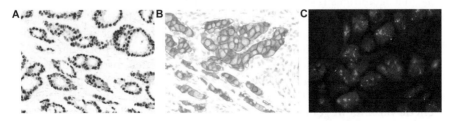

Figure 11.5 Breast cancer biomarkers. (A and B) Immunohistochemistry. (A) ER⁺ positive. (B) *HER-2* overexpression. (C) FISH (*HER-2* gene amplification).

SPOT-Light HER2 CISH kit (Invitrogen, Carlsbad, CA), and Inform Dual ISH (Ventana Medical Systems, Roche Diagnostics, Tucson, AZ) are FDA-cleared tests. These tests are performed on formalin-fixed paraffin-embedded (FFPE) tissue affixed to a slide. The kits have a DNA probe that hybridizes to the *HER2* gene and another probe that hybridizes to the centromere of the chromosome 17 (except for SPOT-Light HER2 CISH kit). The analysis is done using a fluorescent or light microscope to count the number of signals from each probe that will represent the number of HER2 gene copies present in the tumor sample.[33] Since the development of trastuzumab, other small molecular drugs have been designed to target the HER2 receptor when it is overexpressed.[34,35]

Unlike the HER2 example in breast cancer, bladder cancer progression is characterized by increased chromosomal instability and aneuploidy of several chromosomes including chromosome 3, 7, 17 and loss of locus 9p21. Since bladder cancer is a chronic disease, the diagnosis of primary and recurrent disease is a significant problem and patients need to be constantly monitored for recurrence of their disease. The UroVysion Bladder Cancer Kit (Abbott Molecular Inc., Des Plaines, IL) is a FDA-cleared assay that uses multicolor fluorescence *in situ* hybritdzation (FISH) to detect aneuploidy for chromosomes 3, 7, 17 and loss of 9q21 locus (P16/CDKN2A gene) in urine specimens. This assay is also available for use, in conjunction with urine cytology and cystoscopy evaluation, as an aid for diagnosing and monitoring recurrence of urothelial carcinoma. Traditional cytological analysis may not be sensitive enough for this application and it has been shown that the combination of FISH with routine cytology evaluation is more sensitive than cytology alone at detecting urothelial carcinoma.[36–40]

Clearly a better understanding of tumor cell biology and the pathways involved in carcinogenesis associated with the advantages of molecular diagnostics has led to the identification of novel biomarkers and therapeutics for human cancers. In the context of patient management, not only will molecular biomarkers be responsible for the reclassification of many of these tumor types, they will also be responsible for a more personalized medicine.[41]

11.7   PHARMACOGENOMICS

Pharmacogenomics (PGX) has been an important molecular application in clinical laboratories that address drug metabolism (PGX$_m$) and targeted therapies (PGX$_t$). PGX$_m$ refers to the characterization of whole genome variation associated with different pharmacokinetics (PK) or pharmacodynamics (PD) of administered drugs. Pharmacokinetics corresponds to drug absorption, distribution, metabolism, and elimination, and pharmacodynamics represents the effects on drug receptors and other drug targets. Genetic variants in genes that encode proteins involved in PK and PD of a drug will significantly contribute to differences in drug response. PGX$_t$ refers to the identification of drug targets or genetic variants that confer resistance to therapeutics in a targeted pathway. The overall aim of PGX testing is to decrease adverse responses to therapy and increase efficacy by ensuring the appropriate selection and dose of therapy.[42,43]

Many genetic variants, such as single nucleotide polymorphisms (SNPs), deletion, insertion, copy number variations (CNVs), and tandem repeats, in genes may lead to changes in protein structure or stability, and therefore to protein activity. Thus, genetic alterations in genes responsible for drug metabolizing enzymes, drug transporters, and drug targets, will alter response to therapeutics. These proteins are essential in the absorption, distribution and elimination of various endogenous and exogenous substances including pharmaceutical agents. Several groups of drug transporters that may be significant in the field of pharmacogenomics exist, including multidrug resistance proteins (MDRs), multidrug resistance-related proteins (MRPs), organic anion transporters (OATs), organic anion transporting polypeptides (OATPs), organic cation transporters (OCTs) and peptide transporters (PepTs).[44–46]

With the knowledge of genetic factors and the rapid advance in molecular technologies, many pharmacogenomics biomarkers (drug metabolizing enzymes, drug transporters and drug targets) have been characterized, FDA approved and applied in clinical diagnostic laboratories.

With respect to PGX_m, most of the enzymes involved in drug metabolism are members of the cytochrome P450 (CYP) superfamily, which encode 57 CYP enzymes and is responsible for 75–80% of all phase I-independent metabolism and for 65–70% of the clearance of clinically used drugs.[47–49] These cytochrome P450 enzymes are mainly located in the liver and gastrointestinal tract and include greater than 30 isoforms. Classification of these enzymes includes nomenclature, for example CYP2D6*1, the name of the enzyme (CYP) is followed by the family (CYP2), subfamily (CYP2D), and gene (CYP2D6) associated with the biotransformation. Allelic variants are indicated by a *, followed by a number (CYP2D6*1). Genetic variants or polymorphisms in these genes can lead to the following phenotypes: poor, intermediate, extensive and ultrarapid metabolizers. Poor metabolizers (PM) have no detectable enzymatic activity; intermediate metabolizers (IM) have decreased enzymatic activity; extensive metabolizers (EM) are considered normal and have at least one copy of an active gene; ultrarapid metabolizers (UM) contain duplicated or amplified gene copies that result in increased drug metabolism.

CYP2D6 and *CYP2C19* genes are drug metabolizing enzymes involved in the metabolism of cholesterol, steroids and other important lipids, such as arachidonic acid.[50] *CYP2D6* gene, localized on chromosome 22q13.1, encodes a highly polymorphic protein which contains 497 amino acids. *CYP2C19* gene, localized on chromosome 10q24, encodes for a protein which contains 490 amino acids. More than 60 and 24 allele variants have been described for CYP2D6 and CYP2C19, respectively. These variants are result of SNPs, deletions, insertions, gene rearrangements, or deletion or duplication of the entire gene, and result in decreased or increased enzymatic activity.[51] The AmpliChip CYP450 test is an FDA-cleared diagnostic test that can be used to predict how an individual will metabolize drugs that are substrates for CYP2D6 and CYP2C19. It uses the Afffymetrix microarray technology that analyses a patient's genotypic profile of the CYP2D6 (27 allelic variations) and CYP2C19 (three allelic variations) genes. This test provides information about the patients' drug metabolizing status (*i.e.* poor, intermediate, extensive, or ultrarapid metabolizer) based on their genetic information, allowing physicians to select relevant medication and doses of medications for some diseases, such as cardiac diseases, pain and cancer.

Another example of PGX$_m$ is the uridine diphosphate (UDP)-glucuronosyltransferase (*UGT1A1*) gene which plays an important role in the metabolism of drugs and endogenous substances through glucuronidation, allowing these compounds to be more easily excreted. A genetic TA repeat length polymorphism in the promoter of the *UGT1A1* gene has been shown to affect its expression leading to serious clinical manifestations. The "wild-type" allele, *UGT1A1*1*, has six tandem TA repeats in the regulatory TATA box of the *UGT1A1* promoter. The most common polymorphism associated with low activity of *UGT1A1* is the *28 variant, which has seven TA repeats (Table 11.6). UGT1A1*28 reduces activity of the *UGT1A1* enzyme which can then decrease the conversion of the active metabolite of irinotecan to its glucuronidated form for excretion from the body. Decreased UGT1A1 activity, therefore, results in more of the active irinotecan metabolite being present in the body with subsequent adverse reactions. Thus, UGT1A1*28 is associated with an increased risk for severe irinotecan toxicity. Irinotecan is a first line therapy approved for patients with metastatic colorectal cancers and is also used in other forms of cancer as indicated. A single FDA approved assay for this polymorphism has recently been removed from the market due to the low volumes of testing. We have developed a PCR-based fragment sizing assay that identifies the different alleles in this gene.[52] This laboratory-developed test (LDT) can identify the 5, 6, 7 and 8 TA repeats (Figure 11.6). Although the clinical significance of the 5 and 8 TA repeat alleles is not well-established, there is clear evidence that TA repeat size inversely affects UGT1A1 expression.[53] Thus, patients with 8 TA repeats might have the lowest UGT1A1 expression level and, therefore, might require

Table 11.6 Frequency of UGT1A1 polymorphisms in the general Caucasian population.

TA Repeat Number	Allele Designation	Frequency
6/6	*1/*1	46%
6/7	*1/*28	39%
6/8	*1/*37	3%
7/7	*28/*28	9%
7/8	*28/*37	1.5%

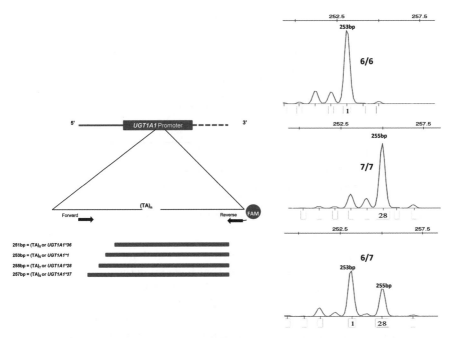

Figure 11.6 (A) Schematic of TA dinucleotide repeat in the UGT1A1 promoter. (B) Electropherograms showing different patterns for detection of TA repeat sizes.

special care if they show the clinical manifestations. Screening for UGT1A1 polymorphisms has been limited to confirmation of a diagnosis of Gilbert's syndrome.

PGX studies in clinical oncology identified many genetic variants as having potential roles in individualizing therapy.[54] This is the so called PGX$_t$ application. Currently, there are 24 FDA-approved drug labels for anticancer agents that include gene variants, functional deficiencies, expressions changes, and chromosomal abnormalities.[54] For example, in breast cancer, HER-2 amplification and overexpression is a positive target for eligibility for treatment with trastuzumab therapy and estrogen receptor (ER) expression is considered an important biomarker as it is a positive target for anti-estrogen therapy (tamoxifen) (Figure 11.5). ER-positive breast cancer patients benefit from endocrine therapy, such as tamoxifen (TAM). TAM is a selective modulator of ER that blocks steroid mechanisms preventing cellular replication and proliferation. CYP2D6 enzyme is responsible for the conversion of TAM to its active metabolites (*i.e.* endoxifen), which is important because they have a greater affinity for the estrogen receptor than TAM itself.[55] It has been postulated that patients with genetic polymorphism in *CYP2D6* have lower or absent CYP2D6 activity, resulting in lower levels of endoxifen. Thus, these patients would have less benefit from tamoxifen therapy than patients with functional copies of *CYP2D6*.[56,57]

ABL1	EGFR	GNAQ	KRAS	PTPN11
AKT1	ERBB2	GNAS	MET	RB1
ALK	ERBB4	HNF1A	MLH1	RET
APC	EZH2	HRAS	MPL	SMAD4
ATM	FBXW7	IDH1	NOTCH1	SMARCB1
BRAF	FGFR1	IDH2	NPM1	SMO
CDH1	FGFR2	JAK2	NRAS	SRC
CDKN2A	FGFR3	JAK3	PDGFRA	STK11
CSF1R	FLT3	KDR	PIK3CA	TP53
CTNNB1	GNA11	KIT	PTEN	VHL

Figure 11.7 The Ion Torrent AmliSeq Cancer Hotspot Panel.

Cetuximab and Panitumumab are two monoclonal antibodies that inhibit the growth and survival of tumor cells that have an activated *EGFR* in colon cancers. However, tumors with *KRAS* codon 12 and 13 mutations may not experience any benefit from these therapies. KRAS is a membrane GTPase that activates many proteins, including *EGFR* signaling pathways, such as c-Raf and PIK3. Only colon cancer patients whose tumors express *EGFR* and whose tumors are negative for *KRAS* mutation will benefit from these drugs.[58–60] A significant number of tumors harbor an activated EGFR pathway and thus therapies targeting the EGFR are being used in numerous tumor types including non-small cell lung cancers (NSCLC). NSCLCs contain activating mutations most commonly found in exons 19 and 21 of the EGFR gene which confer sensitivity to small molecule anti-EGFR therapies to the tumor cells.[61]

Many laboratories have reverted somatic mutation testing to next generation sequencing panels that allow for the simultaneous detection of multiple mutations in many genes (Figure 11.7).

Crizotinib is another example of a targeted therapy. It is an *ALK* inhibitor which is FDA-approved to treat NSCLC that has a distinct rearrangement within this gene. Approximately 2–5% of lung adenocarcinomas harbor this novel EML4-ALK gene rearrangement that is mutually exclusive from EGFR and KRAS mutations. Tumors with ALK rearrangement are eligible for treatment with Crizotinib. Therefore, it is important to identify these patients. The EML4-ALK rearrangement may occur at various breakpoints resulting in multiple possible fusion transcripts. Currently, most laboratories detect ALK rearrangements with an FDA-approved break-apart FISH assay.

11.8 CONCLUSION

Advances in technologies have led to a better understanding the molecular basis of infectious, genetic, and oncologic diseases. Highly complex technologies such as microarrays and next generation sequencing have resulted in biomarker discoveries that have led to the development of novel therapies and companion diagnostics. Qualitative and quantitative molecular testing continues to be critical in the management of patients with infectious disease, genetic disease and cancer. Drug development strategies that incorporate genomics have required clinical laboratories to routinely test for pharmacogenomic markers that address metabolic and targeted aspects of drug delivery. Clearly, the transition to high complexity and high data throughput technologies will once again revolutionize how we approach diagnostic medicine with respect to patient care.

REFERENCES

1. W. E. Grizzle, S. Srivastava and U. Manne, *Cancer Biomarkers*, 2010, **9**, 7.
2. G. J. Tsongalis and W. B. Coleman, *Clin. Chim. Acta*, 2006, **363**, 127.
3. C. T. Wittwer, M. G. Herrmann, A. A. Moss and R. P. Rasmussen, *Biotechniques*, 1997, **22**, 130.
4. S. B. Parks, B. W. Popovich and R. D. Press, *Am. J. Clin. Pathol.*, 2001, **115**, 439.
5. C. T. Wittwer, G. H. Reed, C. N. Gundry, J. G. Vandersteen and R. J. Pryor, *Clin. Chem.*, 2003, **49**, 853.
6. J. Wilhelm and A. Pingoud, *ChemBioChem*, 2003, **4**, 1120.
7. K. H. Allison, *Am. J. Clin. Pathol.*, 2012, **138**, 770.
8. S. H. Katsanis and N. Katsanis, *Nat. Rev. Genet.*, 2013, **14**, 415.
9. K. L. Muldrew, *Curr. Opin. Pediatr.*, 2009, **21**, 102.
10. D. J. Speers, *Clin. Biochem. Rev.*, 2006, **27**, 39.
11. M. S. Watson, G. R. Cutting, R. J. Desnick, D. A. Driscoll, K. Klinger, M. Mennuti, G. E. Palomaki, B. W. Popovich, V. M. Pratt, E. M. Rohlfs, C. M. Strom, C. S. Richards, D. R. Witt and W. W. Grody, *Genet. Med.*, 2004, **6**, 387.
12. The American College of Obstetricians and Gynecologists Committee Opinion, *Obstet Gynecol.* 2011, **486**, 1.
13. C. Bagni and B. A. Oostra, *Am. J. Med. Genet. A*, 2013, **161**, 2809.
14. H. Kiyoi, M. Yanada and K. Ozekia, *Intnl. J. Hematol*, 2005, **82**, 85.
15. J. M. Scandura, *Curr. Oncol. Rep.*, 2005, 7, 323.
16. M. C. Stubbs and S. A. Armstrong, *Curr. Drug Targets*, 2007, **8**, 703.
17. C. H. Lawrie, *Br. J. Haematol.*, 2007, **137**, 503.
18. Y. Ishikawa, H. Kiyoi, A. Tsujimura, S. Miyawaki, Y. Miyazaki, K. Kuriyama, M. Tomonaga and T. Naoe, *Eur. J. Haematol.*, 2009, **83**, 90.
19. R. P. Hasserjian, *Int. J. Lab. Hematol.*, 2013, **35**, 358.
20. T. Naoe and H. Kiyoi, *Int. J. Lab. Hematol.*, 2013, **97**, 165.
21. J. W. Vardiman, J. Thiele, D. A. Arber, R. D. Brunning, M. J. Borowitz, A. Porwit, N. L. Harris, M. M. Le Beau, E. Hellstrom-Lindberg, A. Tefferi and C. D. Bloomfield, *Blood*, 2009, **114**, 937.
22. S. A. Parikh and A. Tefferi, *Am. J. Hematol.*, 2012, **87**, 610.

23. C. G. Mullighan, *Hematology Am. Soc. Hematol. Educ. Program*, 2012, **2012**, 389.
24. H. Inaba, M. Greaves and C. G. Mullighan, *Lancet*, 2013, **381**, 1943.
25. N. Chiorazzi, *Hematology Am. Soc. Hematol. Educ. Program*, 2012, **2012**, 76.
26. A. J. Bench, W. N. Erber, G. A. Follows and M. A. Scott, *Intnl. J. Lab. Hematol.*, 2007, **29**, 229.
27. Y.-L. Kwong, *Br. J. Haematol.*, 2007, **137**, 273.
28. L. J. van't Veer, H. Dai, M. J. van de Vijver, Y. D. He, A. A. Hart, M. Mao, H. L. Peterse, K. van der Kooy, M. J. Marton, A. T. Witteveen, G. J. Schreiber, R. M. Kerkhoven, C. Roberts, P. S. Linsley, R. Bernards and S. H. Friend, *Nature*, 2002, **415**, 530.
29. N. Snoj, P. Dinh, P. Bedard and C. Sotiriou, in *Molecular Biology of breast cancer*, ed. W. B. Coleman and G. J. Tsongalis, Elsevier Press, San Diego, 2012, 26, p. 341.
30. S. Paik, S. Shak, G. Tang, C. Kim, J. Baker, M. Cronin, F. L. Baehner, M. G. Walker, D. Watson, T. Park, W. Hiller, E. R. Fisher, D. L. Wickerham, J. Bryant and N. Wolmark, *N. Engl. J. Med.*, 2004, **351**, 2817.
31. C. Sotiriou, P. Wirapati, S. Loi, A. Harris, S. Fox, J. Smeds, H. Nordgren, P. Farmer, V. Praz, B. Haibe-Kains, C. Desmedt, D. Larsimont, F. Cardoso, H. Peterse, D. Nuyten, M. Buyse, M. J. van de Vijver, J. Bergh, M. Piccart and M. Delorenzi, *J. Natl. Cancer Inst.*, 2006, **98**, 262.
32. L. J. Tafe and G. J. Tsongalis, *Clin. Chem. Lab. Med.*, 2011, **50**, 23.
33. A. C. Wolff, M. E. Hammond, D. G. Hicks, M. Dowsett, L. M. McShane, K. H. Allison, D. C. Allred, J. M. Martlett, M. Bilous, P. Fitzgibbons, W. Hanna, R. B. Jenkins, P. B. Mangu, S. Paik, E. A. Perez, M. F. Press, P. A. Spears, G. H. Vance, G. Viale, D. F. Hayes and American Society of Clinical Oncology, College of American Pathologists, *J. Clin. Oncol.*, 2013, **31**, 3997.
34. A. Ocana and A. Pandiella, *Curr. Pharm. Des.*, 2013, **19**, 808.
35. G. Tinoco, S. Warsch, S. Gluck, K. Avancha and A. J. Montero, *J. Cancer*, 2013, **4**, 117.
36. K. C. Halling, W. Kine, I. A. Sokolova, R. G. Meyer, H. M. Burkhardt, A. C. Halling, J. C. Cheville, T. J. Sebo, S. Ramakumar, C. S. Stewart, S. Pankratz, D. J. O'Kane, S. A. Seelig, M. M. Lieber and R. B. Jenkins, *J. Urol.*, 2000, **164**, 1768.
37. M. Skacel, J. D. Pettay, E. K. Tsiftsakis, G. W. Procop, C. V. Biscotti and R. R. Tubbs, *Anal. Quant. Cytol. Histol.*, 2001, **23**, 381.
38. M. Skacel, M. Fahmy, J. A. Brainard, J. D. Pettay, C. V. Biscotti, L. S. Liou, G. W. Procop, J. S. Jones, J. Ulchaker, C. D. Zippe and R. R. Tubbs, *J. Urol.*, 2003, **169**, 2101.
39. V. Vit, D. Pacik, A. Cermak, I. Falkova, J. Smardova, R. Hrabalkova, M. Svitakova, Z. Pavlovsky and M. Votava, *Cesk. Patol.*, 2009, **45**, 46.
40. I. Karaoglu, A. G. van der Heijden and J. A. Witjes, *World J. Urol.*, 2014, **32**, 651.

41. J. M. Pipas, L. D. Lewis, M. S. Ernstoff and G. J. Tsongalis, *Oncol. Hematol. Rev.*, 2013, **9**, 75.
42. J. Lazarou, B. H. Pomeranz and P. N. Corey, *JAMA*, 1998, **279**, 1200.
43. B. S. Shastry, *Pharmacogenomics J.*, 2006, **6**, 16.
44. A. C. Lockhart, R. G. Tirona and R. B. Kim, *Mol. Cancer Ther.*, 2003, **2**, 685.
45. C. Marzolini, R. G. Tirona and R. B. Kim, *Pharmacogenomics*, 2004, **5**, 273.
46. B. L. Urquhart, R. G. Tirona and R. B. Kim, *J. Clin. Pharmacol.*, 2007, **47**, 566.
47. M. Ingelman-Sundberg, *Trends Pharmacol. Sci.*, 2004, **25**, 193.
48. M. Ingelman-Sundberg, *Pharmacogenomics J.*, 2005, **5**, 6.
49. S. J. Gardiner and E. J. Begg, *Pharmacol. Rev.*, 2006, **58**, 521.
50. D. W. Nebert and D. W. Russel, *Lancet*, 2002, **360**, 1155.
51. S. Bernard, K. A. Neville, A. T. Nguyen and D. A. Flockhart, *Oncologist*, 2006, **11**, 126.
52. A. N. Abou Tayoun, F. B. de Abreu, J. A. Lefferts and G. J. Tsongalis, *Clin. Chim. Acta*, 2013, **422**, 1.
53. E. Beutler, T. Gelbart and A. Demina, *Proc. Natl. Acad. Sci. U. S. A.*, 1998, **95**, 8170.
54. L. Weng, L. Zhang, Y. Peng and R. S. Huang, *Pharmacogenomics*, 2013, **14**, 315.
55. D. W. Lum, P. Perel, A. D. Hingorani and M. V. Holmes, *PLoS One*, 2013, **8**, 76648.
56. Y. Jin, Z. Desta, V. Stearns, B. Ward, H. Ho, K. H. Lee, T. Skaar, A. M. Storniolo, L. Li, A. Araba, R. Blachard, A. Nguyen, L. Ullmer, J. Hayden, S. Lemler, R. M. Weinshilboum, J. M. Rae, D. F. Hayes and D. A. Flockhart, *J. Natl. Cancer Inst.*, 2005, **97**, 30.
57. K. Sideras, J. N. Ingle, M. M. Ames, C. L. Loprinzi, D. P. Mrazek, J. L. Black, R. M. Weinshilboum, J. R. Hawse, T. C. Spelsberg and M. P. Goetz, *J. Clin. Oncol.*, 2010, **28**, 2768.
58. C. J. Allegra, J. M. Jessup, M. R. Somerfield, S. R. Hamilton, E. H. Hammond, D. F. Hayes, P. K. McAllister, R. F. Morton and R. L. Schilsky, *J. Clin. Oncol.*, 2009, **27**, 2091.
59. B. Chibaudel, C. Tournigand, T. Andre and A. de Gramont, *Ther. Adv. Med. Oncol.*, 2012, **4**, 75.
60. J. M. Caretheres. *Clin. Gastroenterol. Hepatol.*, 2014, **12**, 377.
61. M. Roengvoraphoj, G. J. Tsongalis, K. H. Dragnex and J. R. Rigas, *Cancer Treat. Rev.*, 2013, **39**, 839.

CHAPTER 12

Molecular Microbial Diagnostics

KARL-HENNING KALLAND,*[a,b] ØYVIND KOMMEDAL[a] AND
ELLING ULVESTAD[a,b]

[a] Department of Microbiology, Haukeland University Hospital, Bergen,
Norway; [b] Department of Clinical Science, University of Bergen, Bergen,
Norway
*Email: Kalland@gades.uib.no

12.1   INTRODUCTION

The suffering and deaths caused by epidemics remain unsurpassed by any
other malady afflicting the human species.[1] Several attempts at explaining
the dramatic outbreaks were made in the prescientific era, and proposals
included mysterious forces associated with witchcraft, earthquakes and
miasmas residing in foul and decaying air. These forces were not dismissed
until the second half of the 19th century when ingenious work by Louis
Pasteur, Robert Koch and other continental researchers made clear that the
infectious diseases were of microbial etiology.[2] This recognition quickly and
forcefully paved the way for the establishment of diagnostic laboratories in
major hospitals, and thus for the coupling of microbiological science to
medical practice.

The microbiologist's repertoire of diagnostic tests was initially rather
limited, consisting predominantly of specimen examination through the
microscope, colony growth on agar plates and serological tests. The
laboratories were nevertheless very successful at their task, and so provided
scientifically based support for hygienic movements attempting at identifying and controlling carriers of venereal, diarrheal and respiratory diseases.
These methods of investigation are still being utilized in diagnostic

Molecular Biology and Biotechnology, 6th Edition
Edited by Ralph Rapley and David Whitehouse
© The Royal Society of Chemistry 2015
Published by the Royal Society of Chemistry, www.rsc.org

laboratories but owing to several shortcomings, including insufficient diagnostic sensitivity and specificity, they are increasingly being sup-plemented with newer methods of detection.

Foremost amongst the new methods are tests based on analyses of nucleic acids—including DNA and RNA—from the causative microbial agents. Not only are these techniques orders of magnitude faster than the older tech-niques, they are also more sensitive and specific, and enable the detection of non-cultivable microbes as well as microbes that have been rendered non-cultivable by antibiotics. Owing to the ubiquitous presence of nucleic acids in all varieties of life, the diagnostic principles can be utilized to analyze any major pathogen afflicting humans, including bacteria, viruses, parasites and fungi. The progress has made diagnosis of microbial agents ever more relevant, especially in the field of virology, where quick results derived from molecular analyses have virtually revolutionized clinical therapy and disease monitoring.

Microbes have traditionally been associated with disease and mal-functioning, but are now increasingly also being recognized as important promoters of health. A lively research activity has provided evidence that absence of symbiotic microbes, or a skewed distribution of microbes in a community of microbes, may be of relevance for a variety of diseases, in-cluding asthma, autoimmunity, malnutrition and obesity. The older meth-ods of detection were way too insensitive to quantitate such skewedness, whereas the newer molecular techniques seem perfect for the task.

The molecular methodology associated with detection of infectious agents is continuously improving, and so is the capacity to handle ever-increasing numbers of patient specimens. However, owing to extreme sensitivity, the same methodology is liable to produce erroneous results if left unattended. Diagnosticians thus need to understand the shortcomings just as well as they understand the opportunities provided by the tests. In the following we will give an overview of the strengths and weaknesses of the most prevalent and useful nucleic acid-based diagnostic tests. First, we give an overview of sampling techniques, thereafter we review relevant techniques for the de-tection of microbial agents, and finally we briefly demonstrate the tests' utility on diagnostically relevant clinical cases.

12.2 SAMPLE COLLECTION AND NUCLEIC ACID PURIFICATION

12.2.1 Sample Collection and Transport

Virtually any human secretion, fluid or tissue sample can be collected and examined for nucleic acids of suspected infectious agents, in addition to bacterial colonies and infected cell cultures. The microbiological diagnostic laboratory provides kits for sampling and shipping patient specimens. Some kits contain buffers that inactivate infectious agents, inhibit nucleases and lyse membranes to facilitate subsequent nucleic acid extraction. Whenever patient samples are collected and sent without additives, speedy delivery to the laboratory must be ensured and great care must be taken to label the sample appropriately and to avoid contamination of the outside of the collecting vessel. Several companies supply swab kits and buffer kits for stabilizing samples during transport.[3]

12.2.2   Extraction of Nucleic Acids

Different nucleic acid extraction protocols are required for different samples and different microbiological agents. RNA is more susceptible than DNA to rapid degradation by nucleases, and the best way to counteract the degradation problem is to use transport buffers that stabilize the intact infectious particle, or alternatively, to dissolve the sample in guanidine isothiocyanate based lysis buffer that inhibits RNase enzymes. Viral nucleic acids can usually be extracted in the same way that host cell nucleic acids are extracted. Sometimes, *e.g.* for faecal (stool) samples, gentle vortexing followed by low speed centrifugation can pellet debris while viral particles remain in the supernatant. Viral particles (virions) require high speed or ultracentrifugation to be pelleted. The thick cell wall of gram-positive bacteria is more difficult to disrupt than the relatively thinner cell wall of gram-negative bacteria and needs special procedures.[4,5]

Heme in blood, bile in stool and components of urine and additives such as heparin may inhibit the enzymes used in nucleic acid amplification, such as Taq polymerase, and must be removed by the purification procedure prior to the amplification step. Purified nucleic acids should be eluted into a small volume of nuclease-free H_2O or 0.1 mM EDTA (ethylenediaminetetraacetate). Ethanol or isopropanol precipitation may be used to achieve smaller volumes, but alcohols must be completely removed by pipetting to avoid subsequent enzymatic inhibition.

12.2.3   Manual and Automated Extraction of Nucleic Acids

Several commercial manual extraction kits are available for use by clinical laboratories. These kits vary regarding method, cost and time required for extraction. Processing by manual methods involves a higher risk for contamination and switching of patient samples. US Clinical Laboratory Improvement Amendments of 1988 (CLIA) regulations provide guidelines regarding manual extraction of nucleic acids for diagnostic purposes (http://www.cms.hhs.gov/clia/).

Automated extraction instruments are manufactured by many different companies, and like manual methods vary in technique, cost and time requirements. Additionally, these instruments vary regarding specimen capacity and sample size. Automated procedures are in several ways superior to manual methods (see references 6 and 7), although this is not always so.[8] Recovery of nucleic acids from automated instruments is consistent and reproducible. Automated extraction systems keep sample manipulation to a minimum, thus reducing the risk for cross-contamination of samples.

They are most economical when instruments are fully loaded, although smaller and more versatile instruments have now become available.

12.3 NUCLEIC ACID AMPLIFICATION TECHNIQUES

12.3.1 Polymerase Chain Reaction (PCR)

PCR represents the prototype of nucleic acid amplification techniques. Excellent reviews on its applicability in clinical microbiology have been published.[6,7,9] Due to its extreme sensitivity nucleic acid amplification has revolutionized identification of infectious agents in patient samples. Several alternative nucleic acid amplification methods have been developed for use in the clinical microbiological laboratory such as isothermal strand displacement amplification (SDA), the isothermal transcription mediated amplification (TMA/NASBA) and loop-mediated isothermal amplification (LAMP).[10]

12.3.2 The Contamination Problem

Owing to the exquisite sensitivity of nucleic acid amplification techniques great care must be taken to minimize the problem of sample contamination. Sterile filter tips must be employed for all manual pipetting steps. The logistics of the laboratory procedures are of fundamental importance. A separate room should be reserved for work with buffer components, oligonucleotide primers, probes and master mixes. No samples or amplified products should be allowed into this room. Higher air pressure than in the surrounding rooms, restricted access, visible written hygienic guidelines on the door and use of separate clothing and shoes are advantageous. A separate room is necessary for receiving and organizing patient samples and a third room for setting up the PCR reactions. Using conventional PCR, a fourth separate room will be necessary for the detection steps of the amplified products, the amplicons. Commonly used methods for the visualization of amplicons are agarose gel electrophoresis or microplate systems with a detection probe and a streptavidin–enzyme conjugate that binds to biotinylated nucleotides incorporated into the amplicon. Following the washing step, the enzyme bound by the biotin–streptavidin interaction will change a colourless substrate into a coloured derivative in a quantitative way. Using real-time PCR (see below) the amplification and detection can be completed in a closed system and for space considerations may be performed in the same room as the reaction setup. If possible, a one-way flow of persons and samples from the first to the last room is desirable. Decontamination of the critical workspace using UV light or 5% bleach represents additional precautions against contamination of samples. If deoxy-uridine-5'-triphosphate (dUTP) has been used in the deoxynucleotide mix of previous amplification reactions, a pre-amplification step using uracil-N-glycosylase will efficiently destroy amplicon contamination of the subsequent reaction.

12.3.3   Reverse PCR—cDNA Synthesis

Recovery of microbial DNA is suitable for the diagnosis of bacteria, parasites and many viruses. In addition, a large proportion of pathogenic viruses carry RNA-genomes instead of DNA-genomes. The genomes of such RNA viruses must be reverse transcribed to complementary DNA (cDNA) before amplification and identification. When PCR DNA amplification is preceded by cDNA synthesis, the combined protocol is named reverse PCR. Reverse PCR may also be useful in order to quantify transcription (mRNA synthesis) of microbial DNA to show the activity of an infectious genome. Using thermostable enzymes that have both reverse transcriptase and DNA polymerase activity (such as rTth), reverse transcription can be achieved in one reaction and one tube. Otherwise, the reverse transcription is often carried out as a separate first step using a thermolabile reverse transcriptase (typically cloned retroviral enzymes). If random primers (hexamers or nonamers) or oligo(dT) primers are used in the initial reverse transcription, then the subsequent PCR can be performed with different specific primer pairs and designed for a variety of RNA viruses. According to the rTth reverse PCR, the same reverse primer sequence will prime both cDNA synthesis and the PCR amplification. This may be a quicker procedure, but the cDNA has reduced versatility compared with cDNA primed by random hexamers or oligo(dT).

12.3.4   Nested PCR

When patient samples are amplified by PCR, the amplicon very often exhibits a smear in contrast to a distinct band in the agarose gel electrophoresis. This is due to non-specific priming and amplification of host cell nucleic acids. Although the first PCR amplification resulted in a smear in the agarose gel, the specific nucleic acids sought for will usually be highly enriched compared with the starting material. If a small amount of the first PCR is used for a second PCR amplification with a new set of internal primers, then a well-defined PCR-fragment typically results if the infectious agent is present in the initial sample. The use of a second round of PCR with a new set of internal primers is called nested PCR. If only one internal primer is used for the second round of thermocycling, it is called semi-nested PCR. Nested and semi-nested PCR increase sensitivity by as much as 1000 times compared with conventional PCR[11] and increase specificity, but the techniques require more work, longer assay time and increase the risk of contamination.

12.3.5   Real-time PCR

Compared with conventional PCR, and in particular compared with nested PCR, real-time PCR has the advantage that the result of the amplification process is recorded in real-time (*i.e.* while the amplification reaction is running). The risk of contamination is very much reduced since there is no need to open the reaction tube once the amplification has started, *i.e.* the post amplification detection step is omitted. Therefore it takes less time and labour to do real-time PCR than nested PCR; real-time PCR is also amenable to automation. For the above reasons the danger of sample switching is less in real-time PCR. The sensitivity is comparable between real-time PCR and nested PCR.

12.3.6 Visualization of Real-time PCR Amplification

Different fluorescence-based principles are employed for real-time detection. SYBR green represents a class of DNA binding compounds that emit increased fluorescence when intercalated into double-stranded DNA as compared with the unbound chemical in solution (Figure 12.1). As a result, there is a proportional increase in SYBR green fluorescence when DNA amplification products increase. The disadvantage using SYBR green is that non-specifically amplified DNA and amplification artefacts such as primer dimers will also generate increased fluorescence, thus reducing the specificity compared with nested PCR. Real-time PCR assays utilizing an amplicon specific probe increases the specificity very much, because the sequence between the two primers must be complementary to the probe in order to obtain a positive assay. A number of smart assays have been invented based upon three oligonucleotides together, *i.e.* two primers and one probe. Several of these assays include TaqMan assays (Figure 12.2), FRET (fluorescence resonance energy transfer) (Figure 12.3) assays, Eclipse probes, Scorpion probes and Molecular beacon probes.

12.3.7 Real-time PCR Equipment

Compared with conventional PCR real-time quantitative PCR (qPCR) needs more equipment. Lasers are needed to excite fluorophores and recording devices are required to collect and digitalize the fluorescent signals. The choice of thermocycler depends upon a large number of variables, *e.g.* the estimated number of samples per run.

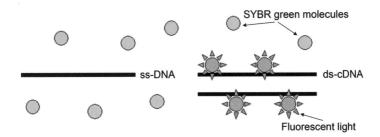

Figure 12.1 The principle of SYBR green PCR assays. The SYBR green molecule has high affinity to double stranded DNA. SYBR green fluorescence is much stronger when it is bound to DNA compared with unbound SYBR green. ss-DNA = single stranded DNA. ds-DNA = double stranded DNA.

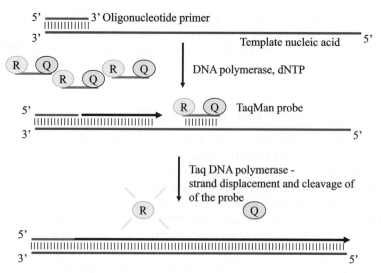

Figure 12.2   The principle of TaqMan real-time PCR assays. The probe is labelled with both a fluorescent chemical group (R) and a quencher (Q). When the Taq polymerase elongates the complementary strand from the primer, its 5′-3′ exonuclease activity cleaves any annealed probe thus separating R from Q. Freed from its quencher, the R fluoresces when exciting light is sent in. The amount of free R, and thus fluorescence, increases exponentially along with amplicons until the PCR reaction approaches the plateau stage.

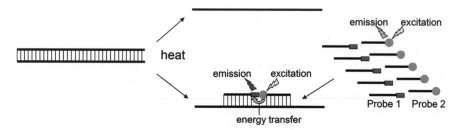

Figure 12.3   The principle of FRET (fluorescence resonance energy transfer). Two different probes are complementary to the target sequence. One probe is labelled with fluorescein and the other with LC Red fluorochrome. The hybridization to the target sequence brings the two fluorochromes physically close enough for the emission energy of fluorescein to excite the LC red so that red fluorescent light is emitted. When PCR generates an increasing concentration of target, an increasing fraction of the probes can be positioned to achieve FRET. This can be monitored as increasing red fluorescence.

12.3.8   Real-time Quantitative PCR

The increasing fluorescence signal plotted on the Y-axis and the increasing cycle numbers along the X-axis generate a sigmoidal curve. The threshold cycle (C_t) is defined as the cycle number when the fluorescent signal exceeds the detection threshold. During the following 4–8 cycles the fluorescence typically approaches exponential increase before approaching a plateau phase.[12] During the exponential phase the logarithm of the fluorescent signal along the Y-axis and the cycle number along the X-axis generate a straight line that can be used for quantification. In this phase of the real-time PCR the fractional threshold cycle number is inversely related to the logarithmic number of starting templates.[13] Parallel PCR-amplification of a dilution series of known amounts of the same template therefore allows the construction of a standard curve from which the exact amount of starting template can be calculated.[9] It must be assumed that the amplification efficiency is equal for all samples and standards. One sign of acceptable quantification is that the slope of the standard curve is between minus 3.3 and minus 3.4. More detailed guidelines to quantification and quality control can be found in user bulletins provided by the manufacturers of thermocyclers and software.

12.3.8.1 Absolute and Relative Quantifications. In order to achieve absolute quantification a known amount of reference standard is required for the dilution series and standard curve generation. Synthetic oligonucleotides, purified PCR fragments, plasmids or *in vitro* transcribed RNA can be quantified using light absorption readings at 260 nM (OD260) wavelength. A dilution series of known amounts of target allow absolute quantification. Relative quantification, relating the amount of template to an accepted standard, is often sufficient, and this can be achieved using the standard curve method or the comparative C_t method (see ref. 9 and references therein).

12.3.9 Determination of "Viral Load" in Clinical Microbiology

Determination of "viral load", which refers to the number of viral genomes in a defined volume of a patient sample, typically in serum or plasma, is becoming increasingly important with the increasing use of antiviral medication. Real-time quantification of viral load is often required before the start of antiviral measures against human immunodeficiency virus (HIV), hepatitis C- and hepatitis B-viruses, against cytomegalovirus (CMV) in patients with compromised immune systems and against Epstein-Barr (EBV) virus in patients following organ transplantations. Reduction of viral load indicates efficient response to treatment, and quantification is of great value to monitor development of viral resistance to the medication or relapse after cessation of therapy. Commercial kits are widely used for quantification of viral load in clinical microbiology. A known amount of synthetic standard is typically included in each reaction to establish a competitive PCR amplification. The resulting amounts of viral and standard amplicons are compared to a standard dilution curve in order to establish the viral load. It is important to be aware that different manufacturers may have different standardization units so that results obtained using different kits may not be comparable. Progress has recently been made to establish internationally accepted standard units for different viruses, *e.g.* for hepatitis B-virus,[14] hepatitis C-virus[15] and parvovirus B19.[16] The WHO provides guidelines and established standards in order to calibrate different methods to obtain comparable International Units per millilitre (IU/mL, http://www.who.int/biologicals/reference_preparations/distribution/en/index.html).

12.3.10   Internal Controls in Microbiological Real-time qPCR

Both negative and positive controls are important in nucleic acid based diagnosis in the microbiological laboratory.[3] The negative control consists of a parallel amplification reaction that does not contain the specific target. There is no rule as to how many negative controls are needed, and there is always a trade-off between the number of negative controls and the cost. Recommendations for a negative control for every fifth tube may come in conflict with available resources.[3] Positive controls can be any target sequence in a separate tube. The best positive control templates are included in the same tube as the diagnostic assay and contain flanking sequences complementary to the primers of the diagnostic assay but with a unique sequence in the region recognized by the probe. Since real-time qPCR assays are typically designed with amplicons shorter than 100 nucleotides it is easy to obtain the desired control template sequence by oligonucleotide synthesis. A more difficult challenge is if the laboratory wants to spike the patient sample with a control sequence to control for the entire process from sample handling, *via* nucleic acid extraction through PCR amplification. For RNA viruses, which represent the greatest challenge, this can be achieved if a known amount of an *in vitro* synthesized RNA is spiked into the patient sample. Any exogenous sequence may be chosen if it can be detected separately from the sequence of the infectious agent, but the optimal control RNA will be transcribed from a vector that can generate an RNA that contains more than 1500 nucleotides and includes a target sequence that can be amplified with the same primers as the diagnostic assay but differs in the region recognized by the probe.

In all same-tube controls it is very important to carefully titrate the amount of control target. Otherwise, the control may out-compete the diagnostic assay and lead to false negatives.

An important example of a positive control RNA is exemplified by human control (endogenous) genes when determining the expression levels from a gene of interest. Typically, the endogenous control gene should be ubiquitously expressed by human cells. Examples include RNAse P, GAPDH or β-actin. When the control PCR reaction for the endogenous gene is positive, it suggests that representative patient material was sampled and that the sample does not contain substances that inhibit the PCR amplification.

12.3.11   Multiplex Real-time PCR

Multiplex PCR refers either to the use of more than one primer pair in one reaction tube in order to amplify more than one target sequence, or to the use of more than one primer/probe set in the case of real-time PCR. The internal control assay described above is one example of multiplex PCR. In microbiological diagnostics the general aim of multiplex PCR is to simultaneously test for different agents in the same reaction. This requires the use of fluorophores and equipment with the ability to distinguish between the different assays. Non-specific amplification products seem to be the most limiting factor of multiplex PCR as the number of oligonucleotide primers and probes increases. Several multiplex assays for the simultaneous detection of two or more different agents are, however, well established. A number of multiplex real-time PCR assays have been published including simultaneous amplification and subtyping of herpes simplex virus 1 and 2 (HSV-1 and HSV-2) and Varicella Zoster virus (VZV) using FRET probes[17,18] or the combination of influenza A subtypes H1N1 and H3N2, influenza B and respiratory syncytial virus (RSV) subtypes A and B[18,19] and additional respiratory infection agents.[20]

12.3.12 Melting Curve Analysis

Melting curve analysis has made possible an increase in the number of agents detected in multiplex real-time PCR, in particular the ability to distinguish between closely related viruses such as HSV-1 and HSV-2 or influenza types A and B. FRET probes (Figure 12.3) are useful for this type of analysis. When one FRET probe is designed to be complementary to one viral subtype but contains one or very few mismatches to the other viral subtype, melting curve analysis is possible due to different melting points (T_m) dependent upon the exact target sequence. Melting curve analysis is performed following completion of the PCR. The probe pair is first allowed to anneal to the target amplicons and thereafter the temperature is increased while the fluorescence is continuously recorded. The viral subtype with mismatch to the FRET probe will exhibit a reduced fluorescence at a lower melting point (T_m) than the viral subtype with perfect homology to the FRET probe. If both subtypes are present in the same mix, this may be visualized as a curve with two different peaks when the fluorescence is plotted on the Y-axis and the temperature is plotted along the X-axis.

12.3.13 Genotyping

Many viruses evolve at high rates after having infected their human hosts. In particular, RNA viruses that replicate by RNA polymerases which lack 5' proof reading in contrast to DNA polymerases, are prone to point mutation errors in the order of 10^{-4} per genome replication, but DNA viruses may also change rapidly upon changes in the selection pressure, owing to their short generation times and enormous number of progeny in each infection. As a consequence, most viral species come in multiple serotypes and genotypes. Some of these types are more efficient at producing disease than others, and it is therefore clinically relevant to distinguish between such genotypes. 12 different main genotypes are presently described for hepatitis C virus.

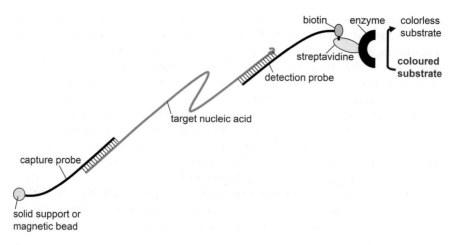

Figure 12.4 The principle of sandwich hybridization.

Both prognosis and the recommended duration of PEG (polyethylene glycol)–interferon α/ribavirin combination therapy vary between HCV genotypes. The prevalent genotypes 1 and 4 usually require 48 weeks of combination therapy while less than half that duration of therapy may be sufficient for genotypes 2 and 3. Also, sustained viral response, in which HCV RNA cannot be detected in serum 6 months following cessation of treatment, is achieved in a higher proportion of cases for genotypes 2 and 3. Likewise, more than 130 genotypes of human papillomaviruses (HPV) are currently recorded, and these types are separated into high-risk types and low-risk types concerning their ability to cause cervical cancer (cancer cervicis uteri). Genotyping can be achieved using DNA sequencing. Due to co-infections with multiple genotypes at different concentrations, other methods have been developed for genotyping in the clinical microbiology laboratory such as sandwich hybridization assays (Figure 12.4) or line probe assays (Figure 12.5).

12.4 OTHER TECHNIQUES USED IN CLINICAL MICROBIOLOGY

12.4.1 Hybridization Techniques

The principle of complementarity, according to which A (adenine) pairs with T (thymine) and C (cytosine) pairs with G (guanine) in a nucleic acid duplex, is the core principle of heredity, all hybridization techniques and the PCR alike. The A–T base pairing involves two and the G–C base pairing involves three hydrogen bonds, respectively. Therefore the relative content of A–T and G–C base pairs contributes strongly to the melting temperature (T_m) of the nucleic acid duplex. The higher the G–C content, the higher the T_m of the nucleic acid duplex. Different types of hybridizations were among the first techniques attempted in laboratory detection of nucleic acids. In

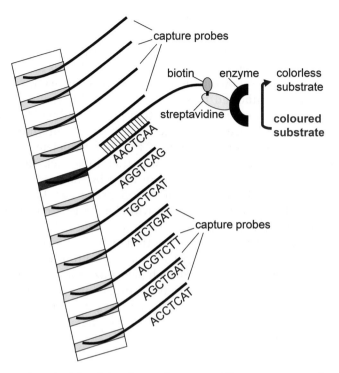

Figure 12.5 The principle of the Line probe assay. Different capture probe sequences are fixed in different bands in a membrane strip. The amplification product hybridizes with the capture probe of complementary sequence. Utilizing biotin labelled PCR primers the position of the complementary capture probe can be visualized as a coloured band following addition of streptavidin-bound enzyme that will change a non-coloured substrate to a coloured product that precipitates in the band.

general, the sensitivity of hybridization techniques was not sufficient for routine detection of infectious agents directly in patient samples. The invention of nucleic acid amplification techniques therefore represented the breakthrough of nucleic acid based laboratory diagnosis. However, in some cases the sensitivity of hybridization was promising and encouraged further development of methods to detect nucleic acids, in particular for bacteria containing multicopy plasmids or repeated genomic sequences.[21–23]

12.4.1.1 Sandwich Hybridization. In the classical hybridization techniques the nucleic acid target was usually fixed on a solid phase (nitrocellulose, nylon membrane, glass slide) and detected by a labelled probe added in the hybridization buffer to the solid phase. These techniques, such as dot blots, colony blots, Southern blots, Northern blots and *in situ* hybridization usually either lacked sufficient sensitivity due to background hybridization or were too laborious for the routine setting. The sandwich hybridization (Figure 12.4) technique turned out, however, to have sufficiently low background and strong enough signal to be employed for some routine laboratory purposes. In this technique the capture probe is fixed to the solid phase and is available to the target in the hybridization buffer. Following annealing between the capture probe and the specific target, unbound material is removed by washing and a new hybridization buffer containing a detection probe is added. The detection probe is typically labelled with either biotin or an enzyme and is designed to bind to a different region of the target nucleic acid than the capture probe. The detection probe will therefore stick to the capture probe and the solid phase only if it is bridged by the specific target. Following a wash step, the retained detection probe can be visualised by addition of a colourless substrate that will change to a coloured product by the enzyme conjugated to the detection probe or by a streptavidin–enzyme complex in the case of a biotin-labelled detection probe. The solid phase may be a magnetic particle or the bottom of a 96-well plate to facilitate automation. The sandwich hybridization technique is used in commercial tests for the detection of human papillomaviruses (HPV) in samples from the female cervix (Digene test), and in the visualization step of conventional PCR assays (*e.g.* several Roche assays). The principle is also utilized in line probe assays.

12.4.1.2   Line Probe Assays.   In line probe assays (Figure 12.5) capture probes are printed in parallel bands (lines) on a membrane. Line probe assays have been developed to distinguish strains or genotypes of infectious agents. Each band or lane contains capture probes that differ only in nucleotides that are characteristic for each strain or genotype. The target nucleic acid is typically first amplified by PCR and next added to the line probe membrane in a hybridization buffer. Following wash, a labelled detection probe is added and visualized as described for the sandwich-hybridization above. The line or band corresponding to one specific genotype will then become coloured (Figure 12.5). Commercial line probe assays have been used for the detection of genotypes of HPV and hepatitis C-virus (HCV), strains of atypical mycobacteria and for detection of HTLV-1 (human T-cell lymphotrophic virus).

12.4.1.3   Peptide Nucleic Acid Fluorescent In Situ *Hybridization (PNA-FISH).* PNA-FISH is one type of *in situ* hybridization utilizing hybridization probes of which the sugar phosphate backbone of DNA is replaced by a polyamide backbone (Figure 12.6).[24,25] Peptide nucleic acids exhibit rapid hybridization kinetics and, when coupled to fluorescent reporter molecules, offer a rapid and sensitive verification of the infectious agent in positive blood cultures.[25] Once a blood culture turns positive, staining of cells on a slide is performed, and based on the results the appropriate PNA-FISH test is selected. Identification results are available within just a few hours and

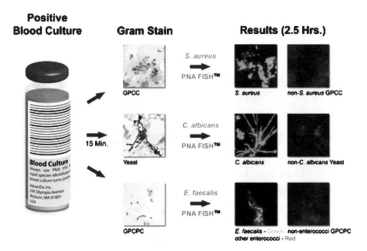

Figure 12.6 Schematics of PNA-FISH.

can be reported to the attending physician, *e.g.* in the case of serious yeast infections.[26]

12.4.2 TaqMan Low Density Arrays (TLDA)

The TaqMan Low Density Array is a 384-well microfluidic card manufactured by Applied Biosystems where up to 384 simultaneous real-time PCR reactions can be carried out without the need for liquid-handling robots or multi-channel pipettes to load samples. Each card allows one to eight samples to be run in parallel with 12 to 380 different TaqMan Gene Expression Assays that are pre-loaded into separate 1–2 µl wells on the card. The TaqMan Low Density Array is customizable for human, mouse and rat genes and has been evaluated for the simultaneous detection of multiple infectious agents.[27] The TaqMan Array is designed for use on the Applied Biosystems 7900HT Fast Real-Time PCR system. The TLDA format increases the number of different real-time PCR assays that are feasible in a given time by one order of magnitude and is suitable for the validation of DNA microarray gene expression results (Figure 12.7).

12.4.3 Nucleic Acid Based Typing of Bacteria

Remarkable progress is currently being achieved in nucleic acid sequencing and associated bioinformatic formatting, storage, analysis and visualization of sequencing data. The progress is variably being referred to as massive parallel sequencing, or next generation sequencing (NGS). It is not possible, at the present time, to foresee exactly how NGS will change microbiological diagnostics, but the potential is formidable. Owing to this potential, several of the nucleic acid based smart techniques covered in the previous edition of this book, will only be listed here. These include Random Amplified

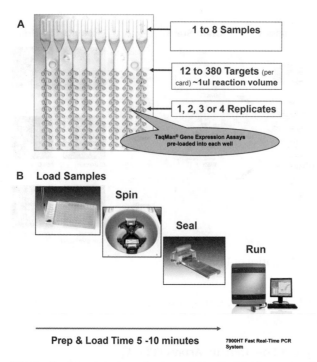

A

1 to 8 Samples

12 to 380 Targets (per card) ~1ul reaction volume

1, 2, 3 or 4 Replicates

TaqMan® Gene Expression Assays pre-loaded into each well

B   Load Samples

Spin

Seal

Run

Prep & Load Time 5 -10 minutes   7900HT Fast Real-Time PCR System

Figure 12.7   (A) The TaqMan Low Density Array (TLDA). (B) About 100 µl master mix and hexamer primed cDNA are loaded into each well and distributed into the reaction wells by low speed centrifugation and the card is sealed and thermocycled for about 40 cycles.

Polymorphic DNA (RAPD) and Amplified Fragment-Length Polymorphism (AFLP).[28,29] Instead, more emphasis will be put on recent progress in bacterial nucleic acid sequencing.

12.4.3.1   The 16S Ribosomal RNA Gene.   The 16S ribosomal RNA (16S rRNA) gene is present in all bacteria, often existing as a multi-gene family and is a nature's gift to bacteriology. Sequence analysis of the bacterial 16S rRNA gene has revolutionized bacterial phylogenetic investigations as well as the clinical microbiologist's ability to routinely provide accurate identification of bacterial isolates. The value of the 16S rRNA gene is linked to its ubiquity, to its patterns of phylogenetic divergence, as well as to its versatility. Owing to accumulating point mutations, bacteria of different species express different 16 rRNA genes, thus allowing the diagnostician to differentiate between the species. The molecule's diagnostic versatility is coupled to the pattern of alternating variable and highly conserved regions along the gene. The conserved regions make possible the design of universal primers that will allow PCR-amplification of the 16S rRNA genes of almost any bacterium. By locating the PCR-primers so that the resulting PCR amplicons contain one or more of the variable 16S regions, identification can be obtained by subsequent DNA sequencing, as exemplified.[30]

The universal PCR as the first step of the sequence analysis also allows for selective amplification of the 16S rRNA gene directly from a sample without the need for cultivating the bacteria on an agar. This enables the detection of microbes that are unable to grow as a consequence of antibiotic treatment prior to sample collection, for the detection of atypical bacteria with special growth requirements and for bacteria that are vulnerable to conditions during sample collection procedures and transportation, *e.g.* anaerobic bacteria. Furthermore, the method has allowed the detection of previously unknown causes of disease, including the up to then uncultured causative organisms for cat-scratch disease (*Bartonella henselae*)[31] and Whipple's disease (*Tropheryma whippelii*).[32]

When mixed bacterial populations are being analysed, sequencing based on the Sanger method is found to be insufficient for identifying individual species. This is partly owing to the generation of mixed sequences and partly due to the phenomenon that signals from DNA belonging to the dominant species completely mask signals from DNA belonging to lower abundance species already at ratios higher than 1 : 10.[33] Although these problems could be reduced to some extent by algorithm based de-convolution of mixed chromatograms[34] or cloning the PCR-product and re-sequencing individual clones,[35] the limitations have seriously hampered effective culture independent investigations of complex microbial communities like those found in the human colon.

Some recently introduced sequencing techniques have resolved these limitations of the Sanger sequencing. In massively parallel sequencing (NGS) the DNA from a reaction tube is distributed on a chip and then each single DNA is re-amplified and sequenced individually in millions of picoliter reactions. Since each of the picoliter sequencing reactions are originating from a single piece of DNA and are being read independently from the neighbouring reactions, the restrictions of the Sanger method are being eliminated.

When massive parallel sequencing is used for 16S rRNA analysis of a sample, the gene will be amplified by PCR prior to sequencing like in Sanger sequencing. This pre-amplification will provide the best sequencing coverage since all the sequencing wells will be used for the gene of interest (deep sequencing) assuring a high sensitivity also for the minor components of complex microbial communities like the human gut microbiome.

In a shotgun approach all DNA in a sample is fragmented and sequenced directly without selective amplification of any specific genes. Although this can allow for culture independent analysis of larger parts of the bacterial genome and even culture-independent whole genome sequencing in some instances, it is not optimal for culture-independent microbial identification from clinical infectious disease samples or description of complex microbial communities since a large number of sequencing wells will be "wasted" on non-microbial DNA and bacterial genes not informative for species identification. In addition to compromising sequencing coverage it also significantly increases the complexity of the subsequent data-analysis.

12.4.3.2   Restriction Endonuclease Analysis—"DNA Fingerprinting".   Restriction enzymes are a group of enzymes that each can recognize one specific short double-stranded DNA sequences (the majority of restriction enzyme recognition sites are between four to nine base pairs) and cleave the DNA at this site. The enzymes work by cleaving the bonds in the phosphate backbone of the DNA molecule. Depending upon the exact DNA sequence of the bacterial genome different mixtures of restriction fragments will result following digestion with one or more defined restriction enzymes. The restriction fragments can be separated using agarose gel electrophoresis with a fluorescent DNA-intercalating compound followed by visualization of the separated restriction fragments (DNA bands) under UV light. A "DNA fingerprint" can thus be obtained for each species of bacterium. The method requires relatively high amounts of purified DNA and the resolution is not high.

12.4.3.3   Pulse Field Gel Electrophoresis (PFGE).   PFGE uses rare-cutting restriction enzymes to fragment the entire bacterial genome into large DNA fragments that are embedded in agarose plugs and placed in wells of the electrophoresis gel where they become part of the gel. The large DNA fragments are separated in the agarose gel by alternate pulses of perpendicularly oriented electrical fields. Following electrophoresis combined with a DNA-intercalating fluorescent compound in the gel, characteristic DNA fragment patterns can be visualized under UV-light (Figure 12.8). The principle of PFGE is that large DNA fragments require more time to reverse direction in an electric field than do small DNA fragments and thus are more retarded in the agarose gel. By alternating the direction of the electric current during gel electrophoresis, it is possible to resolve DNA fragments of 100–1000 kilobases. PFGE has been used extensively for typing of bacterial species, and is often considered to be the "gold standard" of genomic typing methods. DNA degradation may represent a problem.

12.4.3.4   Multilocus Sequence Typing (MLST).   The MLST technique is described in reference.[36] There are three elements to the design of a new MLST system: the choice of the isolates to be used in the initial evaluation; the choice of the genetic loci to be characterized; and the design of primers for gene amplification and nucleotide sequence determination. It is advisable to assemble a diverse isolate collection on the basis of existing typing information or epidemiological data. This should comprise around 100 isolates (95 is a good number for high-throughput sequencing in 96-well microtitre plates) to ensure that the primers developed will be

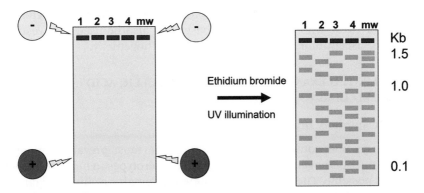

Figure 12.8 Schematics of pulse field electrophoresis. Restriction enzyme digested DNA is loaded into wells and separated in the agarose gel by electrical fields that alternate between the poles every 90 seconds for 24 hours. UV illumination visualizes DNA with an intercalating fluorescent compound. Different samples were loaded in wells 1–4 with a fragment size marker in the rightmost lane.

applicable to as many isolates as possible and to establish the levels of diversity present at each of the loci to be examined. Furthermore, the collection will ideally be representative of the bacterial population, rather than comprising a subset, such as human disease isolates. Housekeeping genes, flanked by genes of similar function, are good targets for MLST and the availability of complete genome sequences has greatly facilitated the identification of candidate loci. Experience with several bacterial species has indicated that PCR fragments of housekeeping genes around 450 base pairs are suitable for MLST and a nested strategy is highly recommended for sequencing.[36] The great advantage of MLST is that sequence data are unambiguous and the allelic profiles of isolates can easily be compared to those in a large central database *via* the Internet (in contrast to most typing procedures which involve comparing DNA fragment sizes on gels). Allelic profiles can also be obtained from clinical material by PCR amplification of housekeeping loci directly from patient samples. Thus isolates can be precisely characterised even when they cannot be cultured from clinical material. MLST can effectively distinguish strains that possess high degrees of homology within the compared gene sequences. This technique is not as laborious as PFGE and it provides balance between sequence-based resolution and technical feasibility.

The MLST-principle is highly compatible with new next-generation sequencing platforms that offer the possibility to sequence an entire bacterial genome in a single run. Whole genome databases for species of interest opens for the possibility to explore and continuously improve MLST as a typing tool for a wide number of species, *e.g.* by changing or increasing the number a targets included (extended MLST). Given the development of user-friendly software that automatically detects and compares the defined genome regions and the inclusion of more variable gene segments, MLST-based typing has the potential to replace all other typing-methods for bacteria in the future. These whole genome databases can also be used to investigate antibiotic susceptibility, vaccine coverage or pathogenicity.

12.5 SELECTED EXAMPLES OF CLINICAL NUCLEIC ACID BASED DIAGNOSIS

12.5.1 Respiratory Infections

Rhinoviruses and coronaviruses are among the most prevalent upper respiratory tract pathogens and exist in too many serotypes to make serological diagnosis possible. The common cold therefore remains a clinical diagnosis.

Lower respiratory tract infections cause a high degree of morbidity and mortality. Traditional culture often fails, is too time consuming or not even possible for many bacterial agents. Serology has limited sensitivity and does not provide information in the acute phase of infection. Real-time PCR and other nucleic amplification techniques targeting core pathogens such as *Chlamydophila pneumoniae*, *Mycoplasma pneumoniae*, *Bordetella pertussis*, *Legionella pneumophila*[37–41] or *Mycobacterium tuberculosis* in respiratory samples have represented a diagnostic revolution for these important infections. In children hospitalized with respiratory difficulties due to bronchiolitis, reverse PCR-tests for human metapneumovirus and respiratory syncytial virus (RSV) have proven to be more sensitive than previous tests based on immunofluorescence or immunochromatography.[42–44] For the diagnosis of influenza A and B, real-time PCR assays have completely replaced viral isolation as the reference standard in the acute phase of the disease.[18,19,45] Real-time PCR assays are available for the specific analyses of avian influenza H5N1.[46,47] swine flu H1N1 and the SARS virus.[48,49]

A recent development is "PCR-in-a-bag" assays: multiplex PCR tests that integrate sample preparation, amplification, detection and analysis all into one system with a turnaround time down to one hour and designed to be performed in the emergency room or on the ward.[50] These easy-to-perform assays have demonstrated sensitivity similar to that obtainable in the lab.[50] The FilmArray (Biofire) respiratory panel offers identification of more than 20 respiratory agents within an hour,[51–53] and GeneXpert (Cepheid) offers *e.g.* simultaneous detection of influenza A, B and the 2009 pandemic H1N1 strain.[54] The cost-effectiveness of these assays is controversial, but they do offer a guide for early treatment as well as isolation of patients. WHO and Cepheid have initiated a programme making the GeneXpert Mtb/RIF assay available at a lower price for a defined group of low-resource countries. This assay offers the simultaneous detection of *M. tuberculosis* and mutations coding for rifampicin resistance.[55]

12.5.2 Central Nervous System (CNS) Disease

Viruses are the most common causes of encephalitis (inflammation of the brain) in humans and more than 100 encephalitogenic viruses are known[56,57] followed by bacterial, parasitic and fungal agents.[58] The aetiology remains unknown in more than half of all cases.[58] Encephalitis is an uncommon disease, but because early start of specific therapy may prevent serious and lasting damage to the brain, early and correct identification of the infectious agent may represent a dramatic difference to each patient. If for example an enterovirus is present in the CSF, the prognosis is usually good without specific treatment. In contrast, when a member of the herpes virus family infects the brain, early antiviral (acyclovir or ganciclovir) treatment is urgent to avoid lasting damage to the brain or even death. In particular, herpes simplex virus type 1 (HSV-1), but also HSV-2 and varicella zoster virus (VZV), account for a large proportion of cases with infectious encephalitis, and may occur in young and otherwise healthy people. As such, amplification of nucleic acids from cerebrospinal fluid (CSF) obtained by lumbar puncture represents a diagnostic breakthrough.

In patients with bacterial meningitis universal amplification of the 16S rRNA gene has been demonstrated to increase diagnostic sensitivity as compared to culture alone, in particular for patients that have received antibiotics prior to lumbar puncture. Since the main proportion of bacterial meningitis is caused by seven to eight bacterial species, multiplex PCR would represent a more sensitive and more rapid diagnostic alternative. There are several new commercial platforms that should have the potential to deliver robust assays at this level of complexity, may be even including the most common viral agents (*e.g.* FilmArray, Luminex, GeneXpert).[59]

12.5.3 Hepatitis

Serological tests detecting specific antibodies and antigens using ELISA and micro-EIA are adequate for the primary diagnosis of hepatitis A, B, C and E viruses (HAV, HBV, HCV, HEV). The detection of HCV RNA is a very useful complement to the standard screening and confirmation test (RIBA) to establish the diagnosis of hepatitis C. In particular, the reverse PCR is useful for the testing whether a newborn child is infected. This diagnosis is difficult to ascertain by serology owing to maternal–foetal transfer of maternal anti-HCV antibodies. Detection of HCV RNA in donor blood or in pools of donor blood may be useful, but the most important role of real-time PCR and other nucleic acid based tests for hepatitis C and B viruses relates to therapy. Commercial HCV reverse PCR tests come in both qualitative and quantitative versions. Before PEG–interferon α/ribavirin therapy is attempted against hepatitis C, an anti-HCV positive patient should in addition test positive in the HCV RNA qualitative test. The next step is then to perform a quantitative reverse PCR-test to establish the HCV RNA IU/mL of serum. The amplification product is employed in a line probe assay or sequenced in order to determine the genotype (see above). Different genotypes require different kinds of treatment, and the quantitative reverse PCR test is used to monitor the response to treatment and to decide whether or not therapy is successful. When therapy is completed, the HCV PCR test is used to make sure that HCV does not relapse in the next 6 months and that a sustained viral response has thereby occurred.

Antiviral treatment is also attempted in hepatitis B in order to prevent accelerating damage to the liver, and quantitative PCR to determine viral load in serum prior to and during treatment is important. Only occasionally is HBV eradicated as a result of treatment. Instead, drug resistant mutants commonly develop and are associated with specific point mutations that can be determined by DNA sequencing.

12.5.4   Gastroenteritis

Acute gastroenteritis causes almost 3 million deaths each year worldwide, mostly among children below 5 years of age in developing countries. The known causes of viral gastroenteritis include the RNA-viruses rotavirus, astrovirus, sapovirus and norovirus and the DNA adenoviruses. Entero-viruses do not usually, despite their name, cause gastroenteritis, but repli-cate in the intestines and can be isolated from faeces in patients with meningitis, myocarditis and symptoms from many other organs. Electron microscopy has been valuable for the discovery and diagnosis of viruses in diarrhoea and the names of rotavirus ("wheel") and astrovirus ("star") refer to the electron microscopy images. Agglutination tests provide rapid diag-nosis of rotavirus and adenovirus gastroenteritis but reverse PCR tests are increasingly coming into use for the rapid diagnosis of viral gastroenteritis. The incubation time of norovirus gastroenteritis may be less than 1 day, and noroviruses commonly cause outbreaks of gastroenteritis in hospitals and in other settings where groups of people live closely together. By means of re-verse PCR and DNA sequencing the hundreds of strains of noroviruses (also known as (caliciviruses/Norwalk agents) have been identified and classified into genogroups and genetic clusters.[60] Reverse PCR tests using nucleic acids extracted from diarrheal faecal specimens have come into broad use to diagnose norovirus infection.[61,62] The genetic variability of viruses causing gastroenteritis represents a great challenge to molecular diagnosis, and new protocols take advantage of new chemistry and computer analysis to design multiplex PCR protocols directed at different strains of each virus and to cover several different viruses in one reaction.[63-65]

12.5.5   Sexually Transmitted Diseases

12.5.5.1   Human Papillomaviruses (HPVs).   HPVs are among the most com-monly sexually transmitted infectious agents. Molecular cloning and DNA sequencing have been crucial for the identification and classification of the more than 130 HPV types presently discovered. The high-risk types, in-cluding the most prevalent types 16 and 18, are the cause of cervical cancer. The low risk types cause benign papillomas and warts and include HPV type 6 and 11 that cause genital warts or condylomas. The cloning and recombin-ant expression of HPV L1 proteins in yeast and insect cells have provided successful vaccines against HPV 16 and 18, and one vaccine additionally protects against HPV 6 and 11. The cytological smear remains the most important screening test for pre-stages of cervical cancer and relies on microscopic evaluation of abnormalities in cells taken from the surface of the cervix. However, microscopy is time-consuming and requires highly skilled and experienced operators. Given that atypical cervical cells are usually caused by HPVs, strategies using HPV nucleic acid detection as a primary screen are being considered. Confirmatory testing using microscopy would then be carried out only on samples that tested positive for HPV DNA. Many commercial nucleic acid based tests are on the market for HPV screening and subtyping. Comparison between these tests is methodologically interesting since different assay principles are used (listed in reference 66). Many issues in HPV nucleic acid testing remain open to the future, including any additional value of focusing the diagnosis on the HPV oncogenes E6 and E7, in particular to monitor their transcriptional activity, the role of next generation sequencing methods and the role of HPV detection in several different human cancer types.[67]

12.5.5.2 Chlamydia Trachomatis and Genital Mycoplasmas and Ureaplasmas.
Nucleic acid detection methods have revolutionized the diagnosis of genital *Chlamydia trachomatis*, one of the most ubiquitously sexually transmitted agents and associated with deep pelvic inflammation and female sterility. Efficient antibiotic treatment is available if the diagnosis is made early. Screening and diagnosis at an early stage is very important. At later stages with established deep pelvic infection the sensitivity of all nucleic acid detection tests become much lower than during early infection.[68] During early stages of infection, self-sampling using vaginal swabs or first pass urine samples provide adequate alternatives to sampling by health professionals.[69,70] Several commercial high capacity *Chlamydia trachomatis* nucleic acid amplification tests streamlined with both positive and negative controls are now available.[71,72]

There is now an increasing awareness that in addition to *Chlamydia trachomatis* several other organisms may be involved in non-gonococcal urethritis, including *Mycoplasma genitalium*, *Mycoplasma hominis*, *Ureaplasma urealyticum* and *Ureaplasma parvum*. In particular, *Mycoplasma genitalium* may be prevalent in some sexually active populations and with many of the same symptoms and complications as *Chlamydia trachomatis*, but, importantly, with different therapeutic requirements.[73] In-house PCR assays have been published by many diagnostic laboratories for the sensitive and specific diagnosis of these agents in urine or semen.[74-76]

12.5.6 HIV Infection and AIDS

The screening (ELISA) and confirmation (Western blot) tests for HIV-1 and HIV-2 infections provide highly reliably laboratory diagnosis of this serious infection. Nucleic acids based amplification tests are therefore not presently important for the establishment of the diagnosis. As opposed to the frustrations encountered in HIV vaccine development, drug treatment of HIV-1 infection and AIDS has been a remarkable success. However, the virus still cannot be eradicated from the body, and life-long combination therapy using three different drugs is necessary to suppress disease progression. By now, multiple drugs have been approved for the treatment of HIV infection and inhibit defined molecular stages of the HIV life cycle, including the reverse transcriptase, the protease, the integrase, the co-receptors (CCR5/CXCR4) and the membrane fusion.[77,78] CD4$^+$ cell counts and HIV viral load in plasma must be determined prior to start and during therapy. When therapy is effective, HIV RNA should be below the detection limit of commercial standardized nucleic acid amplification platforms.[79] If the HIV RNA viral load increases during therapy, it is either a sign that the patient does not take the medication properly or that treatment-resistant HIV strains are evolving. HIV drug resistance is associated with characteristic point mutations that are specific for each drug; therefore sequence analysis of the HIV genome is helpful when selecting a different combination of drugs that may continue to suppress infection.

12.5.7 Bacterial Antibiotic Resistance and Virulence Factor Genes

Detection of bacterial genes that confer antibiotic resistance is important for the choice of proper treatment, to help decide which patients need to be isolated and to trace epidemics. At the same time it may be necessary to identify virulence (disease causing) genes. The following key examples have been selected.

12.5.7.1 Methicillin-Resistant Staphylococcus aureus. *Staphylococcus aureus* has long been a notorious problem among hospitalized patients. Whenever bacteraemia (bacteria in the blood) is suspected, the typical procedure is to identify the type of bacteria using automated blood culture systems followed by microscopy of positive blood cultures. When the microscopy shows Gram-positive cocci in clusters it is a main goal to differentiate between *S. aureus* and coagulase-negative staphylococci and to verify methicillin-resistant *S. aureus* (MRSA).[80] MRSA strains are resistant to all beta-lactam antibiotics. A number of different nucleic acid based techniques have been used to identify the *mecA* gene (the methicillin resistance gene) from putative MRSA colonies and at the same time identify *S. aureus* based upon specific genes such as the *nuc* or *orfX* genes.[25,81,82] Since this verification may take more than 2 days, it is common to start treatment using glycopeptide antibiotics until the results of the *mecA* test is available; this represents a selection pressure for the increasing vancomycin resistance. Glycopeptide-resistant MRSA strains is an emerging problem.[83,84] One important prophylactic measure is to screen and identify asymptomatic carriers of the *mecA* gene among hospital workers and patients. Colonies grown from nasal swab samples are then analysed for *mecA* and *S. aureus* specific genes using PCR, real-time PCR or several other nucleic acid based techniques.[25,81,82] MRSA-detection was one of the first targets for commercial "PCR-in-a-bag" developments. Today sensitive screening of patients for MRSA can be performed in the emergency room and provide results in about 1 hour.[85,86]

More recently, serious MRSA infections have appeared outside hospitals; these so called community-associated MRSA strains harbour a set of phage PVL genes in contrast to the hospital MRSA strains. It is unclear why certain MRSA strains predominate in hospitals and others in the community, but nucleic acid based detection of PVL genes represents a very important tool for epidemiology and surveillance.[87–89] It has also been shown that different MRSA strains differ in their virulence-gene profiles and that this contributes to the complexity of defining the disease-causing abilities of *S. aureus*.[87]

12.5.7.2 Antibiotic Resistance and Virulence in Gram-negative Intestinal bacteria. Extended-spectrum β-lactamase (ESBL)-producing Gram-negative bacteria (such as *Escherichia coli*) are resistant to both penicillins and cephalosporins. The responsible bacterial β-lactamases, *e.g.* TEM, SHV and CTX-M enzymes, can be typed by molecular techniques in order to track epidemics and monitor outbreaks.[90] PFGE is the current "gold standard" for molecular typing, but multilocus sequence typing (MLST) may offer advantages.[91] Real-time PCR combined with deep sequencing may increase speed and capacity for the identification of the antibiotic resistance genes.[92,93] *Klebsiella pneumoniae* and other bacteria containing carbapenemases (KPC) is a further extension of the antibiotic resistance problem.[94]

Although *E. coli* is prevalent in the normal intestinal flora, some strains are highly pathogenic and cause diarrhoea or systemic disease due to acquisition of virulence genes encoded by plasmid, chromosome or bacteriophage DNA. A large number of virulence factor genes separate pathogenic *E. coli* strains into enterotoxigenic (ETEC), enteroinvasive (EIEC), enteropathogenic (EPEC), enterohemorrhagic (EHEC) and enteroaggregative (EAEC).[95] The development of multiplex PCR and other molecular assays has been very useful for the survey of such virulence factors.[96–99]

12.6 CONCLUSIONS AND FUTURE DEVELOPMENT

Nucleic acid based methods have revolutionized routine laboratory diagnosis of infectious agents. In particular, the DNA amplification techniques have contributed to this by providing sufficient sensitivity for the direct examination of infected patient samples. As a result, the clinician can rely much more on the microbiology laboratory in the acute differential diagnosis of each patient. Nucleic acid based techniques have offered much quicker and reliable diagnosis of the individual patient, improved the tracing and monitoring of epidemic outbreaks and the discovery of emerging infectious agents. Nevertheless, there are many problems, such as contamination during nucleic acid amplification and the continuous biological selection and evolution in the microbiological world that without warning may cause a previously well-designed test to fail. The potential for improvement of nucleic acid based laboratory diagnostics is therefore vast. There is a need for back up-tests directed to different regions of the target nucleic acids. We presently see only the beginning of automation and reduction of reaction volumes. Costs severely restrict the repertoire of new tests in the clinical microbiological laboratory—even in developed countries. The future roles of nucleic acid amplification, next generation sequencing and novel technologies incorporating nanotechnology, microfluidics or biosensors remain to be determined. Great challenges endure in microbiology laboratories regarding logistics and quality control of diagnostic procedures. It is therefore evident that nucleic acid based techniques will continue to revolutionize the way we diagnose and treat infections.

REFERENCES

1. M. Oldstone, *Viruses, plagues, and history*, Oxford University Press, Oxford New York, 2010.
2. M. Karamanou, G. Panayiotakopoulos, G. Tsoucalas, A. A. Kousoulis and G. Androutsos, *Infez. Med.*, 2012, **20**, 58–62.
3. M. J. Espy, J. R. Uhl, L. M. Sloan, S. P. Buckwalter, M. F. Jones, E. A. Vetter, J. D. Yao, N. L. Wengenack, J. E. Rosenblatt, F. R. Cockerill, 3rd and T. F. Smith, *Clin. Microbiol. Rev.*, 2006, **19**, 165–256.
4. K. Rantakokko-Jalava and J. Jalava, *J. Clin. Microbiol.*, 2002, **40**, 4211–4217.
5. T. Schuurman, R. F. de Boer, A. M. Kooistra-Smid and A. A. van Zwet, *J. Clin. Microbiol.*, 2004, **42**, 734–740.
6. K. B. Barken, J. A. Haagensen and T. Tolker-Nielsen, *Clin. Chim. Acta*, 2007, **384**, 1–11.
7. M. A. Espy, H. Sandin, C. Carr, C. J. Hanson, M. D. Ward and R. H. Kraus, Jr., *Cytometry A*, 2006, **69**, 1132–1142.
8. T. Schuurman, A. van Breda, R. de Boer, M. Kooistra-Smid, M. Beld, P. Savelkoul and R. Boom, *J. Clin. Microbiol.*, 2005, **43**, 4616–4622.
9. B. Kaltenboeck and C. Wang, *Adv. Clin. Chem.*, 2005, **40**, 219–259.
10. P. J. Asiello and A. J. Baeumner, *Lab Chip*, 2011, **11**, 1420–1430.
11. J. Ikewaki, E. Ohtsuka, R. Kawano, M. Ogata, H. Kikuchi and M. Nasu, *J. Clin. Microbiol.*, 2003, **41**, 4382–4387.
12. J. L. Vaerman, P. Saussoy and I. Ingargiola, *J. Biol. Regul. Homeost. Agents*, 2004, **18**, 212–214.
13. R. G. Rutledge and C. Cote, *Nucleic Acids Res.*, 2003, **31**, e93.
14. J. C. Servoss and L. S. Friedman, *Clin. Liver Dis.*, 2004, **8**, 267–281.
15. J. Saldanha, N. Lelie and A. Heath, *Vox Sang.*, 1999, **76**, 149–158.
16. J. Saldanha, N. Lelie, M. W. Yu and A. Heath, *Vox Sang.*, 2002, **82**, 24–31.

17. J. Burrows, A. Nitsche, B. Bayly, E. Walker, G. Higgins and T. Kok, *BMC Microbiol.*, 2002, **2**, 12.
18. R. M. Ratcliff, G. Chang, T. Kok and T. P. Sloots, *Curr. Issues Mol. Biol.*, 2007, **9**, 87–102.
19. B. Stone, J. Burrows, S. Schepetiuk, G. Higgins, A. Hampson, R. Shaw and T. Kok, *J. Virol. Methods*, 2004, **117**, 103–112.
20. S. Leekha, C. L. Irish, S. K. Schneider, E. C. Fernholz, M. J. Espy, S. A. Cunningham, R. Patel, Y. J. Juhn, B. Pritt, T. F. Smith and P. Sampathkumar, *Diagn. Microbiol. Infect. Dis.*, 2013, **75**, 169–173.
21. K. H. Kalland, L. S. Havarstein and H. Sommerfelt, *Tidsskr. Nor. Laegeforen.*, 1987, **107**, 2510–2512, 2543.
22. K. H. Kalland and G. Haukenes, *Tidsskr. Nor. Laegeforen.*, 1999, **119**, 802–809.
23. K. H. Kalland, H. Myrmel and S. A. Nordbo, *Tidsskr. Nor. Laegeforen.*, 2005, **125**, 3110–3114.
24. M. Sogaard, H. Stender and H. C. Schonheyder, *J. Clin. Microbiol.*, 2005, **43**, 1947–1949.
25. F. C. Tenover, *Clin. Infect. Dis.*, 2007, **44**, 418–423.
26. N. R. Stone, R. L. Gorton, K. Barker, P. Ramnarain and C. C. Kibbler, *J. Clin. Microbiol.*, 2013.
27. M. Kodani, G. Yang, L. M. Conklin, T. C. Travis, C. G. Whitney, L. J. Anderson, S. J. Schrag, T. H. Taylor, Jr., B. W. Beall, R. F. Breiman, D. R. Feikin, M. K. Njenga, L. W. Mayer, M. S. Oberste, M. L. Tondella, J. M. Winchell, S. L. Lindstrom, D. D. Erdman and B. S. Fields, *J. Clin. Microbiol.*, 2011, **49**, 2175–2182.
28. S. Bensch and M. Akesson, *Mol. Ecol.*, 2005, **14**, 2899–2914.
29. F. Saghrouni, J. Ben Abdeljelil, J. Boukadida and M. Ben Said, *J. Appl. Microbiol.*, 2013, **114**, 1559–1574.
30. U. Vogel, R. Szczepanowski, H. Claus, S. Junemann, K. Prior and D. Harmsen, *J. Clin. Microbiol.*, 2012, **50**, 1889–1894.
31. R. A. Tapp, A. F. Roy, R. E. Corstvet and V. L. Wilson, *J. Vet. Diagn. Invest.*, 2001, **13**, 219–229.
32. D. A. Relman, T. M. Schmidt, R. P. MacDermott and S. Falkow, *N. Engl. J. Med.*, 1992, **327**, 293–301.
33. O. Kommedal, B. Karlsen and O. Saebo, *J. Clin. Microbiol.*, 2008, **46**, 3766–3771.
34. O. Kommedal, K. Kvello, R. Skjastad, N. Langeland and H. G. Wiker, *J. Clin. Microbiol.*, 2009, **47**, 3562–3568.
35. M. Al Masalma, F. Armougom, W. M. Scheld, H. Dufour, P. H. Roche, M. Drancourt and D. Raoult, *Clin. Infect. Dis.*, 2009, **48**, 1169–1178.
36. R. Urwin and M. C. Maiden, *Trends Microbiol.*, 2003, **11**, 479–487.
37. S. Kumar and M. R. Hammerschlag, *Clin. Infect. Dis.*, 2007, **44**, 568–576.
38. M. Khanna, J. Fan, K. Pehler-Harrington, C. Waters, P. Douglass, J. Stallock, S. Kehl and K. J. Henrickson, *J. Clin. Microbiol.*, 2005, **43**, 565–571.
39. D. R. Murdoch, *Clin. Infect. Dis.*, 2005, **41**, 1445–1447.

40. Y. R. Chan and A. Morris, *Curr. Opin. Infect. Dis.*, 2007, **20**, 157–164.
41. K. A. Beynon, S. A. Young, R. T. Laing, T. G. Harrison, T. P. Anderson and D. R. Murdoch, *Emerg. Infect. Dis.*, 2005, **11**, 639–641.
42. I. Lee and T. D. Barton, *Drugs*, 2007, **67**, 1411–1427.
43. L. C. Jennings, T. P. Anderson, A. M. Werno, K. A. Beynon and D. R. Murdoch, *Pediatr. Infect. Dis. J.*, 2004, **23**, 1003–1007.
44. C. Deffrasnes, M. E. Hamelin and G. Boivin, *Semin. Respir. Crit. Care Med.*, 2007, **28**, 213–221.
45. M. Petric, L. Comanor and C. A. Petti, *J Infect Dis*, 2006, **194**(Suppl 2), S98–S110.
46. M. D. Curran, J. S. Ellis, T. G. Wreghitt and M. C. Zambon, *J. Med. Microbiol.*, 2007, **56**, 1263–1267.
47. D. L. Suarez, A. Das and E. Ellis, *Avian Dis.*, 2007, **51**, 201–208.
48. J. K. Louie, J. K. Hacker, J. Mark, S. S. Gavali, S. Yagi, A. Espinosa, D. P. Schnurr, C. K. Cossen, E. R. Isaacson, C. A. Glaser, M. Fischer, A. L. Reingold and D. J. Vugia, *Emerg. Infect. Dis.*, 2004, **10**, 1143–1146.
49. C. Drosten, W. Preiser, S. Gunther, H. Schmitz and H. W. Doerr, *Trends Mol. Med.*, 2003, **9**, 325–327.
50. A. Endimiani, K. M. Hujer, A. M. Hujer, S. Kurz, M. R. Jacobs, D. S. Perlin and R. A. Bonomo, *Clin. Infect. Dis.*, 2011, **52**(Suppl 4), S373–383.
51. M. R. Couturier, T. Barney, G. Alger, W. C. Hymas, J. B. Stevenson, D. Hillyard and J. A. Daly, *J. Clin. Lab. Anal.*, 2013, **27**, 148–154.
52. S. P. Hammond, L. S. Gagne, S. R. Stock, F. M. Marty, R. S. Gelman, W. A. Marasco, M. A. Poritz and L. R. Baden, *J. Clin. Microbiol.*, 2012, **50**, 3216–3221.
53. M. Xu, X. Qin, M. L. Astion, J. C. Rutledge, J. Simpson, K. R. Jerome, J. A. Englund, D. M. Zerr, R. T. Migita, S. Rich, J. C. Childs, A. Cent and M. A. Del Beccaro, *Am. J. Clin. Pathol.*, 2013, **139**, 118–123.
54. S. Miller, M. Moayeri, C. Wright, L. Castro and M. Pandori, *J. Clin. Microbiol.*, 2010, **48**, 4684–4685.
55. G. Meyer-Rath, K. Schnippel, L. Long, W. MacLeod, I. Sanne, W. Stevens, S. Pillay, Y. Pillay and S. Rosen, *PLoS One*, 2012, **7**, e36966.
56. R. J. Whitley and J. W. Gnann, *Lancet*, 2002, **359**, 507–513.
57. R. L. Debiasi and K. L. Tyler, *Clin. Microbiol. Rev.*, 2004, **17**, 903–925, table of contents.
58. C. A. Glaser, S. Honarmand, L. J. Anderson, D. P. Schnurr, B. Forghani, C. K. Cossen, F. L. Schuster, L. J. Christie and J. H. Tureen, *Clin. Infect. Dis.*, 2006, **43**, 1565–1577.
59. J. K. Moller, *Methods Mol. Biol.*, 2012, **799**, 37–53.
60. R. Goodgame, *Curr. Infect. Dis. Rep.*, 2007, **9**, 102–109.
61. M. Hoehne and E. Schreier, *BMC Infect. Dis.*, 2006, **6**, 69.
62. W. Hymas, A. Atkinson, J. Stevenson and D. Hillyard, *J. Virol. Methods*, 2007, **142**, 10–14.
63. E. Royuela, A. Negredo and A. Sanchez-Fauquier, *J. Virol. Methods*, 2006, **133**, 14–19.

64. A. Stals, E. Mathijs, L. Baert, N. Botteldoorn, S. Denayer, A. Mauroy, A. Scipioni, G. Daube, K. Dierick, L. Herman, E. Van Coillie, E. Thiry and M. Uyttendaele, *Food Environ. Virol.*, 2012, **4**, 153–167.

65. H. Yan, T. A. Nguyen, T. G. Phan, S. Okitsu, Y. Li and H. Ushijima, *Kansenshogaku Zasshi*, 2004, **78**, 699–709.

66. M. Torres, L. Fraile, J. Echevarria, B. Hernandez Novoa and M. Ortiz, *Open Virol. J.*, 2012, **6**, 144–150.

67. H. M. Wood, R. Bolt and K. D. Hunter, *Expert Rev. Mol. Diagn.*, 2012, **12**, 663–665.

68. C. Rodriguez-Cerdeira, E. Sanchez-Blanco, A. Molares-Vila and A. Alba, *ISRN Obstet. Gynecol.*, 2012, **2012**, 581725.

69. I. J. Bakken and S. A. Nordbo, *Tidsskr. Nor. Laegeforen.*, 2007, **127**, 3202–3205.

70. V. Levy, C. S. Blackmore and J. D. Klausner, *Methods Mol. Biol.*, 2012, **903**, 407–418.

71. A. Cheng, Q. Qian and J. E. Kirby, *J. Clin. Microbiol.*, 2011, **49**, 1294–1300.

72. A. L. Harkins and E. Munson, *ISRN Obstet Gynecol*, 2011, **2011**, 279149.

73. S. A. Weinstein and B. G. Stiles, *Sex. Health*, 2011, **8**, 143–158.

74. S. R. Lee, J. M. Chung and Y. G. Kim, *J. Microbiol.*, 2007, **45**, 453–459.

75. M. Unemo, P. Olcen, I. Agne-Stadling, A. Feldt, M. Jurstrand, B. Herrmann, K. Persson, P. Nilsson, T. Ripa and H. Fredlund, *Euro. Surveill.*, 2007, **12**, E5–E6.

76. S. Yokoi, S. Maeda, Y. Kubota, M. Tamaki, K. Mizutani, M. Yasuda, S. Ito, M. Nakano, H. Ehara and T. Deguchi, *Clin. Infect. Dis.*, 2007, **45**, 866–871.

77. E. De Clercq, *Nat. Rev. Drug Discovey*, 2007, **6**, 1001–1018.

78. L. Menendez-Arias, *Antiviral Res.*, 2013.

79. S. Palmer, *Curr. Opin. HIV AIDS*, 2013, **8**, 87–92.

80. A. van der Zee, W. D. Hendriks, L. Roorda, J. M. Ossewaarde and J. Buitenwerf, *Am. J. Infect. Control*, 2013, **41**, 204–209.

81. K. Levi and K. J. Towner, *J. Clin. Microbiol.*, 2003, **41**, 3890–3892.

82. Y. Misawa, A. Yoshida, R. Saito, H. Yoshida, K. Okuzumi, N. Ito, M. Okada, K. Moriya and K. Koike, *J. Infect. Chemother.*, 2007, **13**, 134–140.

83. P. C. Appelbaum, *Int. J. Antimicrob. Agents*, 2007, **30**, 398–408.

84. G. M. Rossolini, E. Mantengoli, F. Montagnani and S. Pollini, *Curr. Opin. Microbiol.*, 2010, **13**, 582–588.

85. J. Davies, C. L. Gordon, S. Y. Tong, R. W. Baird and J. S. Davis, *J. Clin. Microbiol.*, 2012, **50**, 2056–2058.

86. D. H. Spencer, P. Sellenriek and C. A. Burnham, *Am. J. Clin. Pathol.*, 2011, **136**, 690–694.

87. B. A. Diep, H. A. Carleton, R. F. Chang, G. F. Sensabaugh and F. Perdreau-Remington, *J. Infect. Dis.*, 2006, **193**, 1495–1503.

88. J. A. McClure, J. M. Conly, V. Lau, S. Elsayed, T. Louie, W. Hutchins and K. Zhang, *J. Clin. Microbiol.*, 2006, **44**, 1141–1144.

89. R. R. McDonald, N. A. Antonishyn, T. Hansen, L. A. Snook, E. Nagle, M. R. Mulvey, P. N. Levett and G. B. Horsman, *J. Clin. Microbiol.*, 2005, **43**, 6147–6149.

90. S. N. Seiffert, M. Hilty, V. Perreten and A. Endimiani, *Drug Resist Updates*, 2013, **16**, 22–45.
91. L. L. Nemoy, M. Kotetishvili, J. Tigno, A. Keefer-Norris, A. D. Harris, E. N. Perencevich, J. A. Johnson, D. Torpey, A. Sulakvelidze, J. G. Morris, Jr. and O. C. Stine, *J. Clin. Microbiol.*, 2005, **43**, 1776–1781.
92. C. I. Birkett, H. A. Ludlam, N. Woodford, D. F. Brown, N. M. Brown, M. T. Roberts, N. Milner and M. D. Curran, *J. Med. Microbiol.*, 2007, **56**, 52–55.
93. T. Naas, C. Oxacelay and P. Nordmann, *Antimicrob. Agents Chemother.*, 2007, **51**, 223–230.
94. P. Nordmann, M. Gniadkowski, C. G. Giske, L. Poirel, N. Woodford and V. Miriagou, *Clin. Microbiol. Infect.*, 2012, **18**, 432–438.
95. M. M. Levine, *J. Infect. Dis.*, 1987, **155**, 377–389.
96. L. T. Brandal, B. A. Lindstedt, L. Aas, T. L. Stavnes, J. Lassen and G. Kapperud, *J. Microbiol. Methods*, 2007, **68**, 331–341.
97. S. L. Foley, A. M. Lynne and R. Nayak, *Infect., Genet. Evol.*, 2009, **9**, 430–440.
98. V. K. Sharma, *Mol. Cell. Probes*, 2006, **20**, 298–306.
99. J. R. Yang, F. T. Wu, J. L. Tsai, J. J. Mu, L. F. Lin, K. L. Chen, S. H. Kuo, C. S. Chiang and H. S. Wu, *J. Clin. Microbiol.*, 2007, **45**, 3620–3625.

CHAPTER 13

Molecular Biomarkers: Overview, Technologies, and Strategies

MUKESH VERMA,[a] DEBMALYA BARH,*[b] SANDEEP TIWARI[b,c] AND VASCO AC AZEVEDO[c]

[a] Epidemiology and Genomics Research Program, Division of Cancer Control and Population Sciences, National Cancer Institute (NCI), National Institutes of Health (NIH), 9609 Medical Center Drive, Rockville, MD 20850, USA; [b] Centre for Genomics and Applied Gene Technology, Institute of Integrative Omics and Applied Biotechnology (IIOAB), Nonakuri, Purba Medinipur, West Bengal-721172, India; [c] Departamento de Biologia Geral, Instituto de Ciências Biológicas (ICB), Universidade Federal de Minas Gerais, Pampulha, Belo Horizonte, Minas Gerais, Brazil
*Email: dr.barh@gmail.com

13.1 INTRODUCTION

Biomarkers can be specific cells, molecules, images, genes, gene products, enzymes, hormones, or receptors.[1] The term "biomarkers" is sometimes replaced by "actionable biomarkers" to describe biomarkers that can inform that clinical practice can be applied for disease diagnosis, prognosis, and treatment. Most of such actionable biomarkers exist in cancer although in other diseases progress has also been made in the last few years. Sometime entirely different kinds of biomarkers exist, for example, temperature is a biomarker of abnormal physiology of the body (fever) possibly due to infection, and blood pressure is a biomarker for stroke. To evaluate the current status of the biomarker field in disease diagnosis and therapy, we analyzed published literature starting from 1985 until April 2013 and summarized

Molecular Biology and Biotechnology, 6th Edition
Edited by Ralph Rapley and David Whitehouse
© The Royal Society of Chemistry 2015
Published by the Royal Society of Chemistry, www.rsc.org

Table 13.1 Publications of biomarker studies indicating investigator interest in the field.

Topic	Number of Publications
Biomarker	601401
Biomarker and diagnosis	373653
Biomarker and therapy	173756
Biomarker and cancer	215529
Biomarker and neurological disorders	46330
Biomarker and cardiovascular disorders	57744
Biomarker and metabolic disorders	521
Biomarker and immunologic disorders	73417
Biomarker and infectious agents related diseases	159
Biomarkers of viruses in diseases	11445
Biomarkers of bacteria in diseases	14538
Biomarker and breast cancer and epigenetics	45
Biomarker and breast cancer and methylation	381
Biomarker and breast cancer and histone	247
Biomarker and breast cancer and microRNA	209
Biomarker and breast cancer and proteomics	419
Biomarker and breast cancer and imaging	1094

The analysis was PubMed-based, and references up to April 2013 were considered.

in Table 13.1. The total number of publications in the biomarker field was 601 401. The number of publications in the biomarker diagnosis field was almost double those in therapeutics. This seems reasonable because more biomarkers exist which can be used for disease diagnosis but it takes time to find biomarkers for therapies. Metabolomics disease biomarkers are fewer compared to cardiovascular disease markers.[2] Another important point in such meta-analysis is the proper use of terms for the analysis. For example, when we used "infectious agent biomarker", the number of publications was much less than using the term "bacterial biomarkers" or "viral biomarker" (see Table 13.1). In some diseases, a single biomarker is sufficient to diagnose a disease whereas in other diseases multiple biomarkers are needed for diagnosis. The utility of multiple biomarkers is determined based on their combined sensitivity and specificity, which should be superior to a single marker. In this chapter biomarkers of immune diseases, cardiovascular diseases (CVDs), metabolic diseases, infectious diseases, neurological diseases, and cancer are discussed (see Tables 13.2–13.7).

13.2 OVERVIEW OF BIOMARKERS

Biomarkers are often used in clinics where they are derived from biological fluids that are easily available. Biomarkers show characteristic biological properties that can be detected and measured in parts of the body such as organs or body fluid (blood, urine, saliva, nipple aspirate, pleural lavage, pancreatic juice).[1,3] They represent the normal or disease state of the body and can be used to follow up treatment in the disease state. A number of assays have been developed to identify biomarkers. These assays include immunohistochemistry, imaging, gene constitution (amplification, mutation, and rearrangement), gene and protein expression analysis such as single gene or protein expression, methylation and histone profiling, miRNA polymorphism and miRNA profiling.[3-6] Biomarkers play a major role in medicinal biology.[4-7] In medicine, a biomarker generally refers to a specific protein concentration in blood that may in turn reflect the presence, progression or severity of a disease and guide the treatment.[8] Biomarkers have been used for more than half a century but their application has increased remarkably since the 21st century. A diseased state indicates a structural change in the proteins or enzymes. The physiological changes between the normal and diseased state is compared to look out for biomarkers. This is because the gene and protein expression profiles along with the metabolic expression profiles change in a diseased state. The researchers study the up-regulated and down-regulated genes, proteins and metabolites and understand the genetic patterns closely associated with a particular type of disease that leads to the discovery of a biomarker. Once identified, the biomarkers are validated in a large number of samples and possibly at two different laboratories. Microarrays—RNA, DNA, protein or antibody—play an important role in analyzing a biomarker. The comparison of plasma protein concentration levels in normal and diseased states is often used in biomarker analysis. For example, various new forms of glycoproteins are formed during glycosylation wherein polysaccharides or sugars are added to the polypeptides (proteins). Any abnormal concentration of glycoproteins can act as a biomarker for various diseases such as muscular dystrophy or acute chronic inflammation. In case of pancreatic cancer, RNAase-1 is used as a biomarker for disease diagnosis. An obvious alteration in the pattern of glycolysation of the enzyme was observed in the urine and blood serum in tumorous pancreatic cells. In molecular terms, a biomarker is the subset of markers that can be discovered using various omics or imaging technologies. We have selected biomarkers in cancer, cardiovascular disease, neurological disorders, infectious agents related diseases, metabolic diseases and immune diseases to cover wide-spectrum of diseases and up to date information about biomarkers associated with these diseases and disorders.

13.3 TYPES OF BIOMARKERS

13.3.1 Based on Utility

Physicians and scientist use various types of biomarkers to study human diseases either to track the progression of a disease, or to detect the effect of a drug. The use of biomarkers in the diagnosis of infections, genetic disorders and cancer is well known. The expansion of lab technology and growth of molecular biology increases the feasibility of biomarkers that are technically advanced. Biomarkers play a very important role in personalized medicine as they facilitate the combination of therapeutics with diagnostics. In contrast, big pharmaceutical and biotechnology companies use biomarkers as a drug discovery tool. Biomarkers are important in drug development as they help to determine the pharmacodynamic effects of the drug being developed and assess the safety and the efficacy of the drug. Safer drugs with better efficacy can be developed in a cost effective manner by using biomarkers.

Biomarkers include tools and technologies that can be used in prediction, progression and outcome (survival, recurrence, multiple diseases) of a disease. Biomarkers can be classified based on detection techniques (imaging and non-imaging), properties (molecular, cellular, and nuclear), and utility (diagnosis, prognosis and therapy). Because of the latest high-throughput omics technologies in genomics, epigenomics, proteomics, metabolomics, and transcrptomics, a large number of potential biomarkers have been identified that can be utilized in various ways for disease management and patient care.[2,3] Biomarkers can be classified as diagnostic, prognostic, and therapeutic biomarkers based on their utility.[8] Early detection of biomarkers are used to detect a particular disease in its early stage. Diagnostic biomarkers are those, which are used to identify the presence or absence of a disease or diagnose a disease. Predictive biomarkers are those that are present prior to an event occurring and that predict the outcome and the efficacy of the drug in a treatment. Prognostic biomarkers give valid information about the outcome without using a therapy. They are used to determine how such a disease may develop (etiology of the disease) in an individual and increase survival of the patient. Disease prognosis biomarkers are related to measures of a disease state. Efficacy biomarkers reflect the effects of a particular treatment using a specific drug. Since they correlate with the desired clinical outcomes, they can be used to obtain provisional regulatory approval of a drug. Surrogate biomarkers are used to measure the clinical outcomes. Toxicity biomarkers, as their name reflects, measure the toxicity of drugs or interventions. Target biomarkers reflect the presence of a specific drug target and indicate the drug target interaction and outcome. Lastly, pharmacodynamic biomarkers (pharmacogenomics and pharmacoepigenomics) belong to the category of biomarkers that are used in drug development for intervention and/or treatment.

13.3.2   Based on Diagnosis Approaches

Based on diagnostic approaches, biomarkers can also be classified as imaging biomarkers or non-imaging biomarkers. Imaging biomarkers are those biomarkers which are detectable by using imaging techniques, such as X-ray (mammograms), electrocardiogram (ECG), ultrasound imaging, computed tomography (CT) and CT scanning, magnetic resonance imaging (MRI) and quantum dots.[9] Imaging biomarkers give reproducible results and have tremendous potential in diagnosis as well as therapy and prognosis.

Non-imaging biomarkers can be molecular biomarkers with biophysical properties. They may include nucleic acid based biomarkers such as gene mutations or polymorphisms, alterations in copy number (copy number variance or CNVs), changes in mitochondrial genome, chromosomal aberrations, gene and protein expression profile, microRNA, metabolites, enzymes, antigens, hormones, *etc.*[1,4-7] Serum levels of IL-27, IL-29, IL-31, BALF were high in lung cancer patients.[10] Higher levels of miR-21 and lower levels of miR-451 and miR-485 were observed in lung cancer.[11]

If a combination of biomarkers (biomarker profiles) can detect a disease early, it provides an opportunity for developing intervention and therapeutic approaches.[3] For example, if biomarkers can discriminate pre-rheumatoid arthritis subjects from normal subjects, intervention and therapeutic approaches can be applied in high-risk individuals because such drugs exist which can treat rheumatoid arthritis. Biomarkers are also useful in detecting secondary cancers (recurrence).

13.3.3 Based on Therapy

Biomarkers are used to follow up therapy and their levels indicate the host's response to therapy. Due the recent development in genomics, it is possible to make therapy personalized. The identification of biomarkers represents a fundamental medical advance that can lead to an improved understanding of a disease, and holds the potential to define surrogate diagnostic and prognostic end points.

Understanding biomarkers in a complex system such as central nervous system (CNS) is extremely valuable to develop therapeutics in brain and neurological disorders. Brain parenchyma is a highly complex microvasuclar structure and it undergoes a variety of changes during tumor formation. Neo-angiogenesis starts in tumors and if anti-angiognesis therapy is planned for such brain tumors, biomarkers should be known which can be used to follow therapy. Some biomarkers in this category are vascular endothelial growth factor (VEGF), phosphatidylinositol glycan biosynthesis class F protein (PIGF), angiopoitin-1, and integrin.[12] In Alzheimer's disease, glutamate levels are increased (in the hippocampus region of the brain) after treatment with galantamine.[13] Therefore, glutamate can be considered a therapeutic biomarker of Alzheimer's disease. Galantamine is a cholinesterase inhibitor that is generally used in the treatment of this disease.[14]

An ideal biomarker should indicate whether a treatment is non-toxic and effective. Molecular biomarkers are required to predict the likelihood of an individual tumor's responsiveness or of toxicity in normal organs and to advise optimized treatments with improved efficacy at reduced side effects for each cancer patient. Biomarkers with prognostic value concerning treatment response and patient survival can then be used as targets to develop optimized drugs. For example: (i) chemo-selective treatment of tumors with 9p21 deletion by L-alanosine; (ii) treatment of multidrug-resistant P-glycoprotein-expressing tumor cells by non-cross-resistant natural products or by inhibitors of P-glycoprotein to overcome multidrug resistance; and (iii) natural products that inhibit the epidermal growth factor receptor (EGFR) in EGFR-over expressing tumor cells.[15]

In cancer cells, cyclic AMP dependent protein kinase (PKA) is secreted into the conditioned medium. This PKA, designated as extracellular protein kinase A (ECPKA), is markedly up-regulated in the sera of patients with cancer. The currently available tumor biomarkers are based on the antigen determination method and lack specificity and sensitivity. An ECPKA auto-antibody detection method for a universal biomarker that detects cancer of various cell types has been reported. The receiver-operating characteristic (ROC) plot showed that autoantibody enzyme immunoassay exhibited 90% sensitivity and 88% specificity, whereas the enzymatic assay exhibited 83% sensitivity and 80% specificity. These results show that the autoantibody method distinguished between patients with cancer and controls better than the antigen method could. Serum biomarker measurement in body fluid immuno assays has been the most widely used approach, generally of established tumor-associated markers as carcinoembryonic antigen (CEA), alpha feto protein (AFP), human chorionic gonadotropin (hCG), prostate specific antigen (PSA), and carcinoembryonic antigen 125 (CA125). These biomarkers have low specificity and sensitivity; therefore, they are either not used in screening and diagnosis or used in combination with other biomarkers. The limitations of the presently available serum tumor biomarkers, based on the antigen determination method, indicate the need for other means of screening.

Plasma and urinary concentrations of two members of the vascular endothelial growth factor (VEGF) family and their receptors as potential response and toxicity biomarkers of bevacizumab with neoadjuvant chemoradiation in patients with localized rectal cancer were evaluated by different groups of investigators.[16] Of all biomarkers, pretreatment plasma sVEGFR-1—an endogenous blocker of VEGF—and PlGF—and a factor linked with vascular normalization—were associated with both primary tumor regression and the development of adverse events after neoadjuvant bevacizumab and chemoradiation. Plasma sVEGFR-1 should be further evaluated to establish as a potential biomarker to stratify patients in future studies of bevacizumab and/or cytotoxics in the neoadjuvant setting.

13.3.4 Imaging Biomarkers and Non-Imaging Biomarkers

Mammography is the process of using low-energy X-rays to examine human breast cancer and is used as a screening and diagnostic tool. Mammography is applied in clinic for breast cancer screening and has helped in reducing breast cancer mortality.[17] We should, however, keep in mind that the radiation exposure associated with mammography has a potential health risk.[18,19]

Imaging is an enabling scientific discipline combining advanced technology and complex computational and analytic methods to provide unique ability to extract spatially and temporally defined information from humans.[20,21] It allows us to investigate intact biological system (without isolating samples or taking biopsies) across the spectrum from sub-cellular to macroscopic and from discovery to clinical decision making. By this technology early breast cancer is detected *via* characteristic masses and/or microcalcification. Thus, mammography is considered as a non-invasive biomarker of cancer diagnosis. For the average woman between the age of 50 to 74 years, mammography is recommended every two years. This helps in avoiding unnecessary surgery, treatment, and anxiety. On a cautionary note, mammography has a false-negative rate of approximately 10% due to dense tissues obscuring the cancer and also due to the fact that the appearance of cancer on the mammogram has a large overlap with the appearance of normal tissues.[22] Quantum dots technology (nanotechnology) is also based on imaging and has been successfully used for cancer diagnosis.[23]

Another technology called positron emission tomography (PET) scan was used to evaluate the treatment response in breast cancer patients.[24] Although reasonable success could be achieved, the main problem with imaging technologies (along with health hazards of multiple exposures) is that tumor heterogeneity interferes with interpretation of results and a combination of other biomarkers and patient related information is needed to infer any clinical value. It is re-emphasized here that multiple biomarkers should be used to achieve high sensitivity and specificity (discussed below in other sections of this article).

13.3.5 Invasive and Non-invasive Biomarkers

Tissues are the best source of material to assay early detection cancer biomarkers because they represent true expression of biomarkers during disease development.[9,10] However, tissue collection is a noninvasive procedure and it is difficult to get healthy tissue for comparison. Therefore, such biomarkers are preferred which can be assayed in samples collected non-invasively.[10] Biofluids (urine, blood, sputum) and exfoliated cells are good examples of noninvasive source of biomarkers for early diagnosis of the disease. After identifying biomarkers, the assay and the biomarker have to be approved by the Food and Drug Administration (FDA) so that these biomarkers can be assayed in clinical samples.[25] The FDA has provided guidelines in this direction (www.fda.gov). If biomarkers, assays or devices are planned for clinical use in patient samples, they should be reviewed by the FDA's Center for Devices and Radiological Health (CDRH) for their ability to analytically measure the biomarker.[26] Biomarkers and devices for quantification are expected to yield equivalent results. Biomarkers should have passed analytical and clinical validation tests specified by the FDA. Analytical validity in this context is defined as the ability of an assay to accurately and reliably measure the analyte in the laboratory as well as in the clinical sample. Clinical validation requires the detection or prediction of the associated disease in specimens from targeted patient. Biomarker qualification by the FDA enables collaboration among stakeholders, reduces costs for individual stakeholders and provides biomarkers that are useful for the general public and private parties.

Sometimes biomarkers fail to show reasonable sensitivity and specificity when validation is conducted.[27] Although noninvasive biomarkers are the best for breast cancer detection but when these biomarkers are validated in large number of samples, some of them do not show reasonable sensitivity and specificity.[28] The traditional treatment options for breast cancer are radiation, chemicals, and surgery (lumpectomy, quadendroctemy, mastectomy). Surgery is usually combined with adjuvant therapy (hormonal and/or chemical therapy). Chemicals used for therapy have considerable toxicity and hormonal treatment also have long lasting adverse effects. Surviving patients generally have poor quality of life. Furthermore, resistance to chemotherapy is another problem observed in breast cancer patients.[29,30] Pharmacogenomics is an area of research which may provide some useful information in these cases.[31] By applying diagnostic tests and knowing the genetic background of an individual, personalized treatments are possible.

Noninvasive biomarkers may also help in guiding the type of therapy that is more beneficial for patients. For example, in cases of breast cancer, women with ductal carcinoma *in situ* (DCIS), the treatment is by tamoxifen instead of aromatase inhibitor.[32] However, in early invasive stage, as judged by a panel of biomarkers, aromatase inhibitor proved better for treatment than tamoxiphen. The use of hormonal therapy changed with the age of the patient and tumor characteristics. Most of these characteristics correlated better with early stage than DCIS. This research led to the development of prevention strategies. Now endocrine therapy is used for preventing new primary breast cancers and invasive recurrence for women with DCIS or early invasive breast cancer. The dose used was higher at the early stage but decreased with the age of the patient.

Epigenomic biomarkers have enormous potential for clinical implication in cancer diagnosis and prognosis.[3] Because of the availability of genome-wide methylation, histone, and miRNA analysis technologies, and our rapidly accumulating knowledge regarding epigenome, the translation of findings discussed in this article may be possible in near future. Epigenetic biomarkers may also help in identifying patients who will benefit from the therapy and will not develop resistance to drugs and ultimately increase survival.[33] Recently developed drugs for disease treatment are based on specific pathways and may be useful for those individuals where those pathways are altered. This approach can be designed for personalized medicine and precision medicine. Epigenetic biomarkers may help in such approaches. All potential treatment biomarkers for diseases discussed in this chapter have by no means been exhausted, and it is expected that additional high penetrance biomarkers will be identified.

13.4 OMICS APPROACHES IN BIOMARKER DISCOVERY

Omics is an emerging and exciting area in the field of science and medicine.[4,5,9,34] In the post-genome era, efforts are focused on biomarker discovery and the early diagnosis of different diseases through the application of various omics technologies.[3] Numerous promising developments have been elucidated using omics (transcriptomics, proteomics, metabonomics/metabolomics, peptidomics, glycomics, phosphoproteomics or lipidomics on tissue samples and body fluids) in different diseases.[32,34] The development of high-throughput technologies that permit the solution of deciphering a disease from higher dimensionality may provide a knowledge base, which changes the face of the disease biology, etiology, and therapeutics. The omics technology that has driven these new areas of research consists of DNA and protein microarrays, mass spectrometry and a number of other instruments that enable high-throughput analyses at relatively low cost.[1,34] High-throughput omic technologies are being established and validated to create better predictive models for diagnosis, prognosis and therapy, to identify and characterize key signaling networks and to find new targets for drug development.[34]

13.4.1 Immune Diseases

Various applications of omics technology include studying changes in tissue specific protein expression in normal and disease samples, evaluating changes in immune response of the proteins in disease states, translating results from the laboratory to bed side delivery of patient care, and developing an artificial neural network (ANN) to distinguish high risk individuals from normal subjects. Interferons and cytokines were identified as a big group of biomarkers in immune diseases.[35] Interferons belong to three families: I, II, and III, with multiple subtypes. In infections and chronic inflammatory diseases, interferons and their stimulating genes could be utilized to follow up outcome and survival. Multiple sclerosis is an immune disease with the characteristic feature of deregulated T lymphocyte apoptosis.[36] Biomarkers identified during MS development are CASP8AP2, IL-23, CD36, ITGAL, OLR1, RNASEL, RTN4RL2, and THBS1.[36] Selected biomarkers are shown in Table 13.2.

13.4.2 Cardiovascular Diseases (CVDs)

Cardiovascular diseases may arise due to psychosocial stress.[37] CVD biomarkers, adhesion and proinflammatory molecules (IL-6, other cytokines, C-reactive proteins, and fibrinogen), and pathogens were evaluated to assess their contribution in socioeconomic position (SEP) and CVDs.[37] Selected biomarkers are shown in Table 13.3.

13.4.3 Metabolic Diseases, Metabolomics and Proteomics

A multitude of factors should be considered in selecting the specific technology to be adopted for metabolomics studies.[2] Two major technologies,

Table 13.2 Biomarkers and their characteristics in immune diseases.

Disease	Biomarker	Characteristics
Autoimmune disease (rheumatoid arthritis)	Antigen arrays	Can be used for disease stratification and planning treatment strategies.[73]
Autoimmune disease (rheumatoid arthritis, systemic and cutaneous lupus erythematosus, SLE)	Annexin-I	Annexin-1 exerts its anti-inflammatory effect by suppressing the generation of inflammatory mediators and anti-annexin I antibodies are present in rheumatoid arthritis and SLE patients.[74]
Autoimmune disease (systemic sclerosis, systemic and cutaneous lupus erythematosus, SLE), polymyositis (PM), dermatomyositis (DM), multiple sclerosis (MS)	Prolactin (PRL), ferritin, vitamin D, tumor marker tissue polypeptide antigen (TPA)	High levels of PRL, ferritin, vitamin D, and TPA were observed in SLE, PM, DM, and MS although levels of these markers were different in different diseases.[76]
Rheumatoid arthritis (RA)	IL-17 and Wnt/beta katenin	In rheumatoid arthritis (RA) bone loss occurs.[99] Transcriptomics approaches were applied to identify genes and pathways that contributed to bone loss in RA patients. Results indicated IL-17 and Wnt/beta katenin as transcriptional biomarkers of RA.[100]
Multiple sclerosis (MS)	CASP8AP2, IL-23, CD36, ITGAl, OLR1, RNASEL, RTN4RL2, and THBS1	In infections and chronic inflammatory diseases, interferons and their stimulating genes can be utilized to follow up outcome and survival. Multiple sclerosis is an immune disease with characteristic feature of deregulated T lymphocyte apoptosis. Biomarkers identified during MS development were CASP8AP2, IL-23, CD36, ITGAL, OLR1, RNASEL, RTN4RL2, and THBS1.[36]
Asthma	Nitric oxide	Measurements of fractional excretion of nitric oxide can be used as a biomarker for diagnosing asthma.[161]
Rheumatoid arthritis (RA)	IL-17 and Wnt/beta katenin	In rheumatoid arthritis (RA) bone loss occurs. Transcriptomics approaches were applied to identify genes and pathways, which contributed to bone loss in RA patients. Results indicated IL-17 and Wnt/beta katenin as transcriptional biomarkers of RA.[100]

Autism	Increased number of mast cells and serotonin in urine	High number of mast cells and serotonin in urine are observed in autism patients. Mast cells produce inflammatory cytokines in large amounts in autism patients and the possibility of autoimmunity has been proposed.[116]
Multiple sclerosis (MS)	HLA DRB1-1*5 haplotype, SHP-1 hypermethylation	MS is an inflammatory disease in which the fatty myelin sheaths around the axons of the brain and spinal cord are damaged. Female specific MS is associated with HLA DRB1-1*5 haplotype.[162] SHP-1 hypermethylation is a biomarker for multiple sclerosis.[144]
Asthma	Arginine kinase, sarcoplasmic calcium binding proteins, and tropomycin	Proteomics approaches identified allergenic proteins, based on their reactivity to patient's sera, using tandem mass spectrometry.[160] The most significant allergens identified were arginine kinase, sarcoplasmic calcium binding proteins, and tropomycin that can be used for screening.
Rheumatoid arthritis (RA)	Aggrecan fragments, C-propeptide of type II collagen	In rheumatoid arthritis (RA) inflammation of joint synovium is persistent, which leads to erosion of cartilage and bone. A biomarker, aggrecan fragments, was identified which was detected in synovial fluid (SF). Levels of aggregant fragments (structural components of cartilage) were higher in RA patients than in healthy individuals. Another biomarker was C-propeptide of type II collagen and its levels were directly proportional to the rate of collagen synthesis. Cartilage oligomeric matrix protein (COMP) is also a component of cartilage and it can be measured in serum and SF.
Multiple sclerosis (MS)	Neurofilament light protein (NFL), Glial fibrillary acidic protein (GFAP)	MS is associated with autoimmune-mediated inflammation of the central nervous system and may lead to demyelination and axonal damage. Biomarker neurofilament light protein (NFL) is a cytoskeleton component in large myelinated axon and it is released into the cerebrospinal fluid (CSF) and can be used to determine the level of axonal damage. Glial fibrillary acidic protein (GFAP) is another biomarker for MS.

Table 13.2 (*Continued*)

Disease	Biomarker	Characteristics
Rheumatoid arthritis and systemic and cutaneous lupus erythematosus (SLE)	Annexin I, II and V	Annexin-I exerted its anti-inflammatory effect by suppressing the generation of inflammatory mediators and anti-annexin I antibodies were present in rheumatoid arthritis and systemic and cutaneous lupus erythematosus (SLE) patients.[74] Annexin II and V were involved in coagulation cascade because they had affinity towards phospholipids.
Autoimmune disease (systemic sclerosis, systemic and cutaneous lupus erythematosus, SLE), polymyositis (PM), dermatomyositis (DM), multiple sclerosis (MS)	Prolactin (PRL), ferritin, vitamin D	Prolactin (PRL), ferritin, vitamin D, tumor biomarker tissue polypeptide antigen (TPA) levels were higher in patients with autoimmune disease (systemic sclerosis, systemic and cutaneous lupus erythematosus, SLE), polymyositis (PM), dermatomyositis (DM), multiple sclerosis (MS).[76] Granin family is a group of acidic proteins present in the secretory granules of a wide variety of endocrine, neuronal, and neuroendocrine cells. Few important granins, chromogranin A and B, secretogrannin II, HISL-19 antigen, NESP55, ProSAAS should be studied further for their clinical implication.[77]
Autism	Serotonin, inflammatory cytokines	High number of mast cells and serotonin in urine were observed in autism patients. Mast cells produce inflammatory cytokines in large amount in autism patients and the possibility of autoimmunity was proposed.[110]

| Asthma | Glutathione S transferase M1 (GSTM1) | In one meta-analysis glutathione S transferase M1 (GSTM1) biomarker was identified which helped in screening a cohort of children who were at high risk of developing asthma.[163] In a population-based study C-reactive protein, fibrinogen, and interleukin-6 were found useful biomarkers for screening.[164] |
| RA | CTX-1, CTX-II, C-telopeptide I and II | Biomarkers of bone and cartilage turnover are collagen C-telopeptides I and II, which are predictors of structural damage in RA patients. C-terminal cross-linked telopeptide of type I collagen (CTX-I) could be measured in serum or urine after it was released during bone resorption. Bone erosions and osteoporosis both occurred as a consequence of RA and CTX-I levels could be used for prognosis. On the other hand, increased CTX-II levels were associated with rapid progression of joint damage. For clinical implication, investigators of these studies recommended that new prognostic biomarkers should supply information beyond that provided by risk factors in the prediction of disease outcome.[165] |

Note: For the same disease more than one row are mentioned because biomarkers shown in column 2 were identified in different studies.

Table 13.3 Biomarkers and Their Characteristics in Cardiovascular Disease (CVD).

Disease	Biomarker	Characteristics
Cardiovascular disease (CVD)	C-reactive protein	In a risk prediction model of cardiovascular disease (CVD) C-reactive protein levels were implemented.[8]
Cardiovascular disease (CVD)	Blood pressure	Biomarkers studies of CVD prediction in elderly indicate that mortality and cardiovascular events are dependent on low peripheral pulse pressure not on high blood pressure.[166]
Cardiovascular disease (CVD)	Lipoprotein associated phospholipase A2 and antiphosphorylcholine IgM	In cardiovascular disease (CVD), the relative contribution of genetic and environmental effect was studied on two inflammatory biomarkers, lipoprotein associated phospholipase A2 and antiphosphorylcholine IgM, in a Swedish population.[167]
Cardiovascular disease (CVD)	TNF alpha, IL-6, plasma vitamin E concentrations, total and LDL cholesterol, and antioxidant profiles	Biomarkers discussed in the diagnostic section can be used for follow up of the treatment. Dietary intervention of CVD by fish oil (salmon, herring, and pompus) and other nutrients has been demonstrated in a number of studies. Some of the participants had higher levels of triacylglycerolaemia. Biomarkers TNF alpha and IL-6 were reduced and level of adiponectin increased in the treated arm. Thus TNF alpha, IL-6, and adiponectin were used as therapeutic biomarkers. In another study, argon oil supplement reduced plasma levels of lipids and antioxidant status. Therapeutic biomarkers used in this study were plasma vitamin E concentrations, total and LDL cholesterol, and antioxidant profiles.[168]
Cardiovascular disease (CVD)	Atherosclerosis	A very well characterized biomarker of CVD.
Cardiovascular disease (CVD)	IL-6, other cytokines, C-reactive proteins, and fibrinogen	CVD biomarkers, adhesion and proinflammatory molecules (IL-6, other cytokines, C-reactive proteins, and fibrinogen), and pathogens were evaluated to assess their contribution in socioeconomic position (SEP) and CVD.[37]
Atherosclerosis	Osteoprotegerin (OPG)	Osteoprotegerin (OPG) is such a marker which is independently associated not only with risk factors of atherosclerosis but also with subclinical peripheral atherosclerosis and clinical atherosclerosis and is recommended as a prognostic biomarker for ischemic heart disease and ischemic stroke.[78]

Disease	Biomarker	Description
Hypertension	Methylated ADD1 gene	In hypertension, global methylation profiling and individual gene methylation status (ADD1 gene) have been used for diagnosis and outcome.[146]
Atherosclerosis	Hypermethylation of monoamine oxidae A (MAOA)	Atherosclerosis can be diagnosed when other biomarkers are combined with epigenetic biomarkers, such as hypermethylation of monoamine oxide A (MAOA).[145]
Atherosclerosis (early diagnosis)	Carotid intima media thickness (IMT), circulating oxidized low density lipoprotein (LDL), and flow-mediated dialation (FMD)	For early diagnosis of carotid atherosclerosis for which obesity is the risk factor, the early biomarkers are carotid intima media thickness (IMT) and circulating oxidized low-density lipoprotein.[169] To evaluate endothelium status, flow-mediated dialation (FMD) and IMT have been used as early markers of atherosclerosis in patients with nonalcoholic fatty liver disease (NAFLD).[170]
Atherosclerosis	CARD8, Ephs, ephrins	Atherosclerosis is a chronic inflammatory disease of the vessel wall. The gene CARD8, which codes proteins involved in innate immunity in atherosclerosis patients, is an excellent biomarker for atherosclerosis. Inflammatory markers also over expressed in this disease. Inflammmosomes produce interleukin 1 beta in response to cholesterol crystal accumulation in macrophages. Ephs and ephrins also were proposed as biomarkers of atherosclerosis.[108]
Atherosclerosis	Hypermethylation of monoamine oxidae A (MAOA) and ADD1	Atherosclerosis can be diagnosed when other biomarkers are combined with epigenetic biomarkers, such as hypermethylation of monoamine oxidae A (MAOA). Along with gene-specific methylation biomarkers, global DNA methylation biomarkers were identified for atherosclerosis.[145] Similarly in other CVDs, such as hypertension, global methylation profiling and individual gene methylation status (ADD1 gene) were used for diagnosis and outcome.[146,147]

Note: For the same disease more than one row is mentioned because biomarkers shown in column 2 were identified in different studies.

mass spectrometry (MS) and nuclear magnetic resonance spectroscopy (NMR), are generally considered as they can measure hundreds to thousands of unique chemical entities. MS is highly sensitive and has the capacity to detect metabolites with concentrations in the picomole range and above, requires small biospecimen volumes, enables metabolites to be individually identified and quantified, and is well-suitable for use in a high-throughput mode. However, MS requires expensive consumables, has relatively lower analytical reproducibility, poorly represents highly polar metabolites when using standard chromatography protocols, and requires more complex software and algorithms for routine data analysis.[2] In contrast, NMR allows for the comprehensive generation of metabolite profiles by a single nondestructive method, is fully automated with high-throughput capacity, inherently quantitative, and highly suitable for metabolite structure elucidation, and has very high analytical reproducibility with a well-established mathematical and statistical toolbox. The disadvantages to using NMR include its relative insensitivity in detecting metabolites with concentrations in the micromole range and below. Furthermore, its validity is dependent on the quality of sample collection and handling, as well as the available metadata. Before applying either of these technologies to population-based studies, investigators must consider the advantages and disadvantages in relation to the design and aims of the study.

Quantitative proteomics can be used for the identification of disease biomarkers that could be used for early detection, serve as therapeutic targets, or monitor response to treatment. Several quantitative proteomics tools are currently available to study differential expression of proteins in a variety of types of samples. Two-dimensional gel electrophoresis (2-DE), which was classically used for proteomic profiling, has been coupled to fluorescence labeling for differential proteomics. Isotope labeling methods such as stable isotope labeling with amino acids in cell culture (SILAC), isotope-coded affinity tagging (ICAT), isobaric tags for relative and absolute quantitation (iTRAQ), and ^{18}O labeling have all been used in quantitative approaches for identification of biomarkers. In addition, heavy isotope labeled peptides can be used to obtain absolute quantitative data. Label-free methods for quantitative proteomics, which have the potential of replacing isotope-labeling strategies, are becoming popular. Other emerging technologies such as protein microarrays have the potential for providing additional opportunities for biomarker identification. Selected biomarkers are shown in Table 13.4.

13.4.4 Infectious Diseases

Bacterial infection occurs in pneumonia. Community acquired pneumonia is very common and for proper diagnosis and treatment the other biomarker which showed promise was procalcitonin.[38] C-reactive protein and inflammatory biomarkers could be used in combination with procalcitonin to make intelligent clinical decisions. Selected biomarkers are shown in Table 13.5.

Table 13.4   Biomarkers and their characteristics in metabolic diseases.

Disease	Biomarker	Characteristics
Diabetes	Alanine amino transferase (ALT), gamma glutamyl transferase (GGT), triglyceride, plasminogen activator inhibitor (PAI-1) antigen, ferritin, C-reactive protein (CRP), sex-hormone binding globulin (SHBG)	Biomarkers of risk for diabetes.
Diabetes	IL-6	Inflammation marker of diabetes (genetic studies on association of IL-6 with diabetes have not been completed yet).
Diabetes	Adiponectin	Some groups of investigators consider adiponectin as an excellent biomarker of diabetes pathogenesis.[171]
Diabetes	Insulin resistance (IR) and blood glucose levels	The most studies biomarkers which are used in clinic routinely.
Diabetes	Hemoglobin A1c	For the management of diabetes hemoglobin A1c (HbA1c) is used which is considered as a reliable indicator of glycemic control. In most of the clinical studies in diabetes, HBA1c biomarker is used to determine the glucose control.
Diabetes	1,5-anhydroglucitol (1,5 A)	Circulating biomarker used to measure hyperglycemic condition
Metabolic Diseases	Alpha-hydroxybutyrate (αHB) and linolyl-glycerophosphocholine (L-GPC)	Alpha-hydroxybutyrate (αHB) and linolyl-glycerophosphocholine (L-GPC) were identified as biomarkers of insulin resistance and glucose intolerance in a large population study of more than 1000 participants from the Relationship between Insulin sensitivity and Cardiovascular Disease (RISC) study.[2]
Chronic kidney disease	Proteomic profiling	Differentially expressed peaks in the spectra are future biomarkers for chronic kidney disease.[172]
Diabetes	Proteomic profiling	Isobaric tags for relative and absolute quantitation (iTRAQ) in combination with two-dimensional gel electrophoresis technologies identified a group of proteins, which were associated with diabetes, pancreatitis, and/or pancreatic cancer.[80]

Table 13.4   (*Continued*)

Disease	Biomarker	Characteristics
Diabetes	RCAN1, perilipin A and G0/G1 switch gene, G0S2	Glucose response gene, RCAN1, was over expressed whereas perilipin A and G0/G1 switch gene, G0S2, were under expressed in diabetes-2.[102,103]
Diabetes (early diagnosis)	Dysglycemia, alphahydroxy-butyrate, linoleyolglycerophos-phocholine, advanced glycation end products (AGE), albumin excretion rate (AR) and SNPS in epidermal growth factor gene intron 2	A number of markers have been described for early diagnosis of diabetes. Few of them are dysglycemia, alphahydroxy-butyrate, linoleyolglycerophos-phocholine, advanced glycation end products (AGE), albumin excretion rate (AR) and SNPS in epidermal growth factor gene intron 2.[2,173,174]

Note: For the same disease more than one row are mentioned because biomarkers shown in column 2 were identified in different studies.

13.4.5   Neurological Diseases

For the diagnosis of seizures, carbohydrate deficient transferrin (CDT) was evaluated and results indicated that this biomarker alone could not diagnose the disease but contribute in better diagnosis when combined with other biomarkers such as GGT, ASAT, ALAT, and ASAT/ALAT ratio.[39] A novel missense mutation at position 134 T to A resulting in amino acid change at codon V45E was identified as a biomarker for Norrie disease (ND) which is a rare X-linked disorder characterized by congenital blindness and sometime mental retardation, and deafness.[40] A novel mutation in ATP7B gene was used as a diagnostic biomarker for neurological impairment in Wilson's disease.[41] Selected biomarkers are shown in Table 13.6.

13.4.6   Cancer

Proteomic profiling of samples from cancer and healthy individuals has identified disease-associated profile, especially in Matrix Associated Laser Ionization/Desorption Time-of-Flight Mass Spectrometry (MALDI TOF MS) in colorectal and other cancers which can be used for diagnosis either alone or in combination with other biomarkers.[42] Methylation biomarkers were useful in diagnosis of almost all major cancers.[7,9,43] Defects in the mismatch repair (MMR) genes such as MLH1, MSH2, or MSH6 or methylation of the MLH1 promoter led to erroneous replication of segments of simple nucleotide repeats which contributed to microsatellite instability (MSI). MSI increases the risk of cancer occurrence. However, MSI is uncommon in cancers of the breast as compared to some other cancers such as colorectal carcinoma (CRC). It was observed that prognosis in MSI-positive breast cancer patients was worse than that of patients with MSI-negative tumors. Selected biomarkers are shown in Table 13.7.

Table 13.5 Biomarkers and their characteristics in infectious diseases.

Disease	Biomarker	Characteristics
Infectious diarrhea	Calprotectin	In cases of infectious diarrheal, fecal calprotectin is a good prediction marker.
Fungal infection	1,3-beta-D-glucan (BG)	1,3-beta-D-glucan (BG) can be used as a biomarker in invasive fungal infections (especially infections involving *Candida*) in patients undergoing treatment of candedemia with anidulafungin
Community acquired pneumonia (CAP)	Procalcitonin, C-reactive protein, inflammatory cytokines, and bacterial infection	Bacterial infection occurs in pneumonia. Community acquired pneumonia (CAP) is very common and for proper diagnosis and treatment the other biomarker which showed promise was procalcitonin.[38] C-reactive protein and inflammatory biomarkers can be used in combination with procalcitonin to make intelligent clinical decisions.
Ulcer diseases and gastritis (and gastric cancer)	*H. pylori*	*H. pylori* is the most common chronic bacterial infection in humans. This bacteria is involved not only in gastric cancer but in ulcer diseases and gastritis also. Its synergistic gastrotoxic interaction with non-steroidal anti-inflammatory drugs, and association with atherosclerotic events is a matter of concern.
Diarrhea in AIDS patients	Tubuloreticulin inclusions (TRIs)	Diarrhea in AIDS patients was treated with specific medications and therapeutic response was measured by levels of tubuloreticulin inclusions (TRIs).
Pneumonia	Procalcitonin (PCT)	For pneumonia therapy on more than 100 patients, biomarker procalcitonin (PCT) levels were very useful and this biomarker has been recommended for future prognosis. Sometime detecting bacteria alone is not sufficient to design therapy.[175]
HIV/AIDS	CRC5-delta32 mutation	In one study conducted in Georgia, where the prevalence of HIV/AIDS is high, polymorphism of CCR5 gene was studied.[50] More than 100 subjects were enrolled in this study. Results identified CRC5-delta32 mutation as a marker of the disease in this population.

Table 13.5 (*Continued*)

Disease	Biomarker	Characteristics
HIV/AIDS	Mitochondrial DNA mutations	Mitochondrial DNA mutations are also biomarkers for AIDS, especially in those individuals who are undergoing therapy.[51]
Liver cancer (HCC)	p16 and c-myc hypermethylation	In early liver cancer, where risk factors are infection by HBV and HCV, clustered DNA methylation changes in polycomb repressor target genes were observed. P16 and c-myc hypermethylation also suggested initiation of hepatocellular carcinoma.[127,128]
Tuberculosis	Vitamin D receptor (VDR) methylation	In tuberculosis, vitamin D receptor (VDR) methylation indicated high risk of developing the disease.[149] VDR gene encodes a transcription factor that alters calcium homeostasis and immune function.
Tuberculosis (early diagnosis)	Region-of-Difference-1 (RD-1) gene product and sputum cytokine levels	Region-of-Difference-1 (RD-1) gene product and sputum cytokine levels are considered a biomarker for early detection of tuberculosis.[176,177]
Hepatocellular carcinoma (early diagnosis)	HSP70, CAP2, glypican 3 and glutamine synthase	Hepatocellular carcinoma (HCC) involves infectious agents and the early diagnostic markers for HCC are HSP70, CAP2, glypican 3 and glutamine synthase.[178] These markers were identified based on gene expression analysis of HCC samples.
Hepatocellular carcinoma (HCC)	HBV, HCV, methylation of polycomb repressor, hypermethylation of p16 and c-myc,	In early liver cancer, where risk factors are infection by HBV and HCV, clustered DNA methylation changes in polycomb repressor target genes were observed. P16 and c-myc hypermethylation also suggested initiation of hepatocellular carcinoma.[127,128]
Tuberculosis	Vitamin D receptor (VDR) methylation	In tuberculosis, vitamin D receptor (VDR) methylation indicated high risk of developing the disease.[149] VDR gene encodes a transcription factor, which alters calcium homeostasis and immune function.

Note: For the same disease more than one row are mentioned because biomarkers shown in column 2 were identified in different studies.

Table 13.6 Biomarkers and their characteristics in neurological diseases.

Disease	Biomarker	Characteristics
Neurological disorders	Granins	The granin family is a group of acidic proteins present in the secretory granules of a wide variety of endocrine, neuronal, and neuroendocrine cells. A few important granins, chromogranin A and B, secretogrannin II, HISL-19 antigen, NESP55, ProSAAS should be studied further for their clinical implication.[77]
Neurological disorders	Microvesicles	Microvesicles (MVs) are used as biomarkers of neurological disorders because their release is increased in these diseases.[83]
Neurological disorders (Alzheimer's disease)	Glutamate	In Alzheimer's disease, glutamate levels increased (in the hippocampus region of the brain) after the treatment with galantamine.[13] Therefore, glutamate can be considered a therapeutic biomarker of Alzheimer's disease.[14]
Neurological disorders (seizures)	Carbohydrate deficient transferrin (CDT)	For the diagnosis of seizures, carbohydrate deficient transferrin (CDT) was evaluated and results indicated that this marker alone couldn't diagnose the disease but contribute in better diagnosis when combined with other biomarkers such as GGT, ASAT, ALAT, and ASAT/ALAT ratio.[39]
Norrie disease (ND) (X-linked disorder)	Missense mutation at position 134 T to A	A novel missense mutation at position 134 T to A resulting in amino acid change at codon V45E was identified as a biomarker for Norrie disease (ND) which is a rare X-linked disorder characterized by congenital blindness and sometime mental retardation, and deafness.[40]
Wilson's disease	Mutation in ATP7B gene	A novel mutation in ATP7B gene can be used as a diagnostic marker for neurological impairment in Wilson's disease.[41]
Alzheimer's disease	CXR4 and CCR3	Genes involved in inflammation and immune system regulatory pathways were identified when profiling of Alzheimer and non-Alzheimer's samples was conducted.[116] Results also indicated a role for chemokines and their receptors (CXR4 and CCR3) in patient samples and these two receptors may be considered biomarkers for the Alzheimer's disease.
Alzheimer's disease	Altered methylation of Alu, Line-1, and SAT-alpha sequences	In Alzheimer disease altered methylation levels of repeat sequences (Alu, Line-1, and SAT-alpha) were observed. This modification induces genomic instability, which contributes to disease initiation and progression.[150]
Autism	MECP2 hypermethylation	Epigenetic regulation in autism was also studied and locus specific hypermethylation of MECP2 was observed.[151]

Note: For the same disease more than one row is mentioned because biomarkers shown in column 2 were identified in different studies.

Table 13.7 Biomarkers and Their Characteristics in Cancer.

Disease	Biomarker	Characteristics
Apoptosis (cancer)	Annexin XI	Anti-annexin XI has been reported in different cancers.[74]
Oral cancer	Mutation in exon 4, codon 63 of the p53 gene	Antibodies against abnormal p53 were found in saliva and serum of oral cancer patients.[53]
Gastric cancer	Metabolites of glycolysis, fatty-acid beta oxidation, and cholesterol and amino acid metabolism	In gastric cancer, alterations in metabolites of glycolysis, fatty-acid beta-oxidation, and cholesterol and amino acid metabolism were observed.[122] These metabolites can be used to follow gastric cancer development and treatment response.
Pancreatic cancer	Serum levels of antibodies against periodontal bacteria *P. gingivalis*	In pancreatic cancer the incidence and mortality rates are the same and the survival is only five years after the diagnosis of the disease. Plasma and serum of pancreatic cancer contains antibodies against periodontal bacteria *P. gingivalis* and their level is associated with pancreatic cancer.[84]
Pancreatic cancer	C-reactive proteins, interleukin-6 (IL6), and soluble receptor of tumor necrosis factor alpha	Inflammatory biomarkers, C-reactive proteins, interleukin-6 (IL6), and soluble receptor of tumor necrosis factor alpha are being used in identifying pancreatic cancer patients.[85]
Ovarian cancer	CA 125, Osteopontin, Kllikrein 6, B7-H4, spondin 2, and DcR3	B7-H4, spondin 2, and DcR3 were identified as early biomarkers in ovarian cancer.[179]
Prostate cancer	Prostate-specific antigen, Alpha methyl CoA-racemase	Results have not been validated for Alpha methyl CoA-racemase.[180]
Colon cancer	APC, CDKNA	Expression analysis was used to identify these biomarkers.[181]
Lung cancer	EGFR, KRAS	Multiple investigators have identified these biomarkers.[182,183]
Breast cancer	BRCA-1, BRCA-2, Let-7	These biomarkers have been used in clinical samples.[184]
Acute myeloid leukemia, Chemotherapy resistant AML, Acute lymphoblastic leukemia, Acute non-promyelocytic leukemia	FLT3, DAPK1, hPer3, DNMT3A, repeat sequences LINE-1	Several investigators have identified these markers. Several methylation markers of AML have also been reported.[185-187]
Renal cell carcinoma (Kidney cancer)	APAF-1, DAPK-1	Hypermathylation of APAF-1 and DAPK-1
Breast cancer	SNAPs in in 1p11.2, 2q35, 3p, 5p12, 8q24, 10q23, 13, 14q24.1, and 16q regions	In GWAS study, SNPs identified in this study were primary located in 1p11.2, 2q35, 3p, 5p12, 8q24, 10q23, 13, 14q24.1, and 16q regions.

Disease	Biomarker	Description
Pancreatic cancer	P. gingivalia	Plasma and serum of pancreatic cancer contained antibodies against periodontal bacteria P. gingivalia and their level is associated with pancreatic cancer.[84]
Pancreatic cancer	Inflammatory biomarkers, C-reactive proteins, interleukin-6 (IL6), and soluble receptor of tumor necrosis factor alpha, cell cycle regulatory proteins (cyclin D1 and Ki67), glycolytic enzyme lactate dehydrogenase (LDH), matrix metalloproteinases (MMPs)	Inflammatory biomarkers, C-reactive proteins, interleukin-6 (IL6), soluble receptor of tumor necrosis factor alpha also were used in identifying pancreatic cancer patients.[85] Some biomarkers identified include increased salivary levels of cell cycle regulatory proteins (cyclin D1 and Ki67), glycolytic enzyme lactate dehydrogenase (LDH), matrix metalloproteinases (MMPs) and reduction in DNA repair enzyme (8-oxoquanine DNA glycosylase) and mapsin in oral cancer patients.[86]
Hepatocellular carcinoma (HCC)	hTERT, alpha-fetoprotein (AFP), des-gamma-carboxy prothrombin (DCP)	A highly sensitive method for serum human telomerase reverse transcriptase (hTERT) mRNA for hepatocellular carcinoma (HCC) was reported.[87] Alpha-fetoprotein (AFP) and des-gamma-carboxy prothrombin (DCP) were found to be good markers for HCC.
Retinoblastoma	Trimethylation of H4K20	Retinoblastoma levels were lower whenever trimethylation of H4K20 was present. A correlation with the tumor stage and grade was also established based on these histone biomarkers.[188]
Breast cancer	HDAC1	Another interesting study reported quantitative expression of HDAC1 and its correlation with breast cancer patient's age, lymph node status, tumor size and her2/neu negative, ER and PR positive status.[146]
Kidney cancer	VHL, MET, FLCN, fumarate hydratase, succinate dehydrogenase, TSC1, TSC2, TFE3	Altered expression of VHL, MET, FLCN, fumarate hydratase, succinate dehydrogenase, TSC1, TSC2, and TFE3 genes in kidney cancer was observed.[189]
Lung cancer	IL-27, IL-29, IL-31, BALF	Serum levels of IL-27, IL-29, IL-31, BALF are high in lung cancer patients.[10]
Lung cancer	miR-21, miR-485, miR-451	Higher levels of miR-21 and lower levels of miR-451 and miR-485 were observed in lung cancer.[11]
Lung cancer	CYFRA 21-1	Increased CYFRA 21-1 levels were observed in samples from lung cancer patients.[190]

Note: For the same disease more than one row are mentioned because biomarkers shown in column 2 were identified in different studies.

13.5 TECHNOLOGIES AND STRATEGIES FOR MOLECULAR BIOMARKER DISCOVERY

Technologies and strategies developed for cancer diagnosis can also be applied for other diseases.[4-6] The detection and treatment of cancer is greatly facilitated by omics technologies. For example, genomics analysis provides clues for gene regulation and gene knockdown for cancer management. The approval of Mammaprint and Oncotype DX indicates that multiplex diagnostic biomarker sets are becoming feasible. The microRNA field in human cancers has opened a new avenue for cancer researchers. Some therapeutic drugs targeting DNA methylation and histone deacetylation are currently undergoing keen studies. Proteomics also plays an important role in cancer biomarker discovery and quantitative proteome-disease relationships provide a mean for connectivity analysis. Fluorescent dye enables a more reliable and quantitative analysis and is facilitated by the progress of biochip and cytomics. The huge amount of information collected by multiparameter single cell flow or slide-based cytometry measurements serves to investigate the molecular behavior of cancer cell populations. Metabolite profiling is such a field which can be applied to cancer and other diseases.

13.5.1 Genomics Based Biomarkers in Immune Diseases, Cardiovascular Diseases (CVD), Metabolic Diseases, Infectious Diseases, Neurological Diseases, and Cancer

Discovery and validation of novel disease-associated biomarkers remain a crucial goal of future patient care. Advanced genomic technologies, such as SNP array and next generation sequencing, help shape the genome and epigenome landscapes. Genome wide association studies (GWAS) as a powerful approach to identify common, low penetrance disease loci have been conducted in several types of diseases such as diabetes and cancer and have identified many novel associated loci, confirming that susceptibility to these diseases is polygenic. Though the creation of risk profiles from combinations of susceptible SNPs is not yet clinically applicable, future, large scale GWAS holds great promise for the individualized screening and prevention. Epigenomic biomarkers like DNA methylation have emerged as highly promising biomarkers and are actively studied in multiple diseases. Validated as being associated with disease risk or drug response, some DNA methylation biomarkers are being transferred into clinical use.[34] Discovery of the genes and pathways mutated in diseased states, especially through large scale genome wide sequencing, has provided key insights into the mechanisms underlying disease process and suggested new candidate biomarkers for diagnosis, clinical intervention as well as prognosis. The comprehensive landscapes of cancer genome point out the convergence of mutations onto pathways that govern the course of disease development and indicate that rather than seeking genomics and epigenomics alterations of specific mutated genes, the combination with dynamic transcriptomics, proteomics and metabonomics of the downstream mediators or key nodal points may be preferable for future disease biomarker discovery.

Most of the genomic biomarkers are mutations in genes or small nucleotide polymorphisms (generally present in the noncoding region).

13.5.1.1 Immune Diseases. Among immune diseases, autoimmune diseases are well characterized. These diseases result when the immune system goes awry and recognizes self-tissues as foreign. The major contribution to these diseases is by autoantibodies and autoreactive cellular responses, which ultimately contribute to the ongoing autoimmune disease process. During the process, inflammatory enzymes are recruited to the affected organ and tissues degrading proteolytic enzymes are released. Therefore, identifying biomarkers that may detect the process early is very significant. An ideal biomarker for these diseases should reach abnormal levels (either higher or lower levels compared to control) in conjunction with disease development, should fluctuate in relation to disease severity and should normalize after treatment. Three groups of biomarkers are being characterized in these diseases. First, degradation products arising from destruction of the affected tissue; second, enzymes that plays a role in tissue degradation, and; third, cytokines and other proteins associated with immune system activation and the inflammatory response. In rheumatoid arthritis (RA), persistent inflammation of joint synovium occurs, which leads to erosion of cartilage and bone. A biomarker, aggrecan fragments, was identified which was detected in synovial fluid (SF). Levels of aggregant fragments (structural components of cartilage) were higher in RA patients than in healthy individuals. Another biomarker was C-propeptide of type II collagen and its levels were directly proportional to the rate of collagen synthesis. Cartilage oligomeric matrix protein (COMP) is also a component of cartilage and it can be measured in serum and SF. Multiple sclerosis (MS) is associated with autoimmune-mediated inflammation of the central nervous system and may lead to demyelination and axonal damage. Biomarker neurofilament light protein (NFL) is a cytoskeleton component in large myelinated axon and it is released into the cerebrospinal fluid (CSF) and can be used to determine the level of axonal damage. Glial fibrillary acidic protein (GFAP) is another biomarker for MS.

A genetic link with HLA-DR4 and related allotypes of MHC class II and T cell associated protein PTPN22 with rheumatoid arthritis was established.[44] The presence of autoantibodies to IgGFc (rheumatoid factor) and antibodies to citrullinated peptide (ACPA) also indicated the presence of disease.

In rheumatoid arthritis, SNPS were reported in several genes which were used as biomarkers for disease diagnosis. These genes include PTPN22, IL23R, TRAF1, CTLA4, IRF5, STAT4, CCR6, and PAD14.[45]

13.5.1.2 Cardiovascular Diseases (CVDs). Pulmonary arterial hypertension (PAH) is influenced by genetic background and may be useful in guiding therapy of the disease.[46] In this disease, increase in blood pressure in the pulmonary artery and pulmonary vein occurs which leads to shortness of breath, dizziness and fainting. GWAS studies indicated that other factors (rare exonic mutations, epigenetic phenomena, and interaction with environmental factors) might also contribute in the development of this disease.[47]

13.5.1.3 Metabolic Diseases, Metabolomics and Proteomics. Like other omics approaches, metabolomics also takes the advantage of non-targeted approach for identifying disease associated biomarkers. The technology has the advantage of using minimum amount of sample and no prior knowledge of the substances to be analyzed by nuclear magnetic resonance (NMR) or mass spectrometry (MS).[34] The complete process includes the acquisition of the experimental data, the multivariate statistical analysis, and the projection of the profiles, which is the acquired information, to construct the patient map (or phenotype). Main diseases where metabolomics has been applied are rheumatoid arthritis, spondyloarthritis, systemic lupus erythrematosus, and osteoarthritis.[48] In spondyloarthritis, association of HLA-A, B, and HLA-DR gene expression was studied and population specific alterations were reported.[49]

13.5.1.4 Infectious Diseases. Regarding genomic biomarkers in infectious diseases, we have selected AIDS (although the topic is so big that we cannot cover in this article). In one study conducted in Georgia, where the prevalence of HIV/AIDS is high, polymorphism of CCR5 gene was studied.[50] More than 100 subjects were enrolled in this study. Results identified CRC5-delta32 mutation as a biomarker of the disease in this population. Mitochondrial mutations could also be used as biomarkers for AIDS, especially in those individuals undergoing therapy.[51]

13.5.1.5 Neurologic Diseases. Because of the anatomical location of the nervous system, it is difficult to get biomarkers of the neurological diseases. The accessibility of the affected organ is a challenge and therefore surrogate biomarkers are generally used to identify and follow these diseases. Gene expression analysis, mutations, and SNPs were the common approaches to identify neurological diseases. Compared to healthy subjects, altered gene expression profiling was observed in schizophrenia and Alzheimer's patient samples.[52]

13.5.1.6 Cancer. Here we describe the specific example of breast cancer with reference to genomic approaches in population science for screening, which may result in early detection of the disease and ultimately long survival. Mortality from breast cancer is very high worldwide. More than half of breast cancer cases occur in Western countries. The cost of treatment is higher when breast cancer is detected late in its development; therefore, detecting this cancer early is the key to success. Mammography has been successful in reducing mortality from this cancer, but it is an expensive technique. The occurrence of breast cancer in the general population can be explained by inherited genetic susceptibility, somatic changes, effects of endogenous and exogenous environments, and interaction of these factors (especially gene–environment interactions). In case of oral cancer, exon 4, codon 63 of the p53 gene was mutated in salivary DNA in patients.[53] Autoantibodies against abnormal p53 were reported in saliva and serum of these patients.

Inherited genes for breast cancer susceptibility can be low- or high-penetrance genes; the few genes with allelic variants that confer a high degree of risk to the individual are known as high-penetrance genes. Other genes confer a small to moderate degree of breast cancer risk to the individual and are known as low-penetrance genes. Relatively few individuals in the population carry risk-increasing genotypes at the loci where high-penetrance genes act; therefore, the population-attributable risk is low. On the other hand, the low-penetrance genes are not associated with syndromic or Mendelian patterns but are associated with sporadic breast cancer. The allelic variation of low-penetrance genes is relatively high, and large breast cancer populations carry low-penetrance genes. Different investigators identified low- and high-penetrance genes in breast cancer in a number of populations. The BRCA1 gene was the first gene identified to represent susceptibility to hereditary breast cancer and later on BRCA2 (located on 17q21) was also confirmed for breast cancer and ovarian cancer.[54] To identify breast cancer associated genetic biomarkers a number of cohorts with exposure and lifestyle data and other details of the participants were used.[55] One of those cohorts, the Collaborative Oncological Gene Environment Study (COGS), which is a large scale genotyping cohort funded by European Commission was utilized to identify disease associated biomarkers. More than 150 000 samples were genotyped in this study. Familial based high penetrance susceptibility genes were identified first and then low penetrance genes by association studies.[56,57] Carriers of such genes and SNPs predispose to breast cancer. A panel of 70 genes were able to predict breast cancer prognosis.[58] Genomic markers include small nucleotide polymorphisms (SNPs), mutations, additions and deletions, recombinations, and change in copy number (altered CNVs).[59,60]

GWAS were conducted by different groups in different cohorts to identify breast cancer susceptibility genes which may be useful for breast cancer screening of high risk populations.[61,62] In one such study, genotyping of 2702 women of European ancestry with invasive breast cancer and 5726 controls was conducted.[61] SNPs identified in this study were primary located in 1p11.2, 2q35, 3p, 5p12, 8q24, 10q23, 13, 14q24.1, and 16q regions. Genes affected by these SNPs are involved in regulation of actin cytoskeleton, glycan degradation, alpha linoleic metabolism, circadian rhythm, and drug metabolism.

13.5.2 Proteomics Based Biomarkers in Immune Diseases, Cardiovascular Diseases (CVD), Metabolic Diseases, Infectious Diseases, Neurological Diseases, and Cancer

Proteomics technologies have emerged as a useful tool in the discovery of diagnostic biomarkers and substantial technological advances in proteomics and related computational science have been made in the last decade.[1,63] These advances overcome in part the complexity and heterogeneity of the human proteome, permitting the quantitative analysis and identification of protein changes associated with tumor development. With the advent of new and improved proteomic technologies, it is possible to discover new biomarkers for the early detection and treatment monitor of different diseases. The contribution of the Human Proteomic Organization is valuable in this regard because this organization has provided the dataset of normal healthy human proteomic profiles which can be used as a reference dataset for identifying disease-associated proteomic changes (http://www.hupo.org/). Protein biomarkers are more related to disease phenotype and are more targetable for therapy in comparison with transcriptomic or genomic biomarkers. Proteomics provides a powerful tool to investigate potential biomarkers in diseases due to its high sensitivity, precise characterization of their interaction, and ability to detect functionally significant post-translational modifications. Proteomic biomarkers have been identified in blood (serum and plasma) as well as in tissue samples by applying approaches such as nuclear magnetic resonance spectroscopy (NMR), mass spectrometry (MS), two-dimensional gel electrophoresis and immunoprecipitation. In one study, investigators identified circulating proteomic biomarkers from different stages of breast cancer using an innovative strategy employing high sensitivity label-free proteomics. The approach was MS based and provided semi-quantitative results and could be applied in preclinical and clinical studies. Furthermore, breast cancer patient serum was analyzed by bidimensional nanoUPLC tandem nano ESI-MS to identify breast cancer biomarkers which are differentially expressed at early stages of cancer development.[64] Higher GRHL3 expression and lower levels of TNF alpha were reported during early stage of the diseases whereas PMS2 expression was high in advanced stages of the disease. These results were validated in a different set of patients although the number of participants was low. These investigators plan to evaluate the impact of such markers in determining survival rates of patients and recurrence of breast cancer or other cancers.

Proteomics with the recent advances in mass spectrometry is considered as a powerful analytical method for deciphering proteins expressions alterations as a function of disease progression.[63,65] Proteomics based analyses of breast serum and tissue lysates have resulted in the finding of a number of potential tumor biomarkers providing, therefore, a basis for a better understanding of the breast-cancer development and progression, and eventually serving as diagnostic and prognostic markers.[64] Probably the most widely used proteomic technology is the identification of alterations in protein expression between two different samples through comparative two-dimensional gel electrophoresis (2-DE) which provides high-resolution separation of proteins and offers a powerful method for their identification and characterization.[66]

Proteomic analysis is an essential component to explain the information contained in genomic sequences in terms of the structure, function, and control of biological processes and pathways.[67] The proteome reflects the cellular state or the external conditions encountered by a cell. In addition, proteomic analysis is a genome-wide assay to differentiate distinct cellular states and to determine the molecular mechanisms that control them.[68] Infection-associated proteomic biomarkers have also been characterized.[69] High-throughput proteomic methodologies have the potential to revolutionize protein biomarker discovery and to allow for multiple proteins biomarkers to be assayed simultaneously. With the significant advances in 2-DE and mass spectrometry (MS), protein biomarker discovery has become one of the central applications of proteomics.[67]

13.5.2.1 Immune Diseases. Antigen arrays are valuable for profiling auto-antibodies in diverse rheumatic autoimmune diseases and can be composed of proteins, peptides, protein complexes, glycoproteins, sugar nucleic acids, and lipids. T and B cells can be isolated from disease subjects and used for disease stratification, which ultimately help in designing treatment and disease management of autoimmune diseases (such as rheumatoid arthritis or RA).[70] The proteomic profile might suggest whether the disease was aggressive or not. Treatment can be selected based on the aggressiveness of the disease.[71] Less than two-thirds of all individuals with rheumatoid arthritis had an adequate response to anti-TNF therapy. Biomarkers could identify those individuals who were less likely to respond to this therapy.[70,72] This would not only reduce the cost associated with therapy but also avoid unwanted adverse reactions due to anti-TNF therapy among non-responders. Technologies such as mass cytometry, peptide and protein arrays and BCR (b cell receptor) and TCR (T cell receptor) sequencing might prove useful to improve the management of autoimmune diseases and represent the state of art in analyzing cells, soluble proteins and genes.[73]

Another group of biomarkers which was characterized for autoimmune diseases was annexins, a group of 12 highly conserved proteins which regulate cell cycle. Their abnormal expression was associated with disease development.[74] Annexin-I exerts its anti-inflammatory effect by suppressing the generation of inflammatory mediators and anti-annexin I antibodies are present in rheumatoid arthritis and systemic and cutaneous lupus erythematosus (SLE) patients.[75] Annexin II and V are involved in coagulation cascade because they have affinity towards phospholipids. Prolactin (PRL), ferritin, vitamin D, tumor biomarker tissue polypeptide antigen (TPA) levels are higher in patients with autoimmune disease (systemic sclerosis, systemic and cutaneous lupus erythematosus, SLE), polymyositis (PM), dermato-myositis (DM), multiple sclerosis (MS).[76] The granin family is a group of acidic proteins present in the secretory granules of a wide variety of endo-crine, neuronal, and neuroendocrine cells. A few important granins, chro-mogranin A and B, secretogrannin II, HISL-19 antigen, NESP55, ProSAAS should be studied further for their clinical implication.[77]

13.5.2.2 Cardiovascular Diseases (CVDs). Atherosclerosis is the main cause of cardiovascular diseases. Diagnosis of subclinical atherosclerosis is a clinical challenge. Furthermore, determining the extent of atheroscler-osis aggressiveness in individual patients is a challenge too. Plasma osteo-protegerin (OPG) turned out to be an excellent proteomic biomarker which could detect preclinical atherosclerosis, as validated in a case-control study.[78] Proteomic profiling in circulating cells and plasma extracellular vesicles could distinguish populations at high risk of developing atherosclerosis.[79]

13.5.2.3 Metabolic Diseases. Isobaric tags for relative and absolute quan-titation (iTRAQ) in combination with two-dimensional gel electrophoresis technologies identified a group of proteins which were associated with ei-ther diabetes, pancreatitis, and/or pancreatic cancer.[80]

13.5.2.4 Infectious Diseases. HIV-associated neurodegenerative disease progression and treatment response could be measured by following levels of biomarkers complement C3, soluble superoxide dismutase and prosta-glandin synthase.[81] Although Infectious disease research had focused more on HIV-related diseases, HIV infection had its effects on the central nervous system (CNS) as this lentivirus could infect brain cells. CNS dysfunction then led to a group of cognitive and behavior changes (called HIV-associated neurocognitive disorders or neuroAIDS) which serve as biomarkers.[82]

13.5.2.5 Neurological Diseases. Because of the anatomical location of nervous system, it is difficult to get biomarkers of neurological disorders. The accessibility of the affected organ is a challenge and therefore surro-gate biomarkers are generally used to identify and follow neurological dis-orders. Microvesicles (MVs) have been used as biomarkers of neurological disorders because their release is increased in these diseases.[83] MVs ori-ginate from exosomes (which are derived from endothelial cells) and could be found in plasma or serum. MVs are linked to neurological pathologies with a vascular or ischemic pathogenic component (sometimes used in diagnosis and follow up of strokes). In another disease, multiple sclerosis, MVs of oligodendroglial origin were reported in the cerebrospinal fluid (CSF).[83] MVs detection should be explored further to gain pathogenic information, identify therapeutic targets, and select specific biomarkers for neurological diseases.

13.5.2.6 Cancer. In pancreatic cancer, the incidence and mortality rates are the same and the survival is only five years after the diagnosis of the disease. Plasma and serum of pancreatic cancer patients contain antibodies against periodontal bacteria *P. gingivalia* and their level is associated with pancreatic cancer.[84] Inflammatory biomarkers, C-reactive proteins, interleukin-6 (IL6), and soluble receptor of tumor necrosis factor alpha have also been used in identifying pancreatic cancer patients.[85] Some biomarkers identified include increased salivary levels of cell cycle regulatory proteins (cyclin D1 and Ki67), glycolytic enzyme lactate dehydrogenase (LDH), matrix metalloproteinases (MMPs) and reduction in DNA repair enzyme (8-oxoquanine DNA glycosylase) and mapsin in oral cancer patients.[86]

A highly sensitive method for serum human telomerase reverse transcriptase (hTERT) mRNA for hepatocellular carcinoma (HCC) was reported.[87] Alpha-fetoprotein (AFP) and des-gamma-carboxy prothrombin (DCP) were found to be good markers for HCC. This group also verified the significance of hTERT mRNA in a large scale multi-centered trial. hTERT mRNA was demonstrated to be independently correlated with clinical parameters: tumor size and tumor differentiation. hTERT mRNA proved to be superior to AFP, AFP-L3, and DCP in the diagnosis and underwent an indisputable change in response to therapy. The detection rate of small HCC by hTERTmRNA was superior to the other biomarkers.[87]

To identify the potential biomarkers involved in Hepatocellular carcinoma (HCC) carcinogenesis, a comparative proteomics approach was utilized to identify the differentially expressed proteins in the serum of 10 HCC patients and 10 controls. A total of 12 significantly altered proteins were identified by mass spectrometry. Of the 12 proteins identified, HSP90 was one of the most significantly altered proteins and its over-expression in the serum of 20 HCC patients was confirmed using ELISA analysis. The observations suggest that HSP90 might be a potential biomarker for early diagnosis, prognosis, and monitoring in the therapy of HCC.[88]

Proteomic analysis with 2-DE and MS was used to identify other potential serum markers for breast cancer.[89,90] Protein extracts expressed in the serum of breast cancer patients after depletion of high abundance proteins were compared to sera from healthy women using proteomic approaches. By comparing 2-DE profiles between tumor and non-tumor samples and using MALDI-TOF mass spectrometry of their trypsinized fragments, the identification of two proteins of interest, haptoglobin precursor and alpha-1-antitrypsin precursor, was observed.[93] Separation and analysis of proteins from cells, tissue samples and breast tumor biopsies proved very successful in identifying novel biomarkers.[91] Using proteomic approaches, 26 immunoreactive proteins (antigens) against which sera from newly diagnosed patients with infiltrating ductal carcinomas exhibited reactivity were detected.[89,92]

Among these antigens, peroxiredoxin-2 (Prx-2) belongs to a family of thiol-specific antioxidant proteins that control intracellular H_2O_2 by reducing reactive oxygen species (ROS) issued from free radicals. Such proteins might have an important role and protect the breast tumor cells against oxidative injury and modulate cell proliferation and apoptosis of malignant cells.[93]

Proteomic approaches are useful to identify protein-protein interaction and in one study estrogen receptor alpha and its interaction with a number of transcription factors was characterized which resulted in clinically useful information about breast cancer therapeutics.[94] Laser-capture microdissected breast cancer and normal tissue cells were analyzed by mass spectrometry to identify proteomic profiles associated with breast cancer.[95] In another study, glyoxalase-1 was found to be expressed in breast cancer.[89] This protein was involved in detoxification of methylglyoxal which is a cytotoxic product of glycolysis. Further analysis of tissue microarray indicated correlation of glyoxalase-1 with tumor grade. Based on results from reversed phase protein array, a model was created to predict pathological response in patients receiving neoadjuvant taxane and anthracyclin taxane based systemic therapy, thus indicating translational significance of proteomic biomarkers in breast cancer.[96]

13.5.3 Transcriptomics Based Biomarkers in Immune Diseases, Cardiovascular Diseases (CVD), Metabolic Diseases, Infectious Diseases, Neurological Diseases, and Cancer

Transcriptomics is the study of how our genes are regulated and expressed in different biological settings. All expressed genes can be quantitatively measured in a tissue at a given time (and this science is termed transcriptomics).[97] Over the last decade, microarray technology based transcriptomic analysis has contributed enormously to our understanding of the molecular basis of a number of diseases. Gene expression profiling offers an unparalleled opportunity to develop biomarkers that are useful in diagnosis and prognosis and in helping to achieve the goal of individualized treatment. However, the limitations of the technology and the danger of inappropriate experimental processes should not be underestimated. In clinical settings, blood transcriptomics of Alzheimer's patients undergoing treatment with EHT 0202 has been conducted.[98] Transcriptomic analysis indicated activation of pathways associated with CNS disorders, diabetes, inflammation, and autoimmunity. Treatment resulted in deactivation of these pathways in patients. Thus transcriptomic biomarker profiling could be used in disease prognosis. Such studies would help us identifying those patient populations who would respond to treatment, and also identifying those individuals who would likely to have recurrence of the disease.

13.5.3.1 Immune Diseases. In rheumatoid arthritis (RA) bone loss occurs.[99] Transcriptomic approaches were applied to identify genes and pathways which contributed to bone loss in RA patients. Results indicated IL-17 and Wnt/beta katenin as transcriptional biomarkers of RA.[100]

13.5.3.2 Cardiovascular Diseases (CVDs).   Atherosclerosis is a pathological process in which the walls of large arteries thicken and lose elasticity as a result of the growth of atheromatous lesions. Transcriptomics based gene expression analysis did not identify specific genes but pathways (especially inflammation related pathways) were identified which were associated with atherosclerosis development.[101] The analysis was conducted in monocytes and macrophages.

13.5.3.3 Metabolic Diseases.   Most gene expression profiling identified diabetes but in a few cases individual gene expression (also called candidate gene expression approach) also provided information about disease-associated transcriptomics. Glucose response gene, RCAN1, was over expressed whereas perilipin A and G0/G1 switch gene, G0S2, were under expressed in diabetes-2.[102,103]

13.5.3.4 Infectious Diseases.   Based on transcriptomics, a regression model was developed which could predict the onset of ventilator-associated pneumonia.[104] In another study, transcriptomics identified a gene-expression pattern which helped in identifying patients who were at high risk of developing trachibronchitis or pneumonia.[105]

13.5.3.5 Neurological Diseases.   In the neurobiology field, neuronal development, function, and the subsequent degeneration of the brain are still serious problems. There is a need to find better targets for developing therapeutic intervention by identifying new biomarkers by applying transcriptomics and other approaches.[106] A high-throughput high-content screening approach is needed to go from genes to gene-networks.

13.5.3.6 Cancer.   To implicate a combination of omics technologies, one investigator performed transcriptomics and metabolomics in breast cancer samples and identified a disease-associated profile, which could be used for diagnosis purposes.[107] Transcriptomics identified genes and pathways whereas the functional significance was evaluated by metabolite characterization.

13.5.4 Immunomics Based Biomarkers in Immune Diseases, Cardiovascular Diseases (CVD), Metabolic Diseases, Infectious Diseases, Neurological Diseases, and Cancer

The huge amount of immunological information hidden in the plasma could be better revealed by combining the characterization of antibody binding target epitomes with improved estimation of effector functions triggered by these binding events. Functional immune profiles can be generated characterizing general immune responsiveness by designing arrays with information about epitope collections from different antibody targets. Immunomics was implicated in searching and identifying proteins of interest in the case of breast or colorectal cancers.[108] Two approaches were developed at their laboratory:[109] the top down SERological Proteome Analysis (SERPA) and the bottom up MAPPing (Multiple Affinity Protein Profiling) strategies. The first one relied on two dimensional electrophoresis (2 DE), immunoblotting, image analysis and mass spectrometry. The second approach dealt with the use of two dimensional immuno affinity chromatography, enzymatic digestion of the antigens, and analyses by tandem mass spectrometry. Using immunoinformatics approach, putative T- and B- cell epitopes of capsid proteins were identified which were conserved in existing serotypes.[109]

13.5.4.1 Immune Diseases. Although autism is a neurological disorder, its regulation involves autoimmunity. This disease is characterized by impaired social interaction and verbal and non-verbal communication. A high number of mast cells and serotonin in urine were observed in autism patients. Mast cells produce inflammatory cytokines in large amount in autism patients and the possibility of autoimmunity was proposed.[110]

13.5.4.2 Cardiovascular Diseases (CVDs). Atherosclerosis is a chronic inflammatory disease of the vessel wall. The gene CARD8, which codes proteins involved in innate immunity in atherosclerosis patients, is an excellent biomarker for atherosclerosis. Inflammatory markers also over expressed in this disease. Inflammosomes produce interleukin 1beta in response to cholesterol crystal accumulation in macrophages. Ephs and ephrins also were proposed as biomarkers of atherosclerosis.[111]

13.5.4.3 Metabolic Diseases, Metabolomics and Proteomics. Chronic low grade inflammation contributes to the pathogenesis of insulin resistance. In diabetes innate immune cells accumulate in metabolic tissues and release inflammatory cytokines, especially IL-1beta and TNF. One investigator proposed cell mediated immunity in diabetes. Regulation of type 1 and type 2 diabetes by immune system occurred differently in diabetic patients under study.[112,113]

13.5.4.4 Infectious Diseases. Tuberculosis (TB) is a common and lethal infectious disease caused by various strains of mycobacteria. The lung is the primary organ which is infected by the bacteria although any part of the body can be infected. For several years antibiotics were able to treat TB but in the past few years antibiotic resistant mycobacteria have been reported. To address this point, the use of anti-mycobacteria antibodies, enhancing the Th1 protective responses by using mycobacterial antigen, or increasing Th1 cytokines, was evaluated and promising results were obtained.[114] Childhood deficiency of vitamin D was proposed to be a susceptible biomarker of TB.[115]

13.5.4.5 Neurological Diseases. Genes involved in inflammation and immune system regulatory pathways were identified when profiling of Alzheimer's and non-Alzheimer's samples was conducted.[116] Results also indicated a role for chemokines and their receptors (CXR4 and CCR3) in patient samples; and these two receptors may be considered biomarkers for the Alzheimer's disease. Another biomarker in the same disease was amyloid beta which is expressed at higher levels in patients compared to age and sex matched controls.

13.5.4.6 Cancer. Genetic variation in innate immunity and inflammation pathway associated with lung cancer risk were proposed in different studies. Some investigators proposed T cell mediated immunity in lung cancer using HLA-DR telomerase derived epitopes. In cervical cancer, cell mediated immunity is the main mechanism of cancer development. Infection-associated cancers, infectious agents and their epitopes were considered excellent biomarkers which could be used for diagnostic purposes.[75,117,118]

13.5.5 Metabolomics Based Biomarkers in Immune Diseases, Cardiovascular Diseases (CVD), Metabolic Diseases, Infectious Diseases, Neurological Diseases, and Cancer

Metabolomics is the study of small molecules of both endogenous and exogenous origin, such as metabolic substrates and their products, lipids, small peptides, vitamins and other protein cofactors, generated by metabolism,[2,119] which are downstream from genes. This approach has received more attention in recent years as an ideal methodology to unravel signals closer to the culmination of the disease process. The compounds identified through metabolomic profiling represent a range of intermediate metabolic pathways that may serve as biomarkers of exposure, susceptibility or disease.[120] Therefore, it is a valuable approach for deciphering metabolic outcomes with a phenotypic change. Inherited metabolic disorders can be measured by following a set of metabolomic biomarkers by directly measuring metabolites using LC-MS or NMR. A few examples of these diseases are: amino acid metabolism (phenylketonuria, maple syrup urine disease, tyrosinemia I, argininemia, homocystinuria, ornithine transcarbamylase deficiency, and nonketotic hyperglycemia), organic acidurias, and mitochondrial defects.[121]

Risk prediction for diabetes and CVD is difficult although a number of biomarkers have been identified (see Table 13.4). So far insulin resistance (IR) and atherosclerosis are the only promising biomarkers.

In gastric cancer, alterations in metabolites of glycolysis, fatty-acid beta-oxidation, and cholesterol and amino acid metabolism were observed.[122] These metabolites can be used to follow gastric cancer development and treatment response.

13.5.5.1 Immune Diseases. Diagnosis of asthma is sometime difficult and respiratory symptoms alone are not sufficient for diagnosing this disease. Measurements of fractional excretion of nitric oxide can be used as a biomarker for diagnosing asthma. There is a considerable effect of the environment on health from pollution, climate change, and epigenetic influences, underlining the importance of understanding gene-environment interactions in the pathogenesis of asthma and response to treatment. Metabolomic and proteomic approaches can be applied to determining levels of nitric oxide.[81,123]

13.5.5.2 Cardiovascular Diseases (CVDs). Metabolites characterized from biofluids of CVD patients serve as biomarkers for CVDs, such as atherosclerosis. Dyslipidemia in HIV patients puts them at high risk of developing CVDs. In another study, intestinal microbiodata and metabolites were suggested to contribute in CVD development.[124,125]

13.5.5.3 Metabolic Diseases, Metabolomics and Proteomics. Serum metabolites from healthy and diabetes-2 patients showed different profiles and disease-associated profiles were suggested biomarkers for diabetes diagnosis. In a large population study diabetes associated metabolites were identified which were generated by seven genes.[126,127]

13.5.5.4 Infectious Diseases. Antibiotic resistance is a common feature of TB where mycobacteria become resistant to antibiotic treatment. GC-MS based technologies were used to identify metabolites, especially metabolites from fatty acid metabolism, from wild type mycobacteria infected and mutant infected samples. These observations laid the groundwork for developing therapeutics for TB treatment. In a recent study, the metabolomic adaptation of bacteria in the host was also evaluated.[128]

13.5.5.5 Neurological Diseases. An epileptic seizure is a transient symptom of abnormal excessive or synchronous neuronal activity in the brain. Errors of metabolism result in abnormal levels of metabolites at the neonatal stage of development resulting in neonatal seizures. One group of investigators demonstrated altered metabolism of astrocytes contributing to seizures.[93,129]

13.5.5.6 Cancer. GC-MS based metabolomics in serum identified colorectal cancer associated biomarkers. Metabolites were successfully used to detect bladder, breast, and pancreatic cancer. Here this is emphasized that pancreatic cancer is such a disease where early detection markers are sought because the incidence and mortality rates are almost the same, therefore, metabolomic biomarkers should be explored for this fetal disease.[130,131]

13.5.6 Epigenomics and miRNOMICS Based Biomarkers in Immune Diseases, Cardiovascular Diseases (CVD), Metabolic Diseases, Infectious Diseases, Neurological Diseases, and Cancer

In addition to genetic code, human cells contain an additional regulatory level predominating the genetic code; this is called epigenetic code.[34] This involves altered gene expression without changing the genomic structure.[33] Due to different chromatin status, condensed or relaxed, the same genetic variants might be associated with different phenotypes, depending on the environment, life-style, and exposure. A rapidly growing number of genes with epigenetic regulation altering their expression by chromatin-remodeling (condensation and relaxation) have been identified.[7,34,43,132] Methylation of cytosines in DNA, histone modifications, alterations of non-coding RNAs (especially miroRNAs) are the mechanisms involved in chromatin remodeling.

The term epigenome is used to define a cell's overall epigenetic state. Basic biological properties of DNA-segments such as gene density, replication timing, and recombination are linked to their GC content. The promoter region is rich in CpG content. A genomic region of about 0.4 kb with more than 50% GC content is called a CpG island. In mammals, CpG islands typically are 200–300 bp long. Promoters of tissue specific genes that are situated within CpG islands are generally unmethylated but during the disease development, these CpG sites start getting methylated. Cytosine methylation can regulate gene expression by hindering the association of some transcriptional factors with their cognate DNA recognition sequences, or methyl CpG binding protein (MBP) can bind to methylated cytosines mediating a repressive signal, or MBPs can interact with chromatin forming proteins modifying the surrounding chromatin, linking DNA methylation with chromatin modification. DNA methylation at position five of cytosine is conducted by DNA methyltransferases (DNMTs). These enzymes are needed for methylation initiation and methylation maintenance.[34]

Alterations due to epigenetic mechanisms can be stably passed over numerous cycles of cell division. A selected epigenetic alteration can be inherited from one generation to another generation.[133] Cancer specific methylation alterations are hallmark of different cancers. Alteration in methylation may cause genomic instability, genomic alterations and change in gene expression.[34,134] A systematic approach to determine epigenetic changes in tumor development may lead to identify biomarkers needed for cancer diagnosis. It has been suggested that an integration of genome and hypermethylome might provide insight into major pathways of cancer development which in turn might help us identify new biomarkers of cancer diagnosis and prognosis.[135] Methylation and microRNA (miR) alterations are the main biomarkers which can be assayed easily in samples non-invasively.[136] The finding that monozygotic twins are epigenetically indistinguishable early in life but with age exhibit substantial differences in the epigenome indicates that environmentally determined alterations in a cell's epigenetic marks are responsible for an altered epigenome.[137] Examples exist to show that environmental factors influence disease initiation, progression, and development.[138,139]

MicroRNAs (miRs), small RNA molecules of approximately 22 nucleotides, have been shown to be up or down regulated in specific cell types and disease states. These molecules have become recognized as one of the major regulatory gatekeepers of coding genes in the human genome. The structure, nomenclature, mechanism of action, technologies used for miR detection, and associations of miRs with human cancer have been evaluated by a number of investigators.[140,141] miRs are produced in a tissue-specific manner, and changes in miR within a tissue type can be correlated with disease status.[142] miRs regulate mRNA translation and their degradation. Among a number of regulators of gene expression, miRs are the key regulators. Tissue specific miRs have been reported by different groups.[28] These RNAs are of small size with distinct stem and loop structure.[143] A number of miRs can be isolated in circulation. Because of their small size and stability (due to secondary structure), these circulating miRs provide a rich source of diagnostic biomarkers.

13.5.6.1 Immune Diseases. Genetic and environmental factors contribute in multiple sclerosis (MS). However, recent research indicate role of epigenetics in MS development. MS is an inflammatory disease in which the fatty myelin sheaths around the axons of the brain and spinal cord are damaged. Female specific MS is associated with HLA DRB1-1*5 haplotype. SHP-1 hypermethylation is a biomarker for multiple sclerosis.[144]

13.5.6.2 Cardiovascular Diseases (CVDs). Atherosclerosis can be diagnosed when other biomarkers are combined with epigenetic biomarkers, such as hypermethylation of monoamine oxide A (MAOA). Along with gene-specific methylation biomarkers, global DNA methylation biomarkers were identified for atherosclerosis.[145] Similarly in other CVDs, such as hypertension, global methylation profiling and individual gene methylation status (ADD1 gene) were used for diagnosis and outcome.[146]

13.5.6.3 Metabolic Diseases, Metabolomics and Proteomics. A combination of biomarkers, genetic and epigenetics, was used for diabetes diagnosis using mutations in duodenal homeobox-1 (PDX-1) and its methylation levels. Hypermethylation sites were reported at multiple sites distributed throughout the genome. At least one investigator reported a combination of histone and methylation biomarkers for diabetes diagnosis.[147,148]

13.5.6.4 Infectious Diseases. In early liver cancer, where risk factors are infection by HBV and HCV, clustered DNA methylation changes in polycomb repressor target genes were observed. P16 and c-myc hypermethylation also suggested initiation of hepatocellular carcinoma. In tuberculosis, vitamin D receptor (VDR) methylation indicated high risk of developing the disease. VDR gene encodes a transcription factor which alters calcium homeostasis and immune function.[149]

13.5.6.5 Neurological Diseases. In Alzheimer's disease, altered methylation levels of repeat sequences (Alu, Line-1, and SAT-alpha) were observed. This modification induced genomic instability that contributed to disease imitation and progression.[150] Epigenetic regulation in autism was also studied and locus specific hypermethylation of MECP2 was observed.[151]

13.5.6.6 Cancer. Both histone modifications and DNA methylation were observed in different cancers, especially breast cancer. When epigentic profiling of MCF-7, MDA-MB-231, and MDA-MB-231 (S30) was followed, decreased trimethylation of H4K20 and hyperacetylation of H4 was observed. Concomitant to a decrease in trimethylation, lower levels of Suv4-20h2 histone methyl transferase were also observed. The effect was more in MDA-MB-231 compared to other cells which suggested that differential expression of histone modifications could represent aggressiveness of the disease. In another study, HDAC6 (one of the histone acetyl transferase) responded to estrogen treatment.[152] Retinoblastoma levels were lower whenever trimethylation of H4K20 was present. A correlation with the tumor stage and grade was also established based on these histone biomarkers. Another interesting study reported quantitative expression of HDAC1 and its correlation with breast cancer patient's age, lymph node status, tumor size and her2/neu negative, ER and PR positive status.[153]

Cancer cells accumulate abnormal DNA methylation patterns that result in malignant phenotypes. The genomic distribution of methylation is not well understood. In this direction, a number of genome-wide association studies have been conducted so that cancer risk associated biomarkers can be identified. Using methylated DNA immunoprecipitation combined with high-throughput sequencing (MeDIP-seq), levels of methylation were compared in samples from normal and cancer cells and global hypomethylation was observed in cancer samples, especially in the CpG rich regions. The location of these CpG rich regions was not related with the transcription start sites of various genes. Using this approach, the methylation patterns during epithelial to mesenchymal transition also was evaluated and used for disease stratification.[154]

In breast cancer, methyl acceptance capacity in malignant breast tissues was approximately 2–3 fold greater compared with matched controls. However, the variation in methyl acceptance capacity among patients varied a lot.[155] Quantitative analysis for 5meC levels showed a substantial decrease compared with normal tissues. Levels of hypomethylation in BRCA1 and BRCA2 cancers were slightly lower but significant.[156] Genome-wide hypomethylation correlated with satellite sequence hypomethylation. Specific regions (Sa2 coding) on chromosome 1 and sat-alpha were specifically hypomethylated.[157] On chromosome 5, the region containing the coding sequence of SATr-1 also showed hypomethylation.

The tissue concentrations of specific miRs have been associated with tumor invasiveness, metastatic potential, and other clinical characteristics for several types of cancers, including chronic lymphocytic leukemia, and breast, colorectal, hepatic, lung, pancreatic, and prostate cancers. By targeting and controlling the expression of mRNA, miRs can control highly complex signal-transduction pathways and other biological pathways. The biologic roles of miRs in cancer suggest a correlation with prognosis and therapeutic outcome. Further investigation of these roles may lead to new approaches for the categorization, diagnosis, and treatment of human cancers. Frequent dysregulation of miR in malignancy highlights the study of molecular factors upstream of gene expression following the extensive investigation on elucidating the important role of miR in carcinogenesis. For example, esophageal carcinogenesis is a multi-stage process, involving a variety of changes in gene expression and physiological structure change. Recent innovation in miRs profiling technology have shed new light on the pathology of esophageal carcinoma (EC), and also showed great potential for exploring novel biomarkers for both EC diagnosis and treatment. A thorough review of the role of miRs in EC, addressing miR functions, their putative role as oncogenes or tumor suppressors and their potential target genes has been explored by different investigators.[158]

In inflammatory breast cancer cells, more than 300 miRs were evaluated for their association with the disease.[159] The most promising miRs were miR-29a, miR-30b, miR-342-5p, and 520a-5p. The functional analysis of these miRs revealed their role in cell proliferation and signal transduction pathways. These markers could be useful to identify inflammatory breast cancer cells. The promoter regions of miR coding region were evaluated and several miR promoters were found hypermethylated, especially those of miR-31, miR-130a, miR-let7a-3/let 7-b, miR-155, and miR-137.[142] In one example, an advantage of using miRs for detecting cancer is demonstrated due to their stability even in fixed tissues.[143] miR-155 predicted prognosis of triple negative breast cancer (higher miR-155 expression correlated with higher angiogenesis and aggressiveness).[160]

13.6 CONCLUSION

In the last few decades, biomarkers have played a very significant biological role in disease diagnosis and therapy and consequently numerous studies have been published so far. Comparatively, most of the work has been carried out on biomarker diagnosis than their usage for therapies where single and multiple biomarkers have been used for diagnosis. In this chapter, a number of important biomarkers for a broad-range of disorders have been described in detail with a further focus on updating informations in this field. Furthermore, different criteria and approaches for the selection and specification of different types of biomarkers have also been described, which are very helpful in ameliorating our knowledge in understanding and distinguishing these biomarkers from one another. The cutting-edge NGS technologies and the OMICS approaches have already contributed a vital role in biomarkers discovery in various diseases and they further provide a platform and an insight into the robustness and progress of this field of knowledge.

REFERENCES

1. M. Verma, J. Kagan, D. Sidransky and S. Srivastava, *Nat. Rev. Cancer*, 2003, **3**, 789–795.
2. E. Ferrannini, A. Natali, S. Camastra, M. Nannipieri, A. Mari, K. P. Adam, M. V. Milburn, G. Kastenmuller, J. Adamski, T. Tuomi, V. Lyssenko, L. Groop and W. E. Gall, *Diabetes*, 2013, **62**, 1730–1737.
3. M. Verma, *Curr. Genomics*, 2012, **13**, 308–313.
4. M. Verma and S. Srivastava, *Recent Results Cancer Res.*, 2003, **163**, 72–84, discussion 264–266.
5. M. Verma, G. L. Wright, Jr., S. M. Hanash, R. Gopal-Srivastava and S. Srivastava, *Ann. N. Y. Acad. Sci.*, 2001, **945**, 103–115.
6. S. Srivastava, M. Verma and D. E. Henson, *Clin. Cancer Res.*, 2001, 7, 1118–1126.
7. M. Verma, *Methods Mol. Biol.*, 2012, **863**, 467–480.
8. S. A. Peters, F. L. Visseren and D. E. Grobbee, *Nat. Rev. Cardiol.*, 2013, **10**, 12–14.
9. C. W. Peng, Q. Tian, G. F. Yang, M. Fang, Z. L. Zhang, J. Peng, Y. Li and D. W. Pang, *Biomaterials*, 2012, **33**, 5742–5752.
10. W. Naumnik, B. Naumnik, K. Niewiarowska, M. Ossolinska and E. Chyczewska, *Exp. Oncol.*, 2012, **34**, 348–353.
11. C. C. Solomides, B. J. Evans, J. M. Navenot, R. Vadigepalli, S. C. Peiper and Z. X. Wang, *Acta Cytol.*, 2012, **56**, 645–654.
12. A. Di Ieva, *Microvasc. Res.*, 2010, **80**, 522–533.
13. J. Penner, R. Rupsingh, M. Smith, J. L. Wells, M. J. Borrie and R. Bartha, *Prog. Neuro-Psychopharmacol. Biol. Psychiatry*, 2010, **34**, 104–110.
14. R. Rupsingh, M. Borrie, M. Smith, J. L. Wells and R. Bartha, *Neurobiol. Aging*, 2011, **32**, 802–810.
15. T. Efferth, *Planta Med.*, 2010, **76**, 1143–1154.
16. D. G. Duda, C. G. Willett, M. Ancukiewicz, E. di Tomaso, M. Shah, B. G. Czito, R. Bentley, M. Poleski, G. Y. Lauwers, M. Carroll, D. Tyler,

C. Mantyh, P. Shellito, J. W. Clark and R. K. Jain, *Oncologist*, 2010, **15**, 577–583.

17. H. Xu, C. Chen, C. M. Liu, J. Peng, Y. Li, Z. L. Zhang and H. W. Tang, *Guangpuxue Yu Guangpufenxi*, 2009, **29**, 3216–3219.

18. S. J. Otto, J. Fracheboud, A. L. Verbeek, R. Boer, J. C. Reijerink-Verheij, J. D. Otten, M. J. Broeders and H. J. de Koning, *Cancer Epidemiol., Biomarkers Prev.*, 2012, **21**, 66–73.

19. H. Allahverdipour, M. Asghari-Jafarabadi and A. Emami, *Women Health*, 2011, **51**, 204–219.

20. M. Heijblom, J. M. Klaase, F. M. van den Engh, T. G. van Leeuwen, W. Steenbergen and S. Manohar, *Technol. Cancer Res. Treat.*, 2011, **10**, 607–623.

21. M. Sentis, *Breast Cancer Res. Treat.*, 2010, **123**(Suppl 1), 11–13.

22. R. Luckmann, *ACP Journal Club*, 2005, **142**, 23.

23. H. Zhang, D. Yee and C. Wang, *Nanomedicine (London, U. K.)*, 2008, **3**, 83–91.

24. L. M. Kenny, A. Al-Nahhas and E. O. Aboagye, *Nucl. Med. Commun.*, 2011, **32**, 333–335.

25. R. Lieberman, *Am. J. Phytomed. Clin. Ther.*, 2012, **19**, 395–396.

26. J. L. Parker, N. Lushina, P. S. Bal, T. Petrella, R. Dent and G. Lopes, *Breast Cancer Res. Treat.*, 2012, **136**, 179–185.

27. C. S. Zhu, P. F. Pinsky, D. W. Cramer, D. F. Ransohoff, P. Hartge, R. M. Pfeiffer, N. Urban, G. Mor, R. C. Bast, Jr., L. E. Moore, A. E. Lokshin, M. W. McIntosh, S. J. Skates, A. Vitonis, Z. Zhang, D. C. Ward, J. T. Symanowski, A. Lomakin, E. T. Fung, P. M. Sluss, N. Scholler, K. H. Lu, A. M. Marrangoni, C. Patriotis, S. Srivastava, S. S. Buys and C. D. Berg, *Cancer Prev. Res.*, 2011, **4**, 375–383.

28. K. Tjensvoll, K. N. Svendsen, J. M. Reuben, S. Oltedal, B. Gilje, R. Smaaland and O. Nordgard, *Biomarkers*, 2012, **17**, 463–470.

29. S. Knappskog, R. Chrisanthar, E. Lokkevik, G. Anker, B. Ostenstad, S. Lundgren, T. Risberg, I. Mjaaland, B. Leirvaag, H. Miletic and P. E. Lonning, *Breast Cancer Res.*, 2012, **14**, R47.

30. E. Rivera and H. Gomez, *Breast Cancer Res.*, 2010, **12**(Suppl 2), S2.

31. W. G. Newman and D. Flockhart, *Pharmacogenomics*, 2012, **13**, 629–631.

32. B. A. Virnig, M. T. Torchia, S. L. Jarosek, S. Durham and T. M. Tuttle, in *Data Points Publication Series*, Agency for Healthcare Research and Quality (US), Rockville (MD), 2011.

33. M. Verma, *Curr. Opin. Clin. Nutr. Metab. Care*, 2013, **16**, 376–384.

34. M. Verma, M. J. Khoury and J. P. Ioannidis, *Cancer Epidemiol., Biomarkers Prev.*, 2013, **22**, 189–200.

35. M. Lucas and S. Gaudieri, *Biomarkers Med.*, 2012, **6**, 133–135.

36. C. Ferrandi, F. Richard, P. Tavano, E. Hauben, V. Barbie, J. P. Gotteland, B. Greco, M. Fortunato, M. F. Mariani, R. Furlan, G. Comi, G. Martino and P. F. Zaratin, *Mult. Scler.*, 2011, **17**, 43–56.

37. A. E. Aiello and G. A. Kaplan, *Biodemography Soc. Biol.*, 2009, **55**, 178–205.

38. G. Lippi, T. Meschi and G. Cervellin, *Eur. J. Intern. Med.*, 2011, **22**, 460–465.

39. G. Brathen, K. S. Bjerve, E. Brodtkorb and G. Bovim, *J. Neurol., Neurosurg. Psychiatry*, 2000, **68**, 342–348.
40. D. Lev, Y. Weigl, M. Hasan, E. Gak, M. Davidovich, C. Vinkler, E. Leshinsky-Silver, T. Lerman-Sagie and N. Watemberg, *Am. J. Med. Genet., Part A*, 2007, **143A**, 921–924.
41. N. Elleuch, I. Feki, E. Turki, M. I. Miladi, A. Boukhris, M. Damak, C. Mhiri, E. Chappuis and F. Woimant, *Rev. Neurol.*, 2010, **166**, 550–552.
42. D. Zhu, J. Wang, L. Ren, Y. Li, B. Xu, Y. Wei, Y. Zhong, X. Yu, S. Zhai, J. Xu and X. Qin, *J. Cell. Biochem.*, 2013, **114**, 448–455.
43. M. Verma, *Antioxid. Redox Signaling*, 2012, **17**, 355–364.
44. S. Viatte, D. Plant and S. Raychaudhuri, *Nat. Rev. Rheumatol.*, 2013, **9**, 141–153.
45. J. Kurko, T. Besenyei, J. Laki, T. T. Glant, K. Mikecz and Z. Szekanecz, *Clin. Rev. Allergy Immunol.*, 2013, **45**(2), 170–179.
46. B. P. Smith, D. H. Best and C. G. Elliott, *Hear Fail. Clin.*, 2012, **8**, 319–330.
47. J. Simino, D. C. Rao and B. I. Freedman, *Curr. Opin. Nephrol. Hypertens.*, 2012, **21**, 500–507.
48. G. Massawi, P. Hickling, D. Hilton and C. Patterson, *Rheumatology (Oxford, U. K.)*, 2003, **42**, 1012–1014.
49. N. Mahfoudh, M. Siala, M. Rihl, A. Kammoun, F. Frikha, H. Fourati, M. Younes, R. Gdoura, L. Gaddour, F. Hakim, Z. Bahloul, S. Baklouti, N. Bargaoui, S. Sellami, A. Hammami and H. Makni, *Clin. Rheumatol.*, 2011, **30**, 1069–1073.
50. G. Kamkamidze, C. Capoulade-Metay, M. Butsashvili, Y. Dudoit, O. Chubinishvili, P. Debre and L. Theodorou, *Georgian Medical News*, 2005, **118**, 74–79.
51. B. J. Grady, D. C. Samuels, G. K. Robbins, D. Selph, J. A. Canter, R. B. Pollard, D. W. Haas, R. Shafer, S. A. Kalams, D. G. Murdock, M. D. Ritchie and T. Hulgan, *J. Acquired Immune Defic. Syndr.*, 2011, **58**, 363–370.
52. J. C. Mar, N. A. Matigian, A. Mackay-Sim, G. D. Mellick, C. M. Sue, P. A. Silburn, J. J. McGrath, J. Quackenbush and C. A. Wells, *PLoS Genetics*, 2011, **7**, e1002207.
53. F. D. Shah, R. Begum, B. N. Vajaria, K. R. Patel, J. B. Patel, S. N. Shukla and P. S. Patel, *Indian J. Clin. Biochem.*, 2011, **26**, 326–334.
54. R. Kirk, *Nat. Rev. Clin. Oncol.*, 2011, **8**, 383.
55. P. Hall, *J. Intern. Med.*, 2012, **271**, 318–320.
56. A. C. Antoniou, A. B. Spurdle, O. M. Sinilnikova, S. Healey, K. A. Pooley, R. K. Schmutzler, B. Versmold, C. Engel, A. Meindl, N. Arnold, W. Hofmann, C. Sutter, D. Niederacher, H. Deissler, T. Caldes, K. Kampjarvi, H. Nevanlinna, J. Simard, J. Beesley, X. Chen, S. L. Neuhausen, T. R. Rebbeck, T. Wagner, H. T. Lynch, C. Isaacs, J. Weitzel, P. A. Ganz, M. B. Daly, G. Tomlinson, O. I. Olopade, J. L. Blum, F. J. Couch, P. Peterlongo, S. Manoukian, M. Barile, P. Radice, C. I. Szabo, L. H. Pereira, M. H. Greene, G. Rennert, F. Lejbkowicz, O. Barnett-

Griness, I. L. Andrulis, H. Ozcelik, A. M. Gerdes, M. A. Caligo, Y. Laitman, B. Kaufman, R. Milgrom, E. Friedman, S. M. Domchek, K. L. Nathanson, A. Osorio, G. Llort, R. L. Milne, J. Benitez, U. Hamann, F. B. Hogervorst, P. Manders, M. J. Ligtenberg, A. M. van den Ouweland, S. Peock, M. Cook, R. Platte, D. G. Evans, R. Eeles, G. Pichert, C. Chu, D. Eccles, R. Davidson, F. Douglas, A. K. Godwin, L. Barjhoux, S. Mazoyer, H. Sobol, V. Bourdon, F. Eisinger, A. Chompret, C. Capoulade, B. Bressac-de Paillerets, G. M. Lenoir, M. Gauthier-Villars, C. Houdayer, D. Stoppa-Lyonnet, G. Chenevix-Trench and D. F. Easton, *Am. J. Hum. Genet.*, 2008, **82**, 937–948.

57. S. A. Gayther, H. Song, S. J. Ramus, S. K. Kjaer, A. S. Whittemore, L. Quaye, J. Tyrer, D. Shadforth, E. Hogdall, C. Hogdall, J. Blaeker, R. DiCioccio, V. McGuire, P. M. Webb, J. Beesley, A. C. Green, D. C. Whiteman, M. T. Goodman, G. Lurie, M. E. Carney, F. Modugno, R. B. Ness, R. P. Edwards, K. B. Moysich, E. L. Goode, F. J. Couch, J. M. Cunningham, T. A. Sellers, A. H. Wu, M. C. Pike, E. S. Iversen, J. R. Marks, M. Garcia-Closas, L. Brinton, J. Lissowska, B. Peplonska, D. F. Easton, I. Jacobs, B. A. Ponder, J. Schildkraut, C. L. Pearce, G. Chenevix-Trench, A. Berchuck and P. D. Pharoah, *Cancer Res.*, 2007, **67**, 3027–3035.

58. P. D. Pharoah and C. Caldas, *Nat. Rev. Clin. Oncol.*, 2010, **7**, 615–616.

59. A. A. Ponomareva, E. Rykova, N. V. Cherdyntseva, E. L. Choinzonov, P. P. Laktionov and V. V. Vlasov, *Mol. Biol. (Mosk.)*, 2011, **45**, 203–217.

60. V. R. Adams and R. D. Harvey, *Am. J. Health-Syst. Pharm.*, 2010, **67**, S3–S9, quiz S15–S16.

61. J. Li, K. Humphreys, T. Heikkinen, K. Aittomaki, C. Blomqvist, P. D. Pharoah, A. M. Dunning, S. Ahmed, M. J. Hooning, J. W. Martens, A. M. van den Ouweland, L. Alfredsson, A. Palotie, L. Peltonen-Palotie, A. Irwanto, H. Q. Low, G. H. Teoh, A. Thalamuthu, D. F. Easton, H. Nevanlinna, J. Liu, K. Czene and P. Hall, *Breast Cancer Res. Treat.*, 2011, **126**, 717–727.

62. C. Turnbull, S. Ahmed, J. Morrison, D. Pernet, A. Renwick, M. Maranian, S. Seal, M. Ghoussaini, S. Hines, C. S. Healey, D. Hughes, M. Warren-Perry, W. Tapper, D. Eccles, D. G. Evans, M. Hooning, M. Schutte, A. van den Ouweland, R. Houlston, G. Ross, C. Langford, P. D. Pharoah, M. R. Stratton, A. M. Dunning, N. Rahman and D. F. Easton, *Nat. Genet.*, 2010, **42**, 504–507.

63. P. R. Srinivas, M. Verma, Y. Zhao and S. Srivastava, *Clin. Chem.*, 2002, **48**, 1160–1169.

64. C. Panis, L. Pizzatti, A. C. Herrera, R. Cecchini and E. Abdelhay, *Cancer Lett.*, 2013, **330**, 57–66.

65. M. Verma, *Methods Mol. Biol.*, 2009, **471**, 197–215.

66. K. Na, M. J. Lee, H. J. Jeong, H. Kim and Y. K. Paik, *Methods Mol. Biol.*, 2012, **854**, 223–237.

67. E. S. Baker, T. Liu, V. A. Petyuk, K. E. Burnum-Johnson, Y. M. Ibrahim, G. A. Anderson and R. D. Smith, *Genome Med.*, 2012, **4**, 63.

68. N. G. Anderson, *Clin. Chem.*, 2010, **56**, 154–160.
69. W. Rozek, J. Horning, J. Anderson and P. Ciborowski, *Proteomics: Clin. Appl.*, 2008, **2**, 1498–1507.
70. E. J. Toonen, C. Gilissen, B. Franke, W. Kievit, A. M. Eijsbouts, A. A. den Broeder, S. V. van Reijmersdal, J. A. Veltman, H. Scheffer, T. R. Radstake, P. L. van Riel, P. Barrera and M. J. Coenen, *PloS One*, 2012, **7**, e33199.
71. P. Emery, *Rheumatology (Oxford, U. K.)*, 2012, **51**(Suppl 5), v22–v30.
72. E. Solau-Gervais, C. Prudhomme, P. Philippe, A. Duhamel, C. Dupont-Creteur, J. L. Legrand, E. Houvenagel and R. M. Flipo, *Jt., Bone, Spine*, 2012, **79**, 281–284.
73. H. T. Maecker, T. M. Lindstrom, W. H. Robinson, P. J. Utz, M. Hale, S. D. Boyd, S. S. Shen-Orr and C. G. Fathman, *Nat. Rev. Rheumatol.*, 2012, **8**, 317–328.
74. L. Iaccarino, A. Ghirardello, M. Canova, M. Zen, S. Bettio, L. Nalotto, L. Punzi and A. Doria, *Autoimmun. Rev.*, 2011, **10**, 553–558.
75. N. Leung, *Med. J. Malaysia*, 2005, **60**(Suppl B), 63–66.
76. H. Orbach, G. Zandman-Goddard, H. Amital, V. Barak, Z. Szekanecz, G. Szucs, K. Danko, E. Nagy, T. Csepany, J. F. Carvalho, A. Doria and Y. Shoenfeld, *Ann. N. Y. Acad. Sci.*, 2007, **1109**, 385–400.
77. A. Bartolomucci, G. M. Pasinetti and S. R. Salton, *Neuroscience*, 2010, **170**, 289–297.
78. R. Mogelvang, S. H. Pedersen, A. Flyvbjerg, M. Bjerre, A. Z. Iversen, S. Galatius, J. Frystyk and J. S. Jensen, *Am. J. Cardiol.*, 2012, **109**, 515–520.
79. O. B. Bleijerveld, Y. N. Zhang, S. Beldar, I. E. Hoefer, S. K. Sze, G. Pasterkamp and D. P. de Kleijn, *Proteomics: Clin. Appl.*, 2013, **7**(7–8), 490–503.
80. W. S. Wang, X. H. Liu, L. X. Liu, D. Y. Jin, P. Y. Yang and X. L. Wang, *J. Proteomics*, 2013, **84C**, 52–60.
81. A. J. Apter, *J. Allergy Clin. Immunol.*, 2010, **125**, 79–84.
82. G. Pendyala and H. S. Fox, *Genome Med.*, 2010, **2**, 22.
83. E. Colombo, B. Borgiani, C. Verderio and R. Furlan, *Front. Physiol*, 2012, **3**, 63.
84. D. S. Michaud, J. Izard, C. S. Wilhelm-Benartzi, D. H. You, V. A. Grote, A. Tjonneland, C. C. Dahm, K. Overvad, M. Jenab, V. Fedirko, M. C. Boutron-Ruault, F. Clavel-Chapelon, A. Racine, R. Kaaks, H. Boeing, J. Foerster, A. Trichopoulou, P. Lagiou, D. Trichopoulos, C. Sacerdote, S. Sieri, D. Palli, R. Tumino, S. Panico, P. D. Siersema, P. H. Peeters, E. Lund, A. Barricarte, J. M. Huerta, E. Molina-Montes, M. Dorronsoro, J. R. Quiros, E. J. Duell, W. Ye, M. Sund, B. Lindkvist, D. Johansen, K. T. Khaw, N. Wareham, R. C. Travis, P. Vineis, H. B. Bueno-de-Mesquita and E. Riboli, *Gut*, 2012, **62**(12), 1764–1770.
85. V. A. Grote, R. Kaaks, A. Nieters, A. Tjonneland, J. Halkjaer, K. Overvad, M. R. Skjelbo Nielsen, M. C. Boutron-Ruault, F. Clavel-Chapelon, A. Racine, B. Teucher, S. Becker, T. Pischon, H. Boeing, A. Trichopoulou, C. Cassapa, V. Stratigakou, D. Palli, V. Krogh, R. Tumino, P. Vineis,

S. Panico, L. Rodriguez, E. J. Duell, M. J. Sanchez, M. Dorronsoro, C. Navarro, A. B. Gurrea, P. D. Siersema, P. H. Peeters, W. Ye, M. Sund, B. Lindkvist, D. Johansen, K. T. Khaw, N. Wareham, N. E. Allen, R. C. Travis, V. Fedirko, M. Jenab, D. S. Michaud, S. C. Chuang, D. Romaguera, H. B. Bueno-de-Mesquita and S. Rohrmann, *Br. J. Cancer*, 2012, **106**, 1866–1874.

86. S. Tejpar, M. Bertagnolli, F. Bosman, H. J. Lenz, L. Garraway, F. Waldman, R. Warren, A. Bild, D. Collins-Brennan, H. Hahn, D. P. Harkin, R. Kennedy, M. Ilyas, H. Morreau, V. Proutski, C. Swanton, I. Tomlinson, M. Delorenzi, R. Fiocca, E. Van Cutsem and A. Roth, *Oncologist*, 2010, **15**, 390–404.

87. N. Miura, Y. Osaki, M. Nagashima, M. Kohno, K. Yorozu, K. Shomori, T. Kanbe, K. Oyama, K. Kishimoto, S. Maruyama, E. Noma, Y. Horie, M. Kudo, S. Sakaguchi, Y. Hirooka, H. Ito, H. Kawasaki, J. Hasegawa and G. Shiota, *BMC Gastroenterol.*, 2010, **10**, 46.

88. Y. Sun, Z. Zang, X. Xu, Z. Zhang, L. Zhong, W. Zan, Y. Zhao and L. Sun, *Int. J. Mol. Sci.*, 2010, **11**, 1423–1433.

89. B. Hamrita, K. Chahed, M. Trimeche, C. L. Guillier, P. Hammann, A. Chaieb, S. Korbi and L. Chouchane, *Clin. Chim. Acta*, 2009, **404**, 111–118.

90. M. A. Fonseca-Sanchez, S. Rodriguez Cuevas, G. Mendoza-Hernandez, V. Bautista-Pina, E. Arechaga Ocampo, A. Hidalgo Miranda, V. Quintanar Jurado, L. A. Marchat, E. Alvarez-Sanchez, C. Perez Plasencia and C. Lopez-Camarillo, *Int. J. Oncol.*, 2012, **41**, 670–680.

91. B. Hamrita, H. Ben Nasr, P. Hammann, L. Kuhn, A. Ben Anes, S. Dimassi, A. Chaieb, H. Khairi and K. Chahed, *Ann. Biol. Clin.*, 2012, **70**, 553–565.

92. B. Hamrita, H. B. Nasr, K. Chahed and L. Chouchane, *Gulf J. Oncolog.*, 2011, **1**, 36–44.

93. Y. M. Chung, Y. D. Yoo, J. K. Park, Y. T. Kim and H. J. Kim, *Anticancer Res.*, 2001, **21**, 1129–1133.

94. F. Cirillo, G. Nassa, R. Tarallo, C. Stellato, M. R. De Filippo, C. Ambrosino, M. Baumann, T. A. Nyman and A. Weisz, *J. Proteome Res.*, 2013, **12**, 421–431.

95. N. Q. Liu, R. B. Braakman, C. Stingl, T. M. Luider, J. W. Martens, J. A. Foekens and A. Umar, *J. Mammary Gland Biol. Neoplasia*, 2012, **17**, 155–164.

96. A. M. Gonzalez-Angulo, B. T. Hennessy, F. Meric-Bernstam, A. Sahin, W. Liu, Z. Ju, M. S. Carey, S. Myhre, C. Speers, L. Deng, R. Broaddus, A. Lluch, S. Aparicio, P. Brown, L. Pusztai, W. F. Symmans, J. Alsner, J. Overgaard, A. L. Borresen-Dale, G. N. Hortobagyi, K. R. Coombes and G. B. Mills, *Clin. Proteomics*, 2011, **8**, 11.

97. D. M. Pedrotty, M. P. Morley and T. P. Cappola, *Prog. Cardiovasc. Dis.*, 2012, **55**, 64–69.

98. L. Desire, E. Blondiaux, J. Carriere, R. Haddad, O. Sol, P. Fehlbaum-Beurdeley, R. Einstein, W. Zhou and M. P. Pando, *J. Alzheimer's Dis.*, 2013, **34**, 469–483.

99. C. Q. Yi, C. H. Ma, Z. P. Xie, Y. Cao, G. Q. Zhang, X. K. Zhou and Z. Q. Liu, *GMR, Genet. Mol. Res.*, 2013, **12**.

100. J. Caetano-Lopes, A. Rodrigues, A. Lopes, A. C. Vale, M. A. Pitts-Kiefer, B. Vidal, I. P. Perpetuo, J. Monteiro, Y. T. Konttinen, M. F. Vaz, A. Nazarian, H. Canhao and J. E. Fonseca, *Clin. Rev. Allergy Immunol.*, 2013.

101. J. C. Laguna and M. Alegret, *Pharmacogenomics*, 2012, **13**, 477–495.

102. H. Peiris, R. Raghupathi, C. F. Jessup, M. P. Zanin, D. Mohanasundaram, K. D. Mackenzie, T. Chataway, J. N. Clarke, J. Brealey, P. T. Coates, M. A. Pritchard and D. J. Keating, *Endocrinology*, 2012, **153**, 5212–5221.

103. T. S. Nielsen, U. Kampmann, R. R. Nielsen, N. Jessen, L. Orskov, S. B. Pedersen, J. O. Jorgensen, S. Lund and N. Moller, *J. Clin. Endocrinol. Metab.*, 2012, **97**, E1348–1352.

104. J. M. Swanson, G. C. Wood, L. Xu, L. E. Tang, B. Meibohm, R. Homayouni, M. A. Croce and T. C. Fabian, *PloS One*, 2012, **7**, e42065.

105. I. Martin-Loeches, E. Papiol, R. Almansa, G. Lopez-Campos, J. F. Bermejo-Martin and J. Rello, *Medicina Intensiva*, 2012, **36**, 257–263.

106. S. Jain and P. Heutink, *Neuron*, 2010, **68**, 207–217.

107. E. Borgan, B. Sitter, O. C. Lingjaerde, H. Johnsen, S. Lundgren, T. F. Bathen, T. Sorlie, A. L. Borresen-Dale and I. S. Gribbestad, *BMC Cancer*, 2010, **10**, 628.

108. J. Hardouin, J. P. Lasserre, L. Sylvius, R. Joubert-Caron and M. Caron, *Ann. N. Y. Acad. Sci.*, 2007, **1107**, 223–230.

109. M. R. Amin, M. S. Siddiqui, D. Ahmed, F. Ahmed and A. Hossain, *Int. J. Bioinf. Res. Appl.*, 2011, **7**, 287–298.

110. M. Careaga and P. Ashwood, *Methods Mol. Biol.*, 2012, **934**, 219–240.

111. S. D. Funk and A. W. Orr, *Pharmacol. Res.*, 2013, **67**, 42–52.

112. J. I. Odegaard and A. Chawla, *Cold Spring Harbor Perspect. Med.*, 2012, **2**, a007724.

113. H. S. Kim and M. S. Lee, *Curr. Mol. Med.*, 2009, **9**, 30–44.

114. M. Gonzalez-Juarrero, *Immunotherapy*, 2012, **4**, 187–199.

115. A. J. Battersby, B. Kampmann and S. Burl, *Clin. Dev. Immunol.*, 2012, **2012**, 430972.

116. A. T. Weeraratna, A. Kalehua, I. Deleon, D. Bertak, G. Maher, M. S. Wade, A. Lustig, K. G. Becker, W. Wood, 3rd, D. G. Walker, T. G. Beach and D. D. Taub, *Exp. Cell Res.*, 2007, **313**, 450–461.

117. R. P. Young and R. J. Hopkins, *Cancer*, 2013, **119**, 1761.

118. Y. Godet, E. Fabre, M. Dosset, M. Lamuraglia, E. Levionnois, P. Ravel, N. Benhamouda, A. Cazes, F. Le Pimpec-Barthes, B. Gaugler, P. Langlade-Demoyen, X. Pivot, P. Saas, B. Maillere, E. Tartour, C. Borg and O. Adotevi, *Clin. Cancer Res.*, 2012, **18**, 2943–2953.

119. H. Wang, V. K. Tso, C. M. Slupsky and R. N. Fedorak, *Future Oncol.*, 2010, **6**, 1395–1406.

120. Y. He, Z. Yu, I. Giegling, L. Xie, A. M. Hartmann, C. Prehn, J. Adamski, R. Kahn, Y. Li, T. Illig, R. Wang-Sattler and D. Rujescu, *Transl. Psychiatry*, 2012, **2**, e149.

121. H. Janeckova, K. Hron, P. Wojtowicz, E. Hlidkova, A. Baresova, D. Friedecky, L. Zidkova, P. Hornik, D. Behulova, D. Prochazkova, H. Vinohradska, K. Peskova, P. Bruheim, V. Smolka, S. Stastna and T. Adam, *J. Chrom. A*, 2012, **1226**, 11–17.

122. H. Song, L. Wang, H. L. Liu, X. B. Wu, H. S. Wang, Z. H. Liu, Y. Li, D. C. Diao, H. L. Chen and J. S. Peng, *Oncol. Rep.*, 2011, **26**, 431–438.

123. L. Pedersen, J. Elers and V. Backer, *Phys. Sportsmed.*, 2011, **39**, 163–171.

124. H. Rose, H. Low, E. Dewar, M. Bukrinsky, J. Hoy, A. Dart and D. Sviridov, *Atherosclerosis*, 2013, **229**(1), 206–211.

125. R. A. Koeth, Z. Wang, B. S. Levison, J. A. Buffa, E. Org, B. T. Sheehy, E. B. Britt, X. Fu, Y. Wu, L. Li, J. D. Smith, J. A. Didonato, J. Chen, H. Li, G. D. Wu, J. D. Lewis, M. Warrier, J. M. Brown, R. M. Krauss, W. H. Tang, F. D. Bushman, A. J. Lusis and S. L. Hazen, *Nat. Med.*, 2013, **19**, 576–585.

126. N. Friedrich, *J. Endocrinol.*, 2012, **215**, 29–42.

127. R. Wang-Sattler, Z. Yu, C. Herder, A. C. Messias, A. Floegel, Y. He, K. Heim, M. Campillos, C. Holzapfel, B. Thorand, H. Grallert, T. Xu, E. Bader, C. Huth, K. Mittelstrass, A. Doring, C. Meisinger, C. Gieger, C. Prehn, W. Roemisch-Margl, M. Carstensen, L. Xie, H. Yamanaka-Okumura, G. Xing, U. Ceglarek, J. Thiery, G. Giani, H. Lickert, X. Lin, Y. Li, H. Boeing, H. G. Joost, M. H. de Angelis, W. Rathmann, K. Suhre, H. Prokisch, A. Peters, T. Meitinger, M. Roden, H. E. Wichmann, T. Pischon, J. Adamski and T. Illig, *Mol. Syst. Biol.*, 2012, **8**, 615.

128. I. du Preez and T. du Loots, *OMICS*, 2012, **16**, 596–603.

129. C. Ficicioglu and D. Bearden, *Pediatr. Neur.*, 2011, **45**, 283–291.

130. T. Kobayashi, S. Nishiumi, A. Ikeda, T. Yoshie, A. Sakai, A. Matsubara, Y. Izumi, H. Tsumura, M. Tsuda, H. Nishisaki, N. Hayashi, S. Kawano, Y. Fujiwara, H. Minami, T. Takenawa, T. Azuma and M. Yoshida, *Cancer Epidemiol., Biomarkers Prev.*, 2013, **22**, 571–579.

131. M. Verma, *Technol. Cancer Res. Treat.*, 2005, **4**, 295–301.

132. S. Khare and M. Verma, *Methods Mol. Biol. (Clifton, N.J.)*, 2012, **863**, 177.

133. M. D. Anway and M. K. Skinner, *Endocrinology*, 2006, **147**, S43–S49.

134. F. Fang, S. Turcan, A. Rimner, A. Kaufman, D. Giri, L. G. Morris, R. Shen, V. Seshan, Q. Mo, A. Heguy, S. B. Baylin, N. Ahuja, A. Viale, J. Massague, L. Norton, L. T. Vahdat, M. E. Moynahan and T. A. Chan, *Sci. Transl. Med.*, 2011, **3**, 75ra25.

135. J. M. Yi, M. Dhir, L. Van Neste, S. R. Downing, J. Jeschke, S. C. Glockner, M. de Freitas Calmon, C. M. Hooker, J. M. Funes, C. Boshoff, K. M. Smits, M. van Engeland, M. P. Weijenberg, C. A. Iacobuzio-Donahue, J. G. Herman, K. E. Schuebel, S. B. Baylin and N. Ahuja, *Clin. Cancer Res.*, 2011, **17**, 1535–1545.

136. L. Yu, N. W. Todd, L. Xing, Y. Xie, H. Zhang, Z. Liu, H. Fang, J. Zhang, R. L. Katz and F. Jiang, *Int. J. Cancer*, 2010, **127**(12), 2870–2878.

137. A. Harder, S. Titze, L. Herbst, T. Harder, K. Guse, S. Tinschert, D. Kaufmann, T. Rosenbaum, V. F. Mautner, E. Windt, U. Wahllander-Danek, K. Wimmer, S. Mundlos and H. Peters, *Twin Res. Hum. Genet.*, 2010, **13**, 582–594.

138. J. Ashley-Martin, J. VanLeeuwen, A. Cribb, P. Andreou and J. R. Guernsey, *Int. J. Environ. Res. Public Health*, 2012, **9**, 1846–1858.
139. J. Qiu, R. Yang, Y. Rao, Y. Du and F. W. Kalembo, *PloS One*, 2012, 7, e36497.
140. M. Alshalalfa, *Adv. Bioinf.*, 2012, **2012**, 839837.
141. K. W. Chang, S. Y. Kao, Y. H. Wu, M. M. Tsai, H. F. Tu, C. J. Liu, M. T. Lui and S. C. Lin, *Oral Oncol.*, 2013, **49**, 27–33.
142. N. Nishida, M. Nagahara, T. Sato, K. Mimori, T. Sudo, F. Tanaka, K. Shibata, H. Ishii, K. Sugihara, Y. Doki and M. Mori, *Clin. Cancer Res.*, 2012, **18**, 3054–3070.
143. P. S. Mitchell, R. K. Parkin, E. M. Kroh, B. R. Fritz, S. K. Wyman, E. L. Pogosova-Agadjanyan, A. Peterson, J. Noteboom, K. C. O'Briant, A. Allen, D. W. Lin, N. Urban, C. W. Drescher, B. S. Knudsen, D. L. Stirewalt, R. Gentleman, R. L. Vessella, P. S. Nelson, D. B. Martin and M. Tewari, *Proc. Natl. Acad. Sci. U. S. A.*, 2008, **105**, 10513–10518.
144. C. Kumagai, B. Kalman, F. A. Middleton, T. Vyshkina and P. T. Massa, *J. Neuroimmunol.*, 2012, **246**, 51–57.
145. J. Zhao, C. W. Forsberg, J. Goldberg, N. L. Smith and V. Vaccarino, *BMC Med. Genet.*, 2012, **13**, 100.
146. L. N. Zhang, P. P. Liu, L. Wang, F. Yuan, L. Xu, Y. Xin, L. J. Fei, Q. L. Zhong, Y. Huang, L. M. Hao, X. J. Qiu, Y. Le, M. Ye and S. Duan, *PloS One*, 2013, **8**, e63455.
147. V. K. Rakyan, H. Beyan, T. A. Down, M. I. Hawa, S. Maslau, D. Aden, A. Daunay, F. Busato, C. A. Mein, B. Manfras, K. R. Dias, C. G. Bell, J. Tost, B. O. Boehm, S. Beck and R. D. Leslie, *PLoS Genetics*, 2011, **B**, e1002300.
148. L. M. Villeneuve, M. A. Reddy, L. L. Lanting, M. Wang, L. Meng and R. Natarajan, *Proc. Natl. Acad. Sci. U. S. A.*, 2008, **105**, 9047–9052.
149. C. Andraos, G. Koorsen, J. C. Knight and L. Bornman, *Hum. Immunol.*, 2011, **72**, 262–268.
150. V. Bollati, D. Galimberti, L. Pergoli, E. Dalla Valle, F. Barretta, F. Cortini, E. Scarpini, P. A. Bertazzi and A. Baccarelli, *Brain, Behav., Immun.*, 2011, **25**, 1078–1083.
151. R. P. Nagarajan, K. A. Patzel, M. Martin, D. H. Yasui, S. E. Swanberg, I. Hertz-Picciotto, R. L. Hansen, J. Van de Water, I. N. Pessah, R. Jiang, W. P. Robinson and J. M. LaSalle, *Autism Res.*, 2008, **1**, 169–178.
152. S. Saji, M. Kawakami, S. Hayashi, N. Yoshida, M. Hirose, S. Horiguchi, A. Itoh, N. Funata, S. L. Schreiber, M. Yoshida and M. Toi, *Oncogene*, 2005, **24**, 4531–4539.
153. Z. Zhang, H. Yamashita, T. Toyama, H. Sugiura, Y. Ando, K. Mita, M. Hamaguchi, Y. Hara, S. Kobayashi and H. Iwase, *Breast Cancer Res. Treat.*, 2005, **94**, 11–16.
154. Y. Ruike, Y. Imanaka, F. Sato, K. Shimizu and G. Tsujimoto, Genome-wide analysis of aberrant methylation in human breast cancer cells using methyl-DNA immunoprecipitation combined with high-throughput sequencing, *BMC Genomics*, 2010, **11**, 137.

155. J. Soares, A. E. Pinto, C. V. Cunha, S. Andre, I. Barao, J. M. Sousa and M. Cravo, *Cancer*, 1999, **85**, 112–118.
156. K. Jackson, M. C. Yu, K. Arakawa, E. Fiala, B. Youn, H. Fiegl, E. Muller-Holzner, M. Widschwendter and M. Ehrlich, *Cancer Biol. Ther.*, 2004, **3**, 1225–1231.
157. A. Narayan, W. Ji, X. Y. Zhang, A. Marrogi, J. R. Graff, S. B. Baylin and M. Ehrlich, *Int. J. Cancer*, 1998, **77**, 833–838.
158. S. L. Zhou and L. D. Wang, *World J. Gastroenterol.*, 2010, **16**, 2348–2354.
159. I. Van der Auwera, R. Limame, P. van Dam, P. B. Vermeulen, L. Y. Dirix and S. J. Van Laere, *Br. J. Cancer*, 2010, **103**, 532–541.
160. A. M. Abdel Rahman, S. D. Kamath, S. Gagne, A. L. Lopata and R. Helleur, *J. Proteome Res.*, 2013, **12**, 647–656.
161. T. L. Mertz, *J. Am. Osteopath. Assoc.*, 2011, **111**(S27–S29), S32.
162. M. J. Chao, S. V. Ramagopalan, B. M. Herrera, M. R. Lincoln, D. A. Dyment, A. D. Sadovnick and G. C. Ebers, *Hum. Mol. Genet.*, 2009, **18**, 261–266.
163. F. Li, S. Li, H. Chang, Y. Nie, L. Zeng, X. Zhang and Y. Wang, *Genet. Test. Mol. Biomarkers*, 2013, **17**(9), 656–661.
164. D. R. Taylor, *Thorax*, 2009, **64**, 261–264.
165. M. G. Tektonidou and M. M. Ward, *Nat. Rev. Rheumatol.*, 2011, **7**, 708–717.
166. G. B. Lim, *Nat. Rev. Cardiol.*, 2012, **9**, 672.
167. I. Rahman, R. Atout, N. L. Pedersen, U. de Faire, J. Frostegard, E. Ninio, A. M. Bennet and P. K. Magnusson, *Atherosclerosis*, 2011, **218**, 117–122.
168. S. Sour, M. Belarbi, D. Khaldi, N. Benmansour, N. Sari, A. Nani, F. Chemat and F. Visioli, *Br. J. Nutr.*, 2012, **107**, 1800–1805.
169. I. Okur, L. Tumer, F. S. Ezgu, E. Yesilkaya, A. Aral, S. O. Oktar, A. Bideci and A. Hasanoglu, *J. Pediatr. Endocrinol. Metab.*, 2013, **26**(7–8), 657–662.
170. M. Kucukazman, N. Ata, B. Yavuz, K. Dal, O. Sen, O. S. Deveci, K. Agladioglu, A. O. Yeniova, Y. Nazligul and D. T. Ertugrul, *Eur. J. Gastroenterol. Hepatol.*, 2013, **25**, 147–151.
171. S. G. Wannamethee, P. H. Whincup, L. Lennon and N. Sattar, *Arch. Intern. Med.*, 2007, **167**, 1510–1517.
172. W. Mullen, C. Delles and H. Mischak, *Curr. Opin. Nephrol. Hypertens.*, 2011, **20**, 654–661.
173. *Duke Medicine Health News*, 2012, **18**, 4–5.
174. K. Nakanishi and C. Watanabe, *Clin. Chim. Acta*, 2009, **402**, 171–175.
175. A. Maisel, S. X. Neath, J. Landsberg, C. Mueller, R. M. Nowak, W. F. Peacock, P. Ponikowski, M. Mockel, C. Hogan, A. H. Wu, M. Richards, P. Clopton, G. S. Filippatos, S. Di Somma, I. Anand, L. L. Ng, L. B. Daniels, R. H. Christenson, M. Potocki, J. McCord, G. Terracciano, O. Hartmann, A. Bergmann, N. G. Morgenthaler and S. D. Anker, *Eur. J. Heart Failure*, 2012, **14**, 278–286.
176. D. P. Dosanjh, M. Bakir, K. A. Millington, A. Soysal, Y. Aslan, S. Efee, J. J. Deeks and A. Lalvani, *PloS One*, 2011, **6**, e28754.

177. R. Ribeiro-Rodrigues, T. Resende, Co, J. L. Johnson, F. Ribeiro, M. Palaci, R. T. Sa, E. L. Maciel, F. E. Pereira Lima, V. Dettoni, Z. Toossi, W. H. Boom, R. Dietze, J. J. Ellner and C. S. Hirsch, *Clin. Diagn. Lab. Immunol.*, 2002, **9**, 818–823.
178. M. Sakamoto, *J. Gastroenterol.*, 2009, **44**(Suppl 19), 108–111.
179. I. Simon, Y. Liu, K. L. Krall, N. Urban, R. L. Wolfert, N. W. Kim and M. W. McIntosh, *Gynecol. Oncol.*, 2007, **106**, 112–118.
180. J. N. Mubiru, A. J. Valente and D. A. Troyer, *The Prostate*, 2005, **65**, 117–123.
181. D. J. Birnbaum, S. Laibe, A. Ferrari, A. Lagarde, A. J. Fabre, G. Monges, D. Birnbaum and S. Olschwang, *Transl. Oncol.*, 2012, **5**, 72–76.
182. F. Al Dayel, *J. Infect. Public Health*, 2012, **5**(Suppl 1), S31–S34.
183. H. Takeda, N. Takigawa, K. Ohashi, D. Minami, I. Kataoka, E. Ichihara, N. Ochi, M. Tanimoto and K. Kiura, *Exp. Cell Res.*, 2013, **319**, 417–423.
184. S. Bernholtz, Y. Laitman, B. Kaufman, S. Shimon-Paluch and E. Friedman, *Breast Cancer Res. Treat.*, 2012, **132**, 669–673.
185. M. Gonen, Z. Sun, M. E. Figueroa, J. P. Patel, O. Abdel-Wahab, J. Racevskis, R. P. Ketterling, H. Fernandez, J. M. Rowe, M. S. Tallman, A. Melnick, R. L. Levine and E. Paietta, *Blood*, 2012, **120**, 2297–2306.
186. R. Claus, B. Hackanson, A. R. Poetsch, M. Zucknick, M. Sonnet, N. Blagitko-Dorfs, J. Hiller, S. Wilop, T. H. Brummendorf, O. Galm, U. Platzbecker, J. C. Byrd, K. Dohner, H. Dohner, M. Lubbert and C. Plass, *Int. J. Cancer*, 2012, **131**, E138–E142.
187. P. A. Ho, M. A. Kutny, T. A. Alonzo, R. B. Gerbing, J. Joaquin, S. C. Raimondi, A. S. Gamis and S. Meshinchi, *Pediatr. Blood Cancer*, 2011, **57**, 204–209.
188. C. A. Krusche, P. Wulfing, C. Kersting, A. Vloet, W. Bocker, L. Kiesel, H. M. Beier and J. Alfer, *Breast Cancer Res. Treat.*, 2005, **90**, 15–23.
189. W. M. Linehan, *Genome Res.*, 2012, **22**, 2089–2100.
190. J. Hur, H. J. Lee, J. E. Nam, Y. J. Kim, Y. J. Hong, H. Y. Kim, S. K. Kim, J. Chang, J. H. Kim, K. Y. Chung, H. S. Lee and B. W. Choi, *BMC Cancer*, 2012, **12**, 392.

CHAPTER 14

Next-generation Molecular Markers: Challenges, Applications, and Future Perspectives

MUKESH VERMA,[a] DEBMALYA BARH,*[b] SYED SHAH HASSAN[c] AND VASCO AC AZEVEDO[c]

[a] Epidemiology and Genomics Research Program, Division of Cancer Control and Population Sciences, National Cancer Institute (NCI), National Institutes of Health (NIH), 9609 Medical Center Drive, Rockville, MD 20850, USA; [b] Centre for Genomics and Applied Gene Technology, Institute of Integrative Omics and Applied Biotechnology (IIOAB), Nonakuri, Purba Medinipur, West Bengal-721172, India; [c] Departamento de Biologia Geral, Instituto de Ciências Biológicas (ICB), Universidade Federal de Minas Gerais, Pampulha, Belo Horizonte, Minas Gerais, Brazil
*Email: dr.barh@gmail.com

14.1 INTRODUCTION

14.1.1 Biological Limitations in Cancer Biomarker Discovery

Metabolomics represent the phenotypic state of a disease.[1] Here we describe the challenges in translational research when metabolomic biomarkers are used for disease diagnosis, risk assessment and prognosis. One barrier is a paucity of commercially available standards for identification and quantification of metabolites for humans, as well as data comparisons across studies.[2] Particular consideration should be given to the quantity needed and which compounds should be included. Additionally, the lack of carefully

Molecular Biology and Biotechnology, 6th Edition
Edited by Ralph Rapley and David Whitehouse
© The Royal Society of Chemistry 2015
Published by the Royal Society of Chemistry, www.rsc.org

selected, well annotated, and easily accessible reference samples greatly limits investigations. Although the standard National Institute of Standards and Technology (NIST) pooled plasma reference sets are available, they are of limited value in determining the individual metabolite variation. Standards for other biological media, such as urine, are needed for investigating the most physiologically plausible pathways that best reflect the etiology of disease. In the case of epidemiologic consortia, different laboratories may be involved in sample analyses; and in turn, inter-laboratory comparisons are problematic without reference samples. An opportunity surrounds the use of archived samples from population-based studies to gain insights into optimizing collection and storage protocols for different media for sample integrity. In addition to high quality samples, quantitative robustness requires provision for quality controls, pooled references, and standard reference materials to control for instrumentation variability drift and allow for comparisons across laboratories. Well-standardized protocols for sample collection, storage, and analysis are also needed.

Cancer cells show progressive heterogeneity at cellular and molecular level. Transient expression of various intermediated components (proteins/ genes/ metabolites *etc.*) during intermediate stages of carcinogenesis affects detection of biomarkers resulting false positive or negative. Similarly, biomarkers may get affected in response to therapy and other internal physiological and pathological factors including age of the subject. Other influential factors in biomarker variation include food habit, nutrition, and, life style.

14.1.2 Clinical and Pathologic Factors in Cancer Biomarker Discovery

Although a considerable amount of knowledge has been obtained in understanding disease biology and identification of biomarkers which can detect a disease, implication of that information in clinic is still challenging. Clinical validation is the main hurdle in the process. Just to make our point clear we are describing an example in cancer. In one case control study of the European Prospective Investigation into Cancer and nutrition (EPIC) where more than 300 breast cancer patients and matched controls were tested for breast cancer over a period of three years using a panel of eight serum markers (osteopontin, haptoglobin, cancer antigen 15-3, carcinoembryonic antigen, cancer antigen-125, prolactin, cancer antigen 19-9, and alpha-fetoprotein), very low specificity (50%) and sensitivity (50%) were observed.[3] This may be due to different subtypes of breast cancer present in collected samples. Such epidemiologic studies should select a broader target set of potential biomarkers which could be enabled by antibody array technologies where profiles of up to 100 antibodies can be followed simultaneously. Making different groups, based on subtypes of cancer based on the status of hormone receptors (estrogen and progesterone), might also be helpful.

Although current biomarkers can differentiate normal versus cancer condition; more accurate detection of each stages of cancer such as initiation, progression, metastasis, early and aggressive stages, and recurrence are required to be developed.

14.1.3   Analytical Limitations in Cancer Biomarker Discovery

Analytical limitation exists in different biomarkers mainly because they have not been validated. Another area that needs progress/attention is the cost and high throughput. Systematic and adequate progress has not been made is the application of biomarkers in clinic. Proper analytical and clinical validation of early markers has not been achieved and the number of biomarkers approved by the FDA for clinical use is very small.[4] Clinical validation of identified biomarkers especially is the key challenge in the field. Here is one example to emphasize our point. The National Cancer Institute has developed guidelines for the analytical and clinical validation of biomarkers but none of the biomarkers have been validated to date.[5] Integration of genomic and proteomic markers with epigenetic markers may help us subtyping different cancers and cancer stages (and this is true for other diseases also).[6] Values of biomarkers may be different in different tissues. For example, results of methylation profiling from blood and tissues are different. Koestler *et al.* conducted a systematic epigenome-wide methylation analysis and demonstrated that shifts in leukocyte subpopulations might account for a considerable proportion of variability in these patterns.[7] Multiplexing of biomarkers may reduce false positive results in screening studies where intention is to identify populations which are at high risk of developing a disease. Quantitative imaging data storage and maintenance have their own challenges as we discussed above. Whether miR expression is localized in a specific part of the diseased tissue has to be carefully evaluated.[2] In a tissue biopsy the local concentration (number of miRs) may be low or high. Determining the accurate level of miRs is very critical.

Although opportunities exist for the use of metabolomic profiling in population-based research, multiple challenges prevent the proper integration of these data into epidemiologic studies for meaningful interpretation.[1,2] Therefore, epidemiologists should strive to understand the principles of metabolomics to better determine and apply their appropriate uses. In addition, it is essential for metabolite profiles to be validated both for analytical purposes and clinical use as biomarkers. Some of the primary challenges to population-based studies include the incorporation of this technology in the initial study design, identifying appropriate sample collection protocols and quality control methods, and the selection of analytical approaches and quantification techniques. New strategies will likely be necessary for combining data from different analytical platforms to allow generalizability of data and interpretation. Furthermore, the need to develop better methods for analyzing large amounts of data remains a critical barrier, including improvement of statistical and bioinformatics methods for data analysis.

Unfortunately, metabolomic profiling is used far too frequently *post hoc* in epidemiology studies. Investigators should strive for well-designed, prospective studies that would establish causal effect, as well as temporal changes. In any case, causality is difficult to establish with a clear biological explanation in association studies, even with particularly well-designed studies. Therefore, it is critical to improve methodologies for integration of metabolomics data with other data, such as genomic and proteomic, to help understand the functional and temporal relationship between a biomarker and an effect.

In general, the concentrations of biomarkers are very low in biospecimens collected from patients therefore quantification of markers requires highly sensitive assays.[2] Standard procedures, reference materials, and quality controls need to be strictly followed to ensure accuracy as well as reproducibility of the assay for a given biomarker. Unfortunately, such good manufacturing/laboratory practice (GMP/GLP) and quality control are not followed in most cases in biomarker discovery.

14.1.4 Intellectual Property in Cancer Biomarker Discovery

The National Institutes of Health (NIH) provides an opportunity to investigators to keep IP rights although the NIH supports the project. Leaving regulatory, financial, intellectual property, and cultural issues aside, developing a diagnostic biomarker often requires expertise or patients that its discoverer may not possess. We know very well that discovery, development, and validation of a biomarker is costly and no assurance that a project will succeed. Therefore, investors are often not interested in investing in such projects. However, investors may agree if they are assured a good return on their investment by means of patent protection of novel biomarkers of high sensitivity, efficacy, and clinical utility.

14.1.5 Health Economy Factors in Cancer Biomarker Discovery

For a better treatment outcome, the procedures should be inexpensive. Most of the novel and useful biomarkers and assay methods are patent protected. Therefore, the cost of diagnostic and prognostic assays using such biomarkers is very high and usually not affordable by common people. Cost effective assays need to be developed to meet the needs of common people.

14.2 NEXT-GENERATION MOLECULAR MARKERS AND THEIR APPLICATIONS

Affinity reagents are considered the most significant tools for biomarker discovery. Affimers are a type of affinity reagents, which are relatively small (13 kDa), non-posttranslationally modified, biophysically stable protein scaffolds.[8] They contain three variable regions into which distinct peptides are inserted. This structure then binds to proteins and other molecules the way an antigen and antibody binds. So far 20 000 Affimer arrays have been tested in serum of RA patients and elevated levels of biomarkers such as C-reactive protein have been identified along with more than 20 new biomarkers. The detection and treatment of different diseases is greatly facilitated by the omics technologies. For example, in cancer, genomics analysis provides clue for gene regulation and gene knockdown for cancer management. Discovery of the involvement of microRNAs in human cancers has opened a new page for cancer researchers.[9,10] The latest technologies are Mammaprint and Oncotype DX, which are being used in clinics confirming the feasibility of multiplexing of techniques.[11,12] Some therapeutic drugs, called epigenetic drugs, target DNA methylation and histone deacetylation in solid tumors and leukemia. These drugs work well with PARP inhibitors.[13] Proteomics also plays an important role in cancer biomarker discovery and quantitative proteome–disease relationships provide a means for connectivity analysis.[14] Fluorescent dye enables a more reliable analysis and it facilitates the progress of biochip and cytomics. The huge amount of information collected by multiparameter single cell flow or slide-based cytometry measurements serves to investigate the molecular behavior of cancer cell populations. Diseases selected in this article are very appropriate for metabolite profiling owing to their unique biochemical properties.

In Figure 14.1, we have presented a schematic approach to demonstrate factors contributing to disease development and approaches to isolate clinical samples and analyze biomarkers for disease detection, diagnosis, prognosis, and drug response. Table 14.1 highlights key biomarkers in different diseases and their application in detection, diagnosis, screening, prognosis, and treatment response.

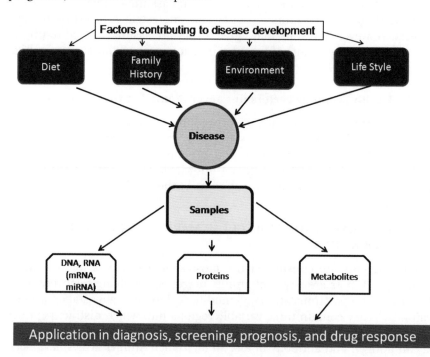

Figure 14.1   Schematic diagram to demonstrate factors contributing to disease development and approaches to isolate clinical samples and analyze biomarkers for disease detection, diagnosis, prognosis, and drug response.

Table 14.1 Next-Generation Molecular Markers for Diagnosis Screening, Early Diagnosis, Risk Assessment, and Prognosis.

Category	Diseases	Markers
Diagnostic Screening		
	Immunologic Diseases	Arginine kinase, sarcoplasmic calcium binding proteins, tropomycin.[15]
	Cardiovascular disease (CVD)	Lipoprotein associated phospholipase A2, antiphosphorylcholine IgM.[16]
	Metabolic Diseases	Glycine, glutamine, glycerophasphatidyl choline.[18]
	Infectious Diseases	Hepatitis B virus (HBV), hepatitis C virus (HCV), certain strains of the human papillomavirus (HPV), Epstein–Barr virus (EBV), human immunodeficiency virus type 1 (HIV-1), human T-cell lymphotropic virus type-1 (HTLV-1), and the gram-negative bacterium *Helicobacter pylori* (*H. pylori*).[24]
	Neurologic Diseases	Proteomic profiling of serum (multiple biomarkers).[29]
	Cancer	Gene promoter hypermethylation in death-associated protein kinase 1 (*DAPK1*), *p16* and *RASSF1A1* could be used as biomarkers for detection of HNSCCs; prostate-specific antigen (PSA) for prostate cancer; serum markers (PAP, tPSA, fPSA, proPSA, PSAD, PSAV, PSADT, EPCA, and EPCA-2), tissue markers (AMACR, methylated GSTP1, and the TMPRSS2-ETS gene rearrangement), and a urine marker (DD3PCA3/UPM-3) for prostate cancer.[35-42] Serum tumor markers like carcinoembryonic antigen (CEA) and cancer antigen (CA) CA 15.3, CA 27.29 for breast and ovarian cancer.[44]
		Using saliva as the starting material, mRNAs specific for esophageal, head and neck, oral and lung cancer also were identified.[48-50] MSI biomarkers were reported in a number of cancers, including bladder, colon, esophageal, and skin cancer.[54-56]
Early Diagnosis		
	Immunologic Diseases	Glutathione S transferase M1 (GSTM1) biomarker was identified which helped in screening a cohort of children who were at high risk of developing asthma;[73] C-reactive protein, fibrinogen, and interleukin-6 for screening.[74]
	Cardiovascular disease (CVD)	Flow-mediated dialation (FMD) and IMT for atherosclerosis.[60]
	Metabolic Diseases	Dysglycemia, alphahydroxybutyrate, linoleyolglycerophosphocholine, advanced glycation end products (AGE), albumin excretion rate (AR) and SNPS in epidermal growth factor gene intron 2.[1,62,63]

Table 14.1 (Continued)

Category	Diseases	Markers
	Infectious Diseases	Region-of-Difference-1 (RD-1) gene product and sputum cytokine levels for early detection of tuberculosis;[64] HSP70, CAP2, glypican 3 and glutamine synthase[65] for HCC.
	Neurologic Diseases	Beta amyloid for early detection of Alzheimer's disease .[66]
	Cancer	B7-H4, spondin 2, and DcR3 were identified as early biomarkers in ovarian cancer.[68–70]
Risk Assessment (screening)		
	Immunologic Diseases	C-reactive protein, fibrinogen, and interleukin-6.[74]
	Cardiovascular disease (CVD)	C-reactive protein.[75]
	Metabolic Diseases	Alanine amino transferase (ALT), gamma glutamyl transferase (GGT), triglyceride, plasminogen activator inhibitor (PAI-1) antigen, ferritin, C-reactive protein (CRP), sex-hormone binding globulin (SHBG), ALT and GGT for diabetes.[78]
	Infectious Diseases	HIV infection for AIDS;[80,81] gamma interferon, p2×7 polymorphism, and mycobacteria antigens for tuberculosis.[82,83]
	Neurologic Diseases	ABCA1 gene polymorphism, and apolypoprotein E genotyping for Alzheimer's disease.[84–86]
	Cancer	Methylation levels of genes NKX-25, CLSTN1, SPOCK2, SLC16A12, DPYS, and NSE1 for screening prostate, colon, and breast cancer.[88]
Prognostic Markers		
	Immunologic Diseases	C-telopeptides I and II (CTX-1 and CTX-2) for rheumatoid arthritis (RA).[92]
	Cardiovascular disease (CVD)	Osteoprotegerin (OPG) for atherosclerosis.[95]
	Metabolic Diseases	Alpha-hydroxybutyrate (Alpha-HB) and linolyl-glycerophosphocholine (L-GPC) for insulin resistance (IR) and glucose intolerance (GI).[1]
	Infectious Diseases	Fecal calprotectin in infectious diarrhea.[96] In a case control study, both bacterial and viral infection was involved. Main bacteria were Salmonella, Campylobacter, Yersinia, and Shigella, and viruses were rotavirus, norovirus, and adenovirus.
	Neurologic Diseases	MRI and proteomic based biomarkers in spinal fluid for Alzheimer's disease.[97]
	Cancer	Comparative genomic hybridization (CHG) identified complex genetic variants associated with adverse prognosis in cancer;[98] cytogenetics biomarkers for breast, head and neck, lung, liver, and ovarian cancers.[99–102]

14.2.1 Diagnostic Screening

14.2.1.1 Immunologic Diseases. Asthma involves inflammation, hyper responsiveness, bronchoconstriction, and symptoms (episodic breathlessness, wheeze, cough, tightness of the chest, and shortness of breath). Proteomics approaches identified allergenic proteins, based on their reactivity to patients' sera, using tandem mass spectrometry.[15] The most significant allergens identified were arginine kinase, sarcoplasmic calcium binding proteins, and tropomycin, which can be used for screening. This study also emphasized that MS analysis is a sensitive and accurate tool in identifying and quantifying aerosolized allergens.

14.2.1.2 Cardiovascular Disease (CVD). In cardiovascular disease (CVD), the relative contribution of genetic and environmental effect was studied on two inflammatory biomarkers, lipoprotein associated phospholipase A2 and antiphosphorylcholine IgM, in a Swedish population.[16] Results indicated that lipoprotein associated phospholipase A2 had low heritability and higher environmental regulation. Therefore, for diagnostic screening of CVD, biomarkers should be selected based on the context (family history, exposure history, lifestyle, *etc.*). Non-invasive technologies transcarnial Doppler, magnetic resonance and computed tomography were used for the diagnosis of intracranial atherosclerosis.[17] In these cases images were considered as biomarkers.

14.2.1.3 Metabolic Diseases. Metabolomic profiles can be used as biomarkers and can indicate the development of obesity, which in turn affects diabetes, cardiovascular disease, liver disease, renal disease and selected cancers. Metabolomic analysis is a valid and powerful tool that helps us understands the mechanism underlying different diseases. Biomarkers identified in obesity included glycine, glutamine, and glycerophasphatidyl choline in serum.[18]

14.2.1.4 Infectious Diseases. Potential applications of biomarkers in infectious diseases include distinguishing bacterial from nonbacterial infection, monitoring response to treatment and predicting survival (outcome). Periodontal disease, where aerobic and anaerobic bacteria infect gums, was proposed as a biomarker and risk factor for CVDs, especially abnormal function of progenitor endothelial cells.[19] Another group demonstrated an association of the oral microbiome with gastrointestinal cancer.[20] In such situations oral microbiodata was considered a biomarker for gastrointestinal cancer. Procalcitonin (PCT), a proinflammatory biomarker, is an excellent early detection biomarker of invasive bacterial infection in febrile children evaluated in emergency departments.[21]

Infectious agents are associated with at least 15% of cancers.[22] Chronic infections are the second most preventable cause of cancer,[23] and about 18% of the global cancer burden has been attributed to infectious agents.[24]

The presence of an infectious agent in tumor tissue is not sufficient to establish it as a causal agent; but the agent is termed carcinogenic if evidence from epidemiologic, clinical, and biologic studies suggests a strong cancer etiology.[25] The International Agency for Research on Cancer (IARC) has identified seven major infectious agents as carcinogenic: hepatitis B virus (HBV), hepatitis C virus (HCV), certain strains of the human papillomavirus (HPV), Epstein–Barr virus (EBV), human immunodeficiency virus type 1 (HIV-1), human T-cell lymphotropic virus type-1 (HTLV-1), and the gram-negative bacterium *Helicobacter pylori* (*H. pylori*).[24] Aflatoxin B1 (AFB1), the product of the common mold *Aspergillus flavus*, was also considered in this category because the IARC has identified AFB1 as a chemical liver carcinogen.[26] Hepatitis C infected individuals can be distinguished from healthy people based on metabolomic profiling (which can be considered a biomarker for screening).[27] It was demonstrated that IL-18 levels in serum could be used as a biomarker for acute Epstein–Barr virus (EBV) infection.[28]

14.2.1.5 Neurologic Diseases. Metabolomics based technologies (liquid chromatography quadrupole time-of-flight mass spectrometry) and chemometrics were used to identify Alzheimer disease (AD) specific profiling.[29] Samples from healthy age-matched controls were also used to identify new biomarkers (profiles). The prediction value for AD was 94–97% in this metabolomic study, which was considered quite remarkable for early detection of AD. The confirmation of these results was done in another set of patients and control and 100% diagnosis was achieved.[29] Such studies are significant to understand etiology, pathophysiology, and treatment of degenerative brain disorders.[30]

14.2.1.6 Cancer. In this section we discuss different tumor types and next generation biomarkers identified in different tumor types. Cancer is a genetic and epigenetic disease and methylation biomarkers have been used in diagnostic screening of cancer.[31,32] Methylation changes in the precarcinoma stage could be used as biomarkers for diagnosis in cancer if these epigenetic changes are not present in normal cells.[32,33] With the development of high throughput next generation sequencing, it has become easier to detect methylation changes qualitatively as well as quantitatively and recently global methylation profiling has also been accomplished.[34] Gene promoter hypermethylation in death-associated protein kinase 1 (*DAPK1*), *p16* and *RASSF1A*1 could be used as biomarkers for detection of HNSCCs. The development of biomarkers for prostate cancer screening, detection, and prognostication revolutionized the management of this disease. During the progression of cancer, it was observed that the levels of certain proteins were elevated. These abnormally increased molecules could be used as biomarkers for gaining an insight on the course of the disease. Prostate-specific antigen (PSA) is a useful, though not specific, biomarker for detecting prostate cancer. Serum markers (PAP, tPSA, fPSA, proPSA, PSAD, PSAV, PSADT, EPCA, and

EPCA-2), tissue markers (AMACR, methylated GSTP1, and the TMPRSS2-ETS gene rearrangement), and a urine marker (DD3PCA3/UPM-3) of prostate cancer are very well characterized biomarkers.[35–42] Disease-specific protein biomarkers also were identified[43] and characterization of such biomarkers in body fluids could aid in the early detection of cancer and help in monitoring cancer progression. It has become easier to identify cancer specific markers with the advancement in the available technologies. In the case of breast cancer, some of the serum tumor markers like carcinoembryonic antigen (CEA) and cancer antigen (CA) CA 15.3, CA 27.29 were not confirmed to be sensitive for early detection; however, their high levels did reflect disease progression and recurrence.[44] Mammoglobin and MASPIN (mammary serine protease inhibitor) were shown to be diagnostic biomarkers.[45] The early detection of circulating breast cancer cells by morphologic methods is currently being challenged by ultrasensitive proteomic and PCR-based methods often enhanced by immunomagnetic bead-based cell capture.[46] The role of MASPIN in diagnosis and prognosis of a variety of cancers is being explored.[47] It has been reported that low or absent MASPIN cytoplasmic expression was frequently observed in oral carcinomas with lymph node metastasis.

Using saliva as the starting material, mRNAs specific for esophageal, head and neck, oral and lung cancer were also identified.[48–50] Seven biomarkers showed 3.5 fold increases in their levels in oral SCC patients compared to healthy individuals. Further validation of values is needed before implementing in clinics.

Microsatellites are repeated sequences of DNA.[51,53] When these repeats are shortened, this is called microsatellite instability (MSI). This stage arises in regions where the mutation rate is very high, mainly due to defective DNA genes such as mismatch repair gene. MSI biomarkers were reported in a number of cancers, including bladder, colon, esophageal, and skin cancer.[54–56] As high as 30% of HNSCC patients had MSI [52] suggesting that MSI is a good biomarker for HNSCC. A number of studies evaluated the potential of detecting these markers in tumor samples as well as in other biospecimens and found these biomarkers useful for diagnosis and prognosis.[57]

14.2.2 Early Diagnosis

14.2.2.1 Immune Diseases. Airways inflammation starts soon after inception of exposure to allergens in asthma. Inflammatory cytokines along with arginine kinase, sarcoplasmic calcium binding proteins are the most practical biomarkers for early diagnosis of asthma.[15,58]

14.2.2.2 Cardiovascular Diseases (CVDs). For early diagnosis of carotid atherosclerosis for which obesity is the risk factor, the early biomarkers are carotid intima media thickness (IMT) and circulating oxidized low density lipoprotein.[59] To evaluate endothelium status, flow-mediated dialation (FMD) and IMT are used as early biomarkers of atherosclerosis in patients with nonalcoholic fatty liver disease (NAFLD).[60]

14.2.2.3 Metabolic Diseases. A number of markers have been described for early diagnosis of diabetes.[1,61] They include dysglycemia, alphahydroxybu-tyrate, linoleyolglycerophosphocholine, advanced glycation end products (AGE), albumin excretion rate (AR) and SNPS in epidermal growth factor gene intron 2.[1,62,63]

14.2.2.4 Infectious Diseases. Region-of-Difference-1 (RD-1) gene product and sputum cytokine levels are considered a biomarker for early detection of tuberculosis.[64] Hepatocellular carcinoma (HCC) involves infectious agents and the early diagnostic biomarkers for HCC are HSP70, CAP2, glypican 3 and glutamine synthase.[65] These biomarkers were identified based on gene expression analysis of HCC samples.

14.2.2.5 Neurological Diseases. A bioinformatic approach has helped in identifying early diagnosis biomarkers for Alzheimer's disease. Based on data from patients and matched controls, the Artificial Neural Network (ANN) identified beta amyloid cascade as a potential biomarker for early detection of Alzheimer's disease.[66]

14.2.2.6 Cancer. The key to successful application of biomarkers is to detect the disease early so that a variety of treatment approaches can be applied.[67] Initially it was thought that genetic changes arise first at the time of initiation of a disease but genome-wide profiling suggests that epigenetic changes occur much earlier than genomic changes.[33] Technologies exist to follow up these changes of early detection. B7-H4, spondin 2, and DcR3 were identified as early biomarkers in ovarian cancer although other investigators have proposed other groups of biomarkers.[68–70]

14.2.3 Risk Assessment (Screening)

Genome-wide association studies (GWAS) are extremely powerful in identi-fying new low-penetrance SNPs (biomarkers) which may have therapeutic implications.[71] Identification of common low-susceptibility alleles is useful because it provides possible insight into the mechanisms of tumor biology in cases of cancer and identifies high risk individuals.[72]

Associations do not necessarily mean causality; the potential for con-founding and reverse causality should always be kept in mind. It will be beneficial to predict complications, in case of diabetes in particular and other diseases in general, and determine subgroups that may be responsive to therapy. Clinical implications should not be exaggerated while charac-terizing prediction biomarkers.

14.2.3.1 Immune Diseases. In one meta-analysis glutathione S transfer-ase M1 (GSTM1) biomarker was identified which helped in screening a cohort of children who were at high risk of developing asthma.[73] In a population based study C-reactive protein, fibrinogen, and interleukin-6 were found useful biomarkers for screening.[74]

14.2.3.2 Cardiovascular Diseases (CVDs). In a risk prediction model of cardiovascular disease, C-reactive protein levels were implemented.[75] However, after completion of studies, it was suggested that evaluation of the potential impact of CRP levels would require studies to quantify the effects of additional CRP assessment on medical decision making. Biomarker studies of CVD prediction in elderly patients indicated that mortality and cardiovascular events were dependent on low peripheral pulse pressure not on high blood pressure.[76] Genetic variants in CVD were studied to evaluate their association with the disease but with limited success.[77]

14.2.3.3 Metabolic Diseases. For diabetes, alanine amino transferase (ALT), gamma glutamyl transferase (GGT), triglyceride, plasminogen activator inhibitor (PAI-1) antigen, ferritin, C-reactive protein (CRP), and sex-hormone binding globulin (SHBG) were identified as prediction markers after a large study "West of Scotland Coronary Prevention Study (WOSCOPS) was conducted. Results from this study were validated in a different population and two biomarkers, ALT and GGT, looked very promising for risk assessment.[78] CRP was not causally related to insulin resistance or obesity. IL-6 upregulation was also associated with diabetes. Early diagnosis of chronic kidney diseases and atherosclerosis in the same subjects could be accomplished by carotid ultra sound technology.[79]

14.2.3.4 Infectious Diseases. HIV/AIDS screening involves behavior biomarkers as well as molecular omics biomarkers and a number of investigators have reported the screening strategies and their outcome in different populations.[80,81] In tuberculosis, gamma interferon, p2×7 polymorphism, and mycobacteria antigens were used as biomarkers for risk assessment and screening.[82,83]

14.2.3.5 Neurological Diseases. The most common biomarkers used for risk assessment of Alzheimer's disease were ABCA1 gene polymorphism, and apolypoprotein E genotyping; although a combination of markers was also used, synergism was lacking.[84–86]

14.2.3.6 Cancer. Transcriptomic and miRNAs biomarkers were used for screening colorectal cancer.[87] For screening prostate, colon, and breast cancer, methylation levels of genes NKX-25, CLSTN1, SPOCK2, SLC16A12, DPYS, and NSE1 were used.[88] Malignant melanoma is one of the most aggressive types of tumor. Because malignant melanoma is difficult to treat once it has metastasized, early detection and treatment are essential. The search for reliable biomarkers of early-stage melanoma, therefore, has received much attention. By using an approach of screening tumor antigens and their auto-antibodies, bullous pemphigoid antigen 1 (BPAG1) was identified as a melanoma antigen recognized by its auto-antibody when anti-BPAG1 auto-antibodies were detected in melanoma patients at both early and advanced stages of disease.[89]

14.2.4 Prognostic Markers

Prognostic biomarkers should be tightly linked to the outcome so that they can be used as surrogate measures of efficacy and treatment response. Prognostic markers can be defined as factors that can predict an outcome in the absence of systemic therapy or predict an outcome different from patients who are devoid of the biomarker, despite empiric therapy (initiating treatment before confirming diagnosis).[90] It is not known whether there is any association between ovarian cyst and estrogen levels during tamoxifen use. A very well designed study was conducted where breast cancer prognostic markers were utilized to evaluate the effect of tamoxifen use in premenopausal women with ovarian cyst and results indicated an association.[91] Hence, prognostic markers can be utilized to classify patients into appropriate groups for treatments.

14.2.4.1 Immune Diseases. Biomarkers of bone and cartilage turnover are collagen C-telopeptides I and II, which are predictors of structural damage in RA patients. C-terminal cross-linked telopeptide of type I collagen (CTX-I) could be measured in serum or urine after it was released during bone resorption. Bone erosions and osteoporosis both occur as a consequence of RA and CTX-I levels and could be used for prognosis. On the other hand, increased CTX-II levels were associated with rapid progression of joint damage. For clinical implication, investigators of these studies recommended that new prognostic biomarkers should supply information beyond that provided by risk factors in the prediction of disease outcome.[92]

European countries restrict anti-tumor necrosis factor therapy for RA and prescription is provided based on evaluation of inflammatory markers, C-reactive protein levels and general health of a patient.[93] Genomic variants of C-reactive proteins play a major role in therapeutics of RA and can be used as biomarkers.[94]

14.2.4.2 Cardiovascular Diseases (CVDs). Osteoprotegerin (OPG) is such a biomarker, which is independently associated not only with risk factors of atherosclerosis but also with subclinical peripheral atherosclerosis and clinical atherosclerosis and is recommended as a prognostic biomarker for ischemic heart disease and ischemic stroke.[95]

14.2.4.3 Metabolic Diseases. Alpha-hydroxybutyrate (Alpha-HB) and linolyl-glycerophosphocholine (L-GPC) were identified as biomarkers of insulin resistance (IR) and glucose intolerance (GI) in a large population study of more than 1000 participants from the Relationship between Insulin sensitivity and Cardiovascular Disease (RISC) study.[1] AHB correlated positively and L-GPC negatively with both diseases (IR and GI) indicating that AHB was a positive predictor and L-GPC, a negative predictor independent of family history of diabetes, sex, age, fasting glucose, and BMI.

14.2.4.4 Infectious Diseases. Diarrhea in children may involve infection. In cases of infectious diarrhea, fecal calprotectin is a good prediction marker.[96] In a case control study, both bacterial and viral infection was involved. The main bacteria were *Salmonella, Campylobacter, Yersinia,* and *Shigella,* and viruses were rotavirus, norovirus, and adenovirus. These studies suggested that fecal calprotectin could be used as a noninvasive biomarker in management of children with infectious diarrhea.

14.2.4.5 Neurological Diseases. A few Alzheimer's diagnosis biomarkers are suitable for prognosis also. One investigator combined MRI and spinal fluid based biomarkers and observed better results than using a single biomarker for prognosis.[97] This investigation was performed in different sites of the brain (for MRI) and spinal fluid biomarker values were used as a reference standard.

14.2.4.6 Cancer. Despite huge efforts into research for studying new biological prognostic markers, only a few out of several hundreds have progressed to clinical use. Comparative genomic hybridization (CHG) identified complex genetic variants associated with adverse prognosis in cancer.[98] Cytogenetics biomarkers turned out to be very useful biomarkers for breast, head and neck, lung, liver, and ovarian cancers.[99–102] In case of HNSCCs, loss of heterozygosity (LOH) on distal arm of 18q was reported to be associated with poorer survival.[103] About 70% of colorectal cancers showed allelic deletions in chromosomes 18q and 17p. The *p53* gene located on 17p is mutated in about 40 to 60% of CRCs and demonstrated association with prognosis and prediction of the disease. CRC patients with chromosome 18q loss showed worse disease-free and overall survival. Markers like Ki-67 staining detecting cell proliferation were significantly correlated with breast cancer outcome. One-fifth of breast cancer patients showed amplification or over-expression of cyclin D1 (*PRAD1* or bcl-1). Germline prognostic markers of bladder cancer were identified and characterized.[104] Stage II and III CRC patients with high microsatellite instability have been reported to show improved survival and better relapse-free survival as compared to microsatellite stable (MSS) patients.

14.2.5 Drug Response

Systems immunology, based on mathematical and computational models, has the potential to predict treatment response.[105] High throughput "omics" technologies generate vast amount of data which may help in identifying immunological processes at high resolution in disease development. Using a systems immunology approach it is possible to construct causal relationships between complex molecular processes and specific disease-associated phenotypes.[106] Drug response and pharmacogenomics in the context of biomarkers in different diseases is shown in Table 14.2.

14.2.5.1 Immune Diseases. The prevalence of autism is 1 in 80 in the USA. Evidence based treatments are practiced in autism[107] although sometimes drug treatment (mGluR antagonist and GABA agonist) is also recommended based on the severity and stage of the disease.[108] The etiology of autism is not completely understood. However, synaptic maturation and plasticity in the pathogenesis of autism spectrum disorder resulting in imbalance of excitation and inhibition has been observed. The drugs mentioned above control excessive excitement.

14.2.5.2 Cardiovascular Diseases (CVDs). Nutritional foods and their biological food components alter cardiovascular disease status and there is very well established evidence of disease modifying effects of these compounds with anti-inflammatory and antioxidant effects in CVD patients. Reduced total LDL cholesterol levels were observed when polyphenolic compounds were given to patients because bioactive phytochemicals play an important therapeutic role in attenuating oxidative damage.

14.2.5.3 Metabolic Diseases. The key players in measuring drug response in type II diabetes mellitus (T2DM) are glycemic index (GI), glucose response curves (GRCs) and daily mean plasma glucose (DMPG). In a trial, a variety of foods were supplied to participants undergoing treatment with oral antidiabetic drugs (OADs).[109] Promising results were obtained in this pilot study and validation of these observations in large population is planned.

14.2.5.4 Infectious Diseases. 1,3-Beta-D-glucan (BG) can be used as a biomarker in invasive fungal infections (especially infections involving *Candida*) in patients undergoing treatment of candedemia with anidulafungin.[110] In those patients who were given anidulafungin followed by fluconazole/voriconazole therapy, decreasing BG concentrations reflected success of the treatment.

In unipolar depression, transcriptomic biomarker approaches were used to identify responders of treatment.[111] The main player in the process was tumor necrosis factor in the inflammatory cytokine pathway. These investigators evaluated genetic background of participants and suggested that SNPs rs1126757 in IL11and rs 7801617 in IL6 play major role in responding

Table 14.2 Drug Response and Pharmacogenomics in the Context of Biomarkers.

Category	Diseases	Markers
Drug Response		
	Immunologic Diseases	Evidence based treatments are practiced in autism [107] although sometime drug treatment (mGluR antagonist and GABA agonist) is also recommended based on the severity and stage of the disease. [108]
	Cardiovascular disease (CVD)	Reduced total LDL cholesterol levels were observed when polyphenolic compounds were given to patients because bioactive phytochemicals play an important therapeutic role in attenuating oxidative damage. Nutritional foods and their biological food components alter cardiovascular disease status and have anti-inflammatory and antioxidant effects in CVD patients.
	Metabolic Diseases	In a trial, a variety of foods were supplied to participants undergoing treatment with oral antidiabetic drugs (OADs). [109] The key players in measuring drug response in type II diabetes mellitus (T2DM) are glycemic index (GI), glucose response curves (GRCs) and daily mean plasma glucose (DMPG).
	Infectious Diseases	In those patients who were given anidulafungin followed by fluconazole/voriconazole therapy, decreasing BG concentrations reflected success of the treatment. 1,3-Beta-ᴅ-glucan (BG) can be used as a biomarker in invasive fungal infections (especially infections involving *Candida*) in patients undergoing treatment of candedemia with anidulafungin. [110]
	Neurologic Diseases	For the treatment of schizophrenia, ionotropic glutamate receptors are targeted. [112] Recent pharmacologic strategies have focused on improving cognition by drugs. [113]
	Cancer	Biomarkers of drug response may be new or the same which are diagnostic or prognostic biomarkers depending on the biology and etiology of the type of cancer studied. In different kinds of cancers pharmacological response of drugs is different and a variety of factors (genetic background, life style, other diseases in the same person, BMI *etc.*) contribute to drug response. [114] Development of epigenetic drugs showed promise in selected cancer treatment. [115]

Table 14.2 (*Continued*)

Category	Diseases	Markers
Pharmacogenomics	Immunologic Diseases	The response of infliximab was favorable in a group of patients where genes regulated by TNF were active compared to other participants of the study indicating a strong evidence of role that genomic background plays in responding to pharmacological drugs.[120]
		Pharmacogenomics of rheumatoid arthritis (RA) has been studied by several investigators[120,121] although RA is a highly heterogeneous disease in all aspects including response to therapy.
	Cardiovascular disease (CVD)	Infliximab treatment was provided to rheumatoid arthritis patients to reduce levels of TNF and results were evaluated based on the genomic background of the participants.[120] Host response due to inflammatory diseases such as arthritis depends on the genomic background and pharmacological agents give response based on the genomic susceptibility and some other yet unidentified factors.[120–122]
	Metabolic Diseases	Genetic background was considered in treatment of type 2 diabetes and an association of ADIPOR2 gene variants with CVDs and type 2 diabetes risk in individuals with defective glucose tolerance (conducted in Finnish population) was observed.[123] Additionally, genetic predisposition and nongenetic risk factors of thioazolidine-related edema in type 2 individuals, the pharmacogenomics of metaformin were also observed.[124]
	Infectious Diseases	Genetic variants at least at three loci, NAT2, CYP2E1, and GSTM1, were involved in pharmacogenomics. Pharmacogenomics of anti-TB drug related hepatotoxicity was studied to understand involvement of genetic background in response to treatment in TB patients.[125]
	Neurologic Diseases	Genetic variants were identified in Alzheimer's disease and demonstrated treatment response.[126, 127] In a recent study of an Italian cohort, response to cholinesaterse inhibitors was evaluated in Alzheimer's patients using 48 SNPs.[126] Results indicated association of two SNPs with the response to treatment.
	Cancer	Promising results were obtained in colon, gastric, breast, ovarian, GI tract and lung cancer.[129–131] Polymorphism in miR encoding genes also was evaluated for its implication in pharmacogenomics.[128]

to antidepressants. This was an excellent example where genomics and transcriptomics biomarkers were used to follow up treatment.

H. pylori is the most common chronic bacterial infection in humans. This bacteria is involved not only in gastric cancer but also in ulcer diseases and gastritis. Its synergistic gastrotoxic interaction with non-steroidal anti-inflammatory drugs and association with atherosclerotic events is a matter of concern. Transmission of *H. pylori* is through oral ingestion, mainly within families in early childhood. Therefore, treatment and prevention approaches are more appropriate for children. One-week proton pump inhibitor based triple therapy with clarithromycin and either amoxicillin or metronidazole is the most common therapy.

14.2.5.5 Neurological Diseases. Schizophrenia is a chronic brain disorder and affects approximately 2.5 million Americans and more than 24 million people worldwide. For the treatment of schizophrenia, ionotropic glutamate receptors are targeted.[112] Recent pharmacologic strategies have focused on improving cognition with drugs.[113] The consensus among investigators in the schizophrenia field is that physiological and pharmacological approaches should be combined for this disease for better outcome of the treatment.

14.2.5.6 Cancer. In different kinds of cancers pharmacological response of drugs is different and a variety of factors (genetic background, life style, other diseases in the same person, BMI *etc.*) contribute to drug response.[114] Biomarkers of drug response may be new or the same which are diagnostic or prognostic biomarkers depending on the biology and etiology of the type of cancer studied. Personalized medicine is recommended for cancer because cancer is a heterogeneous disease and the rate of recurrence after the treatment is high for some cancer types. Development of epigenetic drugs showed promise in selected cancer treatment.[115]

14.2.6 Pharmacogenomics

Pharmacogenomics is the study of interindividual genetic variability that plays a significant role in treatment response and toxicity due to the drug. Based on a better understanding of biological variability, attempts are being made to customize treatment at the personal level.[116] In pharmacogenomic approaches the following points are considered critical: pharmacokinetic related genes and phenotypes, pharmacodynamic targets, genes and products, risk of disease related with metabolomic cycle, physiological variations, and environment interaction. When targeted agents matched with tumor molecular aberrations were implied in a phase I clinical trial, encouraging results were observed.[117] Since pharmacogenomics science is new, success is not guaranteed, as happened in case of gastric cancer where efficacy of Trastuzumab was checked in patients with advanced gastric cancer and the drug was not effective.[118] Other investigators have identified challenges in the field including poor coordinated diagnostic testing and current models of implication in disease stratification.[119]

14.2.6.1 Immune Diseases. Compared to other diseases, the pharmacogenomics of rheumatoid arthritis (RA) has been studied by several investigators.[120,121] Although RA is a highly heterogeneous disease in all aspects including response to therapy. The response of infliximab was favorable in a group of patients where genes regulated by TNF were active compared to other participants of the study indicating strong evidence of the role that genomic background plays in responding to pharmacological drugs.[120]

14.2.6.2 Cardiovascular Diseases (CVDs). Host response due to inflammatory diseases such as arthritis depends on the genomic background and pharmacological agents give response based on the genomic susceptibility and some other yet unidentified factors.[120–122] Infliximab treatment was provided to rheumatoid arthritis patients to reduce levels of TNF and results were evaluated based on the genomic background of the participants.[120] Pharmacogenomics field in CVDs is still in infancy and next few years research should focus on accurately identify, from available therapies, which is the optimal therapy for any particular individual.[121]

14.2.6.3 Metabolic Diseases. In few studies the role of genetic background in treatment of type 2 diabetes was demonstrated, such as association of ADIPOR2 gene variants with CVDs and type 2 diabetes risk in individuals with defective glucose tolerance (conducted in the Finnish population),[123] genetic predisposition and nongenetic risk factors of thioazolidine-related edema in type 2 individuals, the pharmacogenomics of metaformin.[124] Validation in large number of participants is still awaited.

14.2.6.4 Infectious Diseases. Pharmacogenomics of anti-TB drug related hepatotoxicity was studied to understand the involvement of genetic background in response to treatment in TB patients.[125] Genetic variants at least at three loci, NAT2, CYP2E1, and GSTM1, were involved in pharmacogenomics. This study was conducted only in one population and it should be validated in populations with different ethnic backgrounds.

14.2.6.5 Neurological Diseases. Due to its serious effects on day to day life, Alzheimer's pharmacogenomics has been studied by a number of groups in the last decade and genetic variants were identified which were associated with treatment response.[126,127] In a recent study of an Italian cohort, response to cholinesaterse inhibitors was evaluated in Alzheimer's patients using 48 SNPs.[126] Results indicated association of two SNPs with the response to treatment.

14.2.6.6 Cancer. The National Cancer Institute has identified pharmacoepidemiology related to pharmaceutical use and cancer risk, recurrence and survival as one of the priority areas of research. Polymorphism in miR encoding genes also was evaluated for its implication in pharmacogenomics.[128] Although a considerable amount of new-targeted agents have been designed based on cancer biology, challenges and gaps exist between pharmacogenomics knowledge and clinical application. Promising results were obtained in colon, gastric, breast, ovarian, GI tract and lung cancer.[129–131]

14.2.7 Therapeutic Biomarkers in Immune Diseases, Cardiovascular Diseases (CVD), Metabolic Diseases, Infectious Diseases, Neurological Diseases, and Cancer

Since there is no measurable disease while undergoing treatment, generally therapeutic biomarkers are difficult to study. Another key point is the adverse reaction of patients due to treatment. Such parameters are not needed for prognostic biomarkers; therefore, it is easier to identify prognostic biomarkers than therapeutic biomarkers. In the following section, examples of therapeutic biomarkers in different diseases are described.

14.2.7.1 Immune Diseases. In the case of multiple sclerosis (MS), treatment with disease-modifying therapies (DMT) with ability to prevent axonal damage resulted in reduction of progression of the disease from clinically isolated syndrome (CIS).[132] In another study, inclusion of information about Max RNA during interferon treatment of MS resulted in a better response than interferon only treatment.[133] Other investigators also observed similar results.[134–135]

14.2.7.2 Cardiovascular Diseases (CVDs). Biomarkers discussed in the diagnostic section can be used for follow up of the treatment. Dietary intervention of CVD by fish oil (salmon, herring, and pompano) and other nutrients was demonstrated in a number of studies.[136] Some of the participants had higher levels of triacylglycerolaemia. Biomarkers TNFalpha and IL-6 were reduced and the level of adiponectin increased in the treated arm. Thus TNFalpha, IL-6, and adiponectin were used as therapeutic biomarkers. In another study, argon oil supplement reduced plasma levels of lipids and antioxidant status.[137] Therapeutic biomarkers used in this study were plasma vitamin E concentrations, total and LDL cholesterol, and antioxidant profiles.

14.2.7.3 Metabolic Diseases. For the management of diabetes hemoglobin A1c (HbA1c) is used which is considered a reliable indicator of glycemic control. In most of the clinical studies in diabetes, HBA1c biomarker is used to determine the glucose control.

14.2.7.4 Infectious Diseases. Diarrhea in AIDS patients was treated with specific medications and therapeutic response was measured by levels of tubuloreticulin inclusions (TRIs).[138] Although this study may be considered an isolated study TRI has the potential to be a therapeutic biomarker after larger studies are conducted. For pneumonia therapy of more than 100 patients, biomarker procalcitonin (PCT) levels were very useful and this biomarker has been recommended for future prognosis.[139]

14.2.7.5 Neurological Diseases. In Parkinson's disease, glial cell-line derived neurotrophic factor (GDNF) and family of ligands (GDFLs) were used as biomarkers to follow up therapy.[140]

14.2.7.6 Cancer. Several cancer biomarkers are currently used to develop targeted anti-cancer drugs. Topomerase I inhibition in colorectal cancer is one example of targeted therapeutics and utilization of biomarkers.[141] In leukemia, an antibody therapeutics approach was applied, using rituximab, and therapeutic biomarkers Ki67 and PIM1 were followed for their response.[142] Results indicated that Ki63 was an independent biomarker. In another study CD20 was used as a biomarker to evaluate the therapeutic potential of rituximab in B-cell lymphoma.[143] In B-cell non-Hodgkin lymphoma (NHL), treatment with the same agent, interleukin-6 plasma levels were followed to evaluate drug response.[144] Eradication of NHL was achieved with a monoclonal antibody therapy combining rituximab with a blocking anti-CD40 antibody.[145] Other examples of treatment are trastuzumab for breast cancer (target p185neu), gemtuzumab for AML (target CD33), ibritumomab for AML (target CD2090Y), edrecolomab for colorectal cancer (target EpCAM), tositumomab for NHL (target CD20), and cetuximab for colorectal cancer (target EGFR). Cisplatin is a widely used chemotherapeutic in head and neck cancer patients. However, the effect of the drug is limited by the resistance observed in these patients; and it has been seen that hypermethylation in some of the genes responsible for cytotoxicity causes the resistance, for example the *S100P* gene. The above description indicates that biological markers that can predict therapeutic outcomes enable guiding the choice of treatment; and depending on the levels of the biomarkers it could be estimated if the patient would respond to a particular therapy or not and finally an appropriate treatment regimen can be selected. In addition, prediction of drug response helps in reducing the treatment cost.

14.3 RECENT TRENDS AND FUTURE DIRECTIONS

Significant progress in a variety of biomarkers has been made in the last two decades. Metabolomic and epigenomic biomarkers are relatively new and have shown promising results in disease diagnosis and prognosis. The monitoring of relative changes in metabolomic profiles in predisposed versus healthy individuals may help identify unique metabolites involved in disease processes.[146] These profiles have been used to predict the risk of diabetes,[147] cardiovascular disease,[148] and lung cancer,[149] diagnose prostate cancer,[150] differentiate benign and malignant ovarian tumors,[151] and identify biomarkers of Crohn's disease.[7] Such shifts also may identify diagnostic biomarkers, which could provide insights into strategies for disease prevention and be used to monitor the response to treatment. In recent years, metabolomics and other post-GWAS platforms, such as proteomics and transcriptomics, have undergone rapid improvement in both their reliability and throughput; as such, it may be an appropriate time for their use in epidemiologic studies.[152] Although metabolomic profiling has been used in some larger-scale population studies,[153] the number of published reports to date remains small. If the successes of genomics and transcriptomics in epidemiology are reliable indicators, there is a large, yet unexplored, potential for metabolomics to contribute to public health research. So far the assessment of geneotypes of candidate biomarkers in metabolomic diseases and CVD in blood samples has not improved prediction of these diseases. Probably multiple markers (transcriptiomics, epigenomics, metabolomics, genomics) should be tested for better prediction of these diseases. Prognostication is one of the promising area in translation of experimental research into clinical practice. In this approach, patterns of altered gene expression in tumors are used to construct classifiers instead of standard indices such as Nottingham Prognostic Index, Adjuvant Online, and Predict.[154]

Multiplexing of biomarkers may reduce false positive results in screening studies where intention is to identify populations which are at high risk of developing a disease.[2] Quantitative imaging data storage and maintenance have their own challenges as we discussed above. In case of miR biomarkers, whether miR expression is localized in a specific part of the tissue has to be carefully evaluated. In a tissue biopsy the local concentration (number of miRs) may be low or high. Determining the accurate level of miRs is very critical.

It has been seen that a combination of genes measured concurrently imparts better information about the clinical effect than a single gene. Oncotype DX™ is the first multigene predicting test of prognosis for breast cancer patients receiving anti-estrogen therapy (Genomic Health, USA). The FDA has also approved the test MammaPrint that predicts relapse in breast cancer patients by analyzing the activity of 70 genes. In case of prostate cancer, the PSA (Prostate-specific antigen) test has been approved, where PSA levels help in detecting prostate cancer and also predict a recurrence in patients suffering from the disease. Alpha-fetoprotein (AFP) has been approved for the diagnosis and monitoring of patients with non-seminoma testicular cancer.

Significant challenges exist regarding IP issues. At times tests are known but their associations with diseases are new. Sometimes patent belongs to smaller companies for new tests but their clinical utility can only be developed with larger companies in large clinical trials. In general, every situation represents unique opportunities and IP related issues can only be solved in a case by case manner.

The presence of estrogen receptor (ER), progesterone receptor (PR), and human epidermal growth factor receptor 2 (Her2, or ERBB2) is used for the clinical and pathological classification of breast cancer.[3] Generally, ER-positive (+) and PR+ are indicators of good prognosis, and Her2+ is an indicator of bad prognosis. In addition, ER-negative (−), PR−, and Her2− (also called triple-negative) status is considered to be an indicator of poor prognosis. Basal cells exhibit triple-negative features. On the basis of oncologic pathway activity analysis, up to 18 subtypes of breast cancer have been suggested.[4] However, the implications of this information for clinical practice remain to be determined. Furthermore, many prognostic gene expression signatures that dichotomize patient populations into treatment-responsive and nonresponsive groups lack specificity.[155,156] Additional biomarkers are needed that are better prognostic indicators than hormone receptor status, and a better understanding of the genetic characteristics of patients is needed to improve current clinical practice. Ideally, a method for preoperative molecular profiling should be developed that can guide treatment strategies.

In recent years metabolites of biofluids have been analyzed for their potential in cancer diagnosis and follow up of treatment. Especially urine analysis for the routine monitoring of metabolomic disorders has attracted a reasonable amount of interest among scientists because the procedure can be done easily, noninvasively and repeatedly for a large number of samples with high precision. Generally volatile organic metabolites (VOM) get enriched in urine and their analysis is easy.[157] The advantage of adopting metabolomic approach lies in the fact that metabolites are much more stable than RNA and proteins and their levels predict those pathways which are affected during disease development. In one small study with urine from controls and breast cancer, VOMs were identified which were differentially expressed in patients only.[158] Higher levels of 4-carene, 3-heptanone, 1,2,4-trimethylbenzene, 2-methoxythiophene, and phenol, and lower levels of dimethyl sulfides were observed in breast cancer patients. Urine metabolites have also been used for diagnosis of colon, lung, liver, and prostate cancer.[157]

Characterizing metabolomic pathways helps in making treatment decisions.[159] Circulating biomarkers, especially diabetes biomarkers discussed in this article, should be used for disease stratification of patients followed by completion of questionnaires such as: (i) how can the group be best defined in terms of biomarker levels? (ii) cause of the unexpected results; (iii) underlying genetic trait; and (iv) will one group respond better than other? This approach may lead us in the direction of personalized medicine. We have summarized potential challenges and research opportunities in

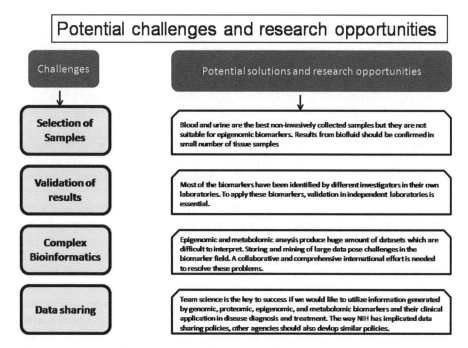

Figure 14.2 Potential challenges and research opportunities in utilizing biomarker information in disease detection, diagnosis, prognosis, and drug response.

utilizing biomarker information in disease detection, diagnosis, prognosis, and drug response in Figure 14.2.

14.4 CONCLUSIONS

Biomarker screening tests face the challenge in transition of tests from the research level in the laboratory to their beneficial use in the clinic. The biomarkers need to be highly specific and sensitive for the detection in clinical samples in order to avoid false positive tests that could lead to misdiagnosis and wrong therapeutic selection. An ideal biomarker should be used for screening in a large population study to validate the effectiveness of the screening and should be proven effective in populations with different genetic background. It has been seen that many biomarkers that correlate with disease statistically may not be useful in the clinic. It is advisable to study patient populations with diversity because the biomarkers might show different responses in different populations.

Considerable amount of knowledge has been obtained in understanding cancer biology and identification of biomarkers which can detect cancer but implication of that information in clinic is still challenging.[160] Clinical validation is the main hurdle in the process. In one case control study of the

Prospect-EPIC (European Prospective Investigation into Cancer and nutrition) where more than 300 breast cancer patients and matched controls were tested for breast cancer over a period of three years using a panel of eight serum biomarkers (osteopontin, haptoglobin, cancer antigen 15-3, carcinoembryonic antigen, cancer antigen-125, prolactin, cancer antigen 19-9, and alpha-fetoprotein), very low specificity (50%) and sensitivity (50%) was observed.[1] This may be due to different subtypes of breast cancer in collected samples. Such epidemiologic studies should select a broader target set of potential biomarkers which could be enabled by antibody array technologies where profiles of up to 100 antibodies can be followed simultaneously. Making different groups, based on the status of hormone receptors (estrogen and progesterone), might also be helpful.

In case of cancer, the need for identification and characterization of early cancer diagnostic biomarkers is high because cancer is a heterogeneous disease and the patient's individual molecular profiling due to tumor microenvironment determines the disease development and response to treatment.[2,5] The tumor microenvironment is affected by several factors including epigenetic factors of the cell.

Reasonable progress has been made in clinic in some cases where biomarkers lead to better efficacy, less toxicity, better diagnosis and predictable prognosis. However, this has been possible only in a few select disease conditions. We anticipate that such success will soon be replicated in several critical diseases. There is an apparent need for new biomarkers and the upcoming technologies promise the development of new biomarkers, which would change the course of disease detection and management. This will be a gain for the medical field with improved patient care and better clinical outcome.

The main areas which need progress/attention are the cost and high throughput. Another area where scope for further progress remains is the application of biomarkers in clinic. Proper analytical and clinical validation of early biomarkers has not been achieved. Clinical validation of identified biomarkers is especially the key challenge in the field. The National Cancer Institute has developed guidelines for the analytical and clinical validation of biomarkers but none of the biomarkers has been validated to date.[161] Integration of genomic and proteomic biomarkers with epigenetic biomarkers may help us subtyping disease stages.[6] Many times results of methylation profiling from blood and tissues are different. Koestler *et al.* conducted a systematic epigenome-wide methylation analysis and demonstrated that shifts in leukocyte subpopulations might account for a considerable proportion of variability in these patterns.[7] The location of miRs in any specific organ should be carefully determined. In a tissue biopsy the local concentration (number of miRs) may be low or high. Determining the accurate level of miRs is very critical.

Association studies are extremely powerful in identifying low-penetrance new SNPs (biomarkers) which may have therapeutic implications. Identification of common low-susceptibility alleles is useful because it provides possible insight into the mechanisms of tumor biology and identify high risk individuals. Since genotyping is not expensive these days, the information from such studies can be utilized in personalized medicine by targeted primary and secondary prevention.

Although individual omics techniques have generated a large number of potential biomarkers for any given cancer; due to heterogeneity of carcinogenesis process, an integrated approach is required to gather all potential biomarkers and to identify key biomarkers for diagnosis, prognosis and therapy.

We emphasize that considerable progress has been made in disease-associated biomarkers which can be used for the complete spectrum of different diseases, from risk assessment to follow up survival. Information discussed in this article may be useful in developing new intervention and therapeutic targets.

REFERENCES

1. E. Ferrannini, A. Natali, S. Camastra, M. Nannipieri, A. Mari, K. P. Adam, M. V. Milburn, G. Kastenmuller, J. Adamski, T. Tuomi, V. Lyssenko, L. Groop and W. E. Gall, *Diabetes*, 2013, **62**, 1730–1737.
2. M. Verma, M. J. Khoury and J. P. Ioannidis, *Cancer Epidemiol., Biomarkers Prev.*, 2013, **22**, 189–200.
3. A. W. Opstal-van Winden, W. Rodenburg, J. L. Pennings, C. T. van Oostrom, J. H. Beijnen, P. H. Peeters, C. H. van Gils and A. de Vries, *Int. J. Mol. Sci.*, 2012, **13**, 13587–13604.
4. C. S. Zhu, P. F. Pinsky, D. W. Cramer, D. F. Ransohoff, P. Hartge, R. M. Pfeiffer, N. Urban, G. Mor, R. C. Bast, Jr., L. E. Moore, A. E. Lokshin, M. W. McIntosh, S. J. Skates, A. Vitonis, Z. Zhang, D. C. Ward, J. T. Symanowski, A. Lomakin, E. T. Fung, P. M. Sluss, N. Scholler, K. H. Lu, A. M. Marrangoni, C. Patriotis, S. Srivastava, S. S. Buys and C. D. Berg, *Cancer Prev. Res.*, 2011, **4**, 375–383.
5. M. Verma, *Methods Mol. Biol.*, 2012, **863**, 467–480.
6. J. M. Yi, M. Dhir, L. Van Neste, S. R. Downing, J. Jeschke, S. C. Glockner, M. de Freitas Calmon, C. M. Hooker, J. M. Funes, C. Boshoff, K. M. Smits, M. van Engeland, M. P. Weijenberg, C. A. Iacobuzio-Donahue, J. G. Herman, K. E. Schuebel, S. B. Baylin and N. Ahuja, *Clin. Cancer Res.*, 2011, **17**, 1535–1545.
7. D. C. Koestler, C. J. Marsit, B. C. Christensen, W. Accomando, S. M. Langevin, E. A. Houseman, H. H. Nelson, M. R. Karagas, J. K. Wiencke and K. T. Kelsey, *Cancer Epidemiol., Biomarkers Prev.*, 2012, **21**, 1293–1302.
8. A. Johnson, Q. Song, P. Ko Ferrigno, P. R. Bueno and J. J. Davis, *Anal. Chem.*, 2012, **84**, 6553–6560.
9. W. Gao, J. Xu and Y. Q. Shu, *Expert Rev. Respir. Med.*, 2011, **5**, 699–709.

10. C. A. Castaneda, M. T. Agullo-Ortuno, J. A. Fresno Vara, H. Cortes-Funes, H. L. Gomez and E. Ciruelos, *Expert Rev. Anticancer Ther.*, 2011, **11**, 1265–1275.

11. J. E. Joh, N. N. Esposito, J. V. Kiluk, C. Laronga, M. C. Lee, L. Loftus, H. Soliman, J. C. Boughey, C. Reynolds, T. J. Lawton, P. I. Acs, L. Gordan and G. Acs, *Oncologist*, 2011, **16**, 1520–1526.

12. E. A. Slodkowska and J. S. Ross, *Expert Rev. Mol. Diagn.*, 2009, **9**, 417–422.

13. B. A. Teicher, *Curr. Opin. Pharmacol.*, 2010, **10**, 397–404.

14. J. Pang, W. P. Liu, X. P. Liu, L. Y. Li, Y. Q. Fang, Q. P. Sun, S. J. Liu, M. T. Li, Z. L. Su and X. Gao, *J. Proteome Res.*, 2010, **9**, 216–226.

15. A. M. Abdel Rahman, S. D. Kamath, S. Gagne, A. L. Lopata and R. Helleur, *J. Proteome Res.*, 2013, **12**, 647–656.

16. I. Rahman, R. Atout, N. L. Pedersen, U. de Faire, J. Frostegard, E. Ninio, A. M. Bennet and P. K. Magnusson, *Atherosclerosis*, 2011, **218**, 117–122.

17. G. Rebovich, E. J. Duffis and L. R. Caplan, *Expert Opin. Med. Diagn.*, 2010, **4**, 267–279.

18. A. Zhang, H. Sun and X. Wang, *Obes. Rev.*, 2012, **14**(4), 344–349.

19. X. Li, H. F. Tse and L. J. Jin, *J. Dent. Res.*, 2011, **90**, 1062–1069.

20. J. Ahn, C. Y. Chen and R. B. Hayes, *Cancer, Causes Control,*, 2012, **23**, 399–404.

21. C. Luaces-Cubells, S. Mintegi, J. J. Garcia-Garcia, E. Astobiza, R. Garrido-Romero, J. Velasco-Rodriguez and J. Benito, *Pediatr. Infect. Dis. J.*, 2012, **31**, 645–647.

22. M. Verma, *Ann. N. Y. Acad. Sci.*, 2003, **983**, 170–180.

23. H. Kuper, H. O. Adami and D. Trichopoulos, *J. Intern. Med.*, 2000, **248**, 171–183.

24. D. M. Parkin, *Int. J. Cancer*, 2006, **118**, 3030–3044.

25. H. zur Hausen, *Eur. J. Cancer*, 1999, **35**, 1174–1181.

26. C. P. Wild and P. C. Turner, *Mutagenesis*, 2002, **17**, 471–481.

27. H. Sun, A. Zhang, G. Yan, C. Piao, W. Li, C. Sun, X. Wu, X. Li, Y. Chen and X. Wang, *Mol. Cell. Proteomics*, 2013, **12**, 710–719.

28. F. L. van de Veerdonk, P. C. Wever, M. H. Hermans, R. Fijnheer, L. A. Joosten, J. W. van der Meer, M. G. Netea and P. M. Schneeberger, *J. Infect. Dis.*, 2012, **206**, 197–201.

29. S. F. Graham, O. P. Chevallier, D. Roberts, C. Holscher, C. T. Elliott and B. D. Green, *Anal. Chem.*, 2013, **85**, 1803–1811.

30. P. Vanek, O. Bradac, P. DeLacy, K. Saur, T. Belsan and V. Benes, *Spine*, 2012, **37**, 1645–1651.

31. M. Verma, *Curr. Opin. Clin. Nutr. Metab. Care*, 2013, **16**(4), 376–384.

32. B. K. Dunn, M. Verma and A. Umar, *Ann. N. Y. Acad. Sci.*, 2003, **983**, 1–4.

33. M. Verma, *Curr. Genomics*, 2012, **13**, 308–313.

34. M. Faryna, C. Konermann, S. Aulmann, J. L. Bermejo, M. Brugger, S. Diederichs, J. Rom, D. Weichenhan, R. Claus, M. Rehli, P. Schirmacher, H. P. Sinn, C. Plass and C. Gerhauser, *FASEB J.*, 2012, **26**, 4937–4950.

35. B. Poudel, A. Mittal, R. Shrestha, A. K. Nepal and P. S. Shukla, *Asian Pac. J. Cancer Prev.*, 2012, **13**, 2149–2152.

36. I. Casanova-Salas, J. Rubio-Briones, A. Fernandez-Serra and J. A. Lopez-Guerrero, *Clin. Transl. Oncol.*, 2012, **14**, 803–811.

37. L. C. Soliman, Y. Hui, A. K. Hewavitharana and D. D. Chen, *J. Chrom. A*, 2012, **1267**, 162–169.

38. L. Murphy and R. W. Watson, *Nat. Rev. Urol.*, 2012, **9**, 464–472.

39. R. Kuner, J. C. Brase, H. Sultmann and D. Wuttig, *Methods*, 2013, **59**, 132–137.

40. S. K. Martin, T. B. Vaughan, T. Atkinson, H. Zhu and N. Kyprianou, *Oncol. Rep.*, 2012, **28**, 409–417.

41. L. Ng, N. Karunasinghe, C. S. Benjamin and L. R. Ferguson, *N. Z. Med. J.*, 2012, **125**, 59–86.

42. J. R. Prensner, M. A. Rubin, J. T. Wei and A. M. Chinnaiyan, *Sci. Transl. Med.*, 2012, **4**, 127rv123.

43. S. Souchelnytskyi, M. Lomnytska, A. Dubrovska, U. Hellman and N. Volodko, *Proteomics*, 2006, **6**(Suppl 2), 65–68.

44. P. D. Hayes, D. A. Payne, N. J. Evans, M. M. Thompson, N. J. London, P. R. Bell and A. R. Naylor, *Eur. J. Vasc. Endovasc.*, 2003, **26**, 665–669.

45. M. P. Endsley and M. Zhang, *Methods Enzymol.*, 2011, **499**, 149–165.

46. E. S. Lianidou, A. Markou and A. Strati, *Cancer Metastasis Rev.*, 2012, **31**, 663–671.

47. A. Alvarez Secord, K. M. Darcy, A. Hutson, Z. Huang, P. S. Lee, E. L. Jewell, L. J. Havrilesky, M. Markman, F. Muggia and S. K. Murphy, *Gynecol. Oncol.*, 2011, **123**, 314–319.

48. Z. J. Xie, G. Chen, X. C. Zhang, D. F. Li, J. Huang and Z. J. Li, *Asian Pac. J. Cancer Prev.*, 2012, **13**, 6145–6149.

49. M. E. Arellano-Garcia, S. Hu, J. Wang, B. Henson, H. Zhou, D. Chia and D. T. Wong, *Oral Dis.*, 2008, **14**, 705–712.

50. C. A. Righini, F. de Fraipont, J. F. Timsit, C. Faure, E. Brambilla, E. Reyt and M. C. Favrot, *Clin. Cancer Res.*, 2007, **13**, 1179–1185.

51. Z. Yalniz, S. Demokan, Y. Suoglu, M. Ulusan and N. Dalay, *Mol. Biol. Rep.*, 2010, **37**, 3541–3545.

52. L. L. Gleich, J. Wang, J. L. Gluckman and C. M. Fenoglio-Preiser, *ORL J. Otorhinolaryngol. Relat. Spec.*, 2003, **65**, 193–198.

53. G. Cuda, A. Gallelli, A. Nistico, P. Tassone, V. Barbieri, P. S. Tagliaferri, F. S. Costanzo, C. M. Tranfa and S. Venuta, *Lung Cancer*, 2000, **30**, 211–214.

54. Z. Q. Yuan, B. Legendre, D. Q. Cai, J. Cao, J. Zhu and T. K. Weber, *Pathology*, 2009, **41**, 393–394.

55. A. I. Shemirani, M. M. Haghighi, S. M. Zadeh, S. R. Fatemi, M. Y. Taleghani, N. Zali, Z. Akbari, S. M. Kashfi and M. R. Zali, *Asian Pac. J. Cancer Prev.*, 2011, **12**, 2101–2104.

56. H. Danaee, H. H. Nelson, M. R. Karagas, A. R. Schned, T. D. Ashok, T. Hirao, A. E. Perry and K. T. Kelsey, *Oncogene*, 2002, **21**, 4894–4899.

57. S. Mourah, O. Cussenot, V. Vimont, F. Desgrandchamps, P. Teillac, B. Cochant-Priollet, A. Le Duc, J. Fiet and H. Soliman, *Int. J. Cancer*, 1998, **79**, 629–633.

58. U. Wahn, R. L. Bergmann and R. Nickel, *Clin. Exp. Allergy*, 1998, **28**(Suppl 1), 20–21discussion 32–26discussion 32–26.

59. I. Okur, L. Tumer, F. S. Ezgu, E. Yesilkaya, A. Aral, S. O. Oktar, A. Bideci and A. Hasanoglu, *J. Pediatr. Endocrinol. Metab.*, 2013, 1–6.

60. M. Kucukazman, N. Ata, B. Yavuz, K. Dal, O. Sen, O. S. Deveci, K. Agladioglu, A. O. Yeniova, Y. Nazligul and D. T. Ertugrul, *Eur. J. Gastroenterol. Hepatol.*, 2013, **25**, 147–151.

61. *Duke Medicine Health News*, 2012, 18, 4–5.

62. E. F. Kern, P. Erhard, W. Sun, S. Genuth and M. F. Weiss, *Am. J. Kidney Dis.*, 2010, **55**, 824–834.

63. A. E. Mehta, *Cleve. Clin. J. Med.*, 1995, **62**, 210–211.

64. D. P. Dosanjh, M. Bakir, K. A. Millington, A. Soysal, Y. Aslan, S. Efee, J. J. Deeks and A. Lalvani, *PloS One*, 2011, **6**, e28754.

65. M. Sakamoto, *J. Gastroenterol.*, 2009, **44**(Suppl 19), 108–111.

66. M. Di Luca, E. Grossi, B. Borroni, M. Zimmermann, E. Marcello, F. Colciaghi, F. Gardoni, M. Intraligi, A. Padovani and M. Buscema, *J. Transl. Med.*, 2005, **3**, 30.

67. M. Verma, G. L. Wright, Jr., S. M. Hanash, R. Gopal-Srivastava and S. Srivastava, *Ann. N. Y. Acad. Sci.*, 2001, **945**, 103–115.

68. I. Simon, Y. Liu, K. L. Krall, N. Urban, R. L. Wolfert, N. W. Kim and M. W. McIntosh, *Gynecol. Oncol.*, 2007, **106**, 112–118.

69. V. Kashuba, A. A. Dmitriev, G. S. Krasnov, T. Pavlova, I. Ignatjev, V. V. Gordiyuk, A. V. Gerashchenko, E. A. Braga, S. P. Yenamandra, M. Lerman, V. N. Senchenko and E. Zabarovsky, *Int. J. Mol. Sci.*, 2012, **13**, 13352–13377.

70. J. Ren, H. Cai, Y. Li, X. Zhang, Z. Liu, J. S. Wang, Y. L. Hwa, Y. Zhang, Y. Yang and S. W. Jiang, *Expert Rev. Mol. Diagn.*, 2010, **10**, 787–798.

71. S. Lindstrom, F. Schumacher, A. Siddiq, R. C. Travis, D. Campa, S. I. Berndt, W. R. Diver, G. Severi, N. Allen, G. Andriole, B. Bueno-de-Mesquita, S. J. Chanock, D. Crawford, J. M. Gaziano, G. G. Giles, E. Giovannucci, C. Guo, C. A. Haiman, R. B. Hayes, J. Halkjaer, D. J. Hunter, M. Johansson, R. Kaaks, L. N. Kolonel, C. Navarro, E. Riboli, C. Sacerdote, M. Stampfer, D. O. Stram, M. J. Thun, D. Trichopoulos, J. Virtamo, S. J. Weinstein, M. Yeager, B. Henderson, J. Ma, L. Le Marchand, D. Albanes and P. Kraft, *PloS One*, 2011, **6**, e17142.

72. K. K. Tsilidis, R. C. Travis, P. N. Appleby, N. E. Allen, S. Lindstrom, F. R. Schumacher, D. Cox, A. W. Hsing, J. Ma, G. Severi, D. Albanes, J. Virtamo, H. Boeing, H. B. Bueno-de-Mesquita, M. Johansson, J. R. Quiros, E. Riboli, A. Siddiq, A. Tjonneland, D. Trichopoulos, R. Tumino, J. M. Gaziano, E. Giovannucci, D. J. Hunter, P. Kraft, M. J. Stampfer, G. G. Giles, G. L. Andriole, S. I. Berndt, S. J. Chanock, R. B. Hayes and T. J. Key, *Am. J. Epidemiol.*, 2012, **175**, 926–935.

73. F. Li, S. Li, H. Chang, Y. Nie, L. Zeng, X. Zhang and Y. Wang, *Genet. Test. Mol. Biomarkers*, 2013.

74. D. R. Taylor, *Thorax*, 2009, **64**, 261–264.

75. S. A. Peters, F. L. Visseren and D. E. Grobbee, *Nat. Rev. Cardiol.*, 2013, **10**, 12–14.

76. G. B. Lim, *Nat. Rev. Cardiol.*, 2012, **9**, 672.

77. J. Madden, C. M. Williams, P. C. Calder, G. Lietz, E. A. Miles, H. Cordell, J. C. Mathers and A. M. Minihane, *Annu. Rev. Nutr.*, 2011, **31**, 203–234.

78. N. Sattar, O. Scherbakova, I. Ford, D. S. O'Reilly, A. Stanley, E. Forrest, P. W. Macfarlane, C. J. Packard, S. M. Cobbe and J. Shepherd, *Diabetes*, 2004, **53**, 2855–2860.

79. A. Betriu-Bars and E. Fernandez-Giraldez, *Nefrologia*, 2012, **32**, 7–11.

80. Y. Shiferaw, A. Alemu, A. Girma, A. Getahun, A. Kassa, A. Gashaw, T. Teklu and B. Gelaw, *BMC Res. Notes*, 2011, **4**, 505.

81. G. B. Gerbi, T. Habtemariam, B. Tameru, D. Nganwa and V. Robnett, *AIDS Care*, 2012, **24**, 331–339.

82. T. Oni, H. P. Gideon, N. Bangani, R. Tsekela, R. Seldon, K. Wood, K. A. Wilkinson, R. T. Goliath, T. H. Ottenhoff and R. J. Wilkinson, *Clin. Vaccine Immunol.*, 2012, **19**, 1243–1247.

83. M. F. Humblet, M. Gilbert, M. Govaerts, M. Fauville-Dufaux, K. Walravens and C. Saegerman, *J. Clin. Microbiol.*, 2010, **48**, 2802–2808.

84. X. F. Wang, Y. W. Cao, Z. Z. Feng, D. Fu, Y. S. Ma, F. Zhang, X. X. Jiang and Y. C. Shao, *Mol. Biol. Rep.*, 2013, **40**, 779–785.

85. M. Silvestrini, G. Viticchi, C. Altamura, S. Luzzi, C. Balucani and F. Vernieri, *J. Alzheimer's Dis.*, 2012, **32**, 689–698.

86. H. M. Schipper, *Alzheimer's Dementia*, 2011, **7**, e118–e123.

87. F. E. Ahmed, P. Vos, S. iJames, D. T. Lysle, R. R. Allison, G. Flake, D. R. Sinar, W. Naziri, S. P. Marcuard and R. Pennington, *Cancer Genomics Proteomics*, 2007, **4**, 1–20.

88. W. Chung, B. Kwabi-Addo, M. Ittmann, J. Jelinek, L. Shen, Y. Yu and J. P. Issa, *PloS One*, 2008, **3**, e2079.

89. T. Shimbo, A. Tanemura, T. Yamazaki, K. Tamai, I. Katayama and Y. Kaneda, *PloS One*, 2010, **5**, e10566.

90. D. J. Sargent, M. B. Resnick, M. O. Meyers, A. Goldar-Najafi, T. Clancy, S. Gill, G. O. Siemons, Q. Shi, B. M. Bot, T. T. Wu, G. Beaudry, J. F. Haince and Y. Fradet, *Ann. Surg. Oncol.*, 2011, **18**, 3261–3270.

91. W. Han, H. Kim, S. Y. Ku, S. H. Kim, Y. M. Choi, J. G. Kim and S. Y. Moon, *Gynecol. Endocrinol.*, 2013, **29**, 16–19.

92. M. G. Tektonidou and M. M. Ward, *Nat. Rev. Rheumatol.*, 2011, **7**, 708–717.

93. D. Plant, I. Ibrahim, M. Lunt, S. Eyre, E. Flynn, K. L. Hyrich, A. W. Morgan, A. G. Wilson, J. D. Isaacs and A. Barton, *Arthritis Res. Ther.*, 2012, **14**, R214.

94. I. Ibrahim, S. A. Owen and A. Barton, *Expert Rev. Clin. Immunol.*, 2012, **8**, 509–511.

95. R. Mogelvang, S. H. Pedersen, A. Flyvbjerg, M. Bjerre, A. Z. Iversen, S. Galatius, J. Frystyk and J. S. Jensen, *Am. J. Cardiol.*, 2012, **109**, 515–520.

96. C. C. Chen, J. L. Huang, C. J. Chang and M. S. Kong, *J. Pediatr. Gastroenterol. Nutr.*, 2012, **55**, 541–547.

97. K. B. Walhovd, A. M. Fjell, J. Brewer, L. K. McEvoy, C. Fennema-Notestine, D. J. Hagler, Jr., R. G. Jennings, D. Karow and A. M. Dale, *AJNR. American journal of neuroradiology*, 2010, **31**, 347–354.

98. K. A. Kolquist, R. A. Schultz, A. Furrow, T. C. Brown, J. Y. Han, L. J. Campbell, M. Wall, M. L. Slovak, L. G. Shaffer and B. C. Ballif, *Cancer Genet*, 2011, **204**, 603–628.

99. A. S. Patel, A. L. Hawkins and C. A. Griffin, *Curr. Opin. Oncol.*, 2000, **12**, 62–67.

100. J. M. Cowan, *Otolaryngol. Clin. North Am.*, 1992, **25**, 1073–1087.

101. Z. Gibas and L. Gibas, *Cancer Genet. Cytogenet.*, 1997, **95**, 108–115.

102. D. Geleick, H. Muller, A. Matter, J. Torhorst and U. Regenass, *Cancer Genet. Cytogenet.*, 1990, **46**, 217–229.

103. R. P. Pearlstein, M. S. Benninger, T. E. Carey, R. J. Zarbo, F. X. Torres, B. A. Rybicki and D. L. Dyke, *Genes, Chromosomes Cancer*, 1998, **21**, 333–339.

104. D. W. Chang, J. Gu and X. Wu, *Urol. Oncol.*, 2012, **30**, 524–532.

105. B. Ludewig, J. V. Stein, J. Sharpe, L. Cervantes-Barragan, V. Thiel and G. Bocharov, *Eur. J. Immunol.*, 2012, **42**, 3116–3125.

106. V. Narang, J. Decraene, S. Y. Wong, B. S. Aiswarya, A. R. Wasem, S. R. Leong and A. Gouaillard, *Immunol. Res.*, 2012, **53**, 251–265.

107. J. D. Bregman, *J. Am. Acad Child. Adolesc. Psychiatry*, 2012, **51**, 1113–1115.

108. L. M. Oberman, *Expert Opin. Invest. Drugs*, 2012, **21**, 1819–1825.

109. P. Karolina, R. Chlup, Z. Jana, K. D. Kohnert, P. Kudlova, J. Bartek, M. Nakladalova, B. Doubravova and P. Seckar, *J. Diabetes Sci. Technol.*, 2010, **4**, 983–992.

110. S. Jaijakul, J. A. Vazquez, R. N. Swanson and L. Ostrosky-Zeichner, *Clin. Infect. Dis.*, 2012, **55**, 521–526.

111. T. R. Powell, L. C. Schalkwyk, A. L. Heffernan, G. Breen, T. Lawrence, T. Price, A. E. Farmer, K. J. Aitchison, I. W. Craig, A. Danese, C. Lewis, P. McGuffin, R. Uher, K. E. Tansey and U. M. D'Souza, *Eur. Neuropsychopharmacol.*, 2012, **23**(9), 1105–1114.

112. R. E. McCullumsmith, J. Hammond, A. Funk and J. H. Meador-Woodruff, *Curr. Pharm. Biotechnol.*, 2012, **13**, 1535–1542.

113. H. M. Ibrahim and C. A. Tamminga, *Curr. Pharm. Biotechnol.*, 2012, **13**, 1587–1594.

114. J. Mullenders, W. von der Saal, M. M. van Dongen, U. Reiff, R. van Willigen, R. L. Beijersbergen, G. Tiefenthaler, C. Klein and R. Bernards, *Clin. Cancer Res.*, 2009, **15**, 5811–5819.

115. S. Maier, C. Dahlstroem, C. Haefliger, A. Plum and C. Piepenbrock, *Am. J. PharmacoGenomicse*, 2005, **5**, 223–232.

116. D. B. Longley, W. L. Allen and P. G. Johnston, *Biochim. Biophys. Acta*, 2006, **1766**, 184–196.
117. A. M. Tsimberidou, N. G. Iskander, D. S. Hong, J. J. Wheler, G. S. Falchook, S. Fu, S. Piha-Paul, A. Naing, F. Janku, R. Luthra, Y. Ye, S. Wen, D. Berry and R. Kurzrock, *Clin. Cancer Res.*, 2012, **18**, 6373–6383.
118. H. Wong and T. Yau, *Oncologist*, 2012, **17**, 346–358.
119. C. B. Weldon, J. R. Trosman, W. J. Gradishar, A. B. Benson, 3rd and J. C. Schink, *J. Oncol. Pract.*, 2012, **8**, e24–e31.
120. L. G. van Baarsen, C. A. Wijbrandts, D. M. Gerlag, F. Rustenburg, T. C. van der Pouw Kraan, B. A. Dijkmans, P. P. Tak and C. L. Verweij, *Genes Immun.*, 2010, **11**, 622–629.
121. S. Marsal and A. Julia, *Pharmacogenomics*, 2010, **11**, 617–619.
122. M. P. Grimaldi, S. Vasto, C. R. Balistreri, D. di Carlo, M. Caruso, E. Incalcaterra, D. Lio, C. Caruso and G. Candore, *Ann. N. Y. Acad. Sci.*, 2007, **1100**, 123–131.
123. N. Siitonen, L. Pulkkinen, J. Lindstrom, M. Kolehmainen, U. Schwab, J. G. Eriksson, P. Ilanne-Parikka, S. Keinanen-Kiukaanniemi, J. Tuomilehto and M. Uusitupa, *Cardiovasc. Diabetol.*, 2011, **10**, 83.
124. G. Ragia and V. G. Manolopoulos, *Pharmacogenomics*, 2012, **13**, 261–264.
125. P. D. Roy, M. Majumder and B. Roy, *Pharmacogenomics*, 2008, **9**, 311–321.
126. F. Martinelli-Boneschi, G. Giacalone, G. Magnani, G. Biella, E. Coppi, R. Santangelo, P. Brambilla, F. Esposito, S. Lupoli, F. Clerici, L. Benussi, R. Ghidoni, D. Galimberti, R. Squitti, A. Confaloni, G. Bruno, S. Pichler, M. Mayhaus, M. Riemenschneider, C. Mariani, G. Comi, E. Scarpini, G. Binetti, G. Forloni, M. Franceschi and D. Albani, *Neurobiol. Aging*, 2013, **34**, 1711.e7–1711.e13.
127. F. Listi, C. Caruso, D. Lio, G. Colonna-Romano, M. Chiappelli, F. Licastro and G. Candore, *J. Alzheimer's Dis.*, 2010, **19**, 551–557.
128. E. Dreussi, P. Biason, G. Toffoli and E. Cecchin, *Pharmacogenomics*, 2012, **13**, 1635–1650.
129. D. T. Merrick, *Chest*, 2012, **141**, 1377–1378.
130. C. Justenhoven, O. Obazee and H. Brauch, *Pharmacogenomics*, 2012, **13**, 659–675.
131. M. Nishiyama and H. Eguchi, *Adv. Drug Delivery Rev.*, 2009, **61**, 402–407.
132. T. Kohriyama, *Rinsho Shinkeigaku*, 2011, **51**, 179–187.
133. S. Malucchi, F. Gilli, M. Caldano, F. Marnetto, P. Valentino, L. Granieri, A. Sala, M. Capobianco and A. Bertolotto, *Neurology*, 2008, **70**, 1119–1127.
134. C. A. Braun Hashemi, Y. C. Zang, J. A. Arbona, J. A. Bauerle, M. L. Frazer, H. Lee, L. Flury, E. S. Moore, M. C. Kolar, R. Y. Washington and O. J. Kolar, *Mult. Scler.*, 2006, **12**, 652–658.
135. A. Miller, L. Glass-Marmor, M. Abraham, I. Grossman, S. Shapiro and Y. Galboiz, *Clin. Neurol. Neurosurg.*, 2004, **106**, 249–254.
136. J. Zhang, C. Wang, L. Li, Q. Man, L. Meng, P. Song, L. Froyland and Z. Y. Du, *Br. J. Nutr.*, 2012, **108**, 1455–1465.

137. S. Sour, M. Belarbi, D. Khaldi, N. Benmansour, N. Sari, A. Nani, F. Chemat and F. Visioli, *Br. J. Nutr.*, 2012, **107**, 1800–1805.

138. L. Ozick, P. Chander, A. Agarwal and A. Soni, *American J. Gastroenterol.*, 1989, **84**, 195–197.

139. A. Maisel, S. X. Neath, J. Landsberg, C. Mueller, R. M. Nowak, W. F. Peacock, P. Ponikowski, M. Mockel, C. Hogan, A. H. Wu, M. Richards, P. Clopton, G. S. Filippatos, S. Di Somma, I. Anand, L. L. Ng, L. B. Daniels, R. H. Christenson, M. Potocki, J. McCord, G. Terracciano, O. Hartmann, A. Bergmann, N. G. Morgenthaler and S. D. Anker, *Eur. J. Heart Failure*, 2012, **14**, 278–286.

140. F. P. Manfredsson, M. S. Okun and R. J. Mandel, *Curr. Gene Ther.*, 2009, **9**, 375–388.

141. D. C. Gilbert, A. J. Chalmers and S. F. El-Khamisy, *Br. J. Cancer*, 2012, **106**, 18–24.

142. E. D. Hsi, S. H. Jung, R. Lai, J. L. Johnson, J. R. Cook, D. Jones, S. Devos, B. D. Cheson, L. E. Damon and J. Said, *Leuk. Lymphoma*, 2008, **49**, 2081–2090.

143. P. C. Tsai, F. J. Hernandez-Ilizaliturri, N. Bangia, S. H. Olejniczak and M. S. Czuczman, *Clin. Cancer Res.*, 2012, **18**, 1039–1050.

144. M. Giachelia, M. T. Voso, M. C. Tisi, M. Martini, V. Bozzoli, G. Massini, F. D'Alo, L. M. Larocca, G. Leone and S. Hohaus, *Leuk. Lymphoma*, 2012, **53**, 411–416.

145. M. P. Chao, A. A. Alizadeh, C. Tang, J. H. Myklebust, B. Varghese, S. Gill, M. Jan, A. C. Cha, C. K. Chan, B. T. Tan, C. Y. Park, F. Zhao, H. E. Kohrt, R. Malumbres, J. Briones, R. D. Gascoyne, I. S. Lossos, R. Levy, I. L. Weissman and R. Majeti, *Cell*, 2010, **142**, 699–713.

146. J. N. Sampson, S. M. Boca, X. O. Shu, R. Z. Stolzenberg-Solomon, C. E. Matthews, A. W. Hsing, Y. T. Tan, B. T. Ji, W. H. Chow, Q. Cai, D. K. Liu, G. Yang, Y. B. Xiang, W. Zheng, R. Sinha, A. J. Cross and S. C. Moore, *Cancer Epidemiol., Biomarkers Prev.*, 2013.

147. X. Zhao, J. Fritsche, J. Wang, J. Chen, K. Rittig, P. Schmitt-Kopplin, A. Fritsche, H. U. Haring, E. D. Schleicher, G. Xu and R. Lehmann, *Metabolomics*, 2010, **6**, 362–374.

148. E. P. Rhee and R. E. Gerszten, *Clin. Chem.*, 2012, **58**, 139–147.

149. J. M. Yuan, Y. T. Gao, S. E. Murphy, S. G. Carmella, R. Wang, Y. Zhong, K. A. Moy, A. B. Davis, L. Tao, M. Chen, S. Han, H. H. Nelson, M. C. Yu and S. S. Hecht, *Cancer Res.*, 2011, **71**, 6749–6757.

150. C. Abate-Shen and M. M. Shen, *Nature*, 2009, **457**, 799–800.

151. T. Zhang, X. Wu, C. Ke, M. Yin, Z. Li, L. Fan, W. Zhang, H. Zhang, F. Zhao, X. Zhou, G. Lou and K. Li, *J. Proteome Res.*, 2013, **12**, 505–512.

152. M. Chadeau-Hyam, T. M. Ebbels, I. J. Brown, Q. Chan, J. Stamler, C. C. Huang, M. L. Daviglus, H. Ueshima, L. Zhao, E. Holmes, J. K. Nicholson, P. Elliott and M. De Iorio, *J Proteome Res.*, 2010, **9**, 4620–4627.

153. C. Gieger, L. Geistlinger, E. Altmaier, M. Hrabe de Angelis, F. Kronenberg, T. Meitinger, H. W. Mewes, H. E. Wichmann, K. M.

Weinberger, J. Adamski, T. Illig and K. Suhre, *PLoS Genetics*, 2008, **4**, e1000282.

154. R. W. Blamey, I. O. Ellis, S. E. Pinder, A. H. Lee, R. D. Macmillan, D. A. Morgan, J. F. Robertson, M. J. Mitchell, G. R. Ball, J. L. Haybittle and C. W. Elston, *Eur. J. Cancer*, 2007, **43**, 1548–1555.

155. S. C. Lee, X. Xu, W. J. Chng, M. Watson, Y. W. Lim, C. I. Wong, P. Iau, N. Sukri, S. E. Lim, H. L. Yap, S. A. Buhari, P. Tan, J. Guo, B. Chuah, H. L. McLeod and B. C. Goh, *Pharmacogenet. Genomics*, 2009, **19**, 833–842.

156. C. Sotiriou and L. Pusztai, *N. Engl. J. Med.*, 2009, **360**, 790–800.

157. G. Ouyang, Y. Chen, L. Setkova and J. Pawliszyn, *J. Chrom. A*, 2005, **1097**, 9–16.

158. C. L. Silva, M. Passos and J. S. Camara, *Talanta*, 2012, **89**, 360–368.

159. D. Y. Wang, S. J. Done, D. R. McCready, S. Boerner, S. Kulkarni and W. L. Leong, *Breast Cancer Res.*, 2011, **13**, R92.

160. S. S. Tang and G. P. Gui, *Biomarkers Med.*, 2012, **6**, 567–585.

161. S. Srivastava, *Gastrointest. Cancer Res. : GCR*, 2007, **1**, S60–63.

CHAPTER 15

Tissue Engineering Design and Application of Synthetic Biology Systems for Therapy

BOON CHIN HENG,[a,c] NILS LINK[b] AND MARTIN FUSSENEGGER*[a]

[a] Department of Biosystems Science & Engineering, ETH-Zurich, Mattenstrasse 26, Basel, Switzerland; [b] Novartis Pharma Stein AG, CH-4332 Stein, Switzerland; [c] Department of Biological Sciences, Sunway University, No. 5 Jalan Universiti, Bandar Sunway 47500, Selangor Darul Ehsan, Malaysia
*Email: fussenegger@bsse.ethz.ch

15.1 INTRODUCTION

15.1.1 Economic Impact of Healthcare

For thousands of years, medical practitioners have attempted to reduce pain, cure disease, extend lifespan and generally improve quality of life. The immense technological progress of the past 30 years has enabled great advances to be made in the treatment of disease. This is reflected statistically in life expectancies, which have increased significantly over the past decades. However, longer life expectancy creates new medical issues, particularly with regard to the longevity of organs and tissues that must now function for longer periods of time. At present, organ transplantation is the only possible therapy for some patients. Thus, as life expectancy increases, so too does the demand for donor organs (Table 15.1). In the USA alone, 112 706 patients were on the waiting list for organ donation in 2011,

Molecular Biology and Biotechnology, 6th Edition
Edited by Ralph Rapley and David Whitehouse
© The Royal Society of Chemistry 2015
Published by the Royal Society of Chemistry, www.rsc.org

Table 15.1   Number of transplants performed in 2011 and their impact on the general public. *Source*: Organ Procurement and Transplantation Network.[1]

Transplant	No. of transplants	Total charge per year (US$ millions)
Bone marrow	20 157	10 378
Cornea	46 081	1124
Heart only	2161	2156
Intestine only	74	89
Kidney only	16 571	4357
Liver only	5898	3404
Lung only	1784	1249
Pancreas	286	82.8
Heart-lung	30	37
Intestine with other organs	107	144
Kidney-heart	66	86
Kidney-pancreas	867	412
Liver-kidney	369	379
Other multi-organ	42	72
Total	94 493	23 970

Table 15.2   Cost of treating the most expensive diseases, 2005. *Source*: www.WebMD.com.

Indication	Total charge per year (US$ billions)
Heart conditions	95.6
Trauma	74.3
Cancer	72.2
Mental disorders	72.1
Joint disorders	57.0
COPD and asthma	53.7
Hypertension	47.4
Diabetes	45.9
Hyperlipidemia	38.6
Back problems	35.0
Total	591.8

while a further 6251 patients passed away while awaiting organ transplants in 2010.[1] Furthermore, the treatment and care of patients undergoing such treatment are costly (almost US$600 billion per year in the USA [Tables 15.1 and 15.2]). Although the use of synthetic materials aids in keeping people alive, many implants function only to a limited extent like living tissues. Unlike real tissues, which are in a constant state of renewal, synthetic implants begin to show signs of decomposition within 10–15 years and must eventually be replaced. Therefore, a key goal of regenerative medicine is to develop artificially generated tissues, which can take over the function of the original tissue without being rejected by the patient's immune system.

15.1.2 Tissue Engineering

As early as 1987, the US National Science Foundation (NSF) approved the proposal submitted by Y. C. Fung to the Granlibakken Workshop to initiate research into tissue engineering to overcome the problem of donor organ shortage. This project is regarded as the birth of tissue engineering, the goal of which, according to the NSF, is 'the application of principles and methods of engineering and life sciences toward fundamental understanding of structure–function relationships in normal and pathological mammalian tissues and the development of biological substitutes to restore, maintain or improve tissue function'. This statement makes it clear that tissue engineering is an interdisciplinary and heterogeneous research field, where biologists, chemists, material scientists and medical practitioners must work together to achieve success.[2]

After 50 years of progress in cell culture, immunology, the cell cycle, proliferation, (trans-)differentiation and extracellular matrices (ECM),[3,4] we are currently in a position to extract, culture and expand, *in vitro*, all the existing cell types of an organism.

The groups of Hay, Heath and Ikehara have performed outstanding investigations of the different growth factors, their effects and the chronology of the markers expressed by cells during differentiation (Table 15.3).[5–8] Nevertheless, the cell model developed thus is far from complete. It is not difficult to keep cells alive in culture, but it is challenging to multiply them without altering their differentiation and phenotype.[9–12] Therefore, research must continue to identify the exact biochemical markers and growth factors that will finally enable the tissue engineer to determine unambiguously whether the cells behave in exactly the same way as cells in living organs. Only then can artificial tissues be engineered which express the same phenotype as the desired organ.[9]

Within the field of tissue engineering, there are four main areas of research: (i) cell therapy, which is based mainly on self-organization of free cells that are injected into damaged tissue,[13,14] (ii) bio-artificial devices, in which cells of a certain type are encapsulated to take on certain functions of the organ, *e.g.* liver cells, pancreatic islet cells and kidney cells,[15–18] (iii) scaffold-assisted tissue engineering, in which cells are grown on a matrix until they reach a certain level of stability,[19,20] and (iv) scaffold-free tissue engineering, in which cells form tissue-like structures *ex vivo* via cell aggregation and expression of their own natural extracellular matrix.[21–23] A discussion of all the topics is beyond the scope of this chapter. Therefore, we will focus on the last two approaches, which involve the growth of real tissues *in vitro*.

15.1.3 Treating Disease Through Tissue Engineering

In the short time since the initiation of tissue engineering programs, immense progress has been made. In principle, all diseases associated with the failure of tissue functions are currently being investigated. Many of the tissues, generated *in vitro* for the replacement of cartilage, articulation,

Table 15.3   Known markers and methods for identification of different cell types.

Cell type	Marker/method
Adipocyte	Adipocyte lipid-binding protein (ALBP) Fatty acid transporter (FAT)
Astrocyte	Glial fibrillary acidic protein (GFAP)
Bone marrow fibroblast	Fibroblast colony-forming unit (CFU-F) Muc-18 (CD146)
Cardiomyocyte	Myosin heavy chain
Chondrocyte	Collagen types II and IV Sulfated proteoglycan
Ectoderm	Neuronal cell-adhesion molecule (N-CAM) Pax6
Ectoderm, neural and pancreatic progenitor	Nestin
Embryoid body (EB)	Vimentin
Embryonic stem (ES), embryonal carci-noma (EC)	Neurosphere Alkaline phosphatase Cripto (TDGF-1) [cardiomyocyte] Cluster designation 30 (CD30) GCTM-2 Germ cell nuclear factor Oct-4 Stage-specific embryonic antigen-3 (SSEA-3) Stage-specific embryonic antigen-4 (SSEA-4) Telomerase TRA-1-60 TRA-1-81 Stem cell factor (SCF or c-Kit ligand)
Endoderm	Alpha-fetoprotein (AFP) GATA-4 Hepatocyte nuclear factor-4 (HNF-4) Fetal liver kinase-1 (Flk1)
Hematopoietic stem cell (HSC)	CD34 Lineage surface antigen (Lin) c-Kit Stem cell antigen (Sca-1) Thy-1
Hepatocyte	Albumin B-1 integrin
Keratinocyte	Keratin
Mesenchymal	CD44
Mesenchymal stem cell (MSC)	Bone morphogenic protein receptor (BMPR) CD34 Sca-1 Lin
Mesoderm	Bone morphogenic protein-4 Brachyury
Myoblast	MyoD Pax7
Neural stem cell	Nestin CD133
Neuron	Microtubule-associated protein-2 (MAP-2)

Table 15.3 (*Continued*)

Cell type	Marker/method
Oligodendrocyte	Myelin basic protein (MPB)
	O4
	O1
Osteoblast	Bone-specific alkaline phosphatase (BAP)
	Hydroxyapatite
	Osteocalcin (OC)
Pancreatic islet	Glucagon
	Insulin
	Insulin-promoting factor-1 (PDX-1) Pancreatic polypeptide
	Somatostatin
Pancreatic progenitor	Nestin
Skeletal myocyte	Myogenin
	MR4
	Myosin light chain
Skeletal muscle	Smooth muscle cell-specific myosin heavy chain
White blood cell (WBC)	Vascular endothelial cell cadherin
	Stro-1 antigen
	CD4
	CD8

heart, liver and pancreatic islets, have already undergone clinical tests. Cartilage has already reached the stage of clinical application[24] due to the robustness and phenotypic stability of the cells and also the fact that cartilage and bone tissues can be grown *in vitro* with relative ease. These tissues do not require strong vascularization—still a major challenge in tissue engineering—since there is sufficient diffusion of nutrients and oxygen through the scaffolds used for this type of cell. Several tests have been conducted with three-dimensional hyaline cartilage to remedy articular damage or abrasions when other treatments fail *e.g.* chondroitin sulfate (Chondrosulf).[25] Other tissues, such as heart valves and heart-muscle tissues, have also been produced *in vitro*.[26,27] However, they have shown poor integration after transplantation into the organ and have not reliably maintained their function. Researchers and medics are working on a cure for liver malfunction (liver cirrhosis and fibrosis), induced by either alcohol abuse or viral infections such as hepatitis. Although the liver can regenerate itself, it can also become so weakened that transplantation is the only solution. To overcome the perpetual shortage of donor organs, an attempt has been made to propagate hepatocytes *in vitro*. However, only the culturing of these cells as scaffold-free microtissues or on three-dimensional scaffolds that mimic the natural extracellular matrix of the liver has been successful; growing hepatocytes as monolayers changes their phenotype and results in the loss of their natural function.[23] In addition, research into the generation of pancreatic islet tissues to combat diabetes, the main cause of hormone dysfunction in developed countries, is ongoing.[28,29]

It is not possible here to list all the organs that are currently being reconstructed in laboratories. We have, therefore, concentrated on the most important diseases, for which a clinical therapy is standard or soon will be. Although researchers have made great advances in the engineering of tissues with simple structures (*e.g.* cartilage and liver), many tissues do not consist of a single cell type but rather a specific architecture of many cell types that enables them to perform their tasks, as is the case, for example, with the kidney, heart and skin. Hence it is very important to understand the structure and cellular composition of native tissue in order to differentially reconstruct them for therapeutic applications.

15.2 CELL TYPES

In recent years, in-depth research has been performed to identify the cells responsible for tissue-/organ-specific functions. There are three broad categories of stem cells: (i) embryonic stem cells (ESC), (ii) induced pluripotent stem cells (iPSC), and (iii) multipotent adult stem cells (ASC). Each category has its advantages and disadvantages for clinical applications. Additionally, tissue engineering can also utilize progenitor and mature somatic cell types that may or may not be derived from trans-differentiation of other somatic lineages.

15.2.1 Embryonic Stem Cells

For over 20 years, researchers have been able routinely to extract and cultivate mouse embryonic stem cells.[30] This cultivation expertise has also permitted the isolation and cultivation of human embryonic stem cell lineages.[31] Embryonic stem cells are non-specialized cells which are found in the inner cell mass of 7–30-day-old blastocysts or embryos.

At this stage of development, the three main cell lineages of ectoderm, mesoderm and endoderm have not yet formed. Thus, embryonic stem cells are pluripotent and are able to differentiate into every cell type of a tissue or organism. This is one of the most important advantages for tissue engineering when considering suitable cell types. Compared with adult stem cells, embryonic stem cells are capable of long-term renewal without differentiation, thus allowing the production of the cell mass required for tissue engineering applications.

As a general rule, a stem cell can do one of the following: divide and generate a new stem cell or differentiate into a particular tissue lineage (Figure 15.1). Symmetric division gives rise to two identical daughter cells, both endowed with stem cell properties, whereas asymmetric division produces only one stem cell and a progenitor cell with limited potential for self-renewal. Progenitors can go through several rounds of cell division before terminally differentiating into a mature cell. The switch between division

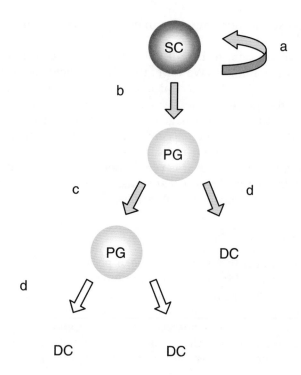

Figure 15.1   Pathways of stem cells (SCs). SCs often divide asymmetrically, thereby generating a daughter SC (a) and a committed progenitor cell (PG) (b). The PG either divides to produce a PG (c) or fully matures to a differentiated cell (DC) (d).

and differentiation is usually triggered easily by external factors such as the cell environment, hormones or signals from neighboring cells. As a result, it is difficult to expand embryonic stem cells in culture without them differentiating spontaneously in one direction. Therefore, embryonic stem cell cultures must be monitored carefully and continuously. Due to lack of knowledge, the research community has not yet established a protocol for the absolute and complete characterization of stem cells, although the following methods are commonly used simultaneously to ensure that the cells display the required stem cell traits:[9,32]

- Visual characterization by microscopy;
- Formation of embryoid bodies;
- Analysis of surface markers, such as the stage-specific embryonic antigen (SSEA)-3, SSEA-4, TRA-l-60, TRA-1-81 and alkaline phosphatase;
- Determination of whether Oct-4—a transcription factor typical of undifferentiated cells is expressed and whether cells display high telomerase activity;
- Teratoma formation in SCID mice;

- Long-term cultivation to ensure long-term self-renewal, a special characteristic of embryonic stem cells;
- Microscopic analysis of chromosome damage;
- Determination of whether sub-culturing is possible after freezing/thawing;
- Differentiation assays leading to a specific cell type;
- Proteomic analysis.[33]

Using these methods, researchers are in a position to vary and improve culture techniques in order to study the fundamental properties of embryonic stem cells, including the precise determination of why embryonic stem cells are not specialized, how they renew themselves over many years and which factors cause stem cells to undergo differ entiation. The determination of these signals and the chronology of their occurrence are of prime importance for the production of the desired phenotype for tissue reconstruction. To trigger differentiation, the sur face of the culture dish, the scaffold, the chemical composition of the culture medium (Table 15.4) or the genetic expression pattern of the cells can all be manipulated.

Embryonic stem cells are the only naturally occurring cells that are able to divide frequently enough without manipulation to generate a sufficient number of cells for therapeutic application. Were it not for ethical concerns, restrictions or bans and also the risk of developing teratomas or rejection by the patient, these cells would be the optimal choice for tissue engineering. Furthermore, we still possess only a rudimentary understanding of most of the differentiation pathways and the culture conditions required to achieve precise lineage control of stem cells. As a result, scientists often obtain differentiated cells, which have a phenotypic resemblance to the cells of a certain tissue but which express not only the usual markers, but also tissue-atypical proteins and do not fulfill all the functions of the native cells in a tissue.

15.2.2   Induced Pluripotent Stem Cells

The clinical applications of human embryonic stem cells were severely hampered by ethical issues,[34,35] as well as immunological hurdles in allogenic transplantation/transfusion therapy.[36,37] Takahashi *et al.*[38] overcame these challenges by successfully reprogramming differentiated somatic cells to a pluripotent embryonic stem cell-like state through the recombinant expression of just four transcription factors: KLF4, OCT4, c-MYC, and SOX2.[38] It is now possible to generate immunocompatible patient-specific cells of any lineage for transplantation/transfusion therapy. Rapid progress has been made in the past few years in the field of cellular reprogramming and derivation of induced pluripotent stem cells (iPSCs).[39,40] Notable breakthroughs in this area include the identification of various small molecules that improve reprogramming efficiency,[41–43] as well as the development of new technology platforms that utilize protein and RNA instead of recombinant DNA for cellular reprogramming.[44–47]

Table 15.4 Mesengenic process, differentiation factors and markers of mesenchymal stem cells (MSCs). The value of MSCs for tissue engineering is apparent after analysis of the different cell types that can be obtained. The diagram displays the stepwise cellular transition upon induction with differentiation factors (italic) from MSCs to highly differentiated phenotypes expressing characteristic cell markers (underlined).

Proliferation	Hematopoietic stem cell CD34, CD45, CD14 Mesenchymal stem cell Integrin ß1, Collagen type I, Fibronectin, CD10, 13, 54, 59, 90, 105, LNGFR, HLA-DR, Stro-1					
	Bone	**Cartilage**	**Muscle**	**Marrow**	**Tendon**	**Connective tissue**
Commitment	Osteogenesis *BMP-2 Vitronectin Dexamethasone FGF Ascorbic acid β-glycerophosphate*	Chondrogenesis *TGF-beta ITS Dexamethasone Praline BMPs*	Myogenesis *Myf5 myogenin MyoD Myf5*	Marrow stroma *Serum PDGF Hydrocortisone*	Tendogenesis *GDF-5, -6, -7 BMP-12*	Adipogenesis *Dexamethasone Isobutylmethyl-xanthine PPARγ2 BMX, Insulin Indomethacin*
Lineage progression	Transitory osteoblast	Transitory chondrocyte	Myoblast	Transitory stromal cell	Transitory fibroblast	
Differentiation	Osteoblast	Chondrocyte	Myoblast fusion			
Maturation	Osteocyte Osteocalcin Alk. phosphatase Alizarin Red	Hypertrophic chondrocyte Type II collagen Chondroadherin GAG	Myotube MyoD1 Myogenin MyHC SH-2,-3 smooth muscle actin CD13, 29, 49	Stromal cell Cytokines (IL-6, IL-11)	Tendon fibroblast CD44 HLA-D	Adipocyte Oil red O Wnt signaling LPL aP2

To date, the vast majority of studies have utilized both recombinant DNA and viral vectors to reprogram somatic cells to iPSCs, a process that carries the risk of permanent genetic modification to the cellular genome. This is a major obstacle for clinical applications of iPSCs, due to overriding safety concerns associated with the introduction of genetically modified cells into the human body.[48] A previous clinical trial reported that genetically modified cells became malignant upon transplantation *in situ* within the patient.[49]

The PiggyBac transposon system[50,51] enables the transient insertion and removal of transgenic elements in a precise and site-specific manner without causing permanent genetic modification to the cellular genome. By utilizing PiggyBac transposons with four different transgene inserts (KLF4, OCT4, c-MYC and SOX2), Woltjen *et al.*[50,51] reprogrammed human fibroblasts into iPSCs and then excised these transgenic elements from the cellular genome, without leaving any permanent trace of genetic alteration.[50,51] Nevertheless, utilizing the PiggyBac transposon system for iPSC generation is tedious and labor-intensive, and requires rigorous screening in order to confirm the absence of genetic modification to the cellular genome.

An alternative approach, which avoids the use of recombinant DNA, is to utilize cell-penetrating peptides and proteins for cellular reprogramming.[52–58] Zhou *et al.*[44] fused a polyarginine (11R) protein transduction domain to four recombinant transcription factors (KLF4, OCT4, c-MYC and SOX2) and utilized these factors to reprogram murine embryonic fibroblasts to iPSCs. Using a similar approach, Kim *et al.*[45] were able to derive iPSC from human fibroblasts with recombinant transcription factors fused to polyarginine protein transduction domains. However, the short half-life of cell-penetrating peptides upon entry into the cell renders this technique relatively inefficient.

Successful cellular reprogramming with cell-penetrating peptides/proteins was quickly followed by the first successful derivation of iPSCs through transfection with chemically modified mRNA.[47] Utilizing mRNA transfection in cellular reprogramming requires overcoming two major technical challenges. The first challenge is the relative instability and short half-life of RNA within the cell, and the second is the innate anti-retroviral response within mammalian cells against foreign RNA that triggers cellular apoptosis. To overcome the first challenge, Warren *et al.*[47] extended the stability and half-life of synthetic mRNA with a 5'-guanine cap,[59,60] while the second challenge was overcome by substituting uridine with pseudouridine[61] and cytidine with 5-methylcytidine.[62] Additionally, B18R protein was supplemented into the culture media during RNA transfection to suppress the interferon-1 pathway that leads to cellular apoptosis upon entry of foreign RNA into the cell.[63] This is achieved by the B18R protein acting as a decoy receptor for type I interferon,[63] and is absolutely crucial for maintaining cell viability during cellular reprogramming with RNA transfection. Warren *et al.*[47] utilized chemically modified mRNA corresponding to the four Yamanaka transcription factors (KLF4, OCT4, c-MYC, and SOX2), together with the supplementation of B18R within the culture medium. This enabled them to reprogram human fibroblasts into iPSC at higher efficiencies than previous techniques that utilized recombinant DNA.

More recently, Miyoshi *et al.*[64] demonstrated that the direct transfection of certain miRNA species alone is able to reprogram somatic cells to pluripotency, without the need for any of the transcription factors that are commonly utilized for iPSC generation. The major advantage of miRNA-based reprogramming is that the innate immunity pathway leading to apoptosis is not activated upon artificial entry of miRNA into mammalian cells, on account of their small size.[65] Additionally, because there is no need for the transfected miRNA to be translated into proteins, its effects are much faster. Furthermore, some miRNA species that are utilized in reprogramming can target various epigenetic factors simultaneously, which leads to global demethylation in target cells.[66–68]

Although the studies on RNA-based cellular reprogramming by Warren *et al.*[47] and Miyoshi *et al.*[64] both reported relatively high efficiency of iPSC generation, it remains unclear which approach is better, due to the relatively few studies and current paucity of experimental data on RNA-based cellular reprogramming.

15.2.3 Adult Stem Cells

Adult stem cells, also referred to as somatic stem cells, represent an ethically acceptable alternative to embryonic stem cells in that they do not require harvesting of embryos. These cells constitute a very small population of undifferentiated cells that can be found amongst the differentiated cells of an organ. They renew themselves and differentiate into the major specialized cells of the organ in which they reside. Their primary tasks are to maintain the proper functioning and repair of the tissue. According to the literature, adult stem cells have been found in the brain, skin, bone marrow, peripheral blood, blood vessels, skeletal muscle and liver. Some of the most important and best characterized adult stem cells include:

- Brain: neural stem cells can develop into four major cell types: (i) neurons, (ii) astrocytes, (iii) oligodendrocytes and (iv) glial cells.[69,70]
- Bone marrow: containing two types of stem cells—first, the hematopoietic stem cells that give rise to all types of blood cells, *i.e.* (i) red blood cells, (ii) B-lymphocytes, (iii) T-lymphocytes, (iv) natural killer cells, (v) neutrophils, (vi) basophils, (vii) eosinophils, (viii) monocytes, (ix) macrophages and (x) platelets; and second, the mesenchymal stem cells (bone marrow stromal cells) which give rise to different cell types, such as (i) osteocytes (bone cells), (ii) chondrocytes (cartilage cells), (iii) adipocytes and (iv) some types of connective tissue cells (Table 15.4).
- Digestive tract: epithelial stem cells differentiate into several cell types such as (i) absorptive cells, (ii) goblet cells, (iii) paneth cells and (iv) entero-endocrine cells.[71]

- Skin: epidermal stem cells can be found in the basal layer of the epidermis. These stem cells give rise to keratinocytes. Follicular stem cells give rise to both the hair follicle and to the epidermis at the base of hair follicles.[72,73]
- *Adipose tissue*: mesenchymal stem cells.[74]
- *Umbilical cord*: hematapoietic and mesenchymal stem cells.[75]

Unfortunately, not all essential human organs contain adult stem cells. As in the case of the heart, the stem cells may not divide or differentiate rapidly enough if the organ has been damaged.[76–78] Most adult stem cells are present only in small numbers and are, therefore, difficult to extract and separate from other cells of the organs. In contrast to embryonic stem cells, adult stem cells have lost their ability to differentiate naturally into all the tissues of an organism. However, recent publications indicate that adult stem cells exhibit plasticity or undergo transdifferentiation under certain conditions.[79–81] The discovery of plasticity among adult stem cells is very important for tissue engineering, as it permits the generation of new types of cells through transdifferentiation. Another advantage of adult stem cells is that it may be possible to use the patient's own cells, to expand them, to generate the desired tissues *in vitro* and to transplant the tissues back into the patient, thereby avoiding the risk of immunological rejection of the transplant. Therefore, adult stem cells, treated to differentiate into specific cell types, could be a renewable source of cells and tissues for the efficient treatment of diseases such as cystic fibrosis, renal failure, stroke, burns, heart disease, diabetes and arthritis.

15.2.4 Mature Cells

In the human body, there are approximately 300 different types of differentiated cells, each with its own specific function. During differentiation, these cells lose the ability to divide and, by adapting to a specific task, can no longer proliferate. Since cell therapy and tissue engineering rely on the implantation of considerable quantities of well-characterized autologous cells that display defined properties, differentiated cells must be manipulated so that they multiply *in vitro*. Taking primary cells from the healthy tissues of choice is mandatory for therapeutic applications. It is unimaginable that immortalized cells could be reintroduced into humans, as they could develop into a cancer. There are different methods for re-initiating the cell division of differentiated cells. One way is to modify the cells genetically and conditionally immortalize them[82] to proliferate for a limited period of time without inducing epigenetic changes or changes in the cell properties.[82,83] Ideal candidate transgenes to control conditional expansion of therapeutic cell populations could be Notch[84] or antisense p27Kip1.[85] However, since these strategies enable proliferation induction in a limited number of cell lines, viral genes such as the simian virus large T antigen, the Herpes virus-16 E6/7 or the ubiquitous telomerase reverse transcriptase are sometimes used generically to trigger proliferation.[86,87] In any case, proliferation-inducing transgenes must be tightly and timely controlled using either state-of-the-art transcription or translation control modalities[88–90] or site-specific recombination technology to excise transgene expression loci prior to implantation.[91–94]

15.2.5 Direct Lineage Conversion and Trans-Differentiated Cells

A number of recent studies have demonstrated direct lineage conversion from one differentiated phenotype to another through the recombinant expression of appropriate lineage-specific genes, particularly transcription factors implicated in upstream control of cell fate.[95] Most notably, murine fibroblasts have been trans-differentiated into neurons,[96,97] cardiomyocytes,[98] and hepatocytes[99,100] while human fibroblasts have been trans-differentiated into neurons[101–103] and hematopoietic progenitors.[104] Nevertheless, it is unlikely that future regenerative medicine applications would bypass the progenitor cell or stem cell stages completely. The limited proliferative capacity of differentiated cells means that direct lineage conversion of one differentiated somatic phenotype to another is unlikely to provide sufficient cell numbers for transplantation/transfusion therapy. It is only at either the stem cell or progenitor cell stages that extensive proliferation *in situ* or *ex vivo* is feasible. Additionally, because of the relatively low number of studies and the current paucity of experimental data, it is unknown whether the process of direct lineage conversion and trans-differentiation could result in cells with aberrant epigenetic reprogramming, and therefore pose a safety challenge in clinical therapy.

15.2.6 Partial Reprogramming to the Transit Amplifying Progenitor Stage

One strategy for overcoming the limited proliferative capacity of differentiated somatic cells is partial reprogramming to the progenitor stage.[105] Instead of reprogramming all the way back to iPSCs, the developmental clock may be reset halfway, to a less immature developmental stage that is more directly applicable to therapeutic applications. Of particular interest is the transit amplifying progenitor stage, which refers to the intermediate stage of cellular differentiation at which stem cells have already committed to a particular lineage and are actively proliferating just before the onset of terminal differentiation. Prominent examples of this stage include hepatic oval cells and keratinocyte progenitors.[105]

The major challenge with this strategy is identifying the appropriate combination of transcription factors and other genes for partial reprogramming, as well as the appropriate somatic cell types that are amendable to reprogramming. Because cell lineages at different states of maturation/differentiation from the same organ/tissue are likely to share some commonalities in gene expression, we hypothesize that partial reprogramming to the progenitor stage would be more efficient than iPSCs derivation, since a common epigenetic and gene regulatory program must differentiated cell types from the same organ/tissue. Moreover, partial re-programming to the progenitor stage will also allay safety concerns regarding the aberrant epigenetic signature that could potentially arise during the reprogramming process. This is because, in partial reprogramming, we implicitly want these cells to be epigenetically predisposed and committed to a particular lineage or subset of lineages. In cardiac tissue engineering, for example, for which transit amplifying progenitors of the cardiomyogenic lineage are required, it will not be of great concern if matured differentiated cardiomyocytes retain much of their original epigenetic signature upon reprogramming to the transit amplifying state of cardiac progenitors, provided that this does not compromise the expected proliferative capacity of such reprogrammed cells.[105] The same cannot be said of iPSCs, which should ideally return to the epigenetic signature of the embryonic state upon reprogramming in order to be considered truly pluripotent.

15.3 EXTRACELLULAR MATRIX

15.3.1 Biological Extracellular Matrices

The second important component in tissue engineering is the extra-cellular matrix (ECM). The ECM represents the secreted product of the resident cells of each tissue and organ. Therefore, the ECM of each tissue and organ has a unique structure and composition, providing structural stability for the tissue. It includes information about the position and alignment of the different cell types and ensures that the relevant growth factors are provided at the right level, time and place to coordinate organ morphogenesis and repair. Although all the cells in an organism are embedded in ECM, sufficient amounts of ECM for use in tissue engineering can only be extracted from a few tissues such as skin, the pericardium, small intestine, urinary bladder, liver and Achilles tendon.[106,107] ECM is by no means static and uniform but is rather a structure that adapts continuously to the requirements of the tissue. The composition and structure of the ECM are directly coupled to its location within the organ, the function of the tissue and the age of the individual.[108–111] For example, the kidney has very little ECM compared with its cellular component, whereas tissue that is primarily structural, such as tendons and ligaments, displays large amounts and a differential composition of ECM.[112,113] ECM is composed mainly of collagen, of which more than 20 types have been identified thus far.[114] The most common is type 1 collagen that has been highly conserved during the course of evolution. Thus, allogenic and xenogenic sources of type 1 collagen are both relevant for tissue engineering, making collagen the most widely used biologic scaffold in therapeutic interventions. The 12 subtypes of collagen are responsible for the distinctive biological activity of the ECM. In combination with laminin they form a three-dimensional mesh-like structure that is adapted to the specific function of a tissue and provides optimal strength, rigidity or plasticity.[115] Laminin, the second most abundant exist between the reprogrammed cells, progenitor, and terminally protein in ECM, is a complex trimeric, cross-linked adhesion protein with separate binding domains for collagen IV, heparin, heparin sulfate and direct cell binding.[116] Laminin exists in different isoforms, depending on the particular mixture of the peptide chains[117,118] and plays an important role in the vascularization and maintenance of vascular structures.[119,120] Since vascularization of scaffolds for tissue repair is the most rate-limiting step, laminin is considered to be an important component of cell-friendly scaffold material.[121]

A very important peptide motif found in most proteins which form the ECM (*e.g.* the glycoproteins fibronectin or vitronectin) is the arginine, glycine and asparaginic acid sequence (RGD).[122,123] This motif binds to cellular adhesion molecules (CAM), known as integrins, thereby anchoring the cells mechanically in the ECM. The different RGD sequences adopt different conformations in the different matrix proteins, so that these sequences are recognized by different integrin subtypes expressed by specific cell types which favors a tissue-specific cellular organization. Furthermore, the generation of focal adhesions generates cell responses such as the polarization of the cells, the production of survival signals and factors for the remodulation of the ECM. Therefore, synthetic scaffolds produced for tissue engineering are often modified with special RGD–peptide sequences to render them biocompatible, to provide better integration and to control the setup of the tissue.[124–126]

Important non-protein components of ECM are the glycosaminoglycans (GAGs), which do not have a structural function. However, they substantially modulate the gel properties of the ECM by retaining water and binding growth factors and cytokines. Also, their ability to mediate ECM–cell interactions makes heparin-rich GAGs a valuable component for tissue engineering-compatible scaffolds.[127] In addition to structural proteins, the ECM also contains trace amounts of a variety of bioactive proteins which functionalize the bio-scaffold. Non-limiting examples include vascular endothelial growth factor, basic fibroblast growth factor, epidermal growth factor, hepatocyte growth factor, keratinocyte growth factor, transforming growth factor beta and platelet-derived growth factor. Due to the presence of such growth factor cocktails, natural ECM grafts are often used to functionalize synthetic scaffolds,[128] which are otherwise biologically inert and fail to degrade and promote infiltration of cells from neighboring tissues. In contrast, the use of unmodified, decellularized ECM promotes rapid cell infiltration, scaffold degradation, deposition of neo-matrix and tissue organization with a minimum of scar tissue.[129] Today, over one million patients have been treated successfully with xenogeneic ECM scaffolds[130] to heal skin lesions,[131] promote vascular reconstruction[132,133] and re-establish the urinary tract,[134–136] the intestine,[137] diaphragm,[138] rotator cuff[139] and muscle structures.[140]

The complex three-dimensional structure and composition of ECM have yet to be fully elucidated, exemplifying how difficult it is to design ECM-containing scaffold mimetics. Bottom-up approaches to functionalize synthetic scaffolds with well-characterized components (*e.g.* laminin, fibronectin, hyaluronic acid, vascular endothelial growth factors)[143] have resulted in some success for very specialized applications, but there is still a long way to go until fully functional ECM mimetics can be used as synthetic scaffolds.

15.3.2   Artificial Extracellular Matrices

It will be extremely challenging to reproduce exactly a biological extracellular matrix through artificial means in the coming decades. To obtain a scaffold resembling biological ECM, many researchers currently use Matrigel, a cocktail of substances extracted from natural ECM.[141] The advantage of Matrigel is that it is a liquid at temperatures below 41 °C, where it can be mixed with desired cell populations to result in a pre-seeded ready-to-implant scaffold upon gelation at physiological temperatures. Despite the superior attributes of ECM, it is sometimes advised to use artificial scaffolds for tissue engineering to achieve reproducibility and control of the individual parameters.

A protocol for the manufacture of a so-called ideal scaffold has yet to be developed. Therefore, special attention must be paid to the essential parameters of ECM, which are to be mimicked. Irrespective of the application, the scaffold should: (i) be bio-compatible, (ii) provide a 3D template for guided growth, (iii) have a highly porous structure for maximum cell load, nutrition and oxygen diffusion, (iv) degradation dynamics should match *de novo* ECM synthesis, (v) be mechanically resistant to biological forces, (vi) withstand sterilization and (vii) production should be economically feasible (Table 15.5). The shape of the starting material, including the fibers,[142] microspheres[143,144] and sheets and films[145,146] must be determined and the generation of the scaffold microstructure studied.[147] For this purpose, different materials, including ceramics, polymers and composites, are under investigation. Ceramics such as hydroxyapatite,[148,149] tricalcium phosphate,[150,151] glass ceramics and glass[152,153] are generally used for the reconstruction of hard tissue, whereas polymers are implemented for the formation of soft tissues.[154,155] Polymers are usually the material of choice as their properties can be modulated with ease. Gelatin,[156] elastin,[19,157] collagen,[158,159] fibrin glue[160,161] and hyaluronic acid,[116–118] all of biological origin, are the most important polymers for production of artificial scaffolds. Synthetic polymers are also available including poly(lactic acid), poly(glycolic acid), polycaprolactone poly(lactic-co-glycolic acid), poly-ethylene, poly(ethylene glycol), poly(ethylene oxide)-block-poly-caprolactone, poly(ethylene terephthalate), polytetrafluoroethylene and polyurethane.[162] Both degradable and non-degradable synthetic polymers exist. Although degradable polymers are usually selected, such as the most widely used poly(lactic acid), the degradation products must be either biologically inert or easily metabolized, but above all non-toxic.

As a result of refining the production process for scaffold material, the spectrum and quality of scaffold types have increased considerably. Latest-generation scaffolds have average pore sizes ranging from a few nanometers[163] to several micrometers.[164] In many cases the pores are interconnected to allow the transport of nutrients and metabolic waste[159] and to provide a high surface-to-mass ratio for the promotion of cell differentiation and proliferation. These scaffolds are produced by various techniques, including: (i) electrospinning, (ii) phase separation, (iii) foaming processes, (iv) microsphere sintering, (v) solid free-form fabrication,

Table 15.5 Ideal parameters for tissue scaffolds.

Characteristics	Comments
Architecture	3D architecture should assist in the guidance and arrangement of the cells
Angiogenic	Scaffold should support fast vascularization of the growing tissue
Biocompatible	Scaffolds should be flexible, rigid, strong and should not induce rejection or immune responses
Biodegradable	Degradation should occur at the same rate as tissue regeneration; degradation products must be non-toxic
Co-casting with cells	It should be possible to produce an optimal scaffold in the presence of the cells
Economically producible	The scaffold should be generally affordable
High surface-to-volume ratio	A critical factor is to generate scaffolds with a high cell load
Homogeneic	Scaffold should be homogeneous and show homogeneous cell distribution
Mechanical strength	Should be adapted to the biological forces in the body
Non-corrosive	Scaffold should withstand body fluids and body temperature
Non-immunogenic	Scaffold should not induce immunogenic responses to the tissue
Non-toxic	Scaffold should not induce toxicity or inhibit tissue development
Porous	Pores should be of correct size to accommodate the infiltrating cells and should be interconnected to allow diffusion of nutrients and oxygen
Specificity	Scaffold should allow growth of different cells types at the same time
Withstands sterilization	Contamination with pathogens should be avoided
Storable	Scaffold should not degrade during storage
Surface chemistry	Surface chemistry should enable easy modification or improved cell adhesion

(vi) shape deposition manufacturing, (vii) fused deposition modeling, (vii) non-fused liquid deposition modeling, (viii) 3D printing, (ix) selective laser sintering, (x) stereo-lithographic processes, (xi) leaching processes and (xii) self-assembly.[165]

Electrospinning is one of the most commonly used techniques to develop scaffolds for tissue engineering, because it is possible to modulate the fiber size from the nanometer to the micrometer scale and to guide the arrangement of fibers in the non-woven 3D fiber network.[166,167] The process consists of forcing a polymer melt through an electrically charged nozzle (10–20 kV), thereby forming a thin polymer string. The solvent quickly evaporates en route to the collector, which can be a simple grounded plate for random orientation of the fibers or a drum for their parallel alignment.[167] Different processing parameters such as the polymer or solvent chosen, viscosity, surface tension, the charge applied, the polymer mass flux through the nozzle, the processing temperature and the humidity, all control the diameter and the ultrastructure of the generated fibers.[168] Another advantage of electrospinning is that fibers of different polymers can be co-deposited into a multilayered structure.[169] Such scaffolds display a range of favorable properties for cell attachment (*e.g.* a wide range of pore sizes, high porosity and parallel fiber orientation).[170,171] Yang *et al.* showed that cells tend to grow in the direction of the fiber and that spatial orientation of the fibers therefore plays a key role in guiding cell growth and the subsequent organization of the tissues.[172]

Phase separation is another valuable and cost-effective process to form scaffolds for tissue engineering. The polymer is usually dissolved in a solvent that is brought into contact with a non-solvent. During the casting process, the solvent diffuses out and the non-solvent diffuses into the polymer solution until the polymer becomes unstable and precipitates, thereby forming the final scaffold after only a few hundred milliseconds.[173,174] The pore size of such scaffolds can be varied from nanometers to several micrometers[175] and can be extended by other processes such as salt leaching.[174] These scaffolds are ideal for the engineering of special tissues such as skin and bone, the pore size of which must be in the range 0.04–0.4 mm.[176]

Self-assembly systems are gaining in importance in the development of artificial scaffolds.[160,177] They utilize a bottom-up approach, whereby molecules undergo self-association to form a higher order structure without external manipulation. Special attention should be paid to the monomer, which must be well designed to induce chemical reactivity and structural compatibility. Given these prerequisites, various building blocks have been used, such as nanofiber scaffolds,[178] peptides[179,180] and dipolar molecules.[181] Although the design of scaffolds by molecular self-assembly is extremely challenging, it is a promising process that can be performed under physiological conditions which allow pre-seeding of the scaffolds with desired cell populations.

The choice of a specific scaffold for tissue engineering is usually based on the type of tissue to be engineered. Several researchers have provided specific information concerning each type of scaffold and their application.[167,182–186] However, this information should be treated with caution because the described scaffolds were, in part, developed by material scientists and chemists and have not been tested in depth with cell cultures. For example, foaming processes may result in inclusion of air bubbles which compromise cell colonization,[176] salt-leaching processes do not completely remove entrapped salt crystals[163] and sterilization radically changes the microstructure of polymer scaffolds. In conclusion, only successful testing in *in vivo* models will finally prove whether a scaffold is suitable for tissue engineering or not. Despite all the obstacles that scientists have encountered in producing natural or synthetic scaffolds for tissue engineering, many scaffolds have already been approved by the FDA and are being used in a clinical setting (Table 15.6).

Table 15.6 A selection of commercially available scaffolds.

Product	Company	Material	Form
AlloDerm	Lifecell	Human skin	Cross-linked Dry sheet
Axis dermis	Mentor	Human dermis	Natural Dry sheet
CuffPatch	Arthrotek	Porcine small intestinal submucosa (SIS)	Cross-linked Hydrated sheet
DurADAPT	Pegasus Biologicals	Horse pericardium	Cross-linked
Dura-Guard	Synovis Surgical	Bovine pericardium	Hydrated sheet
Durasis	Cook SIS	Porcine SIS	Natural Dry sheet
Durepair	TEI Biosciences	Fetal bovine skin	Natural Dry sheet
Fortaderm	Organogenesis	Highly purified Collagen	Dry sheet
FortaFlex	Organogenesis	Highly purified collagen	Cross-linked Dry sheet
FortaPerm	Organogenesis	Highly purified collagen	High-level cross- linked, Resistant to degradation
Gracilis tendon	Regeneration Technologies	Patellar tendon	Frozen tendon
Graft Jacket	Wright Medical Tech	Human skin	Cross-linked Dry sheet
Grafton	Osteotech	Demineralized bone	Natural Dry sheet
Hydrix XM	Caldera Medical	Bovine pericardium	Natural Hydrated sheet
Oasis	Healthpoint	Porcine SIS	Natural Dry sheet
OrthADAPT	Pegasus Biologicals	Horse pericardium	Cross-linked
Peri-Guard	Synovis Surgical	Bovine pericardium	Cross-linked
Permacol	Tissue Science Laboratories	Porcine skin	Hydrated sheet
PriMatix	TEI Biosciences	Fetal bovine skin	Natural Dry sheet
Repliform	Boston Scientific	Human dermal allograft	Natural
Restore	DePuy	Porcine SIS	Natural Sheet
Stratasis	Cook SIS	Porcine SIS	Natural Dry sheet
Strauman Bone Ceramic	Stauman	Hyroxyapatite and tricalcium phosphate	Powder
SurgiMend	TEI Biosciences	Fetal bovine skin	Natural Dry sheet
Surgisis	Cook SIS	Porcine SIS	Natural Dry sheet
Suspend	Mentor	Human fascia lata	Natural Dry sheet
TissueMend	TEI Biosciences	Fetal bovine skin	Natural Dry sheet
VeriCart	Histogenics	Collagen matrix	
Veritas	Synovis Surgical	Bovine pericardium	Hydrated sheet
Xelma	Molnlycke	protein, PGA, water	Gel
Xenoform	TEI Biosciences	Fetal bovine skin	Natural Dry sheet

15.4 TISSUE ENGINEERING CONCEPTS

In addition to an optimal scaffold, a suitable bioreactor and bioprocess are required to produce the cell mass required to assemble artificial tissues. A variety of disposable plates and flasks are commercially available for standard cell culture but only a few systems are available for cultivating artificial tissues.

15.4.1 Cultivation of Artificial Tissues

Although many publications have reported on novel bioreactor configurations,[187,188] only very few have covered the production of life-sized tissues such as those intended for the reconstruction the liver, kidney and heart. *In vitro* fabrication of mammalian tissues for human therapeutic use has become standard practice for small-sized prototype tissues. However, considerable improvements in culture techniques will be required to produce artificial tissues beyond the cubic millimeter size, since larger sized tissue suffers from limited oxygen diffusion, which induces hypoxia and compromises cell viability in the center of the tissue.[189] Most tissue engineers have reached the conclusion that oxygen supply is the most critical factor in limiting tissue growth.[190,191] Therefore, the bioreactor should be optimized to modulate mass transfer into the tissue, which is essential both for the nutrient supply and the elimination of metabolites if optimal tissue viability is to be maintained. Once these limitations have been overcome, it should become possible to produce life-sized tissues for clinical use.

Growing mammalian tissues under *in vitro* conditions is particularly challenging because of their nutrient requirements, their sensitivity to metabolic waste and their susceptibility to shear stresses.[192] The required nutrients and growth parameters and the cells' susceptibility to stress vary considerably, depending on the type of tissue.[193,194] These differences must be accounted for when designing a bioreactor and a bioprocess for a particular type of tissue.

Here we present several cultivation techniques that are currently being applied in tissue engineering. When scaffold design is the primary research interest, it is best to cultivate the artificial tissues in a Petri dish. The scaffold is placed in the dish, covered with culture medium and the cells are seeded on to the scaffold where they quickly migrate and attach. However, the static milieu in the dish may rapidly lead to localized oxygen limitation and insufficient removal of metabolic waste products from the tissue. The thickest bone tissue obtained by this method has been 0.5 mm.[195] Static culture conditions have always resulted in a shell of cells around the scaffold with poor migration into the interior of the scaffold.[196,197] Diffusion can be improved by cultivating tissue samples in medium-containing spinner flasks. Spinner flasks allow constant mixing of the culture and therefore provide better supply of nutrients and oxygen to the tissue. Nevertheless, the culture medium still becomes depleted over time and 50% of it must be exchanged every 3 days.[198] For this kind of bioreactor, the typical mixing rate is 50–80 rpm, a compromise between optimal mass transfer and minimal shear stress.[198] Using spinner flasks, cartilage tissue has been grown to a thickness of 0.5 mm, which is almost five times thicker than isogenic tissues grown in Petri dishes.[193] However, typical cartilage implants used today are 2–5 mm thick.[193]

Although hollow-fiber bioreactors are not suitable for the production of implantable tissues, they have been extensively studied for the design of extracorporeal devices which could provide liver and kidney function in a dialysis-like therapy.[199,200] The hollow-fiber bioreactor consists of a closed container filled with a cell-containing matrix, into which a bundle of semi-permeable hollow fibers is inserted. A constant flow of culture medium through the fiber provides nutrients and eliminates metabolic waste products. With this method, the interface-to-tissue mass ratio is very high and provides a more homogeneous nutrient supply throughout the tissue. This type of bioreactor more closely resembles the situation in a vertebrate body, where cells are usually never more than 200 µm away from the next blood vessel. Studies with hepatocytes cultured in a hollow-fiber bioreactor revealed that, if the distance between the fibers exceeds 250 µm, then a perpendicular flow to the fibers is necessary to achieve a sufficient stream of nutrients to the cells.[201]

An important recent development in the construction of hollow-fiber bioreactors is the use of degradable fiber materials. After degradation of the hollow fibers, the tissue is in principle ready for implantation.[202] Poly(D,L-lactide-co-glycolide) fibers maintain their structural integrity for 4 weeks and degrade homogenously until they disappear completely by 8 weeks, thereby maintaining the structure of the tissue.[202] Although artificial tissues produced by hollow-fiber bioreactors have not yet been used in clinical trials, the bioreactor itself is already in clinical use as cartridges for dialysis/plasma separation. These cartridges contain 50–200 g of primary hepatocytes and are connected for 6–8 h per day in a typical dialysis setting.[17]

One of the most commonly applied reactors for the generation of tissues is the rotating-wall bioreactor. The tissue floats freely in the chamber and the rotation speed, usually 15–30 rpm, is adjusted so that the tissue remains in a state of zero gravity. Cartilage tissues up to 5 mm thick[203] and liver tissues with a thickness of up to 3 mm have been produced.[204] Rotating-wall bioreactors have been used to produce many other tissues, of which the most important has been myocardial tissues.[205,206]

Arguably, the best bioreactor system to cultivate artificial tissues is the perfusion bioreactor, which typically contains a small chamber through which a flow of fresh and defined medium is pumped at a constant rate. The scaffold is usually fixed to a porous support in the middle of the chamber. The flow of medium through the scaffold enhances cell growth inside the scaffold and provides mechanical stimulation in the form of shear force as the media is forced through the scaffold.[207] A disadvantage of this bioreactor is that the orientation of the cells follows the direction of the liquid flow. Although alignment of the cells is desired in tissue engineering, engineers would prefer the cells to align perpendicular rather than parallel to the liquid flow.[208] Perfusion bioreactors are of particular interest for growing tissues for skin replacement and consist of two chambers, through one of which is pumped medium optimized for the growth of epithelial cells and through the other is pumped medium optimized for the growth of connective tissue. This configuration permits the production of two-layer skin tissues.[209]

The perfusion system can even be employed to simulate biological forces. For instance, Watanabe *et al.* applied intermittent hydrostatic pressure (0–5 MPa) when cultivating cartilage,[210] whereas Seidel *et al.* applied mechanical compression forces.[211] Other perfusion bioreactors with a modified configuration have also been used for the tissue engineering of skeletal muscle and osteochondralcomposites.

15.4.2 Design of Scaffold-Free Tissues

Although state-of-the-art artificial scaffolds allow an adequate flow of nutrients to the cells residing at the center, and growth of specific cell types, the physical, chemical and biological properties of the scaffold material are far from being optimal for every tissue.[212] Here are summarized the latest trends in designing scaffold-free artificial tissues, which, in our opinion, represent a valuable extension of current scaffold-based approaches.

The principles of generating scaffold-free microtissue spheroids are straightforward and have been applied for decades to test anticancer drugs in a more realistic tissue-like model[213–215] and also for the analysis of cell differentiation.[216,217] Monodispersed cells aggregate as spheroids whenever intercellular adhesion forces, most often mediated by homotypic interactions between surface proteins of the cadherin family, exceed those of cell–surface interactions.[218] Therefore, microtissue spheroids are produced by: (i) cultivating the cells in culture dishes, spinner flasks and roller bottles with non-adhesive surfaces,[219–221] (ii) centrifugation-based pelleting of the cells,[222] (iii) growth of the cells in small containers[223] and (iv) gravity-enforced re-aggregation of the cells in hanging drops.[23] The hanging drop technology is by far the gentlest aggregation strategy and permits positioning of individual cells within a microtissue without coming in contact with any synthetic material. Using gravity-enforced self-assembly of cells in hanging drops, Kelm and co-workers have successfully designed heart, liver, neuronal and cartilage tissues with unmatched *in vivo* characteristics.[21,23,224–226]

For example, hepatic microtissues generated by the hanging drop technology have shown increased levels of detoxifying enzymes and a more perfect hepatic ultrastructure, including formation of correct polarity and bile canaliculi, compared with hepatocyte monolayer or other 3D cultures.[23] Although it was possible to generate microtissues from a variety of cell types, not all of them will produce the correct extracellular matrix and the desmosome-based intercellular communication network required for correct positioning of individual cells within a microtissue and for the formation of fully functional microtissues.[23,218]

Although cell movement during development has been studied in great detail, the precise positioning of individual cells during the formation of artificial tissues remains largely elusive. Kelm's group has successfully used microtissue spheroids assembled from different cell types to study the relative positioning of different cell populations inside a forming microtissue. For example, gravity-enforced self-assembly of a cell mixture mimicking the natural composition of the heart suggested the presence of molecular forces which position cardiomyocytes preferentially at the periphery of beating 'microhearts'.[21] Microhearts stimulated by addition of phenylephrine or by ectopic expression of bone morphogenetic protein 2 (BMP2) reproduced electrogenic profiles reminiscent of fully functional hearts.[224]

Repositioning of individual cells and assembly of tissue substructures could also be observed in microtissues produced by co-cultivation of human hepatocytes (HepG2) and umbilical vein endothelial cells (HUVECs). In these, HUVECs migrate from the surface of the spheroids to the center thereby forming tubular structures reminiscent of vascular structures.[227] After implantation into a chicken embryo, these vascular structures successfully connected to the chicken vasculature and chicken hemoglobin managed oxygen supply for the implant, which was seamlessly integrated into the embryo tissue without showing any scar structures.[227] Using vascularized microtissues as minimal building blocks, Kelm and co-workers also succeeded in producing fully functional larger sized tissues in the cubic millimeter range. These macrotissues could be assembled into custom shapes by cultivating and fusing prevascularized microtissues in agarose moulds.[22] The design of custom-shaped, scaffold-free, fully vascularized tissues of implantable size will significantly advance tissue engineering in the not-so-distant future.

15.5 CONCLUSIONS

Tissue engineering has grown rapidly as a research discipline in recent years and the number of engineered tissues used in a clinical setting is constantly growing. However, the challenges remain significant. Non-limiting examples include: (i) precise reprogramming of lineage control in different cell types, (ii) increased knowledge of differentiation circuits and, more importantly, de-differentiation networks, (iii) control of position and molecular cross-talk of different cell types within a tissue, (iv) engineering of a fully functional vasculature into larger-sized tissues and (v) industrial-scale production of precursor cells for the assembly of artificial tissues. Given the dramatic progress made in recent years, tissue engineering may well revolutionize medical treatment in the new millennium.

REFERENCES

1. Organ Procurement and Transplantation Network. Accessible at the following website: http://www.statisticbrain.com/organ-donor-statistics/ (Date accessed: 25 September 2012).

2. R. Langer and J. P. Vacanti, *Tissue Eng. Sci.*, 1993, **260**(5110), 920–926.
3. M. Ehrbar, A. Metters, P. Zammaretti, J. A. Hubbell and A. H. Zisch, *J. Control Release*, 2005, **101**(1–3), 93–109.
4. C. Fux, D. Langer and M. Fussenegger, *J. Gene Med.*, 2004, **6**(10), 1159–1169.
5. D. C. Hay, D. Zhao, A. Ross, R. Mandalam, J. Lebkowski and W. Cui, *Cloning Stem Cells*, 2007, **9**(1), 51–62.
6. J. K. Heath and A. G. Smith, *J. Cell Sci. Suppl.*, 1988, **10**, 257–266.
7. J. K. Heath, A. G. Smith, L. W. Hsu and P. D. Rathjen, *J. Cell Sci. Suppl.*, 1990, **13**, 75–85.
8. S. Ikehara, *Proc. Soc. Exp. Biol. Med.*, 2000, **223**(2), 149–155.
9. A. Alhadlaq and J. J. Mao, *Stem Cells Dev.*, 2004, **13**(4), 436–448.
10. I. Jasmund, S. Schwientek, A. Acikgöz, A. Langsch, H. G. Machens and A. Bader, *Biomol. Eng.*, 2007, **24**(1), 59–69.
11. S. H. Khoo and M. Al-Rubeai, *Biotechnol. Appl. Biochem.*, 2007, **47**(Pt 2), 71–84.
12. L. Yao, C. S. Bestwick, L. A. Bestwick, N. Maffulli and R. M. Aspden, *Tissue Eng.*, 2006, **12**(7), 1843–1849.
13. A. I. Caplan, *Tissue Eng.*, 2005, **11**(7–8), 1198–1211.
14. N. Kimelman, G. Pelled, G. A. Helm, J. Huard, E. M. Schwarz and D. Gazit, *Tissue Eng.*, 2007, **13**(6), 1135–1150.
15. W. H. Fissell, *Expert Rev. Med. Devices*, 2006, **3**(2), 155–165.
16. M. Miyamoto, *Hum. Cell*, 2001, **14**(4), 293–300.
17. J. K. Park and D. H. Lee, *J. Biosci. Bioeng.*, 2005, **99**(4), 311–319.
18. A. Saito, *Nephrology (Carlton)*, 2003, **8**(Suppl), S10–S15.
19. W. H. Zimmermann and T. Eschenhagen, *Heart Fail. Rev.*, 2003, **8**(3), 259–69.
20. V. L. Tsang and S. N. Bhatia, *Adv. Biochem. Eng. Biotechnol.*, 2007, **103**, 189–205.
21. J. M. Kelm, E. Ehler, L. K. Nielsen, S. Schlatter, J. C. Perriard and M. Fussenegger, *Tissue Eng.*, 2004, **10**(1–2), 201–214.
22. J. M. Kelm, V. Djonov, L. M. Ittner, D. Fluri, W. Born, S. P. Hoerstrup and M. Fussenegger, *Tissue Eng.*, 2006, **12**(8), 2151–2160.
23. J. M. Kelm and M. Fussenegger, *Trends Biotechnol.*, 2004, **22**(4), 195–202.
24. S. Marlovits, P. Zeller, P. Singer, C. Resinger and V. Vécsei, *Eur. J. Radiol.*, 2006, **57**(1), 24–31.
25. D. W. Hayes Jr., R. L. Brower and K. J. John, *Clin. Podiatr. Med. Surg.*, 2001, **18**(1), 35–53.
26. P. V. Kochupura, E. U. Azeloglu, D. J. Kelly, S. V. Doronin, S. F. Badylak, I. B. Krukenkamp, I. S. Cohen and G. R. Gaudette, *Circulation*, 2005, **112**(9Suppl), I144–I149.
27. T. C. Flanagan, C. Cornelissen, S. Koch, B. Tschoeke, J. S. Sachweh, T. Schmitz-Rod and S. Jockenhoevel, *Biomaterials*, 2007, **28**(23), 3388–3397.
28. A. Cohen and E. S. Horton, *Curr. Med. Res. Opin.*, 2007, **23**(4), 905–917.
29. M. Nair, *Br. J. Nurs.*, 2007, **16**(3), 184–188.

30. M. J. Evans and M. H. Kaufman, *Nature*, 1981, **292**(5819), 154–156.
31. J. A. Thomson, J. Itskovitz-Eldor, S. S. Shapiro, M. A. Waknitz, J. J. Swiergiel, V. S. Marshall and J. M. Jones, *Science*, 1998, **282**(5391), 1145–1147.
32. I. Singec, R. Jandial, A. Crain, G. Nikkhah and E. Y. Snyder, *Annu. Rev. Med.*, 2007, **58**, 313–28.
33. D. Van Hoof, C. L. Mummery, A. J. Heck and J. Krijgsveld, *Expert Rev. Proteomics*, 2006, **3**(4), 427–437.
34. A. Mauron and M. E. Jaconi, *Clin. Pharmacol. Ther.*, 2007, **82**(3), 330–333.
35. L. Waite and G. Nindl, *Biomed, Sci, Instrum.*, 2003, **39**, 567–572.
36. C. J. Taylor, E. M. Bolton and J. A. Bradley, *Philos. Trans. R. Soc. Lond. B Biol. Sci.*, 2011, **366**(1575), 2312–2322.
37. J. E. Lee, M. S. Kang, M. H. Park, S. H. Shim, T. K. Yoon, H. M. Chung and D. R. Lee, *Cell Transplant*, 2010, **19**(11), 1383–1395.
38. K. Takahashi, K. Tanabe, M. Ohnuki, M. Narita, T. Ichisaka, K. Tomoda and S. Yamanaka, *Cell*, 2007, **131**(5), 861–872.
39. G. Mostoslavsky, *Stem Cells*, 2012, **30**(1), 28–32.
40. K. S. Sidhu, *Expert Opin. Biol. Ther.*, 2011, **11**(5), 569–579.
41. J. Hao, D. B. Sawyer, A. K. Hatzopoulos and C. C. Hong, *Rec. Pat. Regen. Med.*, 2011, **1**(3), 263–2274.
42. T. Lin, R. Ambasudhan, X. Yuan, W. Li, S. Hilcove, R. Abujarour, X. Lin, H. S. Hahm, E. Hao, A. Hayek and S. Ding, *Nat. Methods*, 2009, **6**(11), 805–808.
43. B. Feng, J. H. Ng, J. C. Heng and H. H. Ng, *Cell Stem Cell*, 2009, **4**(4), 301–312.
44. H. Zhou, S. Wu, J. Y. Joo, S. Zhu, D. W. Han, T. Lin, S. Trauger, G. Bien, S. Yao, Y. Zhu, G. Siuzdak, H. R. Schüler, L. Duana and S. Ding, *Cell Stem Cell*, 2009, **4**(5), 381–384.
45. D. Kim, C. H. Kim, J. I. Moon, Y. G. Chung, M. Y. Chang, B. S. Han, S. Ko, E. Yang, K. Y. Cha, R. Lanza and K. S. Kim, *Cell Stem Cell*, 2009, **4**(6), 472–476.
46. H. Zhang, Y. Ma, J. Gu, B. Liao, J. Li, J. Wong and Y. Jin, *Biomaterials*, 2012, **33**(20), 5047–5055.
47. L. Warren, P. D. Manos, T. Ahfeldt, Y. H. Loh, H. Li, F. Lau, W. Ebina, P. K. Mandal, Z. D. Smith, A. Meissner, G. Q. Daley, A. S. Brack, J. J. Collins, C. Cowan, T. M. Schlaeger and D. J. Rossi, *Cell Stem Cell*, 2010, **7**, 618–630.
48. S. R. Burger, *Cytotherapy*, 2003, **5**(4), 289–298.
49. S. Hacein-Bey-Abina, C. Von Kalle, M. Schmidt, M. P. McCormack, N. Wulffraat, P. Leboulch, A. Lim, C. S. Osborne, R. Pawliuk, E. Morillon, *et al.*, *Science*, 2003, **302**(5644), 415–419.
50. K. Woltjen, I. P. Michael, P. Mohseni, R. Desai, M. Mileikovsky, R. Hämäläinen, R. Cowling, W. Wang, P. Liu, M. Gertsenstein, *et al.*, *Nature*, 2009, **458**(7239), 766–770.
51. K. Woltjen, R. Hämäläinen, M. Kibschull, M. Mileikovsky and A. Nagy, *Methods Mol Biol.*, 2011, **767**, 87–103.

52. A. K. Haas, D. Maisel, J. Adelmann, C. von Schwerin, I. Kahnt and U. Brinkmann, *Biochem. J.*, 2012, **442**(3), 583–593.
53. A. van den Berg and S. F. Dowdy, *Curr. Opin. Biotechnol.*, 2011, **22**(6), 888–893.
54. M. Grdisa, *Curr. Med. Chem.*, 2011, **18**(9), 1373–1379.
55. H. Noguchi, M. Matsushita, N. Kobayashi, M. F. Levy and S. Matsumoto, *Cell Transplant*, 2010, **19**(6), 649–654.
56. B. Münst, C. Patsch and F. Edenhofer, *J. Vis. Exp.*, 2009, **28**(34), 1627.
57. H. Noguchi and S. Matsumoto, *Acta Med. Okayama*, 2006, **60**(1), 1–11.
58. S. El-Andaloussi, T. Holm and U. Langel, *Curr. Pharm. Des.*, 2005, **11**(28), 3597–3611.
59. X. Jiao, S. Xiang, C. Oh, C. E. Martin, L. Tong and M. Kiledjian, *Nature*, 2010, **467**(7315), 608–611.
60. J. K. Yisraeli and D. A. Melton, *Methods Enzymol.*, 1989, **180**, 42–50.
61. K. Kariko, H. Muramatsu, F. A. Welsh, J. Ludwig, H. Kato, S. Akira and D. Weissman, *Mol. Ther.*, 2008, **16**, 1833–1840.
62. M. S. Kormann, G. Hasenpusch, M. K. Aneja, G. Nica, A. W. Flemmer, S. Herber-Jonat, M. Huppmann, L. E. Mays, M. Illenyi, A. Schams, *et al.*, *Nat. Biotechnol.*, 2011, **29**(2), 154–157.
63. J. A. Symons, A. Alcamí and G. L. Smith, *Cell*, 1995, **81**(4), 551–560.
64. N. Miyoshi, H. Ishii, H. Nagano, N. Haraguchi, D. L. Dewi, Y. Kano, S. Nishikawa, M. Tanemura, K. Mimori, F. Tanaka, *et al.*, *Cell Stem Cell*, 2011, **8**(6), 633–638.
65. K. Drews, G. Tavernier, J. Demeester, H. Lehrach, S. C. De Smedt, J. Rejman and J. Adjaye, *Biomaterials*, 2012, **33**(16), 4059–4068.
66. S. L. Lin, D. C. Chang, S. Chang-Lin, C. H. Lin, D. T. Wu, D. T. Chen and S. Y. Ying, *RNA*, 2008, **14**(10), 2115–2124.
67. S. L. Lin, D. C. Chang, C. H. Lin, S. Y. Ying, D. Leu and D. T. Wu, *Nucleic Acids Res.*, 2011, **39**(3), 1054–1065.
68. S. L. Lin, *Stem Cells*, 2011, **29**(11), 1645–1649.
69. D. Mondal, L. Pradhan and V. F. LaRussa, *Cancer Invest.*, 2004, **22**(6), 925–943.
70. K. Nakashima and T. Taga, *Mol. Neurobiol.*, 2002, **25**(3), 233–44.
71. S. J. Leedham, M. Brittan, S. A. McDonald and N. A. Wright, *J. Cell. Mol. Med.*, 2005, **9**(1), 11–24.
72. L. Alonso and E. Fuchs, *Proc. Natl. Acad. Sci. U. S. A.*, 2003, **100**(Suppl 1), 11830–11835.
73. G. Cotsarelis, *J. Invest. Dermatol.*, 2006, **126**(7), 1459–1468.
74. P. Gir, G. Oni, S. A. Brown, A. Mojallal and R. J. Rohrich, *Plast Reconstr Surg.*, 2012, **129**(6), 1277–1290.
75. E Pelosi, G Castelli and U. Testa, *Blood Cells Mol. Dis.*, 2012, **49**(1), 20–28.
76. P. Anversa, A. Leri, M. Rota, T. Hosoda, C. Bearzi, K. Urbanek, J. Kajstura and R. Bolli, *Stem Cells*, 2007, **25**(3), 589–601.
77. A. Leri, J. Kajstura and P. Anversa, *Physiol. Rev.*, 2005, **85**(4), 1373–1416.

78. K. Urbanek, D. Cesselli, M. Rota, A. Nascimbene, A. De Angelis, T. Hosoda, C. Bearzi, A. Boni, R. Bolli, J. Kajstura, P. Anversa and A. Leri, *Proc. Natl. Acad. Sci. U.S.A.*, 2006, **103**(24), 9226–9231.
79. K. A. Jackson, S. M. Majka, G. G. Wulf and M. A. Goodell, *J. Cell Biochem. Suppl.*, 2002, **38**, 1–6.
80. A. Vescovi, A. Gritti, G. Cossu and R. Galli, *Cells Tissues Organs*, 2002, **171**(1), 64–76.
81. H. E. Young, C. Duplaa, M. Romero-Ramos, M. F. Chesselet, P. Vourc'h, M. J. Yost, K. Ericson, L. Terracio, T. Asahara, H. Masuda, S. Tamura-Ninomiya, K. Detmer, R. A. Bray, T. A. Steele, D. Hixson, M. el-Kalay, B. W. Tobin, R. D. Russ, M. N. Horst, J. A. Floyd, N. L. Henson, K. C. Hawkins, J. Groom, A. Parikh, L. Blake, L. J. Bland, A. J. Thompson, A. Kirincich, C. Moreau, J. Hudson, F. P. Bowyer, T. J. Lin and A. C. Black, *Cell Biochem. Biophys.*, 2004, **40**(1), 1–80.
82. M. Heyde, K. A. Partridge, R. O. Oreffo, S. M. Howdle, K. M. Shakesheff and M. C. Garnett, *J. Pharm. Pharmacol.*, 2007, **59**(3), 329–350.
83. T. May, P. P. Mueller, H. Weich, N. Froese, U. Deutsch, D. Wirth, A. Kröger and H. Hauser, *J. Biotechnol.*, 2005, **120**(1), 99–110.
84. B. Varnum-Finney, L. Xu, C. Brashem-Stein, C. Nourigat, D. Flowers, S. Bakkour, W. S. Pear and I. D. Bernstein, *Nat. Med.*, 2000, **6**(11), 1278–1281.
85. C. Fux, S. Moser, S. Schlatter, M. Rimann, J. E. Bailey and M. Fussenegger, *Nucleic Acids Res.*, 2001, **29**(4), E19.
86. M. B. Goldring, *Methods Mol. Med.*, 2004, **100**, 23–36.
87. C. Priesner, F. Hesse, D. Windgassen, R. Klocke, D. Paul and R. Wagner, *In vitro Cell Dev. Biol. Anim.*, 2004, **40**(10), 318–330.
88. M. Fussenegger, R. P. Morris, C. Fux, M. Rimann, B. von Stockar, C. J. Thompson and J. E. Bailey, *Nat. Biotechnol.*, 2000, **18**(11), 1203–1208.
89. M. Gossen and H. Bujard, *Proc. Natl. Acad. Sci. U. S. A.*, 1992, **89**(12), 5547–5551.
90. W. Weber, B. P. Kramer, C. Fux, B. Keller and M. Fussenegger, *J. Gene Med.*, 2002, **4**(6), 676–686.
91. M. Fussenegger, J. E. Bailey, H. Hauser and P. P. Mueller, *Trends Biotechnol.*, 1999, **17**(1), 35–42.
92. Z. Ivics, A. Katzer, E. E. Stüwe, D. Fiedler, S. Knespel and Z. Izsvák, *Mol. Ther.*, 2007, **15**(6), 1137–1144.
93. C. Miskey, Z. Izsvák, R. H. Plasterk and Z. Ivics, *Nucleic Acids Res.*, 2003, **31**(23), 6873–6781.
94. M. Narushima, N. Kobayashi, T. Okitsu, Y. Tanaka, S. A. Li, Y. Chen, A. Miki, K. Tanaka, S. Nakaji, K. Takei, A. S. Gutierrez, J. D. Rivas-Carrillo, N. Navarro-Alvarez, H. S. Jun, K. A. Westerman, H. Noguchi, J. R. Lakey, P. Leboulch, N. Tanaka and J. W. Yoon, *Nat. Biotechnol.*, 2005, **23**(10), 1274–1282.
95. T. Vierbuchen and M. Wernig, *Nat Biotechnol.*, 2011, **29**(10), 892–907.
96. T. Vierbuchen, A. Ostermeier, Z. P. Pang, Y. Kokubu, T. C. Südhof and M. Wernig, *Nature*, 2010, **463**, 1035–1041.

97. M. Caiazzo, M. T. Dell'Anno, E. Dvoretskova, D. Lazarevic, S. Taverna, D. Leo, T. D. Sotnikova, A. Menegon, P. Roncaglia, G. Colciago, G. Russo, P. Carninci, G. Pezzoli, R. R. Gainetdinov, S. Gustincich, A. Dityatev and V. Broccoli, *Nature*, 2011, **476**(7359), 224–227.

98. M. Ieda, J. D. Fu, P. Delgado-Olguin, V. Vedantham, Y. Hayashi, B. G. Bruneau and D. Srivastava, *Cell*, 2010, **142**(3), 375–386.

99. P. Huang, Z. He, S. Ji, H. Sun, D. Xiang, C. Liu, Y. Hu, X. Wang and L. Hui, *Nature*, 2011, **475**(7356), 386–389.

100. S. Sekiya and A. Suzuki, *Nature*, 2011, **475**, 390–393.

101. Z. P. Pang, N. Yang, T. Vierbuchen, A. Ostermeier, D. R. Fuentes, T. Q. Yang, A. Citri, V. Sebastiano, S. Marro, T. C. Südhof and M. Wernig, *Nature*, 2011, **476**(7359), 220–223.

102. U. Pfisterer, A. Kirkeby, O. Torper, J. Wood, J. Nelander, A. Dufour, A. Björklund, O. Lindvall, J. Jakobsson and M. Parmar, *Proc. Natl. Acad. Sci. U. S. A.*, 2011, **108**(25), 10343–10348.

103. A. S. Yoo, A. X. Sun, L. Li, A. Shcheglovitov, T. Portmann, Y. Li, C. Lee-Messer, R. E. Dolmetsch, R. W. Tsien and G. R. Crabtree, *Nature*, 2011, **476**(7359), 228–231.

104. E. Szabo, S. Rampalli, R. M. Risueño, A. Schnerch, R. Mitchell, A. Fiebig-Comyn, M. Levadoux-Martin and M. Bhatia, *Nature*, 2010, **468**(7323), 521–526.

105. B. C. Heng, M. Richards, Z. Ge and Y. Shu, *J. Tissue Eng. Regen. Med.*, 2010, **4**(2), 159–162.

106. T. W. Gilbert, T. L. Sellaro and S. F. Badylak, *Biomaterials*, 2006, **27**(19), 3675–3683.

107. J. C. Myers, P. S. Amenta, A. S. Dion, J. P. Sciancalepore, C. Nagaswami, J. W. Weisel and P. D. Yurchenco, *Biochem. J.*, 2007, **404**(3), 535–544.

108. A. M. Pizzo, K. Kokini, L. C. Vaughn, B. Z. Waisner and S. l. Voytik-Harbin, *J. Appl. Physiol.*, 2005, **98**(5), 1909–1921.

109. P. Schedin, T. Mitrenga, S. McDaniel and M. Kaeck, *Mol. Carcinog.*, 2004, **41**(4), 207–20.

110. M. Maatta, A. Liakka, S. Salo, K. Tasanen, L. Bruckner-Tuderman and H. Autio-Harmainen, *J. Histochem. Cytochem.*, 2004, **52**(8), 1073–1081.

111. A. C. Bellail, S. B. Hunter, D. J. Brat, C. Tan and E. G. Van Meir, *Int. J. Biochem. Cell Biol.*, 2004, **36**(6), 1046–1069.

112. P. Lin, W. C. Chan, S. F. Badylak and S. N. Bhatia, *Tissue Eng.*, 2004, **10**(7–8), 1046–1053.

113. J. H. Miner, *Kidney Int.*, 1999, **56**(6), 2016–2014.

114. K. Gelse, E. Poschl and T. Aigner, *Adv. Drug Deliv. Rev.*, 2003, **55**(12), 1531–1546.

115. J. Huxley-Jones, D. L. Robertson and R. P. Boot-Handford, *Matrix Biol.*, 2007, **26**(1), 2–11.

116. J. Schwarzbauer, *Curr. Biol.*, 1999, **9**(7), R242–R244.

117. R. Timpl, *Curr. Opin. Cell Biol.*, 1996, **8**(5), 618–624.

118. R. Timpl and J. C. Brown, *Bioessays*, 1996, **18**(2), 123–132.

119. R. Folberg and A. J. Maniotis, *Apmis*, 2004, **112**(7–8), 508–525.

120. Z. Werb, T. H. Vu, J. L. Rinkenberger and L. M. Coussens, *Apmis*, 1999, **107**(1), 11–18.

121. T. Neumann, S. D. Hauschka and J. E. Sanders, *Tissue Eng.*, 2003, **9**(5), 995–1003.

122. M. D. Pierschbacher and E. Ruoslahti, *Nature*, 1984, **309**(5963), 30–33.

123. K. M. Yamada and D. W. Kennedy, *J. Cell Biol.*, 1984, **99**(1 Pt 1), 29–36.

124. S. Miyamoto, B. Z. Katz, R. M. Lafrenie and K. M. Yamada., *Ann. N. Y. Acad. Sci.*, 1998, **857**, 119–129.

125. M. Mochizuki, N. Yamagata, D. Philp, K. Hozumi, T. Watanabe, Y. Kikkawa, Y. Kadoya, H. K. Kleinman and M. Nomizu, *Biopolymers*, 2007, **88**(2), 122–130.

126. J. E. Schwarzbauer, *Curr. Opin. Cell Biol.*, 1991, **3**(5), 786–791.

127. J. P. Hodde, S. F. Badylak, A. O. Brightman and S. L. Voytik-Harbin, *Tissue Eng.*, 1996, **2**(3), 209–217.

128. W. M. Elbjeirami, E. O. Yonter, B. C. Starcher and J. L. West, *J. Biomed. Mater. Res. A*, 2003, **66**(3), 513–521.

129. M. A. Cobb, S. F. Badylak, W. Janas and F. A. Boop, *Surg. Neurol.*, 1996, **46**(4), 389–394; discussion 393–394.

130. S. F. Badylak, *Cell Dev. Biol*, 2002, **13**(5), 377–383.

131. S. Badylak, S. Meurling, M. Chen, A. Spievack and A. Simmons-Byrd, *J. Pediatr. Surg.*, 2000, **35**(7), 1097–1103.

132. S. Nemcova, A. A. Noel, C. J. Jost, P. Gloviczki, V. M. Miller and K. G. Brockbank, *J. Invest. Surg.*, 2001, **14**(6), 321–330.

133. K. J. Zehr, M. Yagubyan, H. M. Connolly, S. M. Nelson and H. V. Schaff, *J. Thorac. Cardiovasc. Surg.*, 2005, **130**(4), 1010–1015.

134. C. Danielsson, S. Ruault, A. Basset-Dardare and P. Frey, *Biomaterials*, 2006, **27**(7), 1054–1060.

135. B. P. Kropp, S. Badylak and K. B. Thor, *Adv. Exp. Med. Biol.*, 1995, **385**, 229–335.

136. P. A. Merguerian, P. P. Reddy, D. J. Barrieras, G. J. Wilson, K. Woodhouse, D. J. Bagli, G. A. McLorie and A. E. Khoury, *BJU Int.*, 2000, **85**(7), 894–898.

137. S. Q. Liu, Gastrointestinal Regenerative Engineering, *Bioregenerative Engineering: Principles and Applications*, Wiley, Hoboken, NJ, 2007.

138. J. R. Fuchs, A. Kaviani, J. T. Oh, D. LaVan, T. Udagawa, R. W. Jennings, J. M. Wilson and D. O. Fauza, *J. Pediatr. Surg.*, 2004, **39**(6), 834–838; discussion 834–838.

139. C. G. Zalavras, R. Gardocki, E. Huang, M. Stevanovic and T. Hedman, *J. Shoulder Elbow Surg.*, 2006, **15**(2), 224–231.

140. P. G. De Deyne and S. M. Kladakis, *Clin. Podiatr. Med. Surg.*, 2005, **22**(4), 521–532.

141. W. L. Rust, A. Sadasivam and N. R. Dunn, *Stem Cells Dev.*, 2006, **15**(6), 889–904.

142. V. Ella, M. E. Gomes, R. L. Reis, P. Törmälä and M. Kellomäki, *J. Mater. Sci. Mater. Med.*, 2007, **18**(6), 1253–1261.
143. J. P. Rubin, J. M. Bennett, J. S. Doctor, B. M. Tebbets and K. G. Marra, *Plast. Reconstr. Surg.*, 2007, **120**(2), 414–424.
144. T. Suciati, D. Howard, J. Barry, N. M. Everitt, K. M. Shakesheff and F. R. Rose, *J. Mater. Sci. Mater. Med.*, 2006, **17**(11), 1049–1056.
145. A. S. Rowlands, S. A. Lim, D. Martin and J. J. Cooper-White, *Biomaterials*, 2007, **28**(12), 2109–2121.
146. T. Takezawa, K. Ozaki and C. Takabayashi, *Tissue Eng.*, 2007, **13**(6), 1357–1366.
147. R. Izquierdo, N. Garcia-Giralt, M. T. Rodriguez, E. Cáceres, S. J. García, J. L. Gómez Ribelles, M. Monleón, J. C. Monllau and J. Suay, *J. Biomed. Mater. Res. A*, 2007, **85**, 25.
148. J. G. Dellinger, J. Cesarano III and R. D. Jamison, *J. Biomed. Mater. Res. A*, 2007, **82**(2), 383–394.
149. I. K. Jun, Y. H. Koh, S. H. Lee and H. E. Kim, *J. Mater. Sci. Mater. Med.*, 2007, **18**(6), 1071–1077.
150. J. O. Eniwumide, H. Yuan, S. H. Cartmell, G. J. Meijer and J. D. de Bruijn, *Eur. Cell Mater.*, 2007, **14**, 30–38; discussion 39.
151. Y. Li and W. Weng, *J. Mater. Sci. Mater. Med.*, 2007, **18**, 2303.
152. C. V. Brovarone, E. Verne and P. Appendino, *J. Mater. Sci. Mater. Med.*, 2006, **17**(11), 1069–1078.
153. J. R. Jones, O. Tsigkou, E. E. Coates, M. M. Stevens, J. M. Polak and L. L. Hench, *Biomaterials*, 2007, **28**(9), 1653–1663.
154. A. R. Boccaccini and J. J. Blaker, *Expert Rev. Med. Devices*, 2005, **2**(3), 303–317.
155. G. C. Engelmayr Jr and M. S. Sacks, *J. Biomech. Eng.*, 2006, **128**(4), 610–622.
156. C. W. Yung, O. Tsigkou, E. E. Coates, M. M. Stevens, J. M. Polak and L. L. Hench, *J. Biomed. Mater. Res. A*, 2007, **83**, 1039.
157. J. B. Leach, J. B. Wolinsky, P. J. Stone and J. Y. Wong, *Acta Biomater.*, 2005, **1**(2), 155–164.
158. K. A. Faraj, T. H. Van Kuppevelt and W. F. Daamen, *Tissue Eng.*, 2007, **13**, 2387.
159. F. J. O'Brien, B. A. Harley, M. A. Waller, I. V. Yannas, L. J. Gibson and P. J. Prendergast, *Technol. Health Care*, 2007, **15**(1), 3–17.
160. E. Alsberg, B. A. Harley, M. A. Waller, I. V. Yannas, L. J. Gibson and P. J. Prendergast, *Tissue Eng.*, 2006, **12**(11), 3247–3256.
161. D. Eyrich, H. Wiese, G. Maier, D. Skodacek, B. Appel, H. Sarhan, J. Tessmar, R. Staudenmaier, M. M. Wenzel, A. Goepferich and T. Blunk, *Tissue Eng.*, 2007, **13**, 2207.
162. J. Velema and D. Kaplan, *Adv. Biochem. Eng. Biotechnol.*, 2006, **102**, 187–238.
163. S. Lin-Gibson, J. A. Cooper, F. A. Landis and M. T. Cicerone, *Biomacromolecules*, 2007, **8**(5), 1511–1518.

164. D. A. Wahl, E. Sachlos, C. Liu and J. T. Czernuszka, *J. Mater. Sci. Mater. Med.*, 2007, **18**(2), 201–209.

165. T. Weigel, G. Schinkel and A. Lendlein, *Expert Rev. Med. Devices*, 2006, **3**(6), 835–851.

166. U. Boudriot, R. Dersch, A. Greiner and J. H. Wendorff, *Artif. Organs*, 2006, **30**(10), 785–792.

167. R. Murugan and S. Ramakrishna, *Tissue Eng.*, 2007, **13**, 1845.

168. L. S. Nair, S. Bhattacharyya and C. T. Laurencin, *Expert Opin. Biol. Ther.*, 2004, **4**(5), 659–668.

169. C. M. Vaz, S. van Tuijl, C. V. Bouten and F. P. Baaijens, *Biomaterials*, 2005, **1**(5), 575–852.

170. Z. Ma, M. Kotaki, R. Inai and S. Ramakrishna, *Tissue Eng.*, 2005, **11**(1–2), 101–109.

171. R. Murugan and S. Ramakrishna, *Tissue Eng.*, 2006, **12**(3), 435–447.

172. F. Yang, R. Murugan, S. Wang and S. Ramakrishna, *Biomaterials*, 2005, **26**(15), 2603–2610.

173. J. Guan, K. L. Fujimoto, M. S. Sacks and W. R. Wagner, *Biomaterials*, 2005, **26**(18), 3961–3971.

174. P. Roychowdhury and V. Kumar, *J. Biomed. Mater. Res. A*, 2006, **76**(2), 300–309.

175. J. Nakamatsu, F. G. Torres, O. P. Troncoso, Y. Min-Lin and A. R. Boccaccini., *Biomacromolecules*, 2006, **7**(12), 3345–3355.

176. P. Sarazin, X. Roy and B. D. Favis, *Biomaterials*, 2004, **25**(28), 5965–5978.

177. S. Zhang, *Biotechnol. Adv.*, 2002, **20**(5–6), 321–339.

178. F. Gelain, A. Lomander, A. L. Vescovi and S. Zhang, *J. Nanosci. Nanotechnol.*, 2007, **7**(2), 424–434.

179. E. Garreta, D. Gasset, C. Semino and S. Borrós, *Biomol. Eng.*, 2007, **24**(1), 75–80.

180. T. C. Holmes, S. de Lacalle, X. Su, G. Liu, A. Rich and S. Zhang, *Proc. Natl. Acad. Sci. U.S.A.*, 2000, **97**(12), 6728–6733.

181. M. Altman, P. Lee, A. Rich and S. Zhang, *Protein Sci.*, 2000, **9**(6), 1095–1105.

182. S. J. Hollister, *Nat. Mater.*, 2005, **4**(7), 518–524.

183. H. H. Lu and J. Jiang, *Adv. Biochem. Eng. Biotechnol.*, 2006, **102**, 91–111.

184. A. G. Mikos, S. W. Herring, P. Ochareon, J. Elisseeff, H. H. Lu, R. Kandel, F. J. Schoen, M. Toner, D. Mooney, A. Atala, M. E. Van Dyke, D. Kaplan and G. Vunjak-Novakovic, *Tissue Eng.*, 2006, **12**(12), 3307–3339.

185. A. S. Mistry and A. G. Mikos, *Adv. Biochem. Eng. Biotechnol.*, 2005, **94**, 1–22.

186. K. Rezwan, Q. Z. Chen, J. J. Blaker and A. R. Boccaccini, *Biomaterials*, 2006, **27**(18), 3413–3431.

187. N. Karim, K. Golz and A. Bader, *Artif. Organs*, 2006, **30**(10), 809–814.

188. Y. Martin and P. Vermette, *Biomaterials*, 2005, **26**(35), 7481–7503.

189. R. M. Sutherland, B. Sordat, J. Bamat, H. Gabbert, B. Bourrat and W. Mueller-Klieser, *Cancer Res.*, 1986, **46**(10), 5320–9.

190. J. M. Piret and C. L. Cooney, *Biotechnol. Adv.*, 1990, **8**(4), 763–783.

191. F. Zhao, P. Pathi, W. Grayson, Q. Xing, B. R. Locke and T. Ma, *Biotechnol. Prog.*, 2005, **21**(4), 1269–80.
192. N. L. Parenteau and J. Hardin-Young, *Ann. N. Y. Acad. Sci.*, 2002, **961**, 27–39.
193. R. P. Lanza, R. Langer and J. P. Vacanti, *Principles of Tissue Engineering*, Academic Press, San Diego, CA, 1997.
194. U. Meyer, U. Joos and H. P. Wiesmann, *J. Oral Maxillofac. Surg.*, 2004, **33**(4), 325–332.
195. I. Martin, R. F. Padera, G. Vunjak-Novakovic and L. E. Freed, *J. Orthop. Res.*, 1998, **16**(2), 181–189.
196. S. L. Ishaug, G. M. Crane, M. J. Miller, A. W. Yasko, M. J. Yaszemski and A. G. Mikos, *J. Biomed. Mater. Res.*, 1997, **36**(1), 17–28.
197. I. Martin, B. Obradovic, L. E. Freed and G. Vunjak-Novakovic., *Ann. Biomed. Eng.*, 1999, **27**(5), 656–662.
198. E. M. Bueno, B. Bilgen and G. A. Barabino, *Tissue Eng.*, 2005, **11**(11–12), 1699–1709.
199. H. D. Humes, S. M. MacKay, A. J. Funke and D. A. Buffington, *Kidney Int.*, 1999, **55**, 2502–2514.
200. H. F. Lu, W. S. Lim, P. C. Zhang, S. M. Chia, H. Yu, H. Q. Mao and K. W. Leong, *Tissue Eng.*, 2005, **11**(11–12), 1667–1677.
201. J. M. Macdonald, M. Grillo, O. Schmidlin, D. T. Tajiri and T. L. James, *NMR Biomed.*, 1998, **11**(2), 55–66.
202. X. Wen and P. A. Tresco, *Biomaterials*, 2006, **27**(20), 3800–3809.
203. L. E. Freed, R. Langer, I. Martin, N. R. Pellis and G. Vunjak-Novakovic, *Proc. Natl. Acad. Sci. U.S.A.*, 1997, **94**(25), 13885–13890.
204. V. I. Khaoustov, G. J. Darlington, H. E. Soriano, B. Krishnan, D. Risin, N. R. Pellis and B. Yoffe, *In vitro Cell Dev. Biol. Anim.*, 1999, **35**(9), 501–509.
205. G. Vunjak-Novakovic and L. E. Freed, *Adv. Drug Deliv. Rev.*, 1998, **33**(1–2), 15–30.
206. L. E. Freed and G. Vunjak-Novakovic, *Cell Dev. Biol. Anim.*, 1997, **33**(5), 381–385.
207. H. L. Holtorf, J. A. Jansen and A. G. Mikos, *J. Biomed. Mater. Res. A*, 2005, **72**(3), 326–334.
208. E. M. Darling and K. A. Athanasiou, *Tissue Eng.*, 2003, **9**(1), 9–26.
209. A. Ratcliffe and L. E. Niklason, *Ann. N. Y. Acad. Sci.*, 2002, **961**, 210–215.
210. S. Watanabe, S. Inagaki, I. Kinouchi, H. Takai, Y. Masuda and S. Mizuno, *J. Biosci. Bioeng.*, 2005, **100**(1), 105–111.
211. J. O. Seidel, M. Pei, M. L. Gray, R. Langer, L. E. Freed and G. Vunjak-Novakovic, *Biorheology*, 2004, **41**(3–4), 445–458.
212. E. R. Ochoa and J. P. Vacanti, *Ann. N. Y. Acad. Sci.*, 2002, **979**, 10–26; discussion 35–38.
213. J. P. Burgues, L. Gómez, J. L. Pontones, C. D. Vera, J. F. Jiménez-Cruz and M. Ozonas, *Eur. Urol.*, 2007, **51**(4), 962–969; discussion 969–970.
214. L. A. Kunz-Schughart, J. P. Freyer, F. Hofstaedter and R. Ebner, *J. Biomol. Screen.*, 2004, **9**(4), 273–285.

215. P. L. Olive and R. E. Durand, *Cancer Metastasis Rev.*, 1994, **13**(2), 121–138.
216. B. J. Conley, J. C. Young, A. O. Trounson and R. Mollard, *Int. J. Biochem. Cell Biol.*, 2004, **36**(4), 555–567.
217. T. Korff, T. Krauss and H. G. Augustin, *Exp. Cell Res.*, 2004, **297**(2), 415–423.
218. R. C. Bates, N. S. Edwards and J. D. Yates, *Crit. Rev. Oncol. Hematol.*, 2000, **36**(2–3), 61–74.
219. H. Okubo, M. Matsushita, H. Kamachi, T. Kawai, M. Takahashi, T. Fujimoto, K. Nishikawa and S. Todo, *Artif. Organs*, 2002, **26**(6), 497–505.
220. A. Rothermel, T. Biedermann, W. Weigel, R. Kurz, M. Rüffer, P. G. Layer and A. A. Robitzki, *Tissue Eng.*, 2005, **11**(11–12), 1749–1756.
221. X. Wang, G. Wei, W. Yu, Y. Zhao, X. Yu and X. Ma, *Biotechnol. Prog.*, 2006, **22**(3), 811.
222. A. Ivascu and M. Kubbies, *J. Biomol. Screen.*, 2006, **11**(8), 922–932.
223. Y. Hasebe, N. Okumura, T. Koh, H. Kazama, G. Watanabe, T. Seki and T. Ariga, *Hepatol. Res.*, 2005, **32**(2), 89–95.
224. J. M. Kelm, L. M. Ittner, W. Born, V. Djonov and M. Fussenegger, *Tissue Eng.*, 2006, **12**(9), 2541–2553.
225. J. M. Kelm, L. M. Ittner, W. Born, V. Djonov and M. Fussenegger, *J. Biotechnol.*, 2006, **121**(1), 86–101.
226. J. M. Kelm, N. E. Timmins, C. J. Brown, M. Fussenegger and L. K. Nielsen, *Biotechnol. Bioeng.*, 2003, **83**(2), 173–180.
227. J. M. Kelm, C. Diaz Sanchez-Bustamante, E. Ehler, S. P. Hoerstrup, V. Djonov, L. Ittner and M. Fussenegger, *J. Biotechnol.*, 2005, **118**(2), 213–229.

CHAPTER 16

Agricultural Biotechnology

KATHLEEN HEFFERON

University of Toronto, Toronto, Ontario, Canada and Cornell University, Ithaca, NY, USA
Email: kathleen.hefferon@utoronto.ca; klh22@cornell.edu

16.1 INTRODUCTION

Agriculture is at the fulcrum of modern society, and our success as a species can be measured relative to our advancements in terms of crop improvement, discovery of bioactive plant compounds, and other applications of plants in medicine. This chapter investigates the use of agricultural biotechnology to generate plants with novel traits, such as resistance to biotic and abiotic stresses, the capacity to remediate polluted soils and watersheds, and to act as production platforms for therapeutic proteins and biodegradable plastics. While plant biotechnology as a discipline covers both molecular plant breeding and transformation technologies, this chapter will focus solely on the latter. Today, transgenic plants are often generated either *via Agrobacterium*-mediated transformation or by particle bombardment; more recently, transplastomic (chloroplast) transgenic plants have been generated as well.[1–3] This chapter describes some of the applications for plant biotechnology with respect to both agriculture and medicine.

16.2 APPLICATIONS IN AGRICULTURE

There are numerous applications of biotechnology in the field of agriculture. Not only have crops been produced that are resistant to plant pathogens and

Molecular Biology and Biotechnology, 6th Edition
Edited by Ralph Rapley and David Whitehouse
© The Royal Society of Chemistry 2015
Published by the Royal Society of Chemistry, www.rsc.org

abiotic stresses, but plants are now utilized for phytoremediation purposes and biofuel production. This next section outlines some of the technologies, which are utilized today, as well as several new technologies that are up and coming.

16.2.1 Herbicide Resistance

One of the most important products generated through agricultural bio-technology has been the introduction of transgenic crop plants that confer resistance to herbicides. Herbicides can be sprayed on transgenic crops without causing damage while simultaneously blocking the growth of neighbouring weeds. The utilization of herbicide-tolerant transgenic crops has therefore greatly reduced the levels of herbicide used on farms for weed management.

Glyphosate [N-(phosphonomethyl)glycine] is the most widely utilized herbicide in the world.[4] Transgenic, glyphosate-resistant crops were first commercialized in 1996, and today almost 90% of all transgenic crops grown across the world possess the glyphosate resistant trait. The mode of action of glyphosate is to inhibit the activity of the plant enzyme 5-enolpyruvyl-shikimate-3-phosphate synthase (Class I EPSPS); this enzyme is involved in the synthesis of amino acids tyrosine, tryptophan and phenylalanine. These amino acids are required for the production of a number of secondary metabolites, including folates, ubiquinones and naphthoquinone[5] As a result, not only is glyphosate highly effective as a broad-spectrum herbicide, it is also considered to be environmentally safe, as its effects on soil and water are very minor. Of note is the feature that glyphosate-resistant crops remove the requirement for tillage, thus reducing the need for labour and preventing soil erosion.[6]

Other strategies have been used to generate herbicide tolerant plants without the use of genetic modification. For example, a nontransgenic plant approach has been used to identify naturally occurring glyphosate resistance genes in wheat germplasm. These newly identified genes can then be used to generate novel herbicide-resistant crops.[7] As another example, a recombinagenic oligonucleobase has been utilized to make a specific mutation in the plant gene encoding glyphosate resistance, EPSPS. The mutated protein enables the plant to exhibit increased resistance or tolerance to this herbicide.[8]

Unfortunately, the evolution of glyphosate-resistant weeds has induced plant scientists to find improved herbicide resistant strategies. For example, herbicide resistance based on a novel set of genes encoding the Class II EPSPS enzymes, which share little homology with Class I EPSPS have been developed.[4]

The herbicide bialaphos, expressed from the *bar* gene, is also used as a herbicide as well as a selectable marker for generating transgenic plants. The *bar* gene was originally identified from *Streptomyces hygroscopicus*, an organism which expresses the secondary metabolite tripeptide bialaphos, which contains phosphinothricin, a glutamate analogue which acts as an inhibitor of glutamine synthetase.[9]

Another type of herbicide tolerance has been developed using aceto-hydroxyacid synthase (AHAS), an enzyme which is utilized by both plants and microorganisms for valine, leucine and isoleucine biosynthesis. Like glyphosate, weeds have emerged which are resistant to AHAS inhibition. As a result, scientists have identified new AHAS variants which exhibit selectively increased resistance to herbicides such as imidazoline herbicides and AHAS inhibiting herbicides.[9]

16.2.2   Crops Resistant to Abiotic Stresses

In an attempt to address issues surrounding climate change and the threat of increased excessive temperature, drought and flooding, agricultural research has been steadily moving toward the development of crop plants which are able to provide high yields yet are more tolerant to adverse environmental conditions (Figure 16.1). The next section describes a few of the technologies under development to develop novel plant varieties that are less susceptible to damage or loss by such stresses.

16.2.3   Drought Tolerance

Since a decrease of water availability will be an expected result of climate change, increasing drought tolerance of crops is paramount. Fortunately, there are an array of genes that have been identified which can be

Figure 16.1   Drought resistant rice.
Source: the International Rice Research Institute.

manipulated through biotechnology to improve drought tolerance.[10] One method of generating drought tolerant plants involves genetically modifying their rates of transpiration. By regulating the stomatal closure responses, more efficient water conservation would be permitted and plants could prosper in areas which were previously unsuitable for growth due to drought conditions.[11]

Many crop plants do not possess the capacity to osmoregulate their internal environment to the extent that stress-tolerant plants are capable of. A number of osmoprotectants are available both from plants and micro-organisms that can help plants exert more drought tolerance, these include compounds such as polyamines and some sugar alcohols.[10–13] Glycine-betaine, for example, which is derived from the *bet A* gene product of *E. coli*, encodes a chlorine dehydrogenase. When transformed into maize, the *bet A* gene confers greater tolerance to drought stress than wild type varieties and can provide higher grain yield than its nontransformed counterpart after three weeks of drought stress. Transgenic plants expressing genes that encode heat shock proteins and molecular chaperones can also increase drought resistance. For example, the cold shock proteins CspA and Csp B, from *E. coli* and *B. subtilis*, respectively, function as RNA chaperones and have been demonstrated to improve drought resistance for several crop species.[14]

16.2.4 High Salinity Stress

High salinity stress can impair crop production from at least 20% of irrigated land worldwide. A highly saline environment can disrupt the ionic and osmotic equilibrium of the cell, and can lead to plant growth inhibition and even death. In certain salt-tolerant plants, a number of gene products are upregulated in response to high salinity stress. Some of these stress responsive genes encode osmolytes, ion channels, receptors, components of calcium signalling, and other regulatory signalling factors or enzymes which have the capacity to confer salinity-tolerant phenotypes when transferred to sensitive plants.[12] Some of the mechanisms by which a plant can confer salinity tolerance involve sequestration of Na^+ and Cl^- in vacuoles of the cells, blocking of Na^+ entry into the cell, and Na^+ exclusion from the transpiration stream.[12,13,15]

The engineering of the regulatory machinery involving transcription factors, such as the overexpression of stress inducible transcription factors, has emerged as a novel strategy for controlling the expression of many stress-responsive genes.[14] Transcription factors in general regulate several different genes at once that may collectively contribute to the stress response. For example, the AREB/ABF transcription factors regulate the abscisic acid-responsive genes, whose products play a role in stress resistance. Overexpression of these genes *via* their transcription factors can therefore lead to increased drought and salt tolerance. As mentioned above, increased stoma closure can improve stress tolerance as well. Control of transcription factors involved in stomatal movement offers a means to increase resistance to stress in general.[11,14]

Antiporter genes have also been identified as sources for novel stress resistant varieties of plants. As an example, the rice antiporter-regulating protein OsARP was overexpressed in transgenic tobacco plants and demonstrated to have superior tolerance to saline conditions. These transgenic plants accumulated more sodium ions in their leaf tissue than their wild-type counterparts.[12] Other transgenic plants which express an ion transporter stress-related polypeptide (ITSRP), as well as a potassium channel stress-related protein (PCSRP) and a pyrophosphatase stress-related protein (PPSRP) have all exhibited increased tolerance to environmental stresses.[13,14] Other methods for inducing stress tolerance in plants exist. For example, transgenic plants that encode enzymes that catalyse proline production have exhibited increased tolerance to environmental stresses.[15] Stress tolerance can also involve reducing a plant's need for nitrogen. In one instance, rice plants were modified to express a nitrogen utilization protein in their roots.[16]

16.2.5 Disease-Resistant Crops

Although many countries throughout the world grow and commercialise transgenic crops, the vast majority of them exhibit herbicide tolerance and insect resistant traits. Today, transgenic disease resistant crops represent roughly 10% of the total number of approved field trials in North America. Of these, resistance to bacteria and fungi *via* the expression of antimicrobial proteins and resistance to viruses *via* coat protein mediated resistance represent most of the disease resistant varieties that have been under construction. Other strategies, such as the regulation of plant defence pathways or the use of RNA interference, are also under development to provide biotechnological solutions to infection by crop pathogens.[17]

Although pathogen resistance can be improved to some extent by traditional plant breeding, many shortfalls exist that can be more readily overcome by the use of plant biotechnology. Pathogen resistance can be increased, the range of resistance broadened, and environmental impacts, such as deleterious effects on soil microbes lessened, all through the use of transgenic plant technologies. The following section lists some of the technologies that have been under development to improve disease resistance in plants.

16.2.6 Insect Resistance

Transgenic crops expressing the Bt (*Bacillus thuringiensis*) toxin gene were the first insect resistant plants commercialized. Insects that are susceptible for this toxin ingest the transgenic crop cultivar expressing the precursor Bt protein, which binds to the insect gut wall and becomes activated in that alkaline environment. The insects soon die as a result of pores that are

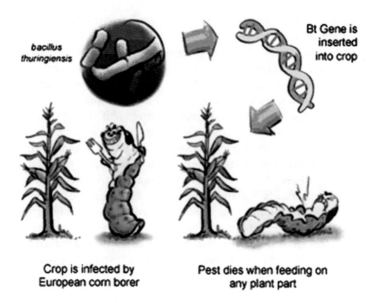

bacillus
thuringiensis

Bt Gene is
inserted
into crop

Crop is infected by
European corn borer

Pest dies when feeding on
any plant part

Figure 16.2   Mechanism of action of Bt. Art by Jiang Long and Jed Philpot.
Source: scq.ubc.ca.

created in the gut cell membrane (Figure 16.2). Although first applied as a
spray in organic farming, crops have been engineered to express this biological pesticide, thus providing a more environmentally friendly method to
protect crops. One of the primary examples of use of this technology is the
generation of Bt corn to combat the corn borer, a pest that resides within the
plant and thus remains undeterred by externally sprayed pesticides. In order
to resist the possibility of Bt-resistant insects, multiple Bt genes are now
included as stacked transgenes and expressed in a single plant. Today, over
20 different genes conferring insect resistance have been genetically engineered into the major crops. There is no question that Bt crops have offered
substantial environmental benefits for global agriculture in general, by
reducing pesticide applications while at the same time improving crop
quality and yield.[18–20]

Another development in insect resistance can be found in the use of the
alarm pheromone enzyme (*E*)-*β*-Farnesene (E*β*F) synthase, which specifically
targets aphids, major agricultural pests that cause significant yield losses of
crop plants each year, both by feeding on crops and by acting as vectors for
harmful plant viruses.[21] (E)-*β*-Farnesene (E*β*F) is a volatile sesquiterpene that
is released from cornicles on the aphid's abdomen when their natural enemies attack them. E*β*F can interrupt aphid feeding and cause other aphids
in the vicinity to become agitated or disperse from the plant. Crop plants
that can synthesize and release the E*β*F pheromone could both repulse
aphids and attract natural enemies that use E*β*F as a foraging cue. Transgenic *Arabidopsis* and tobacco plants which express E*β*F synthase genes from
peppermint and sweetworm wood have been demonstrated to repel aphids

and attract their natural enemies, including ladybugs and parasitoid wasps. To date, few plant resistance (R) genes and their homologues associated with aphid resistance have been identified. The use of this technology would remove the practice of applying insecticides, which are undesirable both for causing environmental harm as well as inviting the possibility of insecticide resistance.

Other insect resistant genes that have been expressed in transgenic plants include protease inhibitors that specifically act against digestive proteinases of insects, such as alpha amylase inhibitors and ribosome-inactivating proteins (RIPs). In addition to this, lectins that bind to specific carbohydrates in the insect gut can be generated in transgenic plants and act as feeding deterrents against insects. Finally, siRNAs have been designed to impair insect feeding and have been expressed in transgenic plants. For example, miRNAs corresponding to the coding sequence of the salivary transcript Coo2 downregulated gene expression and impaired aphid feeding on faba beans. Furthermore, crops that express a transgene that encodes a hairpin RNA targeting an essential gene such as cytochrome P450 of the insect pest have resulted in retarding the growth of feeding larvae.[22]

16.2.7 Virus-Resistant Plants

Unlike the development of virus-resistant plants by conventional molecular breeding techniques, virus-resistance in plants is relatively easy to achieve using a transgenic approach. It has been well established that virus resistance can be conferred by expressing virus-derived genes, such as those encoding the coat protein, movement protein, protease and replicase. Expression of the viral coat protein, for example, is believed to confer resistance by altering the equilibrium of the virus gene products within the virus-infected cell, thereby preventing any incoming related viruses from uncoating and undergoing their replication cycles.[23] In addition to this, some noncoding plant RNA virus sequences can also be incorporated into susceptible plants. RNA interference, using short sequences, which correspond to the plant virus, has also been routinely utilized to produce virus resistant plants.[24] Other non-viral genes, including plant resistant genes and plant-derived microRNAs, ribosome-inactivating proteins, cellular protease inhibitors, and even antibody fragments have been demonstrated to generate virus resistant plants.[23]

A well-established success story of a crop plant which has been engineered for virus resistance is the example of transgenic papayas which are resistant to papaya ringspot virus (PRSV), now considered to be responsible for saving Hawaii's papaya industry which at the time was under extreme threat due to virus-contaminated farmers' fields. Gonsalves *et al.* expressed PRSV coat protein in papaya plants which conferred resistance to the virus.[25] Many other examples exist, using other virus components. For instance, genetically-engineered resistance in plants to tomato yellow leaf curl geminivirus (TYLCV) was demonstrated by expressing a truncated version of the replication associated protein, or Rep, a viral gene product which is essential for virus replication. It is likely that in this case, resistance is due to the overexpression of this truncated version of Rep, which then disrupts the balance required for virus replication.[25]

16.2.8 Bacterial Resistance in Plants

Plants that are resistant to bacteria have also been developed. A well-known example is the use of the *Pseudomonas syringae* gene product harpin to create disease resistance by inducing the hypersensitive response (HR) of higher plants, characterized by the rapid, localized death of plant cells at the site of invasion by bacteria and other pathogens.[26] Similarly, a novel antimicrobial protein obtained from a fraction of an aqueous extract of the fungus *Lyophyllum shimeji* has been demonstrated to inhibit the growth of plant pathogens including *Pyricularia oryzae* and *Rhizoctonia solani*, which can cause damage to rice crops.[27]

16.2.9 Genetic Engineering for Fungal Resistance in Plants

Plants have been engineered for fungal resistance. Plants expressing chitinase and gluconase polypeptides, proteins that are involved in the defence response, have been shown to exert enhanced resistance to fungal infection. These antifungal genes may act by degrading the fungal cell wall, composed primarily of chitin, and in other self-defence pathways. It is the combined action of chitinase and glucanase co-expressed in transgenic plants that provide maximum resistance to fungal infection by acting in a synergistic fashion.[28] Other genes involved in the plant immune response to fungal infection have also been identified. For example, genes conferring late blight disease resistance have been characterized in some potato varieties and can be transferred to susceptible varieties.[29]

Another means of conferring resistance to fungal infection involves the linking of a single chain antibody gene against a fungal (*Fusarium graminearum*) cell wall protein to an antimicrobial gene such as a defensin or a chitinase. These antibody-linked antimicrobial proteins could attach to the cell wall of a fungal pathogen and block infection. The result of producing this synthetic construct was a reduction in disease symptoms to a number of fungal pathogens of the *Fusarium* species, unfortunately a more broad spectrum resistance to a wide variety of fungal pathogens was not possible. In addition to this, overexpression of antifungal lipid transfer proteins (LTPs) has also been demonstrated to confer fungal disease resistance in transgenic plants. LTPs are known to act as a signal for systemic acquired resistance (SAR) in plants.[30]

16.2.10 Nematode Resistance in Plants

Nematodes cause approximately $118 billion in losses annually to world crops and are difficult to manage. Nemasticides are environmentally harsh and do not offer adequate control for nematode infestations, since these pathogens can be difficult to reach deep within the soil. As only a few crops exist which are naturally resistant to nematodes, transgenic crops have been designed which confer nematode resistance.[31] Approaches that have been used to control nematodes include limiting their ability to uptake dietary protein from the crop or by preventing root invasion, as well as the use of RNA interference strategies. Tissue specific promoters can be used to confer transgenic resistance to root tissues alone, where the vast majority of nematodes invade and feed. Several nematode resistance traits can be added in a single plant by gene stacking, including any natural resistance genes that are available.[32]

16.2.11 Disease Resistance Against a Broad Spectrum of Pathogens

Transfer of resistance genes from one crop species to another has proven to be successful and examples are abundant in plant biotechnology. These include the introduction of the R-gene *Rxo1* from maize into rice, which conferred resistance against bacterial streak disease caused by *Xanthomonas oryzae* pv. Oryzicola, and the introduction of *RPI-BLB2* from wild potato *Solanum bulbocastanum,* which conferred resistance to *Phytophtohora infestans* in potato.[33] While it is indeed possible to confer resistance to transgenic plants by introducing functional R genes from unrelated plant species, this technology is limited by the fact that resistance is often conferred to only a single pathogen. Therefore, the employment of stacking or pyramiding multiple resistance genes through a single plant *via* a combination of genetic engineering approaches has proven to be successful for providing resistance to a broad spectrum of pathogens in several crop species. Problems do exist using this technology, including reduced overall fitness of the plant and the development of pathogens which are resistant to the particular resistance gene that has been introduced.

Other approaches routinely used for engineering resistance against a number of pathogens involve the expression of antimicrobial peptides or proteins involved in the production of antimicrobial metabolites. As mentioned above in the fungal resistance section, chitinases and β1–3 glucanases from a wide variety of organisms have been investigated extensively in a variety of crops, as they can break down the main structural components of fungal cell walls. In addition to this, the use of harpins to evoke a generalized defence reaction for the plant has been utilized as a strategy for broad based pathogen resistance. Similarly, genes that encode enzymes derived from the cytochrome P450 family have been identified and used to generate plants that exhibit resistance to a broad spectrum of pathogens.[34] Defensins, which belong to a family of proteins that retard the growth of insects, fungi, several plant viruses and gram-positive bacteria, have been transformed into plants.[35] Many plant defensins are believed to inhibit growth by interacting with the cell membrane of the pathogen, and possess other activities useful in defence, such as α-amylase activity, proteinase activity and translation inhibition activity. Finally, transgenic plants developed which express plant fatty acid amide hydrolase (FAAH) coding sequences have been demonstrated to exhibit an enhanced response to a variety of pathogens.[36]

16.3 OTHER APPLICATIONS FOR PLANTS

16.3.1 Phytoremediation

The pollution of soils and water by heavy metals and other toxic chemicals are of great concern for both the environment and for human health. Heavy metals cannot readily be degraded; rather, they become transformed from one oxidation state or organic complex to another. Phytoremediation refers to plant-based decontamination of polluted environments. Plants can be utilized to clean polluted soils and waters, by collecting and thereby inactivating metals in the rhizosphere or translocating them in the aerial parts. At least 400 plant species have now been identified and tested for the potential to uptake and accumulate heavy metals. Other pollutants such as PCBs, TNT, pharmaceuticals, textile dyes, phenolics, heavy metals, and radionuclides can be taken up by plants.[37] The ability of plants to accumulate heavy metals varies significantly between species and even among different cultivars within species. Phytoremediation can be subdivided into different categories; these are phytoextraction, phytofiltration, phytostabilization, phytovolatization and phytodegradation. Phytoextraction, for example, involves the use of plants to remove contaminants such as heavy metals from the soil. Heavy metal ions can be taken up by the plant to accumulate in the aerial parts, where they can be removed by defoliation to dispose of or burnt to recover the metal (Figure 16.3). Phytofiltration, on the other hand, involves using plant roots or seedlings for removal of metal contaminants from water basins. For phytostabilization, the plant roots absorb the pollutants from the soil and render them harmless by keeping them in the rhizosphere, thus preventing them from leaching back into the soil. Phytovolatization uses plants to volatilize pollutants such as Se and Hg from their foliage. Phytodegradation involves using plants and their associated microorganisms to degrade organic pollutants. Advances in agricultural biotechnology now make it possible for the genes required for phytoremediation to be transferred to another, more desirable plant.[38]

Recently, hairy roots have demonstrated potential for applications in phytoremediation. Plant roots have been shown in certain instances to stimulate the degradation of contaminants by the release of different enzymes associated with the removal of some organic pollutants. The application of genetic engineering techniques and the further elucidation of microbe-assisted phytoremediation may in the future enhance plants' ability to clean up the environment.[39,40]

There are many examples of phytoremediation development in a broad spectrum of plants today. For example, a novel set of *Pcr* genes have been identified that confer tolerance to cadmium in plants. Heavy metals can be removed from contaminated soils and water using plants transformed with

Figure 16.3 Phytoremediation system.
  Source: Federal Remediation Technologies Roundtable (frtr.gov).

the *Pcr* genes. This same research group were able to generate plants with enhanced resistance and more pronounced growing abilities than wild type plants in environments contaminated with heavy metals. Since these transgenic plants have the ability to pump out heavy metals *via* ZntA-like heavy metal pumping ATPases, they may constitute a safer food crop with a lower heavy metal content than their conventional counterparts.[41]

Transgenic plants have been generated which express a fungal ATP-binding cassette (ABC) transporter protein which confers resistance to heavy metal and herbicide accumulation.[42] Plants that express this fungal protein could be utilized to remove contaminants from polluted soil or water. Arbaoui *et al.* demonstrated that kenaf (*Hibiscus cannabinus* L.) and corn (*Zea mays* L.) could decontaminate sludge of cadmium and zinc by tolerating the metals and bioaccumulating them, indicating that both could be used for phytoremediation purposes.[43] Alternatively, plants can be transformed with proteins that have biocidal properties, so that they can be used to degrade organically polluted areas.

The uses for plants in phytoremediation could be substantial. For example, a recent study identified the extent and type of contamination various heavy metals such as Pb, As, Sb, Zn and Cu of a former lead smelting site in the area of Marseille, France, dating from the Industrial Revolution.[44] Two perennial native plants, *Globularia alypum* L. and *Rosmarinus officinalis* L., were selected for their potential to remove pollutants from the soil. The plants were shown to be metal-tolerant and to accumulate low amounts of the heavy metals, indicating that both species may not be used for phytoextraction (plants that take up contaminants from the soil and store them in harvestable tissue), but may be optimal for phytostabilization (a process in which plants are used to immobilize metals and radionuclides in the soil and minimize their mobility in water or dust).

An experimental remediation program in Kuwait serves as a second example. The Gulf War created some of the worst environmental pollution on the planet in the form of oil spills. Phytoremediation programs have been established and implemented since 1995 in an effort to restore the environment.[45] Over time, ornamental shrubs and trees grown in oil-contaminated soil were shown to have a positive impact and reduce pollution.

Plants have been induced to collect higher levels of metal chelates. Phytochelatin produced in plants, for example, sequesters lead ions and can detoxify plants while at the same time remove lead from the environment.[46] Plants can also be used to remove herbicides from the environment. For example, clofibric acid (CA) was demonstrated to be taken up and translocated into the shoots of *Scirpus validus* (soft stemmed bulrush) growing hydroponically. The authors found that *S. validus* could account for the removal of 28–62% of the total mass loss of CA from the test system, further implying that phytoremediation technology could have great potential for the removal of herbicides and other pharmaceutical proteins from inflowing waters.[47]

16.3.2   Biofuel Production

Biofuels, or fuels based on conversion of plant biomass to ethanol, have become an important component of the energy revolution that is taking place around the globe. Cellulose is a complex carbohydrate that is an important constituent of all plant cell walls. Cellulosic ethanol can be produced from wood, grasses, maize and other non-food or feedstock crop plants. Cellulose can be broken down by hydrolysis using enzymes from microorganisms such as bacteria and fungi into simple sugars such as glucose, which can then be fermented into ethanol.[50] Ethanol generated from biofuel is generally used as a gasoline additive to improve vehicle emissions.

The issue of overcoming biomass recalcitrance to enzyme hydrolysis is a critical hurdle for making biofuels into a reality. Much focus has been on the generation of recombinant cellulose enzymes by genetically engineering the bacteria and fungi that they are derived from. More recently, attempts have been made to modify plants to be more conducive for biofuel production. For example, transgenic plants with altered cellulose biosynthetic properties have been developed.[48] These traits will be carried to several plant species that have potential as biofuels, including rice, wheat, barley, maize, a number of Brassica species, cotton and Eucalyptus. Other researchers have focused on expressing all or some of the cellulolytic enzymes required for hydrolysis in the plants themselves, thus reducing both enzyme and thermochemical pretreatment costs.[49]

Wood that is highly lignified represents a good raw material for biofuel production, since lignin yields more energy when burned than cellulose. Some researchers have learned how to control lignin and cellulose content in plants, by simultaneously transforming plants with multiple genes from the phenylpropanoid pathway. For example, transgenic trees can be generated which express these gene combinations and thus produce wood with greater lignin content, making them more attractive for biofuel production.[50]

Cellulosic ethanol can also be produced using algae as the plant biomass. Algal cellulose can be hydrolyzed using fungi from the phylum Neocallimastigomycota. The sugars produced can be fermented and the ethanol collected.[51] Synthetic polynucleotides have also been generated in transgenic plants, these encode processing enzymes that will generate substrates for fermentation and the production of ethanol.[52]

Another plant system that has been genetically engineered for biofuel production is Alamo switchgrass (*Panicum virgatum* L.), with reduced expression of 4-coumarate-CoA ligase (4CL). Transgenic plants that were generated were determined to have reduced lignin content reductions of up to 5.8% and ratios of acid soluble lignin (ASL) to acid insoluble lignin (AIL) and syringyl/guaiacyl (S/G) of 21.4–64.3% and 11.8–164.5%, respectively, a ratio higher than those of conventional plant biomass. These plants were demonstrated to promote enzymatic hydrolysis of cell walls by pretreatment with a mild alkali than conventional, nontransgenic plants.[53]

16.4 APPLICATIONS OF PLANT BIOTECHNOLOGY IN MEDICINE

Agricultural biotechnology plays a significant role in the field of medicine as well. Crop plants with improved nutritional qualities, including biofortified foods, are soon to become available, as are vaccines, monoclonal antibodies and other therapeutic proteins that are generated using plant production platforms. The use of plant compounds in drug discovery is also discussed in this section.

16.4.1 Improved Nutritional Qualities of Plants

A large variety of biologically active compounds have been identified in vegetables, fruits and grains that act in a specific fashion to improve human health. Some of these, including vitamins and minerals, can be introduced to or increased in crop plants which are consumed by the world's malnourished, who have poor access to the essential vitamins and minerals which are critical to human health.

16.4.1.1 Malnutrition and Biofortified Food. Today, approximately 1 billion, or 1 out of 7 people on this planet, are undernourished. Moreover, at least 40% of the world's population are suffering from malnutrition or "hidden hunger" as a result of the lack of essential micronutrients in their daily diets. The majority of these people consume meals that centre around a staple crop such as rice, maize or cassava, and lack access to the variety of fruits and vegetables required in a healthy diet. Nutrient deficiencies such as iron, zinc and vitamin A are responsible for almost two-thirds of childhood deaths worldwide. These statistics are augmented by the fact that the world's population is expected to reach 9.5 billion by 2050, with most of this population increase predicted to take place in the developing world. Such a rapid population growth will present even greater challenges with respect to achieving global food security. One possible step toward alleviating hunger and malnutrition will be the introduction of more self-sustainable, nutrient-rich, and biofortified staple crops.

The generation of biofortified plants requires some knowledge of the micronutrients themselves. Micronutrients are composed of vitamins and minerals. While vitamins are organic molecules that are synthesized by the plant, minerals are inorganic compounds that must be removed by the plant from the soil and stored in their edible tissues. Biofortified plants can be generated using molecular plant breeding or transgenic technologies; both procedures have been outlined in a previous section. Even if the plant itself contains high concentrations of a particular micronutrient, the micronutrient also has to be readily bioavailable, or absorbed and utilized by the body as well. What follows are examples of biofortified crops that are currently under development.

16.4.1.2 Vitamin A (β-carotene) Biofortification. Three million preschool-aged children have visible eye damage as a result of vitamin A deficiency, and approximately half a million of these will go blind. Two-thirds of these children will die shortly afterwards. β-carotene, the precursor molecule required for vitamin A biosynthesis, is not found in cereal grains such as rice. Golden Rice provides a biotechnological solution to reduce diseases related to vitamin A deficiency. Golden Rice is a transgenic crop, which has been designed by inserting two genes which reconstitute the carotenoid biosynthetic pathway from other organisms into the rice genome. This β-carotene accumulates, rendering the rice grain a golden colour (Figure 16.4).[54]

Golden Rice is the first nutrient-rich crop that could become a powerful tool against malnutrition. Biofortified crops such as Golden Rice could reach remote rural populations and provide essential nutrients for those who lack access to supplementation programs. It is hoped that eventually Golden Rice would provide the recommended daily allowance of vitamin A for children in rice-based societies, such as India and Vietnam. Similarly, high beta-carotene maize, has been generated to combat vitamin A deficiency in Africa.

16.4.1.3 Iron Biofortified Crops. Iron deficiency, the most common micronutrient deficiency in the world, affects more than two billion people. Iron deficiency impairs physical growth, mental development and learning capacity, and increases the risk of women dying during delivery or during the postpartum period. Iron biofortified staple crops can be produced using conventional plant breeding strategies or biotechnology. Genetically engineered *japonica* rice plants have been developed with six times the iron content of their nontransgenic counterparts. This was accomplished by inserting plant genes encoding the iron transport protein,

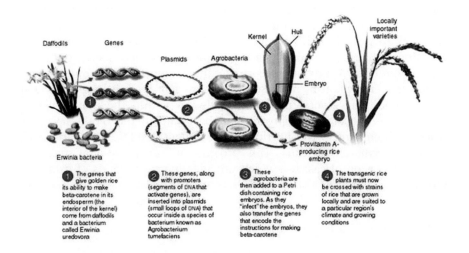

Figure 16.4   Golden Rice.
Photo courtesy of ISAAA.

nicotianamine synthase, and ferritin, a storage centre for iron into the rice genome. Together, these two genes allow the rice plant to absorb more iron from the soil, where it can accumulate within the rice kernel. A third gene, encoding phytase, was also engineered into this rice line to prevent the plant anti-nutrient phytate from inhibiting iron absorption into the intestine, thus increasing the bioavailability of the iron in the transgenic plant.[55]

16.4.1.4   Calcium Biofortified Crops.   Calcium deficiency is a prominent contributor to osteoporosis. Recently, genetically modified carrots have been developed that express increased levels of a plant calcium transporter; transgenic carrot roots were shown to accumulate two-fold-higher calcium levels. A series of randomized clinical trials demonstrated that calcium derived from these transgenic carrots could be absorbed in the gut and incorporated directly into the bones of both mice and humans. Recently, tomatoes expressing a similar calcium biofortified transgene exhibited increased fruit firmness and prolonged shelf life, suggesting additional commercial benefits exist for calcium-biofortified fruits and vegetables. This technology may be eventually applied to a wide variety of crops because it involves the over-expression of a gene that is present in all plants.[56]

16.4.1.5   Biofortified Corn with Multivitamins.   A potential short term solution for those who are most vulnerable to malnutrition is the generation of a nutritionally complete single staple crop, such as rice or corn, that is biofortified with all of the micronutrients that are necessary to maintain one's health. While not attractive in the long term, crops such as these could maintain the health and nutrition status of subsistence farmers as they seek to provide more a conventional means of achieving nutritional diversity in their diet. Recently, a triple-vitamin fortified maize containing high amounts of β-carotene, ascorbate, and folate that is commonly consumed as a crop plant by rural South Africans has been developed, by genetically modifying several metabolic pathways within the maize plant.[57] Another example of this approach can be found in the BioCassava Plus project, which targets the nutritionally deficient staple of a quarter of a billion sub-Saharan Africans. This program uses conventional means to produce a version of cassava, which provides increased levels of iron, zinc and vitamin A, as well as pathogen resistance.[58]

16.4.2 Plants with Improved Health Benefits

While biofortified plants produce vitamins and minerals essential to human health, functional foods also contain other bioactive compounds that have health benefits or can reduce the risk of chronic diseases (Table 16.1). Although functional foods are most often available in the form of conventional foods as part of a regular diet, a number of plants can be altered using biotechnology to increase their nutritional benefits. One example is the increased expression of the antioxidant anthocyanin through metabolic engineering. Anthocyanins are red, purple, or blue flavonoids and are found at high levels in blueberries, blackberries and raspberries. Anthocyanins can protect plant cells from sunlight damage, and are associated with protection against a broad range of human diseases. Recently, a transgenic tomato plant, which produced anthocyanin was generated by expressing two transcription factors from snapdragons in the plants. The fruits produced from these tomato plants were an intense deep purple colour and possessed an

Table 16.1 Examples of plants with improved health benefits through biotechnology.

Attribute	Health Benefit	Reference
Omega-3 fatty acids in seed storage oils	Protection against cardiovascular disease	60
Anthocyanin in tomatoes	Antioxidant, protects against chronic diseases such as cancer	59
Phytic acid in various food crops	Antioxidant, anticancer agent, an anti-oxidant, prevents Parkinson's disease	81
Erucic acid expressed in oil crop seeds	Skin health	82
Acetyl L-carnitine (ALCAR) and L-carnitine (LCAR) in seeds	Alzheimer's disease, depression, and schizophrenia	83
Tocopherol (vitamin E) in lettuce	Heart disease, Parkinson's disease, Alzheimer's disease and cancer	84

enhanced level of antioxidants. Cancer susceptible mice that were fed the high-anthocyanin tomatoes demonstrated an extension of life span of up to 30%.[59]

The metabolic engineering of fatty acids in plant seed storage oils has also been investigated. For example a "designer oilseed" transgenic plant has been developed which synthesizes omega-3 fatty acids found in routinely in fish oils.[60]

16.4.3 Therapeutic Proteins Produced in Plants

Production of plant-derived biopharmaceuticals, or molecular farming, involves the generation of pharmaceutical compounds such as vaccine proteins and monoclonal antibodies in plant tissue. The number of therapeutic proteins produced in plants has extended greatly over the past two decades, and ranges from human monoclonal antibodies against HIV to vaccine proteins against smallpox and other potential biological warfare threats, and even to an assortment of anticancer therapeutic agents for the newly emerging field of personalized medicine (Table 16.2). Molecular farming originally developed as a response to the urgency for safe, effective and inexpensive therapeutic proteins in developing countries. These vaccines needed to be easily transportable so that they were accessible to remote regions while retaining their stability for great lengths of time without refrigeration. Plant-derived vaccines can be directly eaten in the form of plant tissues such as tomatoes, corn or bananas, and have been demonstrated to effectively elicit an immune response to a particular pathogen. Unlike conventional vaccines which have limited shelf lives, vaccine proteins produced in plants are generally quite stable and can be stored as seed at room temperature for months or even years without substantial losses. Plants expressing vaccine proteins can be raised using local farming techniques and do not require sophisticated instrumentation for processing or trained medical personnel.

Diarrhoea due to diseases such as cholera, Norwalk virus and rotavirus, three of most devastating childhood diseases found in developing countries, represents one of the focus areas for the design of plant-derived vaccines. Vaccines generated from plants are also being produced as a defence against poorly funded "orphan" infectious diseases such as hookworm or dengue fever. In the future, it is possible that several different vaccine antigens can be expressed at the same time in a single plant.

A variety of methods are now used to produce biopharmaceuticals from plant tissue. The original form of molecular farming took place using transgenic plants. One of the earliest examples is the use of transgenic potato tubers to express the surface antigen of Hepatitis B virus (HBVsAg).[61] A preliminary clinical trial involving individual volunteers who were fed transgenic potato tubers expressing the HBV surface antigen found that the volunteers developed an increased antibody response to HBV, as opposed to those who were fed nontransformed potato tubers, who served as negative controls.[61]

Table 16.2   Recent therapeutic proteins produced in plants.

Antigen	Expression method	Disease/condition
Interleukin 10 (IL-10)	Transgenic Arabidopsis plants	Immune-regulatory protein
Apolipoprotein AI Milano (ApoAI Milano)	Transgenic safflower seeds	High-density lipoprotein
Lewis Y specific monoclonal antibody (MB314)	Moss expression system	Tumor targeting antibody
α-L-iduronidase	Transgenic maize kernels	Lysosomal storage diseases
anti-epidermal growth factor receptor (EGFR) antibody	Transgenic tobacco plants	Head and neck squamous cell carcinoma, paediatric and adult glioma, and nasopharyngeal and oesophageal cancers
Bacillus anthracis (PA antigen)	Transgenic chloroplasts	Anthrax
CTB-ama1 & CTB-msp1	Transgenic chloroplasts	Malaria
HIV C4V3 polypeptide	Transgenic chloroplasts	HIV
Tuberculosis ESAT-6 (6 kDa early secretory antigenic target)	Transgenic chloroplasts	Tuberculosis
VP8 of rotavirus	Transgenic chloroplasts	Rotavirus
Hepatitis B core antigen	Cowpea Mosaic Virus	Hepatitis B virus
Human papillomavirus L1 and L2	Potato Virus X	Human Papillomavirus
Influenza virus HN epitopes	Tobacco Mosaic Virus	Influenza virus
Non Hodgkin lymphoma idiotype	Tobacco Mosaic Virus	Non-Hodgkin lymphoma
Ebola Virus Monoclonal Antibody	Bean Yellow Dwarf Virus	Ebola virus

Chloroplasts have also been engineered to express vaccine proteins. Plastid engineering, or transplastomy, mitigates the issue of low expression levels of recombinant proteins, which are a frequent drawback of nuclear transformations. Plastid genomes resemble those of prokaryotes, and so the constructs used to transform plastids more resemble those used in bacterial transformations. Since a single plant cell may possess tens of thousands of plastids, high levels of biopharmaceutical proteins can be produced using this approach. An example of therapeutic proteins produced in tobacco chloroplasts are the malarial antigens CTB-ama1 and CTB-msp.[62]

Plant viruses can also be utilized as production platforms for therapeutic proteins. Plant positive-sense RNA viruses, such as tobacco mosaic virus, potato virus X and cowpea mosaic virus, have been engineered into cDNA infectious clones and used to express both full length polypeptides (such as the L1 and L2 proteins of human papillomavirus) or short epitopes which are fused to the virus capsid protein and displayed on the surface of the virus particle.[63] The foot and mouth disease virus epitope has been expressed in this manner on the surface of cowpea mosaic virus.[64] In general, plant viruses can express much larger amounts of foreign proteins than transgenic plants, and under a shorter time frame. A drawback to the use of plant virus expression vectors is that there are size restrictions with respect to the foreign insert that can be tolerated by the virus; larger inserts tend to render the virus unstable.

As an alternative to plant RNA viruses, geminiviruses, such as bean yellow dwarf virus are small single stranded DNA viruses which can amplify large therapeutic proteins to extremely high levels in plants. For example, a monoclonal antibody to Ebola virus has been generated using the bean yellow dwarf virus vector and shown to be biologically functional.[65]

Magnifection refers to a more recent technique for producing plant-derived biopharmaceuticals. In this instance, virus expression vectors are deconstructed into 'modules', one of which contains the therapeutic protein and the others containing the machinery necessary for virus replication. These deconstructed vectors frequently lack movement and coat proteins, thus mitigating biosafety concerns about recombinant virus escape, rather, they are introduced into the host plant by vacuum infiltration, in such a way that every plant cell becomes infected and replication is synchronous. An example of the use of magnifection is the expression of influenza virus H1N1 epitopes using a TMV-based vector in tobacco plants.[66]

Besides fully grown mature plants, therapeutic proteins can be continuously expressed in cell lines, which would provide more controlled environmental conditions. Aquatic plant cells, such as laemna cells, or hairy roots, which can release the therapeutic protein into the media where it can easily be collected and purified, are two examples of the innovative ways that plants can be used to produce biopharmaceuticals.[67]

The fact that in some instances proteins require only partial purification from plant tissue provides another advantage by reducing the cost of protein production even further. For example, vaccines produced in maize kernels can be ground and stored in the form of cornmeal, and the fruit of tomatoes expressing vaccines can be lyophilized into a powder and stored at room temperature for months, before being reconstituted into a juice. Human lactoferrin, a milk protein with multiple health benefits and required for iron absorption, has been produced in transgenic rice with a protein accumulation of about 5% of the rice flour dry weight.[68] Large-scale extraction of lactoferrin is straight forward, requires simple purification steps, and has been estimated to be economically feasible. Alternatively, therapeutic proteins can be produced in seeds such as safflower, then easily purified as oil bodies from other cellular components in the form of fusion proteins with oleosins, for example, or by attachment of the desired protein to the oilbody *via* an affinity ligand. The therapeutic proteins of interest can then be recovered using flotation centrifugation.[69]

16.5 EXPRESSION OF BIODEGRADABLE PLASTICS AND OTHER POLYMERS

16.5.1 Biodegradable Plastics

While plants themselves can produce natural polymers, such as rubber, starch and cellulose, they can also be engineered to synthesize novel types of polymers, including biodegradable plastics. Biodegradable plastics, such as polyhydroxyalkanoates (PHAs), can be completely degraded into CO_2 and water in composters, landfills, or sewage treatment plants by the action of naturally occurring micro-organisms.[70] PHAs are used in the making of films, coated paper, compost bags and disposable food utensils, and water bottles. Plant-based biodegradable polymers such as PHAs can be generated in crops such as switchgrass (*Panicum virgatum* L.), the predominant tall grass of the North American prairie, on land that is marginal or unsuitable for food crops. Besides not competing as a food source, biodegradable plastic from plants such as switchgrass offer an alternative to petroleum-based plastics as an environmentally friendly and carbon-neutral source of polymers. There are a number of species of bacteria that naturally produce PHA, and their biosynthetic pathways have been well characterized and are reproducible in plants. For example, polyhydroxybutyrate (PHB), another bacterial polyester that resembles petrochemically produced plastics. This was accomplished by transforming *A. thaliana* plants with genes derived from the bacteria *R. eutropha* that encode acetoacetyl-CoA reductase and PHB synthase. Synthesis of the polymer took place using a cytosolic β-oxothiolase naturally present in *A. thaliana* along with the additional bacterial enzymes that were introduced to the plant. The result was the production of PHB polymer in the cytosol, nucleus or vacuoles. PHB has also been synthesized in transgenic plants such as sugarcane, to levels as great as 4.8% of the dry leaf weight.[71] Plastics can be produced in a number of crop plants, including alfalfa (*Medicago sativa*), cotton, potato, canola, tobacco, sugar crops (beet [*Beta vulgaris*]) and cane (*Saccharum officinarum*) and fibre crops such as flax (*Linum usitatissimum*).[72,73] Indeed, switchgrass is considered to have a future role in bioethanol production for biofuels, and it is perceived that transgenic switchgrass could produce a double crop of biodegradable plastics and cellulosic ethanol in the future. Although promising, there are still a few stumbling blocks to the use of plants as synthesizers of biodegradable plastics. An environmentally friendly and cost-effective scale up procedure for extracting the product from plants is still a major consideration and efforts are currently underway to resolve this problem.

16.5.2 Spider Silk Produced in Plants

Spider silk proteins used by spiders for building spider webs have also been generated in transgenic tobacco, potato and *Arabidopsis* plants. Spider silk is composed of proteins that assemble in the form of elastic Lycra®-like fibres, and their innate strength makes them attractive for a number of industrial and medical applications. A silk–elastin fusion protein has been produced in the endoplasmic reticulum of transgenic tobacco at a concentration of 80 mg of pure protein/kg of tobacco leaves and extracted using a simple buffer-extraction procedure. While extremely heat stable, the resulting extracted spider silk proteins have lost many of their structural properties, suggesting that they have not assembled correctly.[74–77] Nonetheless, the growth of anchorage-dependent mammalian cells on culture plates coated with spider silk-elastin was shown to be comparable with conventionally coated plates (using collagen, fibronectin and poly-D-lysine), offering the potential of plant-derived spider silk for new applications.

16.6 PUBLIC PERCEPTION, INTELLECTUAL PROPERTY ISSUES

While much has been accomplished in terms of the generation of novel crops by agricultural biotechnology, considerable effort will be required to see these new technologies become a reality. A major stumbling block is the concern regarding public perception of genetically modified food. Despite advances that have been made to date, in certain parts of the world concerns over the safety of "Frankenfood" have continued to prevent the increased production of genetically modified, or GM food, for those who need it most. For example, The 'Golden Rice' project is considered to be the model prototype for the potential of agricultural biotechnology to produce bio-fortified foods that could have enormous humanitarian benefits. In terms of intellectual property issues, 'freedom to operate' has been maintained from the very start and Golden Rice is offered free of charge with no attached conditions to subsistence farmers in developing countries. While one may think that such a technology made available would be lauded by the world as an affordable solution to what many would consider one of the largest causes of malnutrition in the world, Golden Rice continues to draw criticism from anti-GM groups and has yet to be distributed. These organizations continue to worry that acceptance of GM technology utilized for humanitarian purposes would facilitate the acceptance of other genetically modified foods. Concerns of these groups range from worries about environmental or human health hazards to the monopolization of crops by multinational corporations. Unfortunately for people who live in suboptimal climates such as sub-Saharan Africa, these issues deter their ability to bring much needed improved crops into their fields and ultimately to the marketplace to interact with their international trading partners, including some European countries. Increased regulation of GM crops due to unwarranted concerns has led in fact to the opposite effect that GM opponents intended, that the only entities who can afford to overcome the regulatory hurdles are indeed the multinational corporations that the anti-GM movement opposes. Regardless of this, biofortification of foods *via* transgenic crop production continues to provide a truly feasible means by which nutritionally complete foods can be delivered to malnourished people who reside in remote locations and offers an attractive alternative to supplementation programmes, which can be sensitive to the whim of international funding.

Another example of negative publicity and public perception on agricultural biotechnology is the potential impact of Bt crops on beneficial insects such as the Monarch butterfly. A highly discredited study suggested that Monarch larvae feeding on leaves covered in pollen shed from Bt maize plants (event Bt176) grew poorly when compared to those larvae which fed on uncontaminated leaves.[78] Regardless of the faults found in the study, which were addressed by the scientific community, this report was used as fodder by biotechnology opponents.[79] Similarly, a recent study has been highly discredited which claims that GM foods resulted in increased numbers of tumours in mice has induced a sort of mass hysteria among GM opponents. Again, the study in question was poorly conducted and provides no statistically significant data, yet is highly cited by biotechnology opposition groups and adds fuel to the fire of the GM controversy.

In regards to the use of transgenic strategies for pathogen resistance, issues related to public perception remain high, partially due to concerns regarding the effect of transgenes on beneficial organisms, and partially because of the limited success of this technology and the ineffective levels of pathogen resistance that have been realized in many cases. Taken together these two issues weaken the argument for the financial and time commitments, which are required for the production of lengthy field trials, required for the commercial production of GM crops.[80]

While opposition to GM crops remain, it is slowly becoming eroded as tensions resulting from the need for food security and the restrictions due to climate change increase across the globe. Public perception of the need for therapeutic proteins made in plants, for example, is considered to be much more acceptable than what is conceived about some other GM crops.[81] The issue of public acceptance for agricultural biotechnology will continue to evolve.

REFERENCES

1. A. Pitzschke and H. Hirt, *EMBO J.*, 2010, **29**, 102.
2. N. J. Taylor and C. M. Fauquet, *DNA Cell Biol.*, 2002, **21**, 963.
3. N. Scotti, M. M. Rigano and T. Cardi, *Biotechnol. Adv.*, 2012, **30**, 387.
4. S. O. Duke and S. B. Powles, *Pest Manag. Sci.*, 2008, **64**, 319.
5. L. Pollegioni, E. Schonbrunn and D. Siehl, *FEBS J.*, 2011, **278**, 2753.
6. A. L. Cerdeira and S. O. Duke, *2010, GM Crops*, 2010, **1**, 16.
7. W. H. Davies, US7087809, 2006.
8. P. R. Beetham, P. L. Avissar, K. A. Walker and R. A. Metz, US6870075, 2005.
9. G. A. Kleter, J. B. Unsworth and C. A. Harris, *Pest Manag. Sci.*, 2011, **67**, 1193.
10. J. Deikman, M. Petracek and J. E. Heard, *Curr. Opin. Biotechnol.*, 2012, **23**, 243.

11. E. Cominelli and C. N. Tonelli, *N. Biotechnol.*, 2010, **27**, 473.
12. T. Hadiarto and L. S. Tran, *Plant Cell. Rep.*, 2011, **30**, 297.
13. M. Reguera, Z. Peleg and E. Blumwald, *Biochim. Biophys. Acta.*, 2012, **1819**, 186.
14. S. S. Hussain, M. A. Kayani and M. Amjad, *Biotechnol. Prog.*, 2011, **27**, 297.
15. S. S. Gill and N. Tuteja, *Plant Signal. Behav.*, 2010, **5**, 26.
16. B. J. Miflin and D. Z. Habash, *J. Exp. Bot.*, 2002, **53**, 979.
17. D. B. Collinge, H. J. Jørgensen, O. S. Lund and M. F. Lyngkjaer, *Annu. Rev. Phytopathol.*, 2010, **48**, 269.
18. G. Sanahuja, R. Banakar, R. M. Twyman, T. Capell and P. Christou, *Plant Biotechnol. J.*, 2011, **9**, 283.
19. M. R. Gatehouse and N. Ferry, *Phil. Trans. R. Soc. B*, 2011, **366**, 1438.
20. A. Raybould, G. Caron-Lormier and D. A. Bohan, *J. Agric. Food Chem.*, 2011, **59**, 5877.
21. X.-D. Yu, J. Pickett, Y.-Z. Ma, T. Bruce, J. Napier, H. D. Jones and L.-Q. Xia, *J. Integr. Plant Biol.*, 2012, **54**, 282.
22. V. Bhatia, P. L. Uniyal and R. Bhattacharya, *Biotechnol. Adv.*, 2011, **29**, 879.
23. D. V. Reddy, M. R. Sudarshana, M. Fuchs, N. C. Rao, G. Thottappilly and V. Venkata, *Adv Virus Res.*, 2009, **75**, 185.
24. C. Simón-Mateo and J. A. García, *Biochim. Biophys. Acta*, 2011, **1809**, 722.
25. D. Gonsalves, *Curr. Top. Microbiol. Immunol.*, 2002, **266**, 73.
26. E. Noris, G. P. Accotto, R. Tavazza, A. Brunetti, S. Crespi and M. Tavazza, *Virology*, 1996, **224**(1), 130.
27. B. H. Kvitko, A. R. Ramos, J. E. Morello, H. S. Oh and A. J. Collmer, *Bacteriol.*, 2007, **189**(22), 8059.
28. S. A. Ceasar and S. Ignacimuthu, *Biotechnol Lett.*, 2012, **34**, 995.
29. P. Smyda, H. Jakuczun, K. Dębski, J. Sliwka, R. Thieme, M. Nachtigall, I. Wasilewicz-Flis and E. Zimnoch-Guzowska, *Plant Cell Rep.*, 2013.
30. O. Wally and Z. K. Punja, *GM Crops*, 2010, **1**, 199.
31. H. J. Atkinson, C. J. Lilley and P. E. Urwin, *Curr. Opin. Biotechnol.*, 2012, **23**, 251.
32. H. J. Atkinson, P. E. Urwin and M. J. McPherson, *Annu. Rev. Phytopathol.*, 2003, **41**, 615.
33. E. A. van der Vossen, J. Gros, A. Sikkema, M. Muskens, D. Wouters, P. Wolters, A. Pereira and S. Allefs, *Plant J.*, 2005, **44**, 208.
34. W. Li, M. Shao, J. Yang, W. Zhong, K. Okada, H. Yamane, G. Qian and F. Liu, *Plant Sci.*, 2013, **207**, 98.
35. M. A. Anderson, F. T. Lay and R. L. Heath, US7544861, 2009.
36. K. D. Chapman, R. Shrestha, E. Blancaflor and R. A. Dixon, US7316928, 2008.
37. A. Bhargava, F. F. Carmona, M. Bhargava and S. Srivastava, *J. Environ. Manage.*, 2012, **30**, 103.
38. P. C. Abhilash, J. R. Powell, H. B. Singh and B. K. Singh, *Trends Biotechnol.*, 2012, **30**, 416.

39. M. I. Georgiev, E. Agostini, J. Ludwig-Müller and J. Xu, *Trends Biotechnol.*, 2012, **30**, 528.

40. E. Agostini, M. A. Talano, P. S. González, A. L. Oller and M. I. Medina, *Appl. Microbiol. Biotechnol.*, 2013, **97**, 1017.

41. S. A. Hasan, Q. Fariduddin, B. Ali, S. Hayat and A. Ahmad, *J. Environ. Biol.*, 2009, **30**, 165.

42. G. Del Sorbo, H. Schoonbeek and M. A. De Waard, *Fungal Genet. Biol.*, 2000, **30**, 1.

43. S. Arbaoui, A. Evlard, M. E. Mhamdi, B. Campanella, R. Paul and T. Bettaieb, *Biodegradation*, 2013, **24**(4), 563–567.

44. E. Testiati, J. Parinet, C. Massiani, I. Laffont-Schwob, J. Rabier, H. R. Pfeifer, V. Lenoble, V. Masotti and P. Prudent, *J. Hazard. Mater.*, 2012, **29**, 248.

45. A. Yateem, *Environ. Sci. Pollut. Res. Int.*, 2013, **20**, 100.

46. D. K. Gupta, H. G. Huang and F. J. Corpas, *Environ. Sci. Pollut. Res. Int.*, 2013, **20**, 2150.

47. D. Q. Zhang, R. M. Gersberg, T. Hua, J. Zhu and W. J. Ng, *Environ. Sci. Pollut. Res. Int.*, 2013, **20**, 4612.

48. S. Saha and S. Ramachandran, *Recent Pat. DNA Gene Seq.*, 2013, **7**, 36.

49. R. Brunecky, J. O. Baker, H. Wei, L. E. Taylor, M. E. Himmel and S. R. Decker, *Methods Mol. Biol.*, 2012, **908**, 197.

50. A. Harfouche, R. Meilan and A. Altman, *Trends Biotechnol.*, 2011, **29**, 9.

51. M. Y. Menetrez, *Environ. Sci. Technol.*, 2012, **44**, 7073.

52. Z. Wang, R. Li, J. Xu, J. M. Marita, R. D. Hatfield, R. Qu and J. J. Cheng, *Bioresour. Technol.*, 2012, **110**, 364.

53. P. Beyer, *N. Biotechnol.*, 2010, **27**, 478.

54. R. A. Sperotto, F. K. Ricachenevsky, A. Waldow Vde and J. P. Fett, *Plant Sci.*, 2012, **190**, 24.

55. S. Gómez-Galera, E. Rojas, D. Sudhakar, C. Zhu, A. M. Pelacho, T. Capell and P. Christou, *Transgenic Res.*, 2010, **19**, 165.

56. J. Jeong and M. L. Guerinot, *Proc. Natl. Acad Sci. U. S. A.*, 2008, **105**, 1777.

57. R. Sayre, J. R. Beeching, E. B. Cahoon, C. Egesi, C. Fauquet, J. Fellman, M. Fregene, W. Gruissem, S. Mallowa, M. Manary, B. Maziya-Dixon, A. Mbanaso, D. P. Schachtman, D. Siritunga, N. Taylor, H. Vanderschuren and P. Zhang, *Annu. Rev. Plant Biol.*, 2011, **62**, 251.

58. M. Giorgio, H. P. Mock, A. Matros, S. Peterek, E. G. Schijlen, R. D. Hall, A. G. Bovy, J. Luo and C. Martin, *Nat. Biotechnol.*, 2008, **26**, 1301.

59. N. Ruiz-López, O. Sayanova, J. A. Napier and R. P. Haslam, *J. Exp. Bot.*, 2012, **63**, 2397.

60. L. J. Richter, Y. Thanavala, C. J. Arntzen and H. S. Mason, *Nat. Biotechnol.*, 2000, **18**, 1167.

61. A. Davoodi-Semiromi, M. Schreiber, S. Nalapalli, D. Verma, N. D. Singh, R. K. Banks, D. Chakrabarti and H. Daniell, *Plant Biotechnol. J.*, 2010, **8**, 223.

62. J. F. Buyel, J. A. Bautista, R. Fischer and V. M. Yusibov, *J. Chromatogr. B*, 2012, **880**, 19.

63. K. Gopinath, J. Wellink, C. Porta, K. M. Taylor, G. P. Lomonossoff and A. van Kammen, *Virology*, 2000, **267**, 159.
64. H. Lai, J. He, M. Engle., M. S. Diamond and Q. Chen, *Plant Biotechnol. J.*, 2012, **10**, 95.
65. G. Roy, S. Weisburg, K. Foy, S. Rabindran, V. Mett and V. Yusibov, *Arch Virol.*, 2011, **156**, 2057.
66. M. A. Talano, A. L. Oller, P. S. González and E. Agostini, *Recent Pat. Biotechnol.*, 2012, **6**, 115.
67. C. Lin, P. Nie, W. Lu, Q. Zhang, J. Li and Z. Shen, *Protein Expr. Purif.*, 2010, **74**, 60.
68. J. Boothe, C. Nykiforuk, Y. Shen, S. Zaplachinski, S. Szarka, P. Kuhlman, E. Murray, D. Morck and M. M. Moloney, *Plant Biotechnol. J.*, 2010, **8**, 588.
69. B. P. Mooney, *Biochem. J.*, 2009, **418**, 219.
70. L. A. Petrasovits, L. Zhao, R. B. McQualter, K. D. Snell, M. N. Somleva, N. A. Patterson, L. K. Nielsen and S. M. Brumbley, *Plant Biotechnol. J.*, 2012, **10**, 569.
71. K. Bohmert-Tatarev, S. McAvoy, S. Daughtry, O. P. Peoples and K. D. Snell, *Plant Physiol.*, 2011, **155**, 1690.
72. J. B. Van Beilen and Y. Poirier, *Plant J.*, 2008, **54**, 684.
73. J. Yang, L. A. Barr, S. R. Fahnestock and Z.-B. Liu, *Transgenic Res.*, 2005, **14**, 313.
74. J. Scheller, D. Henggeler, A. Viviani and U. Conrad, *Transgenic Res.*, 2004, **13**, 51.
75. J. Huang, C. W. P. Foo and D. L. Kaplan, *Polymer Rev.*, 2007, **47**, 29.
76. J. Scheller, K.-H. Gührs, F. Grosse and U. Conrad, *Nat. Biotechnol.*, 2001, **19**, 573.
77. A. M. Gatehouse, N. Ferry and R. J. Raemaekers, *Trends Genet.*, 2002, **18**, 249.
78. R. Edwards, Demand for executive to ban crop trials until effects of GM food on health are studied, *Sunday Herald*, December 8, 2002.
79. M. P. Oeschger and C. E. Silva, *Adv. Biochem. Eng. Biotechnol.*, 2007, **107**, 57.
80. A. Einsele, *Adv. Biochem. Eng. Biotechnol.*, 2007, **107**, 1.
81. S. M. Boue, T. E. Cleveland, C. Carter-Wientjesd, B. Y. Shih, D. Bhatnagar, J. M. McLachlan and M. E. Burow, *J. Agric. Food Chem.*, 2009, **57**, 2614.
82. X. Li, E. N. van Loo, J. Gruber, J. Fan, R. Guan, M. Frentzen, S. Stymne and L. H. Zhu, *Plant Biotechnol. J.*, 2012, **10**, 862.
83. P. D. E. Fraser, M. Enfiled and P. M. Bramley, *Arch. Biochem. Biophys.*, 2009, **483**, 196.
84. Y. Yabuta, H. Tanaka, S. Yoshimura, A. Suzuki, M. Tamoi, T. Maruta and S. Shigeoka, *Transgenic Res.*, 2013, **22**, 391.

CHAPTER 17

Vaccine Design Strategies: Pathogens to Genomes

NIALL McMULLAN

School of Life Sciences, University of Hertfordshire, Hatfield, AL10 9AB, UK
Email: n.m.mcmullan@herts.ac.uk

17.1 INTRODUCTION

Prophylactic vaccines have provided extensive control of several infectious diseases. Most notably, the twentieth century witnessed the global eradication of smallpox. Polio outbreaks are limited to a few countries and effective vaccines against measles, mumps, rubella, diphtheria, tetanus and meningitis have controlled or eliminated them in countries where mass immunisation programmes have been established. This has been achieved largely through use of classical vaccine designs, in particular, live vaccines, inactivated or killed vaccines and isolated macromolecule subunit vaccines. The success of these vaccines has been primarily against pathogens that cause acute self-limiting diseases and often where humans are the only or primary host. Despite these successes, there is a real need for new vaccines for recently emerging, chronic diseases, such as HIV/AIDS and hepatitis C where no effective vaccine is available. The World Health Organisation (WHO) data for 2011 show over 1.5 million deaths due to HIV/AIDS and over one million deaths due to tuberculosis. In 2012, there were between 3–4 million new cases of hepatitis C infections annually and 350 000 deaths due to hepatitis C-associated liver disease, with an estimated 150 million infected individuals worldwide. While socio-economic factors contribute to

Molecular Biology and Biotechnology, 6th Edition
Edited by Ralph Rapley and David Whitehouse
Published by the Royal Society of Chemistry, www.rsc.org

the success or failure of vaccination programmes, the variable or limited efficacy of some vaccines contribute to the persistence of infectious diseases such as tuberculosis and typhoid. The advent of recombinant DNA technology saw the introduction of a highly effective recombinant protein-based vaccine against hepatitis B. This represented a paradigm shift in vaccine design strategy. No longer was vaccine design dependent on the ability to isolate and culture the pathogen. A similar strategy saw the introduction of two new vaccines against human papilloma virus, the causative agent of cervical cancer. Significant advances in genome sequencing and bioinformatics technologies provided the tools for the development of a novel recombinant protein vaccine against meningitis B, the first vaccine developed through genome-based vaccine design. This century has also seen the development of experimental vaccines using vaccine designs such as plasmid DNA vaccines and recombinant vector vaccines for protein expression in the host.

This chapter aims to provide an overview of classical vaccine design strategies and new strategies using recombinant DNA-based technologies and bioinformatics and the key immunological issues underpinning these strategies. The methods referred to are described in more detail elsewhere in this volume.

17.2 IMMUNOLOGICAL CONSIDERATIONS IN VACCINE DESIGN

Active immunisation is a complex process involving components of both innate and adaptive immunity. The primary objective of vaccination is activation of antigen-specific B and T cells which facilitate effective immunity in the form of antibody-mediated (humoral) immunity driven by antibody-producing B cells, and/or cell-mediated immunity involving activated CD8 + cytotoxic T lymphocytes (CTLs). Another key aspect of immunity is the concept of immunological memory, essential to the success of prophylactic vaccines in controlling infectious diseases. Long-term immunity is achieved through the generation and maintenance of long-lived memory B and T cells and this is dependent largely on TH cells. The central role of CD4 + helper T (TH) cell subsets to the above processes cannot be understated. As the name implies, TH cells orchestrate many key aspects of immune responses by producing an array of cytokines that collectively influence the direction, magnitude and duration of the immune response. These effects include: differentiation of antigen-primed B cells into antibody-secreting plasma cells, differentiation of antigen-primed cytotoxic T cells into effector cytoxic T lymphocytes (CTLs) and formation of memory B and T cells.

T cells recognise only peptide sequences derived from intracellular processing of antigenic proteins and presented on host cell surfaces by major histocompatibility complex (MHC) molecules. MHC presentation of antigenic peptides leads to activation of CD4 + and/or CD8 + T cells. The above events are dependent on the initial interactions that occur between the vaccine and cells of the innate immune system.

17.2.1 Antigen Recognition and Activation of Innate Immunity

In the normal course of infection, pathogen-associated molecular patterns (PAMPs) expressed on microbes are recognised by pattern recognition receptors (PRRs) expressed on cells of the innate immune system. This recognition lacks the subtle specificity of adaptive immunity but permits the initial recognition of a broad range of pathogens. Upon binding to the pathogen, PRRs provide the activation signals that trigger the innate immune response. The major PRRs involved are the toll-like receptors (TLRs) and collectively they bind a range of PAMPs.[1] Some TLRs are expressed on the plasma membrane and these bind external PAMPs, for example TLR4 binds lipopolysaccharide (LPS) found on Gram-negative bacteria and other TLRs bind peptidoglycans, lipopeptides and other PAMPs expressed on the external surface of pathogens. These interactions trigger internalisation and phagocytic killing. Some TLRs are located on internal membranes, notably endosomal membranes, and recognise genomic material released following internalisation of the pathogen. For example, TLR3 binds to viral double-stranded RNA and TLR9 binds to bacterial DNA (discussed further under DNA vaccines) which provide further cellular activation signals. During the initial immune response, tissue macrophages are activated *via* TLR ligation and secrete inflammatory cytokines and chemokines that direct other leukocytes into the infected tissues. Among these are neutrophils which provide the main phagocytic killing defence, and immature dendritic cells both of which interact with the antigenic material *via* PRRs. Dendritic cells (DCs) are the key antigen presenting cells involved in eliciting T cell responses. Upon activation *via* TLR signals, immature DCs internalise antigen and undergo maturation. During this phase, DCs degrade the antigenic material, breaking protein components into peptides which in turn are bound by MHC molecules for presentation to T cells. During maturation, DCs also change chemokine receptor expression which permits their migration to regional lymph nodes where they encounter naïve T cells.

17.2.2 Antigen Processing and the Major Histocompatibility Complex

T cells are specific for antigenic peptides bound to host MHC molecules and expressed on the surface of host cells. The presence of MHC molecules is essential for T cell responses, a phenomenon known as MHC-restriction. In addition, different subsets of T cells respond to different classes of MHC molecules; CD4 + T cells *e.g.* Tн cells, are MHC class II-restricted, *i.e.* they recognise only MHC class II presented peptides, whereas CD8 + cytotoxic T cells are MHC class I-restricted. The MHC displays extensive genetic polymorphism. In humans, the MHC is known as the human leukocyte antigen (HLA) system located on chromosome six. HLA class I alleles are located at three loci, HLA-A, -B and -C. HLA class II molecules are encoded at three loci, HLA-DP, -DP and -DR. Thousands of allelic variations in HLA genes are present in the population. Polymorphism is exhibited in the peptide binding region permitting binding and presentation of many different antigenic peptides. Each MHC molecule may bind several different peptides depending on the amino acid composition of the peptide binding site and the peptide sequence itself. HLA class I molecules preferentially bind nonamers with containing hydrophobic residues at key positions 2, 3 and 9. These are known as anchor residues, and permit binding of several peptides to a single class I molecule. Class II molecules optimally bind peptides of 13–18 amino acids with binding determined primarily by an internal core sequence of 7–10 residues that provide the main interacting residues. In general, these core residues are aromatic or hydrophobic amino acids.

The two classes of MHC molecules differentially present peptides from one of two antigen processing pathways, the cytosolic pathway and the endocytic pathway each of which processes antigen from a different source (Figure 17.1). Endogenous antigens, for example newly-synthesised viral proteins or tumour proteins, are degraded in proteasomes of the cytosolic pathway. Peptides generated by the proteasome are bound by a transporter protein, TAP (transporter associated with antigen processing) that has affinity for peptides containing 8–16 amino acids. TAP translocates these peptides to the rough endoplasmic reticulum where they are trimmed by endoplasmic reticulum aminopeptidase (ERAP) into nonamers. MHC class I molecules are assembled in the RER by chaperone molecules, calnexin and ERp57. Once the class I molecule is assembled, the antigenic peptides are bound by MHC class I molecules in a process involving chaperone molecules, tapasin (TAP-associated protein) and calreticulin. The resultant complex is transferred to the plasma membrane for presentation to CD8 + T cells. By contrast, CD4 + T cell responses are to exogenous antigens, for example, phagocytosed microbes or soluble proteins which are processed through the endocytic pathway. This involves sequential proteolysis in different endosomal compartments of decreasing pH, resulting in peptides of 13–18 amino acids. These peptides associate with MHC class II molecules in the late endosomes. At this stage, the MHC class molecule loses the invariant chain, a short sequence that blocks the peptide-binding site and prevents binding of peptides in the RER, and the MHC–peptide complex transferred to the plasma membrane for presentation to CD4 + T cells.

Differential processing and presentation of antigen is a key consideration in vaccine design. Live vaccines are able to provide antigen for both processing pathways but inactivated vaccines and subunit vaccines do not provide a source of endogenously-synthesised protein thus there is little or no CTL responses to these vaccines. The need for CTL responses against intracellular pathogens causing chronic infections has been the driving force behind the development of recombinant vector vaccines and plasmid DNA vaccines, considered later in this chapter.

MHC class I molecules are expressed by most nucleated cells therefore any virally-infected cell or abnormal self-cell is a potential target for CTLs. By contrast, MHC class II expression is limited to antigen presenting cells (APCs) namely dendritic cells, activated macrophages and activated B cells.

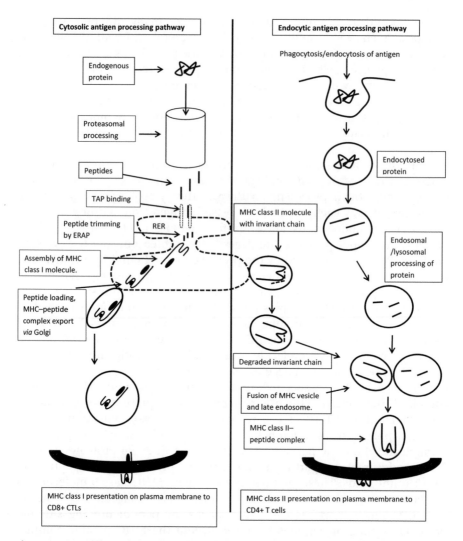

Figure 17.1 Differential processing of cytosolic and endocytic antigens leads to MHC class I-restricted presentation and MHC class II-restricted presentation, respectively. See text for details.

These cells are referred to as professional APCs as they possess other co-stimulatory molecules such as CD80/86 required for T cell activation. Mature dendritic cells and macrophages activated *via* TLRs express MHC class II molecules, whereas resting B cells become APCs following antigen binding.

17.2.3 Cross-Presentation of Antigen by Dendritic Cells

Cross-presentation describes the presentation of peptides derived from exogenous antigen uptake on MHC class I molecules. The ability of DCs to present exogenous antigen on MHC class I molecules to CTLs was first shown by Bevan in the 1970's.[2] While *in vitro* studies indicate that other professional APCs demonstrate cross-presentation, DCs are the most likely cells to perform cross-presentation *in vivo*.[3] Different models have been proposed for how proteins derived from exogenous sources can encounter MHC class I molecules.[4] Proteins from the endocytic processing pathway may enter the cytosolic pathway using TAPs, where they are loaded onto MHC class I molecules in the endoplasmic reticulum. Alternatively, MHC class I molecules and TAP proteins may be translocated to phagosomal and endosomal membranes where peptides derived from endocytic processing are loaded onto class I molecules.[5]

Dendritic cells display considerable heterogeneity in terms of morphology, anatomical location and functionality. in humans, there are two main lineages: plasmacytoid DCs (pDCs) normally resident in secondary lymphoid tissues, in particular lymph nodes, and conventional DCs (cDCs-formerly myeloid derived DCs) which populate the blood and extra-lymphoid tissues and migrate to lymph nodes following antigen uptake in infected tissues. These cells are further subclassified on the expression of blood DC antigen (BDCA) markers: pDCs express BDCA2 (CD303 + DCs) and BDCA4 (CD304 + DCs) whereas cDCs express BDCA1 (CD1c + DCs) and BDCA3 (CD141 + DCs). Despite these differences, *in vitro* evidence suggests that all DC populations are capable of cross-presenting exogenous antigens and thus may present antigen both to CD4 + T cells and CD8 + T cells.

In addition to their phagocytic properties, DCs also express surface receptors that are able to endocytose antigen into different endosomal compartments which may influence the efficiency and extent of cross-presentation.[4,6] Antigen uptake into early endosomes by targeting CD40 or the mannose receptor (MR) showed greater efficiency in cross-presentation that targeting late endosomes *via* DEC205 (CD205).

In addition to antigen presentation, dendritic cells also secrete cytokines that influence the immune response. Most notably activated cDCs secrete IL-12 that promotes differentiation of TH cells into TH1 cells. This TH subtype produces Il-2 and interferon-γ and are essential in promoting CTL development from naïve cytotoxic T cells. This is a key element in developing cellular immunity to chronic intracellular infections.

Dendritic cell subpopulations also vary in their expression of TLRs and therefore respond to different PAMPs. Ligation of different TLRs influences cytokine production from DCs that in turn can influence the type of immune response elicited. Variations in TLR expression may also influence antigen processing. cDCs typically express TLRs1-8 and TLR10 and so recognise a wide range of micro-organisms, in turn secreting IL-12, tumour necrosis factor -alpha (TNF-α) and IL-6 primarily through TLR2 and 4 signalling whereas pDCs express TLR7 and TLR9 and secrete high levels of interferon-alpha. Due to common signalling molecules, there can be synergistic signals between different TLRs.

17.2.4   Induction of B and T Cells in Lymph Nodes

Movement of antigen from the tissues to lymph nodes is an essential step in initiating adaptive immune responses. Prior to encountering antigens, naïve B and T cells circulate between the blood and secondary lymphatic tissues. B cells bind specifically to epitopes on the surface of unprocessed pathogens or secreted proteins, whereas T cells bind specifically to MHC-presented peptides on the surface of DCs. Upon antigen binding, known as antigen-priming, B cells undergo rapid clonal expansion to form clonal populations of B cells and establish germinal centres within the lymph nodes. Differentiation of the activated B cells gives rise to antibody-secreting plasma cells, the source of soluble antibodies, and long-lived memory cells. The first antibodies produced are IgM but B cells also undergo isotype switching to express IgG isotypes. This is an important event as IgG isotypes display a wider range of functional activities, such as opsinisation and neutralisation. These events require essential cytokine co-signals from TH cells for B cell responses to protein antigens (T-dependent antigens). By contrast, poly-saccharide antigens, T-independent antigens, can activate B cells directly in the absence of T cell signals, by cross-linking the Ig antigen receptors. The antibodies tend to be IgM with little or no isotype switching.

T cell responses require engagement of the TCR with the MHC–peptide complex and additional co-stimulatory signals provided by interactions between molecules on the APC surface and counter ligands on the T cell, for example MHC class I or class II binding to CD8 or CD4, respectively, and very importantly, between CD80 or CD86 on the APC and CD28 on T cells. Finally, cytokines provide a key third activation signal which has profound effects on the development of TH cells into different TH cell subsets, described below. Upon activation, TH cells differentiate into cytokine-secreting effector TH cells and memory TH cells and cytotoxic T cells differentiate into functional cytotoxic T lymphocytes and memory CTLs. The memory cells may reside in the lymph nodes conferring central memory, or migrate to peripheral tissues where they establish peripheral memory. The memory cells provide specific, long-lived protection and are essential to the protective efficacy elicited by vaccines.

17.2.5   Polarisation of T-helper Cells

TH cells can be classified into two main functional subsets, TH1 or TH2 cellsm, that differ in the cytokines that they produce. In turn, this results in different types of immune responses. As referred to above, polarisation involves additional cytokines that determine the subset. Following antigen-priming, naïve TH cells develop into TH0 cells, that are driven into functional TH cells by cytokine signals from the APC or auxillary immune cells. In the presence of IL-12, and interferon-gamma (IFN-γ), TH cells polarise into TH1 cells, whereas IL-4 promotes TH2 development. TH1 cells are characterised by IFN-γ production which promotes cellular immunity through activation of macrophages and CTLs and also promote production of opsonising and neutralising IgG subclasses by B cells. By contrast, TH2 cells secrete IL-4, IL-5 and IL-13 which promote antibody-mediated immunity, in particular, some IgG subclasses, but importantly, IgA and IgE isotypes associated with immunity to mucosal pathogens (note that IgE antibodies are also the drivers of type I hypersensitivity reactions in allergic individuals). The cytokines produced by the TH subsets also cross-regulate the two subsets, tending to promote development of their own subset while inhibiting the other. In general, the TH response is mixed, with the extent of different TH involvement determining the effectiveness of the immune response to a given type of pathogen.

17.2.6 Protective Immunity against Acute and Chronic Infections

The primary effectors of immunity are antibodies that bind specifically to the pathogen and mark it for destruction. In the case of intracellular pathogens this may also involve CTL responses to control the spread of the pathogen within the host. Antibodies promote phagocytic uptake and killing of the pathogen by interactions between the bound antibodies and receptors on phagocytes, and activation of the complement system. This is the basis of immunity induced by subunit vaccines against bacterial meningitis. Neutralising antibodies that bind to exotoxins are effective in preventing diseases where the bacterial exotoxins are the primary pathogenic feature, for example, tetanus and diphtheria. Vaccine-induced neutralising antibodies are also effective in conferring immunity against several viral diseases by preventing viral entry into host cells. The success of vaccine-induced antibody-mediated immunity has been limited mainly to pathogens that cause acute, self-limiting infections. Immunity to intracellular pathogens that establish chronic infections require cellular immunity in the form of CTL responses to eliminate the infected host cells. These pathogens possess different features that permit survival within the host. In the case of HIV infection, antibody responses are effectively rendered ineffective by the high mutation rate in key B cell epitopes. While some variants may be susceptible to antibody-mediated immunity, escape mutants persist and spread the infection. Furthermore, HIV can disrupt normal antigen processing and presentation so presenting additional challenges. Infection with Hepatitis C is often asymptomatic reflecting little innate immune response. The pathogen also displays significant antigenic variation. Vaccines against parasitic diseases have proved elusive mainly due to the complex life cycle of the pathogens.

17.3 VACCINE DESIGN STRATEGIES

This section considers the relative merits of vaccine designs both in current use in human immunisation programmes and those under development. The various designs are summarised in Table 17.1.

Table 17.1 Examples of vaccine designs in current use in humans and experimental designs in development.

Vaccine type	Disease	General characteristics (see text for details)
WHOLE ORGANISM VACCINES		
Live, attenuated	Influenza (intranasal vaccine) Measles Mumps Polio (Sabin vaccine) Rotavirus Rubella Tuberculosis Varicella Yellow fever	High levels of protective serum IgG antibodies, mucosal IgG tends to be lower. Mucosal IgA significant with polio vaccine. CTL responses likely due to *de novo* synthesis of proteins, prolonged stimulation due to replication of vaccine. Stimulate innate immune receptors, typically one or two boosters required. Extensive control of disease. Risk of reversion.
Inactivated/ killed	Influenza (killed/subunit) Hepatitis A Polio (Salk vaccine) Rabies	High levels of protective antibodies, reduced or absent CTL responses due to absence of de novo protein synthesis. Seasonal influenza vaccine reformulated for different viral strains. Less immunogenic than live vaccines, more boosters required than live vaccines, stable and no risk of reversion.
SUBUNIT VACCINES		
Polysaccharide/ conjugate	Bacterial meningitis *(Haemophilus influenzae* type B, *Neisseria Meningitidis* serogroups A, C, W-135, Y) Streptococcal pneumonia	Use isolated surface polysaccharides as main target for opsonising antibodies. Immune responses enhanced significantly by conjugation to carrier proteins that provide TH cell responses to increase IgG production and B cell memory. Multiple boosters required. Tetravalent vaccine used for *N. meningitides* protects against four serogroups. Multivalent vaccines available for *S. pneumoniae* disease.
Toxoid	Diphtheria Tetanus pertussis	Inactivated exotoxins (toxoids) or toxoid with other subunits (pertussis), protective antibody responses, multiple boosters, may require later boosters. Good disease control.
Recombinant proteins	Hepatitis B HPV-associated cancers Meningitis B	Stimulate neutralising antibodies to prevent viral entry into host cells. Recombinant proteins form virus-like particles (VLPs). HepB in use for three decades, good safety profile and disease control.

Table 17.1 (*Continued*).

Vaccine type	Disease	General characteristics (see text for details)
		Recently approved multivalent vaccine, initial data suggest good safety and efficacy.
EXPERIMENTAL VACCINE DESIGNS IN DEVELOPMENT		
Plasmid DNA vaccines Recombinant vector vaccines		Heterologous proteins expressed in host, several pathogen targets under study. *In vivo* expression increases likelihood of cytosolic processing for CTL responses. Good safety profiles in human humans and other species. Immunogenicity an issue with plasmid DNA vaccines, but several adjuvant systems under review. Live Bacterial and viral-based vector vaccines mimic live, attenuated vaccines. Several in development.

17.3.1   Live Vaccines

The success of live vaccines in the control of several infectious diseases is one of the great achievements of medical science. These types of vaccines have been particularly successful against acute viral infections as seen in the global eradication of smallpox, near-total eradication of polio and the extensive control of measles, mumps and rubella. These vaccines exploit viral strains which closely resemble the target pathogen but have reduced pathogenicity in the human host. In most cases live viral vaccines use attenuated strains of the pathogen, for example, the Sabin polio vaccine, measles, mumps, rubella (MMR) vaccine, rotavirus vaccine, the yellow fever vaccine and some influenza vaccines. One notable exception is the smallpox vaccine which uses non-attenuated, vaccinia virus. Vaccinia is a member of the *Poxviridae* family and closely related to the variola virus, the causative agent of smallpox, but is much less virulent. Due to its success as a vaccine, vaccinia is being used to develop recombinant vaccinia vector vaccines, discussed later in this chapter. Live vaccines mimic the target pathogen in several key ways. They possess shared epitopes with the pathogen which elicit specific immune responses that 'cross-react' with the pathogen and importantly interact with host innate receptors and are processed by host cells in a similar manner to the pathogen. Upon immunisation, live viral vaccines replicate in host cells and are disseminated through the lymphatics. Specific B cells bind epitopes on the viral particles and proteins on the virus are processed through the endocytic processing pathway and presented to TH cells. Once activated, the TH cells drive the antigen-primed B cells into antibody secreting plasma cells and memory B cells. Furthermore, newly synthesised viral proteins are processed through the cytosolic processing pathway leading to CTL responses. The replicative ability of these vaccines also provides prolonged stimulation of the host immune system and as a result live vaccines require and a single booster. Live vaccines induce long-term immunity. Antibody and CTL responses to the vaccinia vaccine can be detected decades after vaccination, in the absence of any exposure to the wildtype smallpox virus.[7]

Viral attenuation is achieved typically by successive passage in different cells from the normal host cells and the viral progeny screened for immunogenicity and reduced pathogenicity. Measles virus was first isolated from primary cultures of kidney cells from an infected patient and subsequently passaged through several different cell types to produce the Edmonston B strain, the first live, attenuated measles vaccine. All the live, attenuated measles vaccine strains in current use are derived by further passage of the Edmonston B strain. The unattenuated vaccinia vaccine is unsuitable for use in immunocompromised individuals. Consequently, an attenuated vaccinia called modified vaccinia virus Ankara (MVA) was produced by over 500 passages in chicken embryo fibroblasts.[8] The BCG vaccine against tuberculosis was first developed by culturing *Mycobacterium bovis* for 13 years under abnormal conditions, resulting in the attenuated vaccine preparation. Since then BCG has undergone further extensive attenuations BCG is not a single strain preparation but rather a mixture of genotypically and phenotypically different strains of *M. bovis*. This may contribute to its variable efficacy in controlling tuberculosis.

These attenuation strategies are laborious and time-consuming. As more pathogen genomes are sequenced and analysed *in silico*, virulence factors can be identified and targeted for attenuation.

17.3.2 Inactivated (Killed) Vaccines

An alternative means of making a pathogen safe for vaccination is to inactivate the pathogen by heat or chemical treatment. This approach is used where there is no safe live strain available. Several inactivated vaccines have been licensed for use in humans to prevent several infectious diseases including cholera, influenza, rabies, polio and hepatitis A infection. The Salk polio vaccine, the first polio vaccine to be developed and widely used in several countries, was produced by inactivating polio virus strains with formaldehyde. Heat inactivation is less used as it causes extensive denaturing of proteins which may alter key epitopes, in particular destruction of conformational epitopes. In this regard chemical inactivation is preferred as the process can be more controlled. Inactivated vaccines induce strong antibody responses coupled with TH responses, however the inability of these vaccines to replicate inside the host imposes some limitations. As there is no *de novo* synthesis of proteins, there is limited or no cytosolic processing resulting in diminished CTL responses compared with live vaccines. Such an approach is not likely to succeed where the pathogen establishes chronic, intracellular infection, although they may prove useful as part of heterologous prime-boost vaccination strategies. The absence of replication also reduces the period of immune-stimulation, so more boosters are required. The advantages of inactivated vaccines compared with live vaccines are that they are more temperature-stable and there is no risk of reversion to virulence.

17.3.3 Subunit Vaccines

Subunit vaccines are composed of antigenic macromolecules derived from pathogenic organisms. Several subunits have been licensed for use in humans and other animals and have proved highly effective in controlling infectious diseases. The most common subunit designs are based on isolated bacterial capsular polysaccharides, inactivated bacterial exotoxins (toxoids) and recombinant proteins. These vaccines overcome several issues that exclude the use of whole-organism vaccines, for example the absence of suitably attenuated strains. In the case of some pathogenic Gram-negative bacteria inactivated vaccines are not suitable due to the high levels of lipo-polysaccharide (LPS) endotoxin.

17.3.3.1 Polysaccharide and Conjugate Vaccines. Polysaccharide vaccines use isolated capsular polysaccharides derived from cultures of the pathogen. This approach has proved successful against the Gram-negative meningococcal pathogens, namely *Haemophilus influenzae* type b (Hib) and *Neisseria meningitides* (which cause life-threatening meninigitis. These pathogens possess anti-phagocytic properties due mainly to their hydrophilic polysaccharide capsules. The *N. meningitides* vaccine MCV4, is a quadrivalent preparation containing polysaccharides from four different serotypes, serogroups A, C, W-135 and Y (the serotype B recombinant protein vaccine is described later). The immunological basis for these vaccines is that they induce opsonising antibodies against the capsular polysaccharides, in turn promoting phagocytic clearance of the pathogens. One key limitation of these vaccines is that polysaccharides are T-independent antigens. Due to their polymeric structure, the polysaccharides are able to induce B cell responses by cross-linking antigen receptors, *i.e.* the membrane-bound immunoglobulins, but this results in limited production of IgG molecules and little or no memory cell production, both of which are dependent on TH cytokine signals. This is overcome by conjugating the polysaccharide by reductive amination to a carrier protein to form a conjugate vaccine, which provides the peptides for TH cell activation. The carrier protein most commonly used is tetanus toxoid, as is used in the tetanus vaccine (see below). This strategy exploits TH responses to the tetanus toxoid vaccine. Polysaccharide-based vaccines are also used against *Streptoccocus pneumoniae*, a respiratory pathogen responsible for many deaths and cases of respiratory disease globally. There are over 90 serotypes of the pathogen. One vaccine preparation contains 23 different polysaccharides covering a range of strains. Three conjugate vaccines are also available; Prevnar, containing seven polysaccharides covering the serotypes 4, 6B, 9V, 14, 18C, 19F and 23F conjugated to CRM197, a nontoxic, recombinant form of diphtheria toxin. Prevnar 13 which contains 13 different polysaccharides expanding the range of protection and Synflorix, a decavalent vaccine which uses a carrier protein from non-pathogenic strains of *H. influenzae.*

17.3.3.2 Toxoid Vaccines. Anti-toxin antibodies provide effective immunity against infectious diseases where bacterial exotoxins are the primary pathogenic feature. Toxoid vaccines contain chemically inactivated exotoxins (toxoids) adsorbed onto an alum adjuvant. The toxin is isolated and extensively purified from cultures of the pathogen. This strategy has proved highly effective against tetanus and diphtheria. *Clostridium tetani,* the causative agent of tetanus, is an anaerobic, Gram-positive organism transmitted through open wounds. Tetanus is due to the action of a powerful exotoxin, tetanospasmin, encoded on a plasmid and expressed under anaerobic conditions. The toxin is a potent neurotoxin that enters the CNS *via* neuromuscular synapses. The toxoid vaccine induces high protective levels of neutralising antibodies and has proved successful in controlling tetanus in many countries where the vaccine is readily available. However, tetanus, in particular neonatal tetanus, is still a major cause of death in several economically underdeveloped countries where the vaccine is not readily affordable. Anti-tetanus toxin antibodies pooled form the sera of immune individuals is also used as a passive immunising agent where there has been a risk of exposure to the pathogen.

Corynebacterium diphtheriae, a Gram-positive, facultative anaerobe, is the causative agent of diphtheria. The pathogen is spread in aerosols and is associated with close contact with infected individuals and contaminated surfaces. The pathogen usually colonises the upper respiratory tract and many infected individuals remain asymptomatic. Diphtheria toxin is the main virulence factor and, unusually, is encoded on a bacteriophage. The toxin inhibits protein synthesis by inactivation of elongation factor-2 (EF-2). As with the tetanus, the toxoid vaccine induces high levels of neutralising IgG antibodies. Toxoid vaccines activate T$_H$ cells which drive the antibody response. Both toxoid vaccines are often administered with the pertussis vaccine as a combined vaccine (DTP), which contains inactivated pertussis toxin and isolated macromolecules from *Bordetella pertussis.*

17.3.3.3 Virus-Like Particle (VLP) Vaccines: The Advent of Recombinant Protein Vaccines. Subunit vaccines based on recombinant viral proteins came to prominence in the 1980's. Sequencing of the HBV genome[9,10] was followed closely by the introduction of the Hepatitis B virus (HBV) vaccine, the first recombinant protein vaccine. The gene encoding the HBV surface antigen (HBsAg) was cloned into a plasmid and expressed in yeast cells.[11,12] The extensively-purified recombinant HBsAg proteins spontaneously assemble into three-dimensional virus-like particles (VLPs). VLPs are highly geometric structures formed by spontaneous assembly of viral capsid proteins, following expression in eukaryotic cells, and range in size from 20–100 nm in diameter.[13] The particulate structure permits uptake by DCs. The recombinant HBsAg protein used in the HBV vaccine assembles into a VLP structure of 22 nm in diameter. The vaccine has proven safety and efficacy and has been at the centre of vaccination against HBV infection for 30 years. Protective immunity is achieved mainly through inducing high levels of protective antibodies which prevent viral entry into hepatocytes. In common with other subunit vaccines, two to three boosters are required to achieve efficacy. The vaccine is often viewed as the first 'anti-cancer vaccine' as HBV is one of the major causes of hepatic cancer. Similarly, VLP-based vaccines against human papilloma virus (HPV) were introduced to prevent cervical cancer. HPVs are a large family of over 100 types of which HPV 16 and HPV18 are the most common causes of cervical cancer. The VLP-based vaccines contain VLPs of HPV antigens reassembled from the L1 capsid proteins of HPV. Two VLP-based vaccines are currently in use in humans. A bivalent vaccine containing VLP antigens for HPV 16 and HPV 18 reassembled from L1 proteins of HPV16 and HPV18,[14] and a quadrivalent vaccine containing VLP antigens for HPV6, HPV11, HPV16 and HPV18 reassembled from L1 proteins of HPV6, HPV11, HPV 16 and HPV18.[15] The bivalent vaccine uses a baculovirus expression system for expression of the recombinant HPV antigens whereas the quadrivalent vaccine was produced in yeast.

Chimaeric VLPs expressing proteins from other viruses than the VLP structural proteins, offer a platform for presenting heterologous antigens from other viruses on the same VLP structure, through conjugating epitopes to the VLP structure or creating structures from fusion proteins. VLPs produced following fusion of the HBV core gene and the HIV Tat protein transduction domain were taken up by DCs and induced HIV-specific CTL activity indicating that the HIV protein was effectively expressed on the core HBV structure.[16] The HBV core structure has been shown to express epitopes derived from malarial antigens of the circumsporozoite protein (CSP) of *Plasmodium falciparum*.[17] This chimaeric VLP approach has significant potential and several candidate vaccines using chimaeric VLPs derived from a range of both plant and animal viruses are being investigated.[18] The empty core structure of VLPs and their uptake by DCs hold potential for their use as delivery vehicles for transfer plasmid DNA into DCs.[19,20]

17.3.3.4 The Recombinant Protein Subunit Vaccine Against Serogroup B Neisseria meningitidis: *The First Genome-Based Vaccine Design.* Serogroup B *Neisseria meningitides* (MenB), a Gram-negative, capsulated bacterium, is one of several serotypes of *N. meningitidis* that can cause life-threatening meningitis and sepsis. Effective vaccines against four other *N. meningitidis* serogroups (A C, Y and W-135), described in a previous section, are based on isolated polysaccharides. However, this vaccine strategy has not proved a viable option for serogroup B *N. meningitidis* disease as the polysaccharide capsule of Men B closely resembles polysialic acid (*N*-acetyl neuraminic acid) that is widely distributed on human cells. Protein subunit vaccines against MenB have been developed that use isolated proteins from the outer membrane vesicles (OMVs) and have proved effective in controlling strain-specific MenB epidemics by inducing protective antibodies against surface-exposed proteins in the OMVs.[21] The major limitation of this approach is the antigenic variability in the OMV surface proteins, thus protective antibodies are strain-specific and do not confer effective immunity against other strains expressing variants of the OMV proteins. The novel MenB vaccine was developed using a genome-based approach referred to as reverse vaccinology, to identify putative protein vaccine candidates.[22] The vaccine is composed of recombinant surface proteins adsorbed onto alum and an OMV from the New Zealand strain of *N. meningitides*. The strategy employed in the development of this vaccine is described in detail below.

17.4   REVERSE VACCINOLOGY: GENOME-BASED VACCINE DESIGN

Just as recombinant DNA technology marked a shift in vaccine design with the introduction of VLP-based vaccines, rapid advances in genomics and proteomics technologies have revolutionised vaccine design strategies. In contrast to classical vaccinology which involved the culture and analysis of the pathogen, genome-based vaccinology starts with *in silico* analysis of the genetic information of the pathogen. This approach has been named reverse vaccinology (RV).[22] In essence, RV uses bioinformatics tools to mine the genome of a pathogen to identify open reading frames (ORFs). These are then analysed to predict protein vaccine candidates based on their likely cellular location, for example, certain sequences indicate the likelihood of a protein being expressed on the surface of the pathogen and therefore may be a suitable for inducing protective antibodies. Identification of candidate proteins is followed by high expression cloning in *E. coli* and the recombinant proteins evaluated, *in vivo,* for immunogenicity and protective efficacy against the wild-type pathogen. The development of the meningococcus B vaccine is used below to illustrate the RV strategy (Figure 17.2).

17.4.1   Reverse Vaccinology Approach to Vaccine Design: The Development of the Serogroup B Meningococcus Vaccine

The RV approach to developing a protein-based vaccine against MenB began with whole-genome sequencing of MC58, a virulent strain of serogroup B *N. meningitides.*[23] *In silico* analysis of unassembled DNA fragments of the MC58 genome identified 570 open reading frames (ORFs) of interest.[24] These ORFs were amplified by polymerase chain reaction(PCR)and cloned in

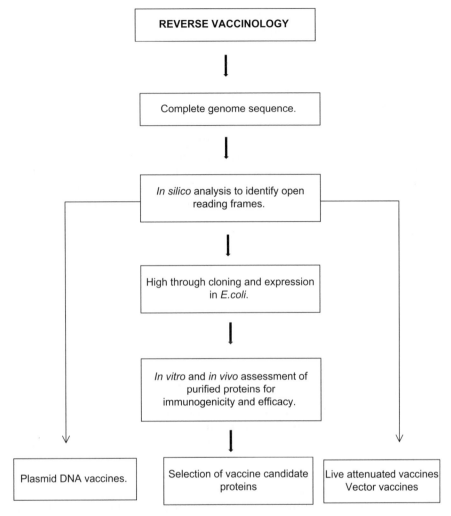

Figure 17.2   Genome-based vaccine design: reverse vaccinology approach to identifying vaccine candidate proteins. In silico analysis can also be applied
for identifying sequences for plasmid DNA vaccines and live vector
vaccines. Identification of virulence genes can lead to live, attenuated
vaccines. Comparative and pan-genomics can help predict vaccine
candidate proteins specific for selected pathogenic strains, and candidate proteins for universal vaccines against multiple strains of a
pathogen.

E. coli. The polypeptides were expressed as either His-tagged or glutathione
S-transferase (GST) fusion proteins to enhance expression and permit ease of
purification. In total, 350 recombinant proteins were produced and tested for
immunogenicity in mice. Immune sera were screened by Western blotting,
enzyme-linked immunosorbent assay (ELISA) and flow cytometry to detect

antibodies to surface-expressed proteins on a range of MenB strains. Finally, sera were screened for complement-mediated bactericidal activity (BCA) which correlates with protective efficacy in humans. On the basis of these studies, seven proteins, referred to as genome-derived *Neisseria* antigens (GNAs) were selected for further studies. Based on BCA data and protective efficacy, three antigens were selected for vaccine development.[25] These included GNA 2132, named Neisseria Heparin binding antigen (NHBA), GNA 1870, a complement factor H binding protein (fHbp) and GNA1994, the Neisseria adhesin protein A (NadA). Two additional antigens, GNA 1030 and GNA 2091, were selected as fusion partners for GNA 2132 (NHBA) and GNA 1870 (fHbp), respectively. This work led to the development of the four component protein subunit MenB vaccine (4CMenB), approved in 2013. The vaccine is composed of three recombinant proteins; the two fusion proteins, NHBA-GNA 1030 and fHbp-GNA 2091, and the individually expressed NadA, each adsorbed onto aluminium hydroxide (alum), and OMVs from the strain used for the New Zealand OMV vaccine. Initial clinical trials data indicate the vaccine is safe and induces protection against several strains of MenB.[26,27]

17.4.2 Pan-Genomics and Comparative Genomics

The RV approach to the MenB vaccine described above, was based on analysis of a single genome. One of the difficulties in developing universal vaccines against pathogens is the strain variability associated with key antigenic determinants. Pan-genomics is the analysis of strains within a single species by direct sequencing or comparative genome hybridization to identify conserved antigen sequences which can form the basis of a vaccine against all strains within the species. This approach has identified four major pili antigens suitable for vaccine development against the group B streptococcus, *S. agalactiae*, a member of the group B streptococci (GBS) and a leading cause of neonatal sepsis.[28] Similarly, comparative genomics involves screening of multiple genomes of pathogenic strains, but includes non-pathogenic commensal strains, to identify antigens that are present in pathogenic strains of *E. coli* but not in commensal organisms.[29] When used to immunise mice, a combination of these pathogen-associated proteins protected mice against sepsis.[30]

Reverse vaccinology may provide other important information useful to understanding the mechanisms of pathogenicity, for example, the identification of the factor H binding protein (fHbp) used in the MenB vaccine described above. In addition to being a suitable vaccine antigen, identifying its ability to disrupt the complement system helps in our understanding of how the pathogen survives in the host. Furthermore, fHbp also shows significant antigenic variation among strains of MenB which has led to development of a potentially 'universal' vaccine for MenB, using an engineered protein expressing antigenic determinants for the major fHbp variants.[31]

17.4.3 Bioinformatics Tools for Identification of B and T cell Epitopes

Several bioinformatics/immunoinformatics tools have been developed for prediction of epitopes from genomic and proteomic data. In the development of the MenB vaccine, the process began with analysis of the MenB genome using BLASTX from the National Centre for Biotechnology Information (NCBI) to identify ORFs and exclude those associated with cytoplasmic functions, as potential B cell epitopes are associated with surface proteins. Most of the tools available, such as PSORT (www.psort.org) allow prediction of these epitopes based on various characteristics, for example, transmembrane domains, anchoring motifs, leader sequences and lipoprotein signals. Similarly, Vaxign (www.violinet.org/vaxign)[32] and NERVE (new enhanced reverse vaccinology environment),[33] which were designed to complement the RV approach, predict potential vaccine candidates using similar criteria, and also permit exclusion of sequences homologous to host proteins so minimises the chances of generating autoantibodies. These tools also allow prediction of MHC class II-restricted T cell epitopes. A recently described webserver, Jenner-Predict, expands the search range to include known functional domains associated with pathogen–host and virulence factors, for example, invasion, porin, flagellin, toxins, choline-binding, penicillin-binding, transferrin-binding, fibronectin-binding and solute binding.[34] This server also runs comparisons with known epitopes from the Immune epitope database (IEDB, www.iedb.org), a manually curated database of published epitope sequences, for predicting immunogenicity and autoimmunity.[35] IEDB offers an extensive range of tools for predicting B and T cell epitopes. Another webserver, CBTOPE, allows prediction of conformational epitopes, which make up a larger part of the B cell epitope repertoire, using linear peptide sequences, without prior knowledge of the tertiary structure.[36]

Several dedicated tools are available for predicting potential CTL epitopes based on MHC class I binding, proteasomal processing and immunogenicity (see IEDB, www.iebd.org) and its companion site (http://tools.immuneepitope.org). NetChop, for example, predicts peptides generated from proteasomal processing of a protein, whereas NetCTL predicts T cell epitopes along a protein sequence. NetCTL was used to predict several putative MHC class I-restricted vaccine epitopes for West Nile Virus.[37]

17.5 DNA VACCINES

The potential use of plasmid DNA as vaccines was realised in the 1990's by the demonstration that, firstly, a marker protein encoded in a simple plasmid DNA could be expressed in mammalian cells.[38] Subsequently, it was demonstrated that plasmids encoding viral proteins elicited protective antibody and CTL responses.[39,40] The basic principle underlying DNA vaccines is that the antigenic proteins of interest are expressed *in vivo* in the cells of the immunised individual. These proteins are then processed through the cytosolic pathway within the host cells leading to MHC class I presentation to CTLs. Secreted proteins can be taken up by dendritic cells and processed for class II presentation to TH cells. DCs may also cross-present these proteins to both TH cells and CTLs. Although no DNA vaccines have been approved for use in humans, four vaccines have been approved for use in veterinary medicine. These include vaccines against West Nile virus (WNV) in horses,[41] infectious haematopoietic necrosis virus in farmed salmon,[42] a therapeutic vaccine for melanoma in dogs[43] and a vaccine encoding growth hormone releasing hormone (GHRH) for preventing foetal loss in pigs.[44] Plasmid DNA offers several advantages over conventional live vaccines in that they are non-replicating, cannot be transmitted and are unlikely to change to a pathogenic state—a risk with live, attenuated vaccines.

Several studies have generated a great deal of information on the safety, immunogenicity, design and formulation of plasmid DNA vaccines.[45–48] Clinical trials have shown that these preparations are well-tolerated and safe for use in humans.[49] A number of plasmid DNA vaccines are in development and have been clinically tested in humans and other species against both infectious pathogens and cancers.[50–54] Although they have an excellent safety profile in humans and show great potential, the immunogenicity and protective efficacy of DNA vaccines is low. The main strategies to improve these factors include modification of the plasmid, use of adjuvants and targeting of APCs, in particular dendritic cells.

17.5.1 Design of Plasmid DNA Vaccines

DNA vaccines consist of bacterial plasmid DNA containing a eukaryotic expression cassette and an open reading frame (ORF) with multiple cloning sites containing the desired antigenic protein. The bacterial region contains a replication of origin site and a selection marker. The eukaryotic expression cassette contains a powerful promoter, typically the cytomegalovirus (CMV) with the associated intron A (CMVIntA) placed upstream of the transgene, which directs transcription of mRNA, and a downstream polyadenylation signal usually from bovine growth hormone, for export of mRNA to the cytoplasm.

Several modifications have been made to the basic design outlined above to improve the immunogenicity, efficacy and safety of plasmid DNA vaccines.[45,55,56] Protein expression may be enhanced by inclusion of a Kozak consensus sequence—gccgccRcc**ATG**G (R may be an A or G, the underlined ATG is the transgene start codon)—immediately upstream of the transgene in the 5′ untranslated region (UTR). Any start codons upstream of the Kozak sequence should be removed. As expression of several genes is dependent on splicing, the inclusion of an intron within the 5′ UTR may enhance transgene expression.[45,57] In addition, modifying the 5′ UTR immediately downstream of the CMV promoter to include the human T cell leukaemia virus type I R region (HTLV-I R) can enhance transgene expression.

Due to differences in tRNA pools between species, optimising codons for the transgene by removal of inhibitory elements to suit expression in host cells has been shown to increase immune responses to Lassa virus antigens,[58] mycobacterial antigens,[59] and HIV antigens.[60] The latter study also demonstrated similar benefits from optimising RNA sequences. Furthermore, enhancing promoter efficiency and modifying secretory leader sequences alongside codon optimisation has produced synergistic enhancement of the response to a DNA vaccine encoding HIV proteins.[61] Inserting additional genes encoding immunomodulatory proteins, in particular cytokines, or coadministered on a separate plasmid, can increase the response to DNA vaccines. Several cytokines have been examined in this context with promising results observed in non-human primates, using a co-plasmid encoding the TH1 cytokines, Il-12 and Il-15.[62] Co-administration of a plasmid encoding Il-15 with a HIV DNA vaccine followed by boosting with a vaccinia-based HIV vaccine has been shown to increase the frequency of memory CD8+ cells compared to responses without the Il-15 plasmid.[63] Another modification is to include CpG motifs on the plasmid backbone. CpGs bind to the intracellular toll-like receptor TLR9, and have been shown to increase the immune response when administered with proteins.[64] However, results were more variable with DNA vaccines and require more investigation. The contribution of CpGs may be secondary to the plasmid itself as the double-stranded DNA structure has been shown to activate TLR-associated signalling pathways through activation of the TANK-binding (TBK-1) kinase even in the absence of TLR9.[65] This immunostimulatory effect is due to recognition of plasmid DNA through the stimulator of interferon genes–TBK-1 (STING–TBK-1) signalling cascade.[66,67]

The immune response to DNA vaccines can be enhanced by delivery using a ballistic gene gun approach, rather than through traditional intramuscular or intradermal injection.[68] Microporation, using hundreds of microneedles carrying the plasmid, also improves the immune response.[69] Electroporation is another method for vaccine delivery that may also increase immunogenicity of DNA vaccines.[70]

17.6 LIVE RECOMBINANT VECTOR VACCINES

Live vaccines have been shown to be highly effective in the control of several infectious diseases, as described earlier in this chapter. The success of these vaccines is due to their ability to replicate inside the host and activate both antibody and cellular immunity. An additional benefit is that vector vaccines may induce immunity to both the target pathogen and the pathogen from which the vector was derived. Although this approach has been limited by the availability of suitably attenuated strains, their ability to infect host cells has encouraged their use as vectors to deliver antigenic genes from other pathogens. Once injected into the host, recombinant vector vaccines enter host cells where the proteins encoded by the transgene are expressed and presented to B and T cells. Several live vector vaccines using attenuated viruses and bacteria are in development against a range of infectious pathogens and cancers.

17.6.1 Live Recombinant Bacterial Vector Vaccines

Bacterial vectors possess the intrinsic immunogenicity of the vector organism with genomic capacity to accept several transgenes. The use of bacteria as vectors has focused on intracellular bacteria as these provide opportunities for intracellular processing of the antigenic protein of interest. Due to the paucity of live bacterial vaccines approved for use in humans, work on bacterial vectors has focused mainly on BCG and attenuated *Salmonella enterica*, serovar Typhimurium. Both infect host cells are taken up by dendritic cells and enter secondary lymphoid tissues.

17.6.1.1 Recombinant BCG Vector Vaccines. The potential of recombinant BCG (rBCG) vectors was first demonstrated in the late 20th century with the use of shuttle vectors to introduce the candidate transgene.[71,72] Since then rBCG vectors have been used to express pathogenic genes from a range of pathogens although the immune responses have been variable.[73,74] Several shuttle vectors have been used for transgene expression in rBCG vectors. Typically these are episomal plasmids combining a mycobacterial replicon derived from the naturally occurring *Mycobacterium fortuitum* plasmid, pAL5000 with an *E. coli* cloning vector. Site-specific integrating vectors under the control of mycobacterial heat shock proteins or inducible promoters, have been developed from mycobacteriophages, which enable integration between the *attP* phage and *attB* mycobacterial attachment sites. Protective immunity against rotavirus was observed in mice following administration of rBCG vector expressing the rotavirus antigen VP6 using an integrating vector under a hsp60 promoter.[75] By contrast, using an episomal vector with the same promoter proved unstable. However, the episomal vector proved protective when the VP6 sequence was linked to the mycobacterial 19 kDa lipoprotein sequence.[76] An rBCG (AFRO-1) expressing perfringolysin has been described.[77] AFRO-1 elicited greater T cell responses than BCG in rhesus macaques, when used in a heterologous prime-boost strategy with a recombinant adenovirus vector booster expressing fusion protein composed of mycobacterial antigens. Expression of perfringolysin in AFRO-1-infected cells permits greater endosomal escape and cytosolic localisation of heterologous proteins compared to BCG. Incorporation of GM-CSF sequences into BCG have been shown to increase cellular expression of mycobacterial antigens and increased protection against both pulmonary tuberculosis and disseminated tuberculosis.[78] The use of cytokine adjuvants encoded in the shuttle vector may increase immunogenicity and thus efficacy of rBCG vectors.

17.6.1.2 Live Recombinant Salmonella Vaccines. As a facultative intracellular pathogen with ability to survive in the gastrointestinal tract,

Salmonella has the potential as a vehicle to deliver heterologous antigens by oral immunisation. Its potential as a live vaccine delivery system has been extensively studied.[79]

Salmonella genus comprises two species *S. enterica* and *S. bongori*, of which *S. enterica* is the main pathogen associated with diseases in humans and other animal species. There are over 2000 serovars of *S. enterica* displaying host-specificity. *S. enterica* serovar Typhi is a human-restricted pathogen responsible for typhoid fever, whereas *S. enterica*, serovar Typhimurium causes typhoid-like disease in mice but has less virulence in humans.

The virulence of *Salmonella* is associated with a complex secretory system known as the type 3 secretion system (T3SS), and found in several other Gram-negative bacterial pathogens. These systems permit the translocation of *Salmonella* effector proteins into the host cell cytosol and are the main mechanism of *Salmonella* pathogenesis. The T3SS involves a multi-component needle-like apparatus, known as the injectisome, which permits delivery of the effector proteins into the host cell. In Salmonella, the virulence genes encoding the T3SS proteins are located in two main *Salmonella* pathogenicity islands (SPIs), SPI1 and SPI2. These give rise to two groups of T3SS effector proteins which play different roles in Salmonella virulence. The SPI1 system (SPI1-T3SS) encodes effector proteins that facilitate invasion of non-phagocytic cells of the gastrointestinal epithelium. Delivery of SPI1 effector proteins inducing membrane-ruffling in the host cells resulting in formation of a modified phagosome known as the Salmonella-containing vacuole (SCV).[80] SPI1 is not required for entry into conventional phagocytes such as macrophages and DCs. The SPI2 system (SPI2-T3SS) is activated intracellularly and exports effector proteins into the host cell cytosol promoting bacterial replication and dissemination. The ability of T3SSs to deliver heterologous antigenic proteins into the host cell cytosol is the primary focus in developing vaccines based on recombinant attenuated *Salmonella* vaccines (RASVs). The potential to deliver heterologous antigens using Salmonella as a vector was first demonstrated using SPI1-T3SS.[81] However, as the SPI1 system is active primarily in the extracellular phase of infection, most focus has been on the SPI2 system which is activated during the intracellular phase and also is active in APCs.[82]

The ability of several potential vaccine candidates have been investigated using live RASVs based mainly on the serovar Typhimurium.[79] One approach is to express heterologous proteins as fusion proteins. The heterologous antigenic gene sequence is fused with a sequence for a Salmonella effector protein as a delivery system, under the control of the SPI2 promoter, and the encoding plasmid is expressed in Salmonella within the host cell. The SPI2 effector protein SseF is able to translocate heterologous fusion proteins containing antigens of Listeria monocytogenes into the cytosol of host cells and elicit protective immune responses in mice both using plasmid-based expression cassettes and chromosomal-expression cassettes.[83] Further studies using several SPI2 effector proteins, SifA, SteC, SseF, SseJ and SseF, demonstrated effective translocation of fusion proteins into the host cell cytosol, however, *in vivo* only SseJ or SifA fusion proteins provoked significant T cell responses.[84] This indicates that intracellular events, post immunisation, differ depending on the effector protein used. Thus the choice of effector protein partner and the choice of promoter is an important factor in designing effective RASVs. Using *Salmonella* to express recombinant pneumococcal antigens, PspA and PspC, using different signal sequences from a type 2 secretion (T2SS) have also shown that RASVs can elicit protective immunity in mice to *Streptococcus pneumoniae*.[85] However, the level of protection varied significantly depending on the signal sequence used. In general, the immune response is dependent on the amount of antigen produced. However, this needs to be balanced against the effect of high level expression which can deplete nutrients and the potential toxicity of the antigen which may prematurely kill the RASV used. Recently, RASVs have been developed that have delayed antigen synthesis.[86] These utilize the strong LacI-repressible P_{trc} promoter and attenuated strains of serovar Typhimurium that synthesise different levels of LacI under transcriptional control of an arabinose-regulated promoter. In the absence of arabinose, LacI is produced which binds to and inhibits the Ptrc promoter thus blocking antigen synthesis. *In vivo*, the arabinose-poor environment causes a progressive decrease in LacI permitting increased antigen synthesis.[86] Ongoing modifications to RASVs holds the promise of extensively regulated RASVs that are safe for use and of sufficient immunogenicity as vector vaccines.[87]

17.6.2 Recombinant Viral Vector Vaccines

17.6.2.1 Live Recombinant Vaccinia Vector Vaccines. The success of the vaccinia viral vaccine has made poxviruses the most common candidates for use as a vectors for vaccine development. Vaccinia virus has a linear, double-stranded DNA genome of approximately 190 kbp and the commonly used attenuated canarypox virus ALVAC, has a genome of approximately 330 kbp. Recombinant vaccinia vectors (rVVs) have been extensively tested as vaccines against a range of infectious diseases[88,89] and cancers.[90–92] Due to the intrinsic immunogenicity of vaccinia virus it is potentially harmful to the aging population and immunocompromised individuals so its use as a vector has been confined mainly to the attenuated modified vaccinia virus Ankara (MVA). MVA possess the necessary infectivity and ability to express transgenes but cannot produce infectious viral particles and has an excellent safety record in human trials.[93] MVA is able to infect isolated DCs and induce cross-presentation of the heterologous proteins.[94–97] CTL responses *in vivo* are predominantly due to cross-priming, with stable proteins being favoured over preprocessed peptides.[98] A recent study described a recombinant vaccinia vector vaccine encoding multiple T cell epitopes which elicited strong T cell responses in mice.[99] This study used an rV vector encoding the influenza nucleoprotein(NP) containing T cell epitopes from other influenza proteins. The rV vector was used as a booster following priming with plasmid DNA encoding the same influenza antigens.

Both MVA and non-attenuated vaccinia induce neutralising antibodies which limits the number of immunisations. That said, the use of adjuvants may increase efficacy while limiting the number of immunisations required nor does it reduce their potential as a component of heterologous prime-boost immunisation regimens. An alternative is the use of attenuated canarypox virus which does not induce neutralising antibodies in humans. Although not as intrinsically immunogenic in humans as vaccinia or MVA, the canarypox virus vector vaccine ALVAC has a shown real potential in a HIV/AIDS clinical trial using ALVAC as a priming dose followed by a protein-based vaccine (AIDSVAX) in a heterologous prime-boost regimen.[100] This was the first HIV vaccine trial to show significant efficacy in humans.

17.6.2.2 Live Recombinant Adenovirus Vector Vaccines. The Adenoviridae are a large, ubiquitous group of viruses composed of at least 55 serotypes which infect humans (Had1-55). They are most commonly associated with respiratory infections but some species may cause conjunctivitis and gastro-enteritis. Adenoviruses are large, non-enveloped viruses, 90–100 nm in diameter with an icosahedral structure and a dsDNA genome ranging in size from 26–45 kbp and infect most animal cell types. Viral replication occurs in the nucleus of infected cells but is extrachromosomal, reducing the risk of mutational insertion when used as vectors. Attenuated Had7 has been used for many years to immunise US military personnel and has a good safety record. Adenovirus is stable and easily manipulated. Replication-incompetent vectors are generated by deletion of the E1 gene. E1 is an essential protein for viral replication upregulates transcription by activation of early adenovirus promoters and drives host cells into S-phase. As with other recombinant viral vectors, recombinant adenovirus vectors (rAdVs) are produced by homologous recombination. The deletion of E1 and E3 viral genes, the latter is not essential for replication, can accept transgene increases the capacity to accept transgenes. HAd-5 vectors encoding HIV-Gag proteins or Ebola virus antigens, have induced protective immunity in non-human primates.[101,102] Adenoviral vectors encoding hepatitis C virus (HCV) antigens, NS4, NS5a and NS5b can induce expression in DCs resulting in significant HCV-specific responses in cultured, autologous T cells.[103]

One potential barrier to the success of rAdVs is the presence of pre-existing neutralising antibodies and memory cells to adenovirus as a result of the prevalence of adenoviruses. This is the likely reason for the withdrawal of a possible rADV –based HIV-1 vaccine following clinical trials, despite encouraging pre-clinical data. A recent study showed significant levels of pre-existing, neutralising antibodies in humans to different serotypes of ade-novirus.[104] By comparison, levels of neutralising antibodies to chimpanzee adenovirus were lower. A study involving the chimpanzee-based adenoviral vector vaccine, ChAdY25 showed very low levels of neutralising antibodies in human sera to the vaccine and the vaccine elicits cellular immune responses.[105] These findings, and other work using bovine adenovirus (BAdV)-based vectors support the view that rAdVs based on non-human adenovirus offer a viable system for vaccine delivery.[106–108]

17.7   VACCINE ADJUVANTS AND TARGETING DENDRITIC CELLS

Adjuvants are used in order to increase the immunostimulatory properties of vaccines. This is achieved mainly by increasing uptake up the vaccine and stimulating cellular processes. While whole organism vaccines, in particular live vaccines, have high intrinsic immunogenicity due to their ability to activate innate immunity and present multiple B and T cell epitopes. In contrast, subunit vaccines often include adjuvants and plasmid DNA are limited in the range of immunostimulatory motifs and the quantity of protein they produce in the host.

A limited number of adjuvants are approved for use in human vaccines. Alum, either aluminium phosphate or aluminium hydroxide, is universally approved. The vaccine antigens are adsorbed onto alum particles. Alum has been shown to interact directly with membrane lipids on DCs.[109] It has been proposed that alum provides additional stimulatory signals by inducing cell death with the release of host cell DNA, which acts as an endogenous stimulus.[110] AS04 is an alum-based adjuvant containing a TLR4 agonist, monophosphoryl lipid A (MPL) used in the hepatitis B vaccine, Fendrix and the human papillomavirus vaccine, Cervarix.

Squalene-based adjuvants have also been approved for use in human vaccines, for example MF59 and AS03, prepared as oil in water emulsions. These have been used in several influenza subunit vaccines against seasonal influenza and pandemic influenza outbreaks. Although the precise mechanism of action remains unclear, squalenes have been shown to promote leukocytes to the vaccination site and promote DC migration to lymph nodes.[111]

Other adjuvants have been used in clinical trials but are not yet fully approved for use in humans. These adjuvants have been tested mainly as adjuvants for plasmid DNA vaccines where low immunogenicity remains an obstacle. The main approach is the use of biodegradable, synthetic polymeric material in the form of nanoparticles, such as poly(vinylpyridine), poly-lactide-co-glycolides (PLGs) and polylactide-co-glycolide acid (PLGA) which have an established safety profile in humans.[112,113] Plasmid DNA may be adsorbed or trapped within the particle. PLG increases plasmid uptake and promotes DC maturation and migration to lymph nodes and increases CTL responses.[114–116]

Plasmid-encoded cytokine adjuvants may also increase immunogenicity and protective efficacy. Co-administration of plasmids encoding interleukin-2 (IL-2)-Ig fusion protein significantly increased the biological half-life of Il-2 resulting in increased memory T cells.[117] Plasmids expressing Il-12 have shown significant enhancement of immune responses, in particular T cell responses and promotion of TH1 cells.[118] Combining cytokines have been examined in this context with promising results observed in non-human primates using a co-plasmid encoding the TH1 cytokines, IL-12 and IL-15.[62] Co-administration of a plasmid encoding Il-15 with a HIV DNA vaccine followed by boosting with a vaccinia-based HIV vaccine has been shown to increase the frequency of memory CD8+ cells compared to responses without the Il-15 plasmid.[63]

Strategies to target DCs using antibody fusion partners have shown increased immune responses to HIV proteins. This strategy uses antibody-targeting of the DEC205 antigen uptake receptor on DCs (see earlier section on cross-presentation). Plasmids encoding a fusion protein comprised of Ig-HIVgagp41 or Ig-HIVgagp24 formulated with synthetic dsRNA elicited significant responses in mice and humans, respectively.[119,120]

17.8 CONCLUDING REMARKS

The 21st century has seen rapid advances in vaccinology. Genome-based vaccine design in conjunction with recombinant DNA technology and the ever-expanding capabilities of immunoinformatics tools have revolutionised vaccine design. Such strategies have realised new vaccines and promise to deliver more at a much faster pace than seen before. Recombinant protein vaccines are but one part of this. Genome-based strategies can be employed in the design of plasmid DNA vaccines and live vector vaccines. Identification of virulence factors also holds promise of developing new live, attenuated vaccines.

REFERENCES

1. H. Kumar, T. Kawai and S. Akira, *Int. Rev. Immunol.*, 2011, **30**, 16.
2. M. J. Bevan, *J. Exp. Med.*, 1976, **143**, 1283.
3. S. Jung, D. Unutmaz, P. Wong, G.-I. Sano, K. de Ios Santos, T. Sparwasser, S. Wu, S. Vuthoori, K. Ko, F. Zavala, E. Pamer, D. Littman and R. A. Lang, *Immunity*, 2002, **17**, 211.
4. O. P. Joffre, E. Segura, A. Savina and S. Amigorena, *Nat. Rev. Immunol.*, 2012, **12**, 557.
5. S. Burgdorf, C. Scholz, A. Kautz, R. Tampe and C. Kurts, *Nat. Immunol.*, 2008, **9**, 558.
6. B. Chatterjee, A. Smed-Sörensen, L. Cohn, C. Chalouni, R. Vandlen, B.-C. Lee, J. Widger, T. Keler, L. Delamarre and I. Mellman, *Blood*, 2012, **120**(10), 2011.
7. W. E. Demkowicz Jr., R. A. Littaua, J. Wang and F. A. Ennis, *J. Virol.*, 1996, **70**(4), 2627.
8. G. Sutter and C. Staib, *Curr. Drug Targets Infect. Disord.*, 2003, **3**, 263.
9. Y. Ono, H. Onda, R. Sasada, K. Igarashi, Y. Sugino and K. Nishioka, *Nucleic Acids Res.*, 1983, **11**(6), 147.
10. P. Valenzuela, P. Gray, M. Quiraga, J. Zaldiver, H. M. Goodman and W. J. Rutter, *Nature*, 1979, **280**, 815.

11. P. Valenzuela, A. Medina, W. J. Rutter, G. Ammerer and B. D. Hall, *Nature*, 1982, **298**, 347.
12. W. J. McAleer, E. B. Buynak, R. Z. Maigetter, D. E. Wampler and M. R. Hilleman, *Nature*, 1984, **307**(5947), 178.
13. C. Ludwig and R. Wagner, *Curr. Opin. Biotechnol.*, 2007, **18**, 537.
14. J. Paavonen, D. Jenkins, F. X. Bosch, P. Naud, J. Salmonerón, C. M. Wheeler, S.-N. Chow, D. L. Apter, H. C. Kitchener and X. Castellsague *et al.*, *Lancet*, 2007, **69**, 2161.
15. S. M. Garland, M. Hernandez-Villa, C. M. Wheeler, G. Perez, D. M. Harper, S. Leodolter, G. W. Tang, D. G. Ferris, M. Steben and J. Bryan, *N. Eng. J. Med.*, 2007, **356**, 1928.
16. X. Chen, Y. Yu, Q. Pan, Z. Tang, J. Han and G. Zang, *Acta Biochim. Biophys. Sin. (Shanghai)*, 2008, **40**, 996.
17. E. H. Nardin, G. A. Oliveira, J. M. Calvo-Calle, K. Wetzel, C. Maier, A. J. Birkett, P. Sarpotdar, M. L. Corado, G. B. Thornton and A. Schmidt, *Infect. Immun.*, 2004, **72**, 6519.
18. G. T. Jennings and M. F. Bachmann, *Biol. Chem.*, 2008, **389**(6), 521.
19. A. Touze and P. Coursaget, *Nucleic Acids Res.*, 1998, **26**, 1317.
20. S. Takamura, M. Niikura, T. C. Li, N. Takeda, S. Kusagawa, Y. Takebe, T. Miyamura and Y. Yasutomi, *Gene Ther.*, 2004, **11**, 628.
21. J. O'Hallahan, D. Lennon, P. Oster, R. Lane, S. Reid, K. Mulholland, J. Stewart, L. Penney, T. Percival and D. Martin, *Vaccine*, 2005, **23**(17–18), 2197.
22. R. Rappuoli, *Curr. Opin. Microbiol.*, 2000, **3**(5), 445.
23. H. Tettelin, N. J. Saunders, J. Heidelberg, A. C. Jeffries, K. E. Nelson, J. A. Eisen, K. A. Ketchum, D. W. Hood, J. F. Peden and R. J. Dodson *et al.*, *Science*, 2000, **287**, 1809.
24. M. Pizza, V. Scarlato, V. Masignani, M. M. Giulani, B. Arico, M. Comanducci, G. T. Jennings, L. Baldi, E. Bartolini and B. Capecchi *et al.*, *Science*, 2000, **287**, 1816.
25. M. M. Giuliani, J. Adu-Bobie, M. Comanducci, B. Arico, S. Savino, L. Santini, B. Brunelli, S. Bambini, A. Biolchi and B. Capecchi *et al.*, *Proc. Natl. Acad. Sci. U. S. A.*, 2006, **103**(29), 10834.
26. J. Donnelly, D. Medini, G. Boccadifuoco, A. Biolchi, J. Ward, C. Frasch, E. R. Moxon, M. Stella, M. Comanducci and S. Bambini *et al.*, *Proc. Natl. Acad. Sci. U. S. A.*, 2010, **107**, 19490.
27. M. E. Santolaya, M. L. O'Ryan, M. T. Valenzuela, V. Prado, R. Vergara, A. Muñoz, D. Toneatto, G. Graña, H. Wang, R. Clemens and P. M. Dull, *Lancet*, 2012, **379**, 617.
28. D. Maione, I. Margarit, C. D. Rinaudo, V. Masignani, M. Mora, M. Scarselli, H. Tettelin, C. Brettoni, E. T. Iacobini and R. Rosini *et al.*, *Science*, 2005, **309**(5731), 148.
29. A. A. Bhagwat and M. Bhagwat, *Pathog. Dis.*, 2008, **5**, 487.
30. D. Moriel, I. Bertoldi, A. Spagnuolo, S. Marchi, R. Rosini, B. Nesta, I. Pastorello, V. A. M. Corea, G. Torricelli and E. Cartocci,*et al.*, *Proc. Natl. Acad. Sci. U. S. A.*, 2010, **107**(20), 9072.

31. M. Scarselli, B. Aricò, B. Brunelli, S. Savino, F. Di Marcello, E. Palumbo, D. Veggi, L. Ciucchi, E. Cartocci and M. J. Bottomley *et al.*, *Sci. Transl. Med.*, 2011, **3**, 91ra62.
32. Y. He, Z. Xiang and H. L. Mobley, *J. Biomed. Biotechnol.*, 2010, DOI 10.1155/2010/297505, Article ID 297505.
33. S. Vivona, F. Bernante and F. Fillipini, *BMC Biotechnology*, 2006, **6**, 35.
34. V. Jaiswal, S. K. Chanumolu, A. Gupta, R. S. Chauhan and C. Rout, *BMC Bioinformatics*, 2013, **14**, 211.
35. R. Vita, L. Zarebski, J. A. Greenbaum, H. Emami, I. Hoof, N. Salimi, R. Damle, A. Sette and B. Peters, *Nucleic Acids Res.*, 2010, **38**, D854.
36. H. R. Ansari and G. P. S. Raghava, *Immunome Res.*, 2010, **6**, 6.
37. M. V. Larsen, A. Lelic, R. Parsons, M. Nielsen, I. Hoof, K. Lamberth, M. B. Loeb, S. Buss, R. Bramson and O. Lund, *PLoS ONE*, 2010, **5**(9), e12697.
38. J. A. Wolff, R. W. Malone, P. Williams, W. Chong, G. Ascadi, A. Jani and P. L. Felgner, *Science*, 1990, **247**, 1465.
39. D. C. Tang, M. DeVit and S. A. Johnson, *Nature*, 1992, **356**, 152.
40. J. B. Ulmer, J. J. Donnelly, S. E. Parker, G. H. Rhodes, P. L. Felgner, V. J. Dwarki, S. H. Gromkowski, R. R. Deck, C. M. DeWitt and A. Friedman *et al.*, *Science*, 1993, **259**(5102), 1745.
41. A. H. Davidson, J. L. Traub-Dargatz, R. M. Rodeheaver, E. N. Ostlund, D. D. Pedersen, R. G. Moorhead, J. B. Stricklin, R. D. Dewell, S. D. Roach, L. E. Long, S. J. Albers, R. J. Callan and M. D. Salman, *J. Am. Vet. Med. Assoc.*, 2005, **226**, 240.
42. K. A. Garver, S. E. LaPatra and G. Kurath, *Dis. Aquat. Organ.*, 2005, **64**, 13.
43. P. J. Bergman, M .A. Camps-Palau, J. A. McKnight, N. F. Leibman, D. M. Craft, C. Leung, J. Liao, I. Riviere, M. Sadelain and A. E. Hohenhaus *et al.*, *Vaccine*, 2006, **24**, 4582.
44. E. L. Thacker, D. J. Holtkamp, A. S. Khan, P. A. Brown and R. Draghia-Akli, *J. Anim. Sci.*, 2006, **84**, 733.
45. J. A. Williams, *Vaccines*, 2013, **1**, 225.
46. F. Saade and N. Petrovsky, *Expert Rev. Vaccines*, 2012, **11**(2), 189.
47. D. M. Klinman, S. Klaschik, D. Tross, H. Shirota and F. Steinhagan, *Vaccine*, 2010, **28**(16), 2801.
48. M. A. Kutzler and D. B. Weiner, *Nat. Rev. Genet.*, 2008, **9**(10), 776.
49. J. A. Schalk, F. R. Mooi, G. A. M. Berbers, L. van Aerts, H. Ovelgonne and T. Kimman, *Hum. Vaccines*, 2006, **2**(2), 45.
50. R. B. Ross, J. Immune Based Ther. Vaccines, 2009, **7**(3). Access online at http://www.jibtherapies.com/content/7/1/3.
51. J. Rice, H. Ottensmeier and F. K. Stevenson, *Nat. Rev. Cancer*, 2008, **8**, 108.
52. D. J. Laddy and D. B. Weiner, *Intl. Rev. Immunol.*, 2006, **25**, 99.
53. K. S. Rosenthal and D. H. Zimmerman, *Clin. Vaccine Immunol.*, 2006, **13**(8), 821.
54. J. J. Donnelly, J. J. Wahren and M. A. Liu, *J. Immunol.*, 2005, **175**, 633.
55. D. R. Gill, I. A. Pringle and S. C. Hyde, *Gene Ther.*, 2009, **16**, 165.

56. J. L. Brandsma, in Methods in Molecular Medicine, DNA Vaccines: Methods *and Protocols*, ed. W. M. Saltzman, H. Shen and J. L. Brandsma, *Humana Press Inc.*, Totowa, NJ, 2nd Edition, 2006, vol. 127.

57. J. A. Williams, A. E. Carnes and C. P. Hodgson, *Biotechnol. Adv.*, 2009, **27**(4), 353.

58. K. A. Cashman, K. E. Broderick, E. R. Wilkinson, C. I. Shaia, T. M. Bell, A. C. Shurtleff, K. W. Spik, C. V. Badger, M. C. Guttieri, N. Y. Sardesai and M. C. Schmaljohn, *Vaccines*, 2013, **1**, 262.

59. H.-J. Ko, S.-Y. Ko, Y.-J. Kim, E.-G. Lee, S.-N. Cho and C.-Y. Kang, *Infect. Immun.*, 2005, **73**(9), 5666.

60. S. Megati, D. Garcia-Hand, S. Capello, V. Roopchand, A. Masood, R. Xu, A. Luckay, S. Y. Chong, M. Rosati and S. Sackitey *et al.*, *Vaccine*, 2008, **26**(40), 5083.

61. S. Wang, D. J. Farfan-Arribas, S. Shen, T. H. Chou, A. Hirsch, F. He and S. Lu, *Vaccine*, 2006, **24**(21), 4531.

62. S. Y. Chong, M. A. Egan, M. A. Kutzler, S. Megati, A. Masood, V. Roopchard, D. Garcia-Hand, D. C. Montefiori, J. Quiroz and M. Rosati *et al.*, *Vaccine*, 2007, **25**, 4967.

63. S. Li, X. Qi, Y. Gao, Y. Hao, L. Cui, L. Ruan and W. He, *Cell. Mol. Immunol.*, 2010, **7**, 491.

64. D. M. Klinman, *Intl. Rev. Immunol.*, 2006, **25**, 135.

65. C. Coban, S. Koyama, F. Takeshita, S. Akira and K .J. Ishii, *Hum. Vaccin.*, 2008, **4**, 453.

66. C. J. Desmet and K. J. Ishii, *Nat. Rev.*, 2012, **12**, 479.

67. K. Kobiyama, N. Jounai, T. Aoshi, M. Tozuka, F. Takeshita, C. Coban and K. J. Ishii, *Vaccines*, 2013, **1**, 278.

68. M. A. Barry and S. A. Johnson, *Vaccine*, 1997, **15**, 788.

69. J. A. Mikszta, J. B. Alarcon, J. M. Brittingham, D. E. Sutter, R. J. Pettis and N. G. Harvey, *Nat. Med.*, 2002, **8**, 415.

70. A. Luckay, M. K. Sidhu, R. Kjeken, S. Megati, S. Y. Chong, V. Roopchand, D. Garcia-Hand, R. Abdullah, R. Braun and J. Montefiori, *Virol.*, 2007, **81**(10), 5257.

71. K. Matsuo, R. Yamaguchi, A. Yamazaki, H. Tasaka, K. Terasaka, M. Totsuka, K. Kobayashi, H. Yukitake and T. Yamada, *Infect. Immun.*, 1990, **58**, 4049.

72. W. R. Jacobs, M. Tuckman and B. R Bloom, *Nature*, 1987, **327**(6122), 532.

73. M. Dennehy and A.-L. Williamson, *Vaccine*, 2005, **23**, 1209.

74. R. Chapman, G. Chege, E. Shephard, H. Stutz and A.-L. Williamson, *Curr. HIV Res.*, 2010, **8**(4), 282.

75. M. Dennehy, W. Bourn, D. Steele and A.-L. Williamson, *Vaccine*, 2007, **25**, 3646.

76. A. A. Ryan, T. M. Wozniak, E. Shklovskaya, M. A. O'Donnell, B. Fazekas de St Groth, W. J. Britton and J. A. Triccas, *J. Immunol.*, 2007, **179**(12), 8418.

77. I. Magalhaes, D. R. Sizemore, R. K. Ahmed, S. Muellar, L. Wehlin, C. Scanga, F. Weichold, G. Schirru, M. G. Pau and J. Goudsmit *et al.*, *PLoS ONE*, 2008, **3**(11), e3790.

78. J. K. Nambiar, A. A. Ryan, C. U. Kong, W. J. Britton and J. A. Triccas, *Eur. J. Immunol.*, 2010, **40**(1), 153.
79. W. A. H. Hegazy and M. Hensel, *Future Microbiol.*, 2012, **7**(1), 111.
80. O. Steele-Mortimer, *Curr. Opin. Microbiol.*, 2008, **11**(1), 38.
81. H. Rüssman, H. Shams, F. Poblete, Y. Fu, J. E. Galen and R. O. Donis, *Science*, 1998, **281**(5376), 565.
82. J. Jantsch, D. Cheminay, T. Chakravortty, J. Lindig, J. Hein and M. Hensel, *Cell. Microbiol.*, 2003, **5**(12), 933.
83. M. I. Husseiny and M. Hensel, *Vaccine*, 2009, **27**, 3780.
84. W. A. H. Hegazy, X. Xu, L. Metelitsa and M. Hensel, *Infect. Immun.*, 2012, **80**, 1193.
85. W. Xin, S. Y. Wanda, W. Li, S. Wang, H. Mo and R. Curtiss 3rd, *Infect. Immun.*, 2008, **76**(7), 3241.
86. S. Wang, Y. Li, G. Scarpellini, W. Kong, H. Shi, C.-H. Baek, B. Gunn, S.-Y. Wanda, K. L. Roland, X. Zhang, P. Senechal-Willis and R. Curtiss 3rd, *Infect. Immun.*, 2010, **78**(9), 3969.
87. R. Curtiss 3rd, W. Xin, Y. Li, W. Kong, S. Y. Wanda, B. Gunn and S. Wang, *Crit. Rev. immunol.*, 2010, **30**(3), 255.
88. N. Goonetilleke, S. Moore, S. Dally, N. Winstone, I. Cebere, A. Mahmoud, S. Pinheiro, G. Gillespie, D. Brown and V. Loach *et al.*, *J. Virol.*, 2006, **80**, 4717.
89. H. McShane, A. A. Pathan, C. R. Sander, S. M. Keating, S. C. Gilbert, K. Huygen, H. A. Fletcher and A. V. Hill, *Nat. Med.*, 2004, **10**, 1240.
90. C. Larocca and J. Schloem, *Cancer J.*, 2011, **17**(5), 359.
91. R. Harrop, N. Connolly, I. Redchenko, J. Valle, M. Saunders, M. G. Ryan, K. A. Myers, N. Drury, S. M. Kingsman, R. E. Hawkins and M. W. Carroll, *Clin. Cancer Res.*, 2006, **12**, 3416.
92. R. G. Meyer, C. M. Britten, U. Siepmann, B. Petzold, T. A. Sagban, H. A. Lehr, B. Weigle, M. Schmitz, L. Mateo and B. Schmidt *et al.*, *Cancer Immunol. Immunother.*, 2005, **54**, 453.
93. E. B. Imoukhuede, T. Berthoud, P. Milligan, K. Bojang, J. Ismaili, S. Keating, D. Nwakanma, S. Keita, F. Njie and M. Sowe *et al.*, *Vaccine*, 2006, **24**, 6526.
94. A. Chahroudi, D. A. Garber, P. Reeves, L. Liu, D. Kalman and M. B. Feinberg, *J. Virol.*, 2006, **80**, 8469.
95. W. Kastenmuller, I. Drexler, H. Ludwig, V. Erfle, C. Peschel, H. Bernhard and G. Sutter, *Virology*, 2006, **350**, 276.
96. M. Di Nicola, C. Carlo-Stella, R. Mortarini, P. Baldassari, A. Guidetti, G. F. Gallino, M. Del Vecchio, F. Ravagnani, M. Magni and P. Chaplin *et al.*, *Clin. Cancer Res.*, 2004, **10**, 5381.
97. I. Drexler, C. Staib and G. Sutter, *Curr. Opin. Biotechnol.*, 2004, **15**, 506.
98. G. Gasteiger, W. Kastenmuller, R. Ljapoci, G. Sutter and I. Drexler, *J. Virol.*, 2007, **18**(21), 11925.
99. A. G. Goodman, P. P. Heinen, S. Guerra, A. Vijayan, C. O. S. Sorzano, C. E. Gome and M. Estaban, *PLoS ONE*, 2011, **6**(10), e25938.

100. S. Rerks-Ngarm, P. Pitisuttithum, S. Nitayaphan, J. Kaekungwal, J. Chiu, R. Paris, M. Premsri, C. Namwat, M. de Souza and E. Adams *et al.*, *N. Eng. J.Med.*, 2009, **361**(23), 2209.

101. D. R. Casimiro, A. Tang, L. Chen, T.-M. Fu, R. K. Evans, M.-E. Davies, D. C. Freed, W. Hurni, J. M. Aste-Amezaga and L. Guan *et al.*, *J. Virol.*, 2003, **77**(13), 7663.

102. N. J. Sullivan, T. W. Geisberg, J. B. Geisberg, L. Xu, Z. Y. Yang, M. Roederer, R. A. Koup, P. B. Jahrling and G. J. Nabel, *Nature*, 2003, **424**, 681.

103. W. Li, K. Krishnadas, R. Kumar, D. Lorne, J. Tyrrell and B. Agrawal, *Intl. immunol.*, 2007, **20**(1), 89.

104. H. Chen, Z. Q. Xiang, Y. Li, R. K. Kurupati, B. Jia, A. Bian, D. M. Zhou, N. Hutnick, N. S. Yuan and C. Gray, *J.Virol.*, 2010, **84**(20), 10522.

105. M. D. J. Dicks, A. J. Spencer, N. J. Edwards, G. Wadell, K. Bojang, S. C. Gilbert, A. V. S. Hill and M. G. Cottingham, *PLoS ONE*, 2012, 7(7), e40385.

106. D. S. Bangari and S. K. Mittal, *Vaccine*, 2006, **24**, 849.

107. S. Moffatt, J. Hays, H. HogenEsch and S. K. Mittal, *Virology*, 2000, **272**(1), 159.

108. U. B. Rasmussen, M. Benchaibi, V. Meyer, Y. Schlesinger and K. Schughart, *Hum. Gene. Ther.*, 1999, **10**(16), 2587.

109. T. L. Flach, G. Ng and A. Hari *et al.*, *Nat. Med.*, 2011, **17**(4), 479.

110. T. Marichal, K. Ohata, D. Bedoret, C. Mesnil, C. Sabatel, K. Kobiyama, P. Lekeux, C. Coban, S. Akira, K. J. Ishii, F. Bureau and C. J. Desmet, *et al.*, *Nat. Med.*, 2011, **17**(8), 996.

111. D. T. O'Hagan, G. S. Ott, E. De Gregorio and A. Seubert, *Vaccine*, 2012, **30**(29), 4341.

112. D. T. O'Hagan, M. Singh and J. B. Ulmer, *Immunol. Rev.*, 2004, **199**, 191.

113. S. D. Xiang, C. Selomulya, J. Ho, V. Apostolopoulos and M. Peblanksi, *Wiley Interdiscip. Rev. Nanomed. Nanobiotechnol.*, 2010, **2**(3), 205.

114. K. Denize-Mize, M. Dupuis, M. Singh, C. Woo, M. Ugozzoli, D. O'Hagan, J. Donnelly, G. Ott and D. M. McDonald, *Cell. Immunol.*, 2003, **225**, 12.

115. G. J. Randolph, K. Inaba, D. F. Robbiani, R. M. Steinman and W. A. Muller, *Immunity*, 1999, **11**, 753.

116. U. McKeever, S. Barman, T. Hao, P. Chambers, S. Song, L. Lunsford, Y. Y. Hsu, K. Roy and M. L. Hedley, *Vaccine*, 2002, **20**, 1524.

117. D. H. Barouch, D. M. Truitt and N. L. Letvin, *Vaccine*, 2004, **22**(23–24), 3092.

118. S. Bhaumik, R. Basu, S. Sen, K. Naskar and S. Roy, *Vaccine*, 2009, **27**(9), 1306.

119. G. Nchinda, J. Koroiwa, M. Oks, C. Trumpfheller, C. G. Park, Y. Huang, D. Hannaman, S. J. Schlesinger, O. Mizenina, M. C. Nussenzweig, K. Uberla and R. Steinman, *J Clin. Invest.*, 2008, **118**(4), 1427.

120. C. Trumpfheller, M. P. Longhi, M. Caskey, J. Idoyaga, L. Bozzacco, T. Keler, S. J. Schlesinger and R. M. Steinman, *J. Intern. Med.*, 2012, **271**(2), 183.

Subject Index